MATHEMATICS
for Technical and
Vocational Students
A Worktext

Richard C. Spangler
Bates Technical College
Tacoma, Washington

John G. Boyce
Consultant in Mathematics to the
Bucks County Technical School
Fairless Hills, Pennsylvania

Prentice Hall
Upper Saddle River, New Jersey / Columbus, Ohio

Library of Congress Cataloging-in-Publication Data
Spangler, Richard C.
   Mathematics for technical and vocational students:
A worktext/Richard C. Spangler, John G. Boyce.
      p.   cm.
   Includes index.
   ISBN 0-13-228842-7 (alk. paper)
   1. Shop mathematics.   I. Boyce, John G.   II. Title.
TJ1165.S7315   1998
513'.14'0246—dc21   96-46967
                                                          CIP

Editor: Stephen Helba
Production Editor: Alexandrina Benedicto Wolf
Design Coordinator: Karrie M. Converse
Text Designer: Rebecca M. Bobb
Cover Designer: Proof Positive/Farrolyne Associates
Production Manager: Deidra M. Schwartz
Marketing Manager: Debbie Yarnell

This book was set in Times Ten by Bi-Comp, Inc. and was printed and bound by Courier/
Kendallville, Inc. The cover was printed by Phoenix Color Corp.

 ©1998 by Prentice-Hall, Inc.
Simon & Schuster/A Viacom Company
Upper Saddle River, New Jersey 07458

Printed in the United States of America

10 9 8 7 6 5 4 3 2 1

ISBN: 0-13-228842-7

Prentice-Hall International (UK) Limited, *London*
Prentice-Hall of Australia Pty. Limited, *Sydney*
Prentice-Hall of Canada, Inc., *Toronto*
Prentice-Hall Hispanoamericana, S. A., *Mexico*
Prentice-Hall of India Private Limited, *New Delhi*
Prentice-Hall of Japan, Inc., *Tokyo*
Simon & Schuster Asia Pte. Ltd., *Singapore*
Editora Prentice-Hall do Brasil, Ltda., *Rio de Janeiro*

■■■■ PREFACE

Based on the ninth edition of *Mathematics for Technical and Vocational Students*, by Boyce et al, this Worktext is a textbook of practical mathematics applied to technical and trade work. It is suitable for use in any self-study situation or standard lecture mode. There are many illustrative examples and problems that will be of value both for self-instruction and for formal classwork. The tables and formulas in the back of the book are more comprehensive than those found in similar textbooks written for technical and vocational students.

The answers given may differ slightly from the answers various users find depending upon the readers' procedures in arriving at a solution to the problem. We did use the calculator and the computer in finding many of the answers to problems, since doing every problem by traditional pencil-and-paper methods presented a formidable task. We would appreciate any comments users might have regarding the presented answers.

Features

- The examples are plentiful and lead the students step-by-step to the solution.
- The practice checks are sequenced periodically with each section. After a new idea or technique is explained, students are directed to work a practice exercise in the margin. The practice exercises are self-paced and allow the student to become actively involved with the material before starting the problem set at the end of the unit. The solutions to the practice checks are conveniently located within the margins of each chapter.
- The section problem sets are graduated in degree of difficulty and afford the students more drill, practice, and reinforcement. The answers to odd-numbered problems appear in the back of the text.
- Space is provided for the students to work out the problems in each problem set.
- Except for Chapter 13, each chapter ends with a self-test and a chapter test to aid the students in their preparation for an instructor-given chapter test. These tests contain a variety of problems representative of those found in the chapter, and all answers are given in the back of the book.

Supplements

For the Instructor

- The Instructor's Manual includes answers to all the exercises, self-tests, and chapter tests.
- The Test Item File contains alternate test forms, including four free-response tests per chapter and two final examinations.

For the Students

- The Student Solutions Manual consists of worked-out, step-by-step solutions to all of the odd-numbered end-of-section exercises. This manual may be purchased as an optional resource by students.

Special Notes to the Learner

How to Be Successful in Mathematics

Most people find reading a mathematics or technical book very difficult. But, with the following tips, it doesn't have to be that way.

Read slowly and carefully. Reading a mathematics or technical book is not like reading a novel. You can read and comprehend a page from an average novel in two or three minutes, but reading and comprehending a page from a mathematics or technical book could take you up to an hour. Don't be shocked if you find yourself rereading something several times, because every word and symbol are important.

Be actively involved. Mathematics is not a spectator sport. Work out all examples step-by-step on paper. Both the practice exercises and problem sets are part of this involvement. Be involved right from the start—the pay-off will be more learning power.

Seek help. Even with the most careful reading and practice, some concepts will remain fuzzy. Don't be afraid to seek help—there are no "dumb" questions.

Take time to review. Allow time for ideas to sink in. It helps to review material from time to time.

Success in mathematics comes with perseverance, patience, and doing.

We wish to thank the many people who have offered suggestions and the reviewers who took the time to offer suggestions chapter by chapter. We would appreciate further comments and suggestions from users of the book.

Richard C. Spangler
John G. Boyce

■■■■ CONTENTS

Chapter 1   Whole Numbers   1

1–1   Introduction   1
1–2   Addition of Whole Numbers   1
1–3   Calculator Addition   6
1–4   Subtraction of Whole Numbers   7
1–5   Calculator Subtraction   11
1–6   Multiplication of Whole Numbers   11
1–7   Calculator Multiplication   17
1–8   Division of Whole Numbers   17
1–9   Calculator Division   22
        Self-Test   23
        Chapter Test   24

Chapter 2   Common Fractions   27

2–1   Introduction   27
2–2   Definitions   27
2–3   Changing Whole or Mixed Numbers to Improper Fractions   28
2–4   Changing Improper Fractions to Whole or Mixed Numbers   30
2–5   Changing a Fraction to Lowest Terms   32
2–6   Multiplication of Fractions   35
2–7   Division of Fractions   38
2–8   Problems Involving Multiplication and Division   41
2–9   Changing a Fraction with a Given Denominator to One with a
        Higher Denominator   42
2–10   Changing Two or More Fractions to Equivalent Fractions Having a
        Common Denominator   44
2–11   Addition of Fractions   46
2–12   Addition of Mixed Numbers   48
2–13   Subtraction of Fractions   50
2–14   Subtraction of Mixed Numbers   52
2–15   Problems Involving Addition, Subtraction, Multiplication,
        and Division   55
2–16   Complex Fractions   57
        Miscellaneous Problems in Common Fractions   59

Self-Test 67
Chapter Test 68

Chapter 3 Decimal Fractions   71
3–1 Introduction 71
3–2 Changing a Decimal Fraction to a Common Fraction 74
3–3 Changing a Common Fraction to a Decimal Fraction 76
3–4 Using a Table of Decimal Equivalents 79
3–5 Conversion of Dimensions 81
3–6 Addition of Decimals 84
3–7 Subtraction of Decimals 88
3–8 Multiplication of Decimals 90
3–9 Division of Decimals 95
3–10 Using the Calculator to Solve Decimal Problems 99
    Miscellaneous Problems in Decimal Fractions 109
    Self-Test 116
    Chapter Test 116

Chapter 4 Percentage   119
4–1 Definitions 119
4–2 Finding the Percentage, Given the Base and the Rate 120
4–3 Finding the Rate, Given the Base and the Percentage 123
4–4 Finding the Base, Given the Percentage and the Rate 127
4–5 Using the Calculator to Find the Percentage, Given the Base and the Rate 131
4–6 Using the Calculator to Find the Rate, Given the Base and the Percentage 133
4–7 Using the Calculator to Find the Base, Given the Percentage and the Rate 134
4–8 The $P–R–B$ Triangle 136
    Self-Test 137
    Chapter Test 137

Chapter 5 Ratio and Proportion   141
5–1 Ratio 141
5–2 Reduction of Ratios to Lowest Terms 141
5–3 Proportion 144
5–4 Using the Calculator to Solve Proportion Problems 150
5–5 Averages 153
5–6 Using the Calculator to Find the Average of a Set of Numbers 155
    Self-Test 157
    Chapter Test 158

Chapter 6 Practical Algebra   161
6–1 Use of Letters 161
6–2 Negative Numbers 161
6–3 Definitions 163
6–4 Substitution 167
6–5 Addition 168
6–6 Subtraction 171
6–7 Symbols of Grouping 173
6–8 Multiplication 175
6–9 Division 178
6–10 Equations 181
    Self-Test 188
    Chapter Test 189

Chapter 7   Rectangles and Triangles   191

7–1   Area of Surfaces and Units of Area   191
7–2   The Perimeter of a Rectangle   197
7–3   Finding the Width or Length of a Rectangle   198
7–4   Squares and Square Roots   201
7–5   Finding the Square Root of a Whole Number   203
7–6   The Square Root of Mixed Numbers   204
7–7   Finding the Square Root of a Fraction   208
7–8   Using the Calculator to Solve Square Root Problems   210
7–9   Applications of Square Root   212
7–10   Triangles   219
7–11   Areas of Isosceles Triangles   222
7–12   Areas of Scalene Triangles Using Hero's Formula   223
7–13   Using the Calculator to Solve Triangle Problems   226
7–14   Angles in Triangles   232
7–15   Using the Calculator for Angle Measure in D.MS   235
        Self-Test   238
        Chapter Test   239

Chapter 8   Regular Polygons and Circles   243

8–1   Definitions   243
8–2   Equilateral Triangles   245
8–3   Squares   247
8–4   The Regular Hexagon   249
8–5   The Regular Octagon   252
8–6   Regular Polygons and the Calculator   259
8–7   Quadrilaterals   262
8–8   Area of a Trapezoid   262
8–9   Areas of Composite Figures   266
8–10   Scale   269
8–11   Finding the Drawing Measure   269
8–12   Finding the Actual Dimension on an Object from the Measured
        Dimension on the Diagram   270
8–13   Finding the Area of a Figure Drawn to a Certain Scale   273
8–14   Circles   275
8–15   Circumference   276
8–16   Using the Calculator to Find Circumference   278
8–17   Finding the Diameter of a Circle   279
8–18   Finding the Area of a Circle: Method 1   280
8–19   Finding the Area of a Circle: Method 2   282
8–20   Using the Calculator to Find the Area of a Circle   284
8–21   Finding the Diameter of a Circle When the Area Is Given:
        Method 1   285
8–22   Finding the Diameter of a Circle When the Area Is Given:
        Method 2   287
8–23   Finding the Diameter of a Circle Equal in Area to the Combined Areas
        of Two or More Circles   289
8–24   Short Method of Comparing Areas of Circles   291
8–25   Areas of Ring Sections   292
8–26   Short Method of Finding the Areas of Ring Sections   294
8–27   Arcs and Sectors of Circles   295
8–28   Circles and Regular Figures   297
8–29   Segments of Circles   301
8–30   The Ellipse   303

8–31   Summary of Formulas for Plane Figures 305
        Self-Test 308
        Chapter Test 309

Chapter 9 Solids   311
9–1   Definitions 311
9–2   Prisms and Cylinders 311
9–3   Volumes of Prisms 311
9–4   Finding the Height of a Prism or a Cylinder 316
9–5   Finding the Area of the Base of a Prism or a Cylinder 318
9–6   Lateral Surfaces of Prisms and Cylinders 319
9–7   Pyramids and Cones 322
9–8   Volumes of Pyramids and Cones 322
9–9   Lateral Surfaces of Pyramids and Cones 324
9–10   Frustums of Pyramids and Cones 327
9–11   Finding the Height of the Frustum of a Pyramid or a Cone 330
9–12   Finding the Lateral Surface of the Frustum of a Cone
        or a Pyramid 332
9–13   Spheres 334
9–14   Using the Calculator to Find Volumes of Spheres 337
9–15   Finding the Surface Area of a Sphere 338
9–16   Volume of a Ring 340
9–17   Volumes of Composite Solid Figures 341
9–18   Weights of Materials 344
9–19   Weights of Castings from Patterns 346
9–20   Board Measure 348
9–21   Flooring 351
9–22   Summary of Formulas for Solids 354
        Self-Test 354
        Chapter Test 355

Chapter 10 Metric Measure   359
10–1   Introduction 359
10–2   Units of Length 363
10–3   Units of Area 366
10–4   Units of Volume 371
10–5   Units of Weight 375
10–6   Converting English Length to Metric Length 378
10–7   Converting English Area to Metric Area 381
10–8   Converting English Area to Metric Area Using Constants 383
10–9   Converting English Volume to Metric Volume 385
10–10   Converting English Weight to Metric Weight 388
10–11   Converting English Temperature and Metric Temperature 390
        Self-Test 392
        Chapter Test 393

Chapter 11 Graphs   397
11–1   Definitions 397
11–2   Types of Graphs 398
11–3   Use of Graphs in Experimental Work 402
11–4   Two or More Graphs Combined 402
11–5   Circle Graphs 409
11–6   Bar Graphs 412
        Self-Test 415
        Chapter Test 417

Chapter 12   Measuring Instruments

421

12–1   The Micrometer   421
12–2   The Ten-Thousandths Micrometer   423
12–3   The Vernier Caliper   424
12–4   The Protractor   426
12–5   The Vernier Protractor   427
12–6   The Planimeter   428
12–7   Use of the Planimeter   429
        Self-Test   430
        Chapter Test   431

Chapter 13   Geometrical Constructions

433

13–1   Applications of Geometry   433
13–2   Bisecting a Line Segment   433
13–3   Bisecting an Angle   434
13–4   Bisecting an Arc   434
13–5   Constructing a Perpendicular to a Line at a Given Point on the Line   435
13–6   Constructing a Perpendicular to an Endpoint of a Line Segment   435
13–7   Constructing a Perpendicular to a Line through a Point Not on the Line   436
13–8   Constructing a Line Parallel to Another Line   437
13–9   Dividing a Line Segment into a Number of Equal Parts   437
13–10   Constructing an Angle Equal to a Given Angle   438
13–11   Constructing an Equilateral Triangle of Given Size   439
13–12   Constructing a Circle through Three Given Points   439
13–13   Finding the Center of a Circle or an Arc   440
13–14   Inscribing a Square in a Circle   440
13–15   Constructing a Square of a Given Size   441
13–16   Constructing a Square Equal in Area to the Sum or Difference of Two Given Squares   441
13–17   Constructing a Circle Equal in Area to the Sum or Difference of the Areas of Two Given Circles   442
13–18   Inscribing a Hexagon in a Circle   443
13–19   Inscribing an Equilateral Triangle in a Circle   443
13–20   Constructing a Hexagon Whose Sides Will Be a Given Length   443
13–21   Constructing a Hexagon with One of the Sides on a Given Line   444
13–22   Inscribing an Octagon in a Circle   445
13–23   Inscribing an Octagon in a Square   445
13–24   Constructing an Octagon of a Given Size   446
13–25   Constructing a Pentagon   446
13–26   Constructing a Tangent to a Circle   447
13–27   Constructing a Tangent to a Circle through a Point Outside the Circle   448
13–28   Constructing a Tangent to Two Circles of Equal Size   448
13–29   Constructing an Internal Tangent to Two Equal Circles   449
13–30   Constructing an External Tangent to Two Circles of Unequal Size   449
13–31   Constructing an Internal Tangent to Two Unequal Circles   450
13–32   Constructing an Ellipse   451

Chapter 14   Essentials of Trigonometry

453

14–1   The Right Triangle   453
14–2   Trigonometric Functions   455
14–3   Use of Tables   456
14–4   Sin, Cos, and Tan on the Calculator   458

14–5   Finding an Angle Corresponding to a Given Function 459
14–6   Using the Calculator to Find an Angle, Given Its Function 461
14–7   Solution of Right Triangles 462
14–8   Isosceles Triangles 468
        Miscellaneous Problems Using Trigonometry 470
        Self-Test 477
        Chapter Test 478

Chapter 15   Strength of Materials   481
15–1   Stress and Strain 481
15–2   Kinds of Stresses 481
15–3   Unit Stress 482
15–4   Elastic Limit 482
15–5   Ultimate Strength 482
15–6   Safety Factor 483
15–7   Working Unit Stresses 483
15–8   Pressure in Pipes 486
15–9   Riveted Joints 487
        Self-Test 492
        Chapter Test 493

Chapter 16   Work and Power   495
16–1   Work and Power 495
16–2   Horsepower of a Steam Engine 497
16–3   Horsepower of Gas Engines 498
16–4   Brake Horsepower 499
16–5   Electrical Power 501
16–6   Mechanical Efficiency of Machines 503
        Self-Test 506
        Chapter Test 506

Chapter 17   Tapers   509
17–1   Definitions 509
17–2   Computing Taper and Diameter 509
17–3   American Standard Self-Holding (Slow) Taper Series 512
17–4   Taper Angle 519
17–5   Taper Turning by Offsetting the Tail Stock 523
17–6   Taper Turning by Using the Compound Rest 527
17–7   Taper Turning by Using the Taper Attachment 531
        Self-Test 533
        Chapter Test 534

Chapter 18   Speed Ratios of Pulleys and Gears   535
18–1   Gear Trains 535
18–2   Idlers 539
18–3   Finding the Number of Teeth for a Given Speed Ratio 540
18–4   Compound Gearing 541
18–5   Worm and Gear 547
18–6   Trains of Spur, Bevel, and Worm Gearing 548
18–7   Pulley Trains 549
        Self-Test 553
        Chapter Test 555

Chapter 19   Screw Threads   559
19–1   Introduction 559
19–2   Pitch 559

19–3 Lead 562
19–4 Definitions Applying to Screw Threads 564
19–5 Sharp V-thread 564
19–6 Double Depth of Sharp V-Thread 565
19 7 Minor Diameter 566
19–8 Tap Drill Sizes 567
19–9 The Unified Thread 568
19–10 Double Depth of the Unified Thread 570
19–11 Minor Diameter of Unified Threads 570
19–12 American National Thread 571
19–13 Double Depth of American National Thread 572
19–14 Minor Diameter of American National Thread 573
19–15 Size of Tap Drill for American National Threads 574
19–16 Width of Point of Tool 575
19–17 Square Thread 576
19–18 Tap Drill for Square Thread 576
19–19 The Acme 29-Degree Screw Thread 578
19–20 Tap Drill and Tap, Acme Thread 579
19–21 The Brown and Sharpe 29-Degree Worm Thread 580
19–22 Metric Standard Screw Threads 580
19–23 Whitworth Standard Threads 582
19–24 Tap Drill Size for Whitworth Thread 583
19–25 Radius of Tool Point for Whitworth Thread 583
19–26 American Standard Taper Pipe Threads 584
19–27 Length of Part Having Perfect Threads and Length of Effective Thread 585
19–28 Thickness of Metal between Bottom of Thread and Inside of Pipe in Straight Pipe Thread 585
19–29 Lathe Gearing for Cutting Screw Threads 587
19–30 Fractional Threads 591
19–31 Use of Compound Gearing 593
19–32 Cutting Metric Threads 596
       Self-Test 597
       Chapter Test 598

Chapter 20 Cutting Speed and Feed 601
20–1 Cutting Speed and Surface, or Rim, Speed 601
20–2 Cutting Speed on the Lathe 601
20–3 Cutting Feed of a Lathe 604
20–4 Drill Press and Milling Machine 606
20–5 Drill Press Feed 607
20–6 Milling Machine Feed 609
20–7 Surface Speed or Rim Speed 612
20–8 Finding the Revolutions per Minute, Given the Diameter and the Cutting Speed 613
20–9 Finding the Diameter, Given the Rim Speed and the Revolutions per Minute 614
20–10 Cutting Speed on the Planer 616
20–11 Time Required for a Job on the Planer 617
20–12 Number of Strokes per Minute 618
20–13 Cutting Speed of Shapers 619
       Self-Test 621
       Chapter Test 622

Contents

Chapter 21   Gears   625
21–1   Spur Gears   625
21–2   Diametral Pitch   629
21–3   Proportions of Gear Teeth   632
21–4   Relation Between Circular Pitch and Diametral Pitch   632
21–5   Clearance   635
21–6   Depth of Tooth   636
21–7   Outside Diameter   638
21–8   Pitch Diameter   639
21–9   Use of Formulas   641
21–10   Racks   642
21–11   Center-to-Center Distance of Gears   644
21–12   Selection of Cutters   646
21–13   Bevel Gears   646
21–14   Definitions Applying to Bevel Gears—Pitch Cones   647
21–15   Tooth Parts   651
21–16   Pitch Cone Radius   651
21–17   Addendum Angle and Turning Angle   652
21–18   Dedendum Angle and Cutting Angle   652
21–19   Outside Diameters   652
21–20   Selecting Cutters for Bevel Gears   653
21–21   Dimensions of Teeth at Small End   654
21–22   Miter Gears   655
        Self-Test   656
        Chapter Test   657

Appendix   659

Answers to Problems   697

Answers to Self-Tests   729

Answers to Chapter Tests   732

Index   735

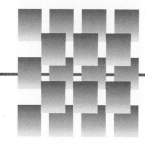

CHAPTER 1

Whole Numbers

1–1   INTRODUCTION

The set of ***whole numbers*** consists of the set of counting numbers (i.e., 1, 2, 3, 4, 5, . . .) and the number zero (0). Many of our everyday mathematical activities use whole numbers exclusively. When someone asks you how old you are or how much you weigh, you usually round the answer off to a whole number. If you buy an item that is priced at two for 25 cents, then the cost of one is usually 13 cents, a whole number. The post office deals in whole numbers when you mail a letter or package. If your package weighs $10\frac{1}{4}$ ounces, you pay for 11 ounces. Can you think of other examples where whole numbers are used?

1–2   ADDITION OF WHOLE NUMBERS

We use the ***decimal system*** with the Arabic numbers 0, 1, 2, 3, 4, 5, 6, 7, 8, and 9 in our everyday work. These ten numerals can be arranged to represent numbers of any size, and in every number the position of each numeral determines its value. For example, in the number 265, the rightmost numeral (the 5) means one times five. The number immediately to the left of the rightmost numeral (the 6) means ten times six, and the numeral immediately to its left (the 2) means one hundred times two. The number 265 then reads two hundred sixty-five. This positioning is called ***place value*** and is summarized in the chart shown in Fig. 1–1 for numbers up to, but not including, ten million.

   Because of the position of the digits in this number, we read the number as five million, four hundred seventy-four thousand, eight hundred twenty-one.

Place Value Chart						
Millions	Hundred thousands	Ten thousands	Thousands	Hundreds	Tens	Ones
5	4	7	4	8	2	1

Figure 1–1

■ **Example**

Add 123, 254, 52, and 964.

Solution   Arrange the addends in vertical format as follows.

$$
\left.\begin{array}{r}
123 \\
254 \\
52 \\
+\ 964
\end{array}\right\} \text{addends}
$$

Add the ones column (the rightmost): $3 + 4 + 2 + 4 = 13$. Write the 3 under the ones column and carry the tens digit (the numeral 1) over to the tens column. Add the tens column, including the carried 1: $1 + 2 + 5 + 5 + 6 = 19$. Write the 9 under the tens column and carry the hundreds digit, 1, over to the hundreds column (the 19 really means 19 tens or 190). Add the hundreds column, including the carried 1: $1 + 1 + 2 + 9 = 13$. Write the 3 under the hundreds column and carry the thousands digit, 1, over to the thousands column. Since there are no thousands in the problem, write the 1 to the left of the 3 in the answer (sum). Your work will look like this:

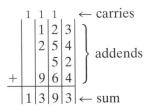

Addition problems are usually added in vertical format as just shown. All of the addends are aligned to the right so that the units, or ones (the rightmost digits), line up one under the other. Carefully aligning the columns is an important part of neat and accurate work! Check your addition by adding the numbers again. If you add downward, then add upward for the check. Your answer should be the same both times.

Practice check: Do exercise 1 on the left.

Add.

1. $\begin{array}{r} 63 \\ 415 \\ 137 \\ +\ 841 \end{array}$

The answer is in the margin on page 4.

■ Problems 1–2

Add the following numbers as indicated. Use the space provided below to work out the answers.

1.		2.		3.		4.		5.		6.	
	278		473		772		225		131		24
	805		380		82		298		887		910
	399		222		993		531		511		802
	258		714		588		759		705		19
	320		26		438		539		760		568
	+ 927		+ 404		+ 406		+ 617		+ 954		+ 625

7.	674	8.	752	9.	570	10.	54	11.	848	12.	223
	273		791		982		592		657		621
	281		32		594		654		346		60
	898		429		172		336		580		206
	750		565		639		293		691		963
	+ 23		+ 492		+ 746		+ 469		+ 386		+ 822

13.	167	14.	847	15.	326	16.	196	17.	537	18.	362
	791		169		430		588		626		121
	427		864		864		914		709		64
	299		71		457		738		715		737
	222		122		362		299		572		393
	+ 543		+ 408		+ 103		+ 7		+ 946		+ 396

19.	436	20.	152	21.	847	22.	727	23.	303	24.	995
	643		476		907		160		710		552
	682		686		562		935		322		768
	588		30		531		458		763		949
	911		813		50		369		36		915
	+ 392		+ 182		+ 413		+ 374		+ 579		+ 718

25.	308	26.	5,741	27.	6,912	28.	4,911	29.	9,002	30.	3,228
	664		3,996		2,498		2,801		9,531		3,713
	29		4,304		5,310		9,601		8,897		5,068
	844		+ 1,580		+ 3,103		+ 9,835		5,448		9,692
	812								4,249		5,938
	+ 764								2,448		32
									+ 34		+ 9,165

31.	1,672	32.	2,996	33.	566	34.	2,130	35.	2,982
	6,052		699		7,945		1,238		1,976
	9,925		8,450		5,450		73		2,100
	483		4,942		9,447		3,544		4,568
	7,179		6,238		69		5,565		96
	4,470		849		4,293		111		828
	+ 964		+ 1,942		+ 7,694		+ 8,109		+ 8,373

36.   7,940,149,875
      + 27,570,978,261

37.   68,067,239,787
      + 22,140,698,112

38.   771,852,260,717
      +   5,841,725,401

39.   250,881,416,443
              2,347,334,289
              3,334,661,784
      +   208,422,380,089

40.   257,380,823,881
              53,327,207,744
                  1,481,544
      +   472,729,139

Rewrite the numbers so that they are in vertical format and add.

41. 6 + 87 + 92 + 34 + 12 + 4 + 197

42. 145 + 465 + 437 + 32 + 96 + 389 + 1,008 + 345

43. 1,987 + 2,543 + 3,678 + 156 + 12 + 5,432 + 98,135 + 6,543

44. 198 + 1,923 + 19,245 + 19,876 + 123 + 9,854 + 2,654 + 198,999

45. 1,009 + 1,000 + 18,763 + 1,965 + 18,639 + 1,859 + 16,846 + 198,871

46. 834 + 8,345 + 37,445 + 24,654 + 2,765 + 24,655 + 29,864 + 468

47. 123 lb + 145 lb + 98 lb + 133 lb + 165 lb + 170 lb

48. 1,678 cars + 1,986 cars + 1,776 cars + 1,492 cars + 2,001 cars

49. 19,876 tons + 18,872 tons + 17,655 tons + 28,775 tons

50. 155 miles + 345 miles + 132 miles + 188 miles + 200 miles

51. If a contractor poured 12 cubic yards of concrete on Monday, 15 cubic yards of concrete on Tuesday, 10 cubic yards of concrete on Wednesday, and 15 cubic yards of concrete on Thursday, and spent the rest of the week removing the forms and cleaning up, how many cubic yards were poured for the week?

52. If a sheetrock mechanic has three jobs that require 120 4-by-8 sheets, 115 4-by-8 sheets, and 130 4-by-8 sheets of sheetrock, respectively, how many 4-by-8 sheets of sheetrock are needed to complete the three jobs?

53. If an electrician needs to install 5 receptacles in the living room, 7 receptacles in the study, 2 receptacles in the bathroom, 6 receptacles in the master bedroom, 4 receptacles in the smaller bedroom, and 10 receptacles in the kitchen, how many receptacles must be installed?

54. The monthly production of cars was as follows: January, 4,356; February, 4,353; March, 4,400; April, 4,290; May, 4,425; June, 4,287; July, 4,456; August, 4,223; September, 4,265; October, 4,365; November, 4,109; December, 4,001. How many cars were produced for the year?

55. The number of people who immigrated to the United States during the 1800s is as follows: 1800–1820, 8,385; 1821–1830, 143,439; 1831–1840, 599,125; 1841–1850, 1,713,251; 1851–1860, 2,598,214; 1861–1870, 2,314,824; 1871–1880, 2,812,191; 1881–1890, 5,246,613; 1891–1900, 3,687,564. How many people immigrated to the United States in the 1800s? (Source: U.S. Immigration and Naturalization Service.)

1–3   CALCULATOR ADDITION

The *scientific calculator* is a great help. However, the danger in using a calculator to solve problems is that the user often becomes dependent upon the answers given by the calculator, and if the wrong sequence of keys is pressed on the calculator, the answer can be incorrect. Care must be taken to enter the correct values, and care must be taken to press the keys in the correct sequence. Each entry must be analyzed by the operator, and the results of the operation must be considered carefully. Always remember that a calculator is a machine and therefore follows your directions blindly. It can respond only to the keys you press and the sequence in which you press them. If you press the wrong key, you'll probably get the wrong answer.

■ **Example**

Using a calculator, do Problem 1 from Problem Set 1–2 again.

$$
\begin{array}{r}
278 \\
805 \\
399 \\
258 \\
320 \\
+\ 927 \\
\end{array}
$$

Solution   Use the calculator in the following manner.

The Display Shows

Turn on the calculator	0.
Enter 278	278.
Press +	278.
Enter 805	805.
Press +	1083.
Enter 399	399.
Press +	1482.
Enter 258	258.
Press +	1740.
Enter 320	320.
Press +	2060.
Enter 927	927.
Press =	2987. (answer)

■ **Problems 1-3**

Do all the rest of Problem Set 1-2 in the same manner. Check your calculator answers against the answers you wrote on pages 2–6. You will notice that you cannot do Problems 36–40 with your calculator. Most calculators have 8-digit or 10-digit capability, and these problems have more than 10 digits in the addends. You might figure out some way of doing these problems with the calculator by doing each addend in two steps.

1-4   SUBTRACTION OF WHOLE NUMBERS

Subtraction can be thought of as the opposite of addition. It is the process of determining the difference between two numbers. The number that is to have another number subtracted from it is called the **minuend,** while the number that is to be subtracted is called the **subtrahend.** The answer to the subtraction process is called the **difference.**

■ **Example**

Subtract 126 from 543.

Solution   The problem indicates that 543 is the minuend and 126 is the subtrahend. Write the numbers in vertical format.

$$543 \leftarrow \text{minuend}$$
$$\underline{-\ 126} \leftarrow \text{subtrahend}$$

Start subtracting in the units column; that is, subtract the 6 from the 3. Since 6 cannot be subtracted from 3, you must borrow a ten from the tens column and subtract 6 from 13, which equals 7. Place the 7 in the units column in the answer. The 4 in the tens column is now 3, since you borrowed a ten from it. Now subtract the 2 from the 3, which gives a remainder of 1. Place this 1 in the tens column in the answer. Lastly, in the hundreds column, subtract the 1 from the 5, which equals

4. Place this difference in the hundreds column of the answer. The work will look like this:

Subtract.

2.   482
     − 291

The answer is in the margin on page 10.

$$\begin{array}{r} {}^{3}{}_{1} \\ 5\not{4}3 \leftarrow \text{minuend} \\ -\ 126 \leftarrow \text{subtrahend} \\ \hline 417 \leftarrow \text{difference} \end{array}$$

To check your answer, add the difference and the subtrahend, and that sum should be the minuend.

Practice check: Do exercise 2 on the left.

■ **Problems 1–4**

Use the space provided below to work out your answers for the problems.

1. Subtract 381 from 1,895.   **2.** Subtract 146 from 850.

3. Subtract 852 from 1,682.   **4.** Subtract 132 from 957.

5. Subtract 665 from 1,084.   **6.** Subtract 670 from 2,064.

7. Subtract 443 from 694.   **8.** Subtract 739 from 1,591.

9. Subtract 154 from 1,284.   **10.** Subtract 754 from 772.

11.   10,327   **12.**   8,851
      − 7,477   − 8,453

13.   7,561
  − 5,339

14.   9,443
  − 4,045

15.   7,609
  − 6,957

16.   8,905
  −1,846

17.   14,652
  − 9,195

18.   8,858
  −   182

19.   19,338
  − 7,353

20.   19,919
  − 8,881

21.   13,488
  − 3,497

22.   8,888
  − 1,015

23.   8,322
  − 4,090

24.   1,224
  −   595

25.   17,431
  − 4,548

26.   13,380
  − 3,121

27.   5,715
  − 4,183

28.   18,446
  − 5,461

29.   13,896
  − 3,849

30.   4,430
  − 1,250

31. In 1992 Democratic candidate William Clinton received 44,908,254 votes, Republican candidate George Bush received 39,102,343 votes, and independent candidate Ross Perot received 19,741,065 votes. How many more votes did President Clinton get than George Bush? How many more votes did George Bush get than Ross Perot? (Source: News Election Service.)

32. In May of 1995 there were 43,068,000 beneficiaries in the Social Security program, while in May of 1994 there were 42,461,000 beneficiaries. What was the increase in that one year? (Source: Social Security Administration.)

Answer to exercise 2

191

33. The net U.S. budget receipts for 1994 were $1,257,187,000; in 1993 they were $1,153,175,000. What was the increase in net receipts from 1993 to 1994?

34. In 1993 South Africa mined 19,907,772 troy ounces of gold, while the United States mined 10,642,314 troy ounces. How many more troy ounces of gold were mined by South Africa than by the United States in 1993? (Source: Bureau of Mines, U.S. Department of the Interior.)

35. In 1994 General Motors had total sales of 2,719,764 units, while the Ford Motor Company had total sales of 1,661,350 units. Compared to Ford, how much greater were the total sales of General Motors in 1994? (Source: Motor Vehicle Manufacturers Association of the United States.)

36. The total number of eggs produced in the United States in 1994 was 73,866,000,000. The total number of eggs produced in the United States in 1993 was 71,936,000,000. How many more eggs were produced in the United States in 1994 than in 1993? (Source: The Economic Research Service of the U.S. Department of Agriculture.)

37. The population of the United States in 1990 was 248,709,873, and of that population, 121,239,418 were males. According to this estimate, how many more females were there than males in 1990? (Source U.S. Bureau of the Census.)

38. In 1945 there were 8,226,373 U.S. Army personnel on active duty. In 1995 there were 521,036 U.S. Army personnel on active duty. How many more people on active duty did the Army have in 1945 than in 1995? (Source: Department of the Army.)

39. In 1994, O'Hare International Airport in Chicago had 66,435,252 passenger arrivals and departures, while John F. Kennedy International Airport in New York City had 28,799,275. How many more passengers arrived at and departed from O'Hare Airport than from J.F.K. Airport in 1994? (Source: U.S. Department of Transportation.)

40. Mount Everest, on the border of Nepal and Tibet, is 29,028 feet high, while Mount McKinley in Alaska is 20,320 feet high. How much higher is Mount Everest than Mount McKinley? (Source: National Geographic Society, Washington, D.C.)

 ## 1–5 CALCULATOR SUBTRACTION

Subtraction on the calculator depends on accurate keying in of the subtrahend and minuend. Make it a practice to read the number in the display after you enter it on the keyboard.

■ **Example**

Using a calculator, do Problem 1 from Problem Set 1–4 again.
Subtract 381 from 1,895.

Solution   Use the calculator in the following manner.

	The Display Shows
Turn on the calculator	0.
Enter 1895	1895.
Press −	1895.
Enter 381	381
Press =	1514. (answer)

■ **Problems 1–5**

Do the rest of Problem Set 1–4 in the same manner. Check your calculator answers against the answers you wrote on pages 8–11. If your answers do not agree, try to analyze why the answers are different so that you do not make the same mistake in the future.

1–6   MULTIPLICATION OF WHOLE NUMBERS

Multiplication can be thought of as addition repeated a given number of times. For example, 3 times 5 can be solved by 5 + 5 + 5. The 3 means that the 5 is to be used a total of three times. The same problem can also be thought of as 5 times 3, or 3 + 3 + 3 + 3 + 3. Written this way, the 3 is used a total of five times. In either case the solution is 15. When two numbers are placed in vertical format for multiplication, the number in the upper position is called the **multiplicand,** and the number in the lower position is called the **multiplier.** The answer to the muliplication problem is called the **product.** Partial products, shown in the following example, are intermediate results of the multiplication process.

■ **Example**

Find the product of 234 and 46.

Solution   Place the numbers in vertical format, such that 234 is the multiplicand and 46 is the multiplier. Starting from the units column, multiply the 6 by the multiplicand's 4. The product of 6 times 4 is 24, which can be thought of as 2 tens and 4 ones. Write the 4 in the answer's unit column and carry over the 2 to the tens column. Now multiply the 6 by the multiplicand's 3, which equals 18. Add the carried-over 2 to the 18 for a sum of 20. Write the 0 in the answer's tens column and carry over the 2 to the hundreds column. Now multiply the 6 by the multiplicand's 2, which equals 12. Add the carried-over 2 to the 12 for a sum of 14. Write 14 into the answer, with the 4 in the hundreds column and the 1 in the thousands column. What now appears is 1,404, the product of 6 times 234, which is the first partial product of this problem.

$$
\begin{array}{ccccc}
 & 2 & 2 & & \leftarrow \text{carries} \\
 & 2 & 3 & 4 & \leftarrow \text{multiplicand} \\
\times & & 4 & 6 & \leftarrow \text{multiplier} \\
\hline
1 & 4 & 0 & 4 & \leftarrow \text{first partial product}
\end{array}
$$

The next step is to multiply the multiplicand, 234, by the multiplier's tens column, 4. Multiply this 4 by the multiplicand's 4, which equals 16. Write the 6 in the answer's ten column (remember, the 16 is really 16 tens) and carry over the 1 to the tens column. Now multiply the 4 by the multiplicand's 3, which equals 12. Add the carried-over 1 to the 12 for a sum of 13. Write the 3 in the answer's hundreds column and carry over the 1 to the hundreds column. Now multiply the 4 by the multiplicand's 2, which equals 8. Add the carried-over 1 for a sum of 9. Write the 9 in the answer's thousands column. Imagining the zero that belongs in the units column, you can see that the second partial product is 9,360.

$$
\begin{array}{ccccc}
 & 1 & 1 & & \leftarrow \text{carries} \\
 & 2 & 3 & 4 & \leftarrow \text{multiplicand} \\
\times & & 4 & 6 & \leftarrow \text{multiplier} \\
\hline
1 & 4 & 0 & 4 & \leftarrow \text{first partial product} \\
9 & 3 & 6 & & \leftarrow \text{second partial product}
\end{array}
$$

The final step to this multiplication process is to add the partial products to get the final product.

Therefore, the product of 46 and 234 is 10,764.

$$
\begin{array}{rl}
234 & \\
\times \quad 46 & \\
\hline
1404 & \leftarrow \text{first partial product} \\
+ \quad 939 & \leftarrow \text{second partial product} \\
\hline
10764 & \leftarrow \text{product}
\end{array}
$$

Practice check: Do exercise 3 on the left.

Multiply.

3.   582
   × 35

The answer is in the margin on page 14.

■ **Problems 1–6**

Use the space provided below to work out the answers for the problems.

1. Find the product of 852 and 10,000.　　**2.** Find the product of 132 and 100.

3. Find the product of 665 and 100.　　**4.** Find the product of 670 and 10,000.

5. Find the product of 694 and 10.　　**6.** Find the product of 739 and 1,000.

7. Find the product of 154 and 1,000.　　**8.** Find the product of 754 and 100.

9. Find the product of 393 and 100.　　**10.** Find the product of 948 and 1,000.

11. Find the product of 446 and 1,000.　　**12.** Find the product of 791 and 100.

13. Find the product of 323 and 1,000.　　**14.** Find the product of 160 and 10.

Answer to exercise 3

20,370

15. Find the product of 1,063 and 10,000.

16. Find the product of 267 and 10,000.

17. Find the product of 527 and 100.

18. Find the product of 322 and 1,000.

19. Find the product of 947 and 10.

20. Find the product of 964 and 10.

21. Find the product of 493 and 496.

22. Find the product of 536 and 743.

23. Find the product of 782 and 688.

24. Find the product of 1,011 and 492.

25. Find the product of 252 and 576.

26. Find the product of 786 and 130.

27. Find the product of 913 and 282.

28. Find the product of 947 and 1,007.

29. Find the product of 662 and 631.

30. Find the product of 150 and 513.

31. Find the product of 827 and 260.

32. Find the product of 1,035 and 558.

33. Find the product of 469 and 474.

34. Find the product of 403 and 810.

35. Find the product of 422 and 863.

36. Find the product of 136 and 679.

37. Find the product of 1,095 and 652.

38. Find the product of 868 and 1,049.

39. Find the product of 1,015 and 818.

40. Find the product of 408 and 764.

41. If a mechanic makes $600 a week, what is the mechanic's yearly salary? (52 weeks = 1 year.)

42. If a bricklayer can lay 165 bricks in 1 hour, how many bricks can this bricklayer lay in an 8-hour day?

43. If 1 cubic yard of concrete costs $55, how much would 13 cubic yards cost?

44. If your car's average speed is 50 miles per hour, how far can you drive in 9 hours?

45. If a worker can make 357 bolts in 1 hour, how many bolts can this worker make in 8 hours?

46. If oak flooring costs 4 dollars a square foot, how much would 1,350 square feet of oak flooring cost?

47. If light travels 186,000 miles per second, how far does light travel in 1 year? (60 seconds = 1 minute; 60 minutes = 1 hour; 24 hours = 1 day; 365 days = 1 year.)

48. If the satellite *Voyager II,* launched September 20, 1977, travels at 17,000 miles per hour, how far does it travel in 1 year? In 10 years? How far has it traveled by today's date?

49. Sound travels approximately 1,100 feet per second in air. How far will sound travel in 2 minutes?

50. One cubic foot contains 1,728 cubic inches. How many cubic inches are contained in 25 cubic feet?

1–7 CALCULATOR MULTIPLICATION

Multiplication on the calculator depends upon accurate keying in of the multiplier and the multiplicand. The operator must be careful not to exceed the display capability of the calculator in use. Many calculators can hold 10-place numbers, that is, numbers up to but not including 10 billion. If the calculator answer to a problem is greater than the display capability, then the calculator may display an error message, or it might display the answer in scientific notation. (Scientific notation will be discussed later in this book.)

■ **Example**

Using a calculator, do Problem 1 from Problem Set 1–6 again.
Find the product of 852 and 10,000.

Solution   Use the calculator in the following manner.

The Display Shows

Turn on the calculator	0.
Enter 852	852.
Press ×	852.
Enter 10000	10000.
Press =	8520000. (answer)

■ **Problems 1–7**

Do the rest of Problem Set 1–6 in the same manner. Check your calculator answers against the answers you wrote on pages 13–17.

1–8   DIVISION OF WHOLE NUMBERS

Division can be thought of as repeated subtraction. For example, 12 divided by 4 can be solved by subtracting the 4 from the 12 for a remainder of 8, subtracting the 4 from the remaining 8 for a new remainder of 4, and subtracting the 4 from the remaining 4 for a final remainder of zero. It took three steps of subtracting 4 from 12 to reach zero, so 12 divided by 4 is 3. The 12 in this problem is called the **dividend,** the 4 is called the **divisor,** and the 3 is called the **quotient.** Signs that indicate division are $\div$, $\overline{)}$, and /. Therefore, dividing 12 by 4 can be indicated by $12 \div 4$ or $4\overline{)12}$ or 12/4 or $\frac{12}{4}$.

Division can be done by repeated subtraction as just demonstrated, but this process can be unwieldy. An algorithm (steps to follow for a solution) is used to speed up the division process.

■ Example

Divide 1,924 by 37.

Solution   Using the division symbol $\overline{)}$, enter 1,924 as the divided and 37 as the divisor.

$$\text{divisor} \rightarrow 37\overline{)1924} \leftarrow \text{dividend}$$

First determine how many times 37 can be divided into the first few digits (from the left) of the dividend. The 37 cannot be divided into (or subtracted from) the 1 or the 19 of the dividend, but it can be divided into the 192 of the dividend. The question that must now be answered is: How many times can 37 be subtracted from 192 before the remainder is less than 37? If the 37 is thought of as 40 and the 192 is thought of as 200, then you can mentally think of the number of subtractions that can be made. In this case you should figure that 40 can be subtracted from 200 five times. (The 40 and 200 were chosen because they are nearly equal to the 37 and the 192 but are much easier to use mentally.) So try 5 as the number of times that 37 can be subtracted from 192.

```
        | | |5| |
37)| 1 |9 |2 |4 |
```

Because the 192 means 192 tens, the 5 means 5 tens, so write the 5 in the quotient's tens place. To check if this is correct, multiply the 5 by the divisor 37. If 5 times 37 is greater than 192, then 5 is too great a number for the quotient's tens place. In this case, 5 is not too great a number because the product of 5 and 37 is 185. This 185 means 185 tens, so align it properly under the 192, as shown.

```
        | | |5|
37)| 1 |9 |2 |4
  −| 1 |8 |5 |
        | |7| |
```

Now subtract the 185 from the 192. If the remainder is greater than the divisor, 37, the 5 is too small a number for the quotient's tens place. In this case, 5 is not too small a number, because the remainder of 185 from 192 is 7. This 7 means 7 tens, so write it in the tens column under the subtraction bar.

To determine the rest of the quotient, bring down the dividend's units amount, 4, to be beside the remainder 7. Now you will be dividing the 37 into the 74. Again, you can make a reasonable guess by thinking of 37 and 74 as 40 and 80. You should be able to figure that 40 can be subtracted from 80 two times.

```
        | | |5| |
37)| 1 |9 |2 |4
  −| 1 |8 |5 |
        | |7 |4|
```

Write the 2 in the quotient's units place. To check if this is correct, multiply the 2 by the divisor, 37. As previously explained, the product must be greater than 74. In this case, 2 times 37 is exactly 74, and when 74 is subtracted from 74, the remainder is zero.

$$
\begin{array}{r}
52 \leftarrow \text{quotient} \\
\text{divisor} \rightarrow 37\overline{)1924} \leftarrow \text{dividend} \\
-185 \\
\hline
74 \\
74 \\
\hline
0 \leftarrow \text{remainder}
\end{array}
$$

Therefore, the quotient of 1,924 divided by 37 is 52.

Practice check: Do exercise 4 on the right.

Divide.

4. $43\overline{)1118}$

The answer is in the margin on page 21.

■ Problems 1–8

Use the spaces provided below to work out the answers for the problems.

1. Divide 4,330 by 10.

2. Divide 52,400 by 100.

3. Divide 17,500 by 100.

4. Divide 7,260,000 by 10,000.

5. Divide 8,790 by 10.

6. Divide 4,450,000 by 10,000.

7. Divide 5,370,000 by 10,000.

8. Divide 18,100 by 100.

9. Divide 7,000 by 100.

10. Divide 9,170 by 10.

11. Divide 6,320 by 10.

12. Divide 29,800 by 100.

13. Divide 6,520 by 10.

14. Divide 55,000 by 100.

15. Divide 6,310 by 10.

16. Divide 150,000 by 1,000.

17. Divide 692,000 by 1,000.

18. Divide 252,000 by 1,000.

19. Divide 875,000 by 1,000.

20. Divide 85,500 by 100.

21. Divide 64,897 by 73.

22. Divide 58,158 by 81.

23. Divide 7,704 by 18.

24. Divide 31,248 by 36.

25. Divide 13,621 by 53.

26. Divide 12,731 by 29.

27. Divide 26,345 by 55.

28. Divide 11,842 by 62.

29. Divide 15,120 by 16.

30. Divide 44,732 by 53.

31. Divide 11,496 by 958.

32. Divide 18,468 by 486.

Answer to exercise 4

26

33. Divide 19,030 by 346.

34. Divide 13,410 by 447.

35. Divide 9,000 by 375.

36. Divide 17,500 by 250.

37. Divide 16,794 by 311.

38. Divide 4,806 by 801.

39. Divide 31,558 by 509.

40. Divide 3,016 by 58.

41. A sheetrock contractor agrees to install and spackle 150 sheets of $\frac{1}{2}$-inch 4-by-8 sheetrock for $1,200. What is the installation cost per sheet?

42. A carpenter is to install a staircase that will have 8-inch risers (the height between steps), and the total height of the staircase is to be 8 feet. How many steps will the staircase have? (12 inches = 1 foot.)

43. How many 15-inch-long pieces of wire can be cut from a 100-foot roll of wire?

44. A contractor agrees to build a 3,500-square-foot house for $140,000. What is the cost per square foot?

45. A motorist drives 376 miles in 8 hours. What is the motorist's average speed per hour?

46. An airplane travels 3,630 miles in 6 hours. What is the airplane's average speed per hour?

47. A gallon of water has a volume of 231 cubic inches. How many gallons are in a full tank that holds 5,775 cubic inches?

48. A car company made 31,025 cars in 1995. What was its average number of cars made per day for the entire year? (365 days = 1 year.)

49. Alabama received $281,060.00 in motor fuel taxes. If the drivers in Alabama bought 2,162,000 gallons of fuel, what was Alabama's tax per gallon in cents?

50. If the area of the United States is 3,618,770 square miles and there are 248,709,813 people in the United States, about how many people are there per square mile?

1–9   CALCULATOR DIVISION

Division on the calculator not only involves careful entry of the dividend and the divisor; it also requires that you enter the dividend before you enter the divisor. Read the problem thoroughly and determine which number is the dividend and which number is the divisor. If the answer seems improbable, you may have entered the numbers in the wrong sequence.

■ **Example**

Using a calculator, do Problem 1 from Problem Set 1–8 again.
Divide 4,330 by 10.

Solution   Use the calculator in the following manner.

	The Display Shows
Turn on the calculator	0.
Enter the dividend, 4330	4330.
Press ÷	4330.
Enter the divisor, 10	10.
Press =	433. (answer)

■ Problems 1–9

Do the rest of Problem Set 1–8 in the same manner. Be watchful of the dividend and the divisor in the word problems. Check your calculator answers against the answers you wrote on pages 19–22.

■ Self-Test

Do the following problems. Use the space provided below to work out the answers. Check your answers with the answers in the back of the book.

1. Add: 231 + 345 + 1,000 + 10,000

2. Add: 3,459 + 8,973 + 3,456

3. Subtract: 3,478 − 2,999

4. Subtract: 3,450,435 − 9,839

5. Multiply: 549 × 10,000

6. Multiply: 5,469 × 3,981

7. Divide: 11,349 ÷ 873

8. Divide: 10,000 ÷ 400

9. The U.S. Department of the Treasury determined that it has in circulation $3,571,913,726 in one-dollar bills; $707,773,472 in two-dollar bills; $4,939,073,340 in five-dollar bills; $11,363,371,940 in ten-dollar bills; $51,586,205,780 in twenty-dollar bills; $21,715,575,800 in fifty-dollar bills; $76,516,974,400 in hundred-dollar bills; $154,472,000 in one-thousand-dollar bills, and $3,470,000 in ten-thousand-dollar bills. How much paper money is in circulation in the United States?

10. Based on the information in Problem 9, how many individual bills are in circulation in the United States?

■ **Chapter Test**

Add the following. Use the space provided below to work out the answers.

1. 152,818 157,699 58,830 + 513,162	**2.** 153,045 254,515 540,678 + 311,581	**3.** 549,400 98,208 416,816 11,533 + 524,095	**4.** 148,509 135,408 565,972 189,720 + 342,160

5. 119,883 62,548 494,262 190,512 8,804 + 420,920	**6.** 111,408 120,560 651,900 237,216 243,089 + 383,959	**7.** 230,868 403,323 104,300 182,535 41,904 9,190 221,628 + 408,285	**8.** 298,073 6,435 57,304 163,784 104,974 865,488 150,378 + 275,010

Subtract the following.

9. 98,362 − 6,052	**10.** 168,997 − 93,755	**11.** 161,211 − 130,441	**12.** 155,465 − 121,907

13. 112,846 − 52,241	**14.** 143,463 − 4,794	**15.** 137,938 − 129,863	**16.** 126,633 − 124,763

Multiply the following.

17. 4,060 × 827	**18.** 1,437 × 913	**19.** 1,127 × 49	**20.** 2,841 × 984

21. 3,640
× 936

22. 2,088
× 358

23. 5,166
× 512

24. 2,805
× 198

Divide the following.

25. 490)34,790

26. 60) 5,820

27. 343)11,662

28. 273)18,291

29. 182) 9,464

30. 96) 5,952

31. 644)16,100

32. 72) 4,968

Answer the following questions.

33. In 1985 the U.S. Mint in Philadelphia minted no silver dollars; $9,353,481 in half dollars; $193,954,740 in quarters; $70,520,096 in dimes; $32,355,748 in nickels; and $49,519,048 in pennies. How much money in coins was minted in Philadelphia in 1985? How many coins did the Philadephia Mint make? (Source: U.S. Department of the Treasury.)

34. In 1993 there were 8,781,080 licensed drivers 19 years old and under. The estimate in 1994 was 8,881,000. How many more licensed drivers 19 and under were there in 1994 than in 1993? (Source: U.S. Department of Transportation.)

35. In 1994 the total units sold of the Ford Taurus were 397,031. In 1993, 360,448 units were sold. How many more Taurus automobiles were sold in 1994 than in 1993? (Source: American Automobile Manufacturers Association.)

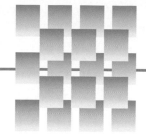

CHAPTER 2

Common Fractions

The changeover to the metric system, as it occurs, will tend to lessen the need for common fractions. Until the changeover is complete, fractions will be used where it is still convenient to use them. Goods produced using English measure will need a set of tools different from the tools made to work on goods produced using metric measure. Thus, mechanics will need two sets of tools during the changeover period. To lessen the cost, some manufacturers are making inserts that convert sockets from English to metric. There are also metric sockets with a $\frac{1}{2}''$ drive so ratchets and torque bars will work with the metric sockets.

Common fractions will be needed for some time, and everyone who works with numbers should know how to use common fractions in solving everyday problems.

2–2   DEFINITIONS

If an inch is divided into eight equal parts, each of the parts is one-eighth of an inch, written $\frac{1}{8}$ of an inch. Two such parts would be two-eighths of an inch, written $\frac{2}{8}$ of an inch; three such parts would be three-eighths of an inch, written $\frac{3}{8}$ of an inch; and so on. In each instance the numeral 8, written below the short horizontal line, shows into how many parts the inch has been divided; the numeral written above the line shows how many of the parts have been taken. Thus, in the expression $\frac{3}{8}$ of an inch, the numeral 8 below the line shows that the inch has been divided into eight equal parts; the numeral 3 above the line shows that three of these equal parts are taken. The inches on the ruler in Fig. 2–1 have been divided into eighths.

To show 15 divided by 3, you may write $15 \div 3 = 5$. Another way to indicate the division of 15 by 3 is to write $\frac{15}{3} = 5$.

Such expressions as $\frac{1}{8}$, $\frac{2}{8}$, $\frac{3}{8}$, and $\frac{15}{3}$ are called **fractions.** The numeral below the fraction line is called the **denominator.** The numeral above the fraction line is called the **numerator.** The numerator and denominator are called the **terms** of the fraction.

$$\text{terms} \longrightarrow \begin{array}{l} \nearrow 3 \leftarrow \text{numerator} \\ \searrow 8 \leftarrow \text{denominator} \end{array}$$

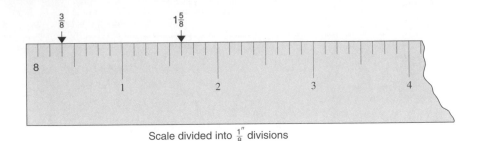

Scale divided into $\frac{1}{8}''$ divisions

Figure 2–1

A fraction like $\frac{5}{8}$, whose value is less than 1, is called a ***proper fraction.*** In a proper fraction the numerator is less than the denominator. A fraction like $\frac{9}{8}$, whose value is greater than 1, is called an ***improper fraction.*** An improper fraction is a fraction whose numerator is equal to or greater than its denominator. A fraction like $\frac{8}{8}$, whose value is equal to 1, is also called an improper fraction.

2–3   CHANGING WHOLE OR MIXED NUMBERS TO IMPROPER FRACTIONS

The ***counting numbers*** are those numbers used to count objects, such as the number of pencils in a box. There is no greatest counting number, but there is a least counting number, namely, the number 1. This collection of numbers is called the set of counting numbers. The word ***set*** implies the collection of all the numbers described.

Whole numbers include the counting numbers and the number zero (0), which is the least number of the set of whole numbers. The whole numbers, then, comprise the set of counting numbers and zero.

$$0, 1, 2, 3, 4, 5, \ldots$$

The changing of a whole number to an improper fraction makes these numbers easier to use in operations involving fractions and mixed numbers. A ***mixed number*** is a number consisting of a whole number and a fraction, such as $5\frac{1}{2}$, read as five and one-half.

Any whole number can be written as an improper fraction by using the whole number as the numerator and the number 1 as the denominator. For instance, $4 = \frac{4}{1}$ and $199 = \frac{199}{1}$.

It is possible to write whole numbers as improper fractions with denominators other than 1. The improper fractions $\frac{8}{1}$ and $\frac{16}{2}$ are equivalent fractions. The bar indicates that the numerator is to be divided by the denominator. In this case, $8 \div 1$ is the same as $16 \div 2$; they are therefore called equivalent fractions. They are also both improper fractions.

As previously stated, changing whole numbers to improper fractions makes these numbers easier to use when working with fractions and mixed numbers. A whole number can be written as an equivalent improper fraction with any counting number as its denominator by writing the whole number as a numerator and 1 as its denominator and then multiplying the numerator and denominator by the desired denominator.

■ Example 1

Change 3 to fifths.

Solution $\frac{3}{1} \times \frac{5}{5} = \frac{15}{5}$

Explanation   Write the whole number as an improper fraction using 1 as a denominator, that is, $\frac{3}{1}$. Multiply the numerator and denominator each by 5, obtaining $\frac{15}{5}$.

■ Example 2

Change $5\frac{3}{8}$ to eighths.

Solution $5\frac{3}{8} = \frac{5}{1} + \frac{3}{8} = \frac{40}{8} + \frac{3}{8} = \frac{43}{8}$

Explanation   In 1 there are 8 eighths, and in 5 there are 5 times 8 eighths, or 40 eighths. In $5\frac{3}{8}$ there are 40 eighths plus 3 eighths or a total of 43 eighths, written $\frac{43}{8}$.

   Quick method: Multiply the whole number by the denominator of the fraction, add the numerator, and write the result over the denominator.

$$5\frac{3}{8} = (8 \times 5 + 3) \text{ eighths} = \frac{43}{8}$$

■ Example 3

Change $6\frac{7}{8}$ to an improper fraction.

Solution $6\frac{7}{8} = (8 \times 6 + 7)$ eighths or $\frac{55}{8}$

Explanation   Multiplying the 6 by the denominator 8 yields 48 (the number of eighths in 6). Adding the 7 eighths yields a total of 55 eighths, written as $\frac{55}{8}$.

   Practice check: Do exercises 1 and 2 on the right.

Change the following mixed numbers to improper fractions.

1. $7\frac{1}{4}$

2. $3\frac{5}{16}$

The answers are in the margin on page 31.

■ Problems 2–3

Answer the following questions involving fractions. Use the space provided below to work out the answers.

1. Is $\frac{5}{8}$ a proper fraction or an improper fraction?

2. Is $\frac{8}{5}$ a proper fraction or an improper fraction?

3. Is $\frac{6}{6}$ a proper fraction or an improper fraction?

4. In the fraction $\frac{3}{4}$ the numeral 3 is called the _____ .

5. In the fraction $\frac{7}{8}$ the numeral 8 is called the _____ .

6. In the fraction $\frac{1}{2}$ the numerals 1 and 2 are called the _____ of the fraction.

7. How many halves in 3? in 4? in 5? in 12?

8. How many quarters in $2\frac{1}{2}$? in $3\frac{1}{2}$? in $7\frac{1}{2}$?

9. How many eighths in 4? in $5\frac{1}{8}$? in $6\frac{7}{8}$?

10. Change the following mixed numbers to improper fractions.

(a) $1\frac{1}{8}$ **(b)** $3\frac{3}{4}$ **(c)** $5\frac{1}{2}$ **(d)** $6\frac{7}{8}$ **(e)** $3\frac{7}{16}$ **(f)** $1\frac{1}{32}$

(g) $5\frac{15}{16}$ **(h)** $7\frac{3}{32}$ **(i)** $5\frac{2}{3}$ **(j)** $8\frac{7}{10}$ **(k)** $12\frac{5}{8}$ **(l)** $15\frac{17}{64}$

(m) $6\frac{3}{10}$ **(n)** $15\frac{9}{10}$ **(o)** $1\frac{1}{10}$ **(p)** $4\frac{31}{100}$

2–4   **CHANGING IMPROPER FRACTIONS TO WHOLE OR MIXED NUMBERS**

Improper fractions are used in multiplying and dividing mixed numbers. The answers to these problems are then changed back to whole or mixed numbers. This changing-back process is also called *reducing.*

■ **Example 1**

Change $\frac{24}{8}$ to a whole number.

Solution $\quad \frac{24}{8} = 24 \div 8 = 3$

Explanation $\quad \frac{24}{8}$ means 24 divided by 8. To change this fraction, carry out the indicated division. The answer is the whole number 3.

■ **Example 2**

Change $\frac{13}{4}$ to a mixed number.

Solution $\quad \frac{13}{4} = 13 \div 4 = 3\frac{1}{4}$

Explanation $\quad \frac{13}{4}$ means that 13 is to be divided by 4. When 13 is divided by 4, there is a remainder of 1. This 1 indicates that the remainder is 1 part of 4 and must be added to the quotient 3. Therefore, $\frac{13}{4} = 3\frac{1}{4}$.

Practice check: Do exercises 3 and 4 on the right.

Answer to exercises 1 and 2

1. $\frac{29}{4}$ 2. $\frac{53}{16}$

Change the following improper fractions to whole or mixed numbers.

3. $\frac{17}{2}$

4. $\frac{43}{16}$

The answers are in the margin on page 33.

■ **Problems 2–4**

Change the following improper fractions to whole or mixed numbers. Use the space provided below to work out the answers.

1. $\frac{4}{2}$ 2. $\frac{8}{4}$ 3. $\frac{16}{8}$ 4. $\frac{8}{5}$ 5. $\frac{4}{9}$ 6. $\frac{8}{3}$

7. $\frac{16}{3}$ 8. $\frac{8}{8}$ 9. $\frac{24}{4}$ 10. $\frac{5}{2}$ 11. $\frac{9}{8}$ 12. $\frac{10}{3}$

13. $\frac{55}{8}$ 14. $\frac{125}{100}$ 15. $\frac{100}{64}$ 16. $\frac{27}{8}$ 17. $\frac{31}{4}$ 18. $\frac{18}{10}$

19. $\frac{15}{3}$ 20. $\frac{27}{12}$ 21. $\frac{10}{4}$ 22. $\frac{35}{16}$ 23. $\frac{17}{8}$ 24. $\frac{33}{32}$

25. $\frac{7}{2}$ 26. $\frac{33}{12}$ 27. $\frac{25}{10}$ 28. $\frac{150}{100}$ 29. $\frac{18}{16}$ 30. $\frac{84}{64}$

31. $\frac{71}{32}$ **32.** $\frac{31}{8}$ **33.** $\frac{62}{32}$ **34.** $\frac{33}{8}$ **35.** $\frac{200}{100}$ **36.** $\frac{37}{12}$

37. $\frac{130}{60}$ **38.** $\frac{37}{24}$ **39.** $\frac{32}{16}$ **40.** $\frac{69}{64}$ **41.** $\frac{27}{10}$ **42.** $\frac{121}{100}$

43. $\frac{400}{100}$ **44.** $\frac{6750}{1000}$ **45.** $\frac{163}{100}$ **46.** $\frac{63}{10}$ **47.** $\frac{6380}{1000}$ **48.** $\frac{47}{10}$

49. $\frac{142}{100}$ **50.** $\frac{880}{100}$

2–5   CHANGING A FRACTION TO LOWEST TERMS

A fraction is in its **lowest terms** when the numerator and denominator are **prime** to each other, that is, when they share no common factor other than 1. For example, 5 and 8 are prime to each other because 1 is the only number that will divide *both* of them without a remainder. The numbers 10 and 12 are *not* prime to each other because both can be divided by 2 without a remainder.

When working with fractions, you will find it helpful to memorize the first seven or eight prime numbers. A **prime number** is a number that can be divided only by itself and 1; the number 1, however, is not a prime number. The first 10 prime numbers are as follows.

2 (the only even prime), 3, 5, 7, 11, 13, 17, 19, 23, 29

Rule _____

To change a fraction to lower terms, divide the numerator and the denominator by a factor common to both. A fraction is in its lowest terms when the numerator and denominator have no common factor other than 1.

_____

■ **Example 1**

Change $\frac{6}{8}$ to lowest terms.

Solution $\dfrac{6}{8} = \dfrac{6 \div 2}{8 \div 2} = \dfrac{3}{4}$

Explanation Dividing both terms of the given fraction by 2 gives the fraction $\frac{3}{4}$. Since 3 and 4 contain no common divisor, the fraction $\frac{3}{4}$ is in its lowest terms.

Example 1 could be also solved by factoring the numerator and denominator into their prime factors and canceling as shown in the following solution.

Solution $\dfrac{6}{8} = \dfrac{\overset{1}{\cancel{2}} \times 3}{\underset{1}{\cancel{2}} \times 2 \times 2} = \dfrac{3}{4}$

Explanation   The prime factors of 6 are 2 and 3, while the prime factors of 8 are 2, 2, and 2. The common 2 in the numerator and denominator can be canceled and the remaining factors multiplied to give $\frac{3}{4}$.

■ **Example 2**

Change $\frac{70}{105}$ to lowest terms.

Solution by Division Method

$$\frac{70 \div 5}{105 \div 5} = \frac{14}{21} \qquad \frac{14}{21} = \frac{14 \div 7}{21 \div 7} = \frac{2}{3} \quad \text{or} \quad \frac{70 \div 35}{105 \div 35} = \frac{2}{3}$$

Explanation   Dividing both terms of the given fraction by 5 gives the fraction $\frac{14}{21}$; dividing both terms of $\frac{14}{21}$ by 7 gives $\frac{2}{3}$. Or dividing both terms of $\frac{70}{105}$ by 35 gives at once $\frac{2}{3}$. Since 2 and 3 contain no common divisor, the fraction $\frac{2}{3}$ is in its lowest terms.

Solution by Prime Factor and Cancel Method   The problem could also be solved by finding all the prime factors of 70 and 105 as shown and canceling the common 5 and 7.

$$\frac{70}{105} = \frac{2 \times \overset{1}{\cancel{5}} \times \overset{1}{\cancel{7}}}{3 \times \underset{1}{\cancel{5}} \times \underset{1}{\cancel{7}}} = \frac{2}{3}$$

Changing a fraction to its lowest terms is also called ***simplification*** of the fraction. You can see that it is easier to read $\frac{2}{3}$ than $\frac{70}{105}$ even though $\frac{2}{3}$ and $\frac{70}{105}$ are equal to each other and represent the same number.

Practice check: Do exercises 5 and 6 on the right.

■ **Problems 2–5**

Simplify the following fractions. (All answers should be given with fractions in their lowest terms and improper fractions changed to whole or mixed numbers.) Use the space provided below to work out the answers.

1. $\frac{3}{6}$ **2.** $\frac{2}{4}$ **3.** $\frac{8}{12}$ **4.** $\frac{12}{8}$ **5.** $\frac{18}{24}$ **6.** $\frac{12}{32}$

7. $\frac{24}{64}$ **8.** $\frac{4}{16}$ **9.** $\frac{8}{32}$ **10.** $\frac{30}{32}$ **11.** $\frac{64}{128}$ **12.** $\frac{125}{1000}$

13. $\frac{100}{10}$ **14.** $\frac{16}{64}$ **15.** $\frac{96}{144}$ **16.** $\frac{375}{1000}$

17. $\frac{157}{314}$ **18.** $\frac{900}{360}$ **19.** $\frac{216}{1728}$ **20.** $\frac{448}{128}$

21. What part of a foot is 6 inches? (12 inches = 1 foot)

22. What part of a foot is 9 inches?

23. Thirty inches is what part of a yard? (36 inches = 1 yard)

24. Six inches is what part of a yard?

25. What part of a pound is 8 ounces? (16 ounces = 1 pound)

26. Four ounces is what part of a pound?

27. What part of a pound is 10 ounces?

28. What part of a mile is 440 yards? (1,760 yards = 1 mile)

29. What part of a yard is 18 inches?

30. What part of a yard is 21 inches?

31. What part of 12 is 3?

32. Fifteen is what part of 100?

33. Three is what part of 8?

34. Nine is what part of 16?

35. What part of 64 is 32?

2–6   MULTIPLICATION OF FRACTIONS

To multiply two or more fractions, multiply the numerators and use that answer as the new numerator; then multiply the denominators and use that answer as the new denominator. It is not necessary to find common denominators when multiplying fractions. The product should be expressed in lowest terms.

Rule

The product of two or more fractions is the product of the numerators over the product of the denominators.

■ Example 1

Multiply $\frac{3}{4}$ by 5.

Solution $\dfrac{3}{4} \times \dfrac{5}{1} = \dfrac{3 \times 5}{4 \times 1} = \dfrac{15}{4} = 3\frac{3}{4}$

or, more briefly,

$$\tfrac{3}{4} \times \tfrac{5}{1} = \tfrac{15}{4} = 3\tfrac{3}{4}$$

Explanation   The multiplier is 5, a whole number. To obtain the product, multiply the numerator of the given fraction by the whole number and place the result over the denominator. Thus, $3 \times 5 = 15$, the numerator of the product that is written over the denominator 4, giving $\frac{15}{4}$ or $3\frac{3}{4}$.

Do the indicated multiplication.

7. $\frac{5}{6} \times 15$

The answer is in the margin on page 37.

Practice check: Do exercise 7 on the right.

■ Example 2

Multiply $\frac{2}{3}$ by $\frac{5}{7}$.

Solution $\dfrac{2}{3} \times \dfrac{5}{7} = \dfrac{2 \times 5}{3 \times 7} = \dfrac{10}{21}$

or, in one step,

$$\tfrac{2}{3} \times \tfrac{5}{7} = \tfrac{10}{21}$$

Explanation   To multiply one fraction by another, multiply the numerators together and place the result over the product of the denominators. In this example, $2 \times 5 = 10$, the numerator of the product, and $3 \times 7 = 21$, the denominator of the product.

■ Example 3

Multiply $\frac{3}{4}$ by $\frac{8}{9}$. (This example is similar to Example 2 and thus may be done in the same way.)

Solution $\dfrac{3}{4} \times \dfrac{8}{9} = \dfrac{3 \times 8}{4 \times 9} = \dfrac{24}{36}$

Explanation   Reduce the answer to lowest terms by dividing the numerator and denominator by 12; thus, $\dfrac{24}{36} = \dfrac{24 \div 12}{36 \div 12} = \dfrac{2}{3}$. It is not necessary to do the multiplication first and then the division to reduce a fraction to its lowest terms. The division can be done before the multiplication, and this often makes the work much easier. Thus,

$$\frac{\overset{1}{\cancel{3}}}{\underset{1}{\cancel{4}}} \times \frac{\overset{2}{\cancel{8}}}{\underset{3}{\cancel{9}}} = \frac{2}{3}$$

The numerator and the denominator each have the common factors 3 and 4. Divide both the numerator and the denominator by 3. Then divide both the numerator and denominator by 4. The answer is simply the product of 1 and 2, that is, 2 divided by the product of 1 and 3, which is 3.

    The numerator and the denominator are divided by factors common to both. Before doing the actual multiplication, always look for common factors in the numerator and the denominator, and divide the numerator and the denominator by them. If all the common factors are found, the answer will be in the lowest terms.

Practice check: Do exercise 8 on the left.

Do the indicated multiplication.

8. $\frac{2}{3} \times \frac{9}{10}$

The answer is in the margin on page 39.

■ Example 4

Multiply $\frac{8}{11} \times \frac{5}{12}$.

Solution $\dfrac{\overset{2}{\cancel{8}}}{11} \times \dfrac{5}{\underset{3}{\cancel{12}}} = \dfrac{2 \times 5}{11 \times 3} = \dfrac{10}{33}$

Explanation   This case is similar to Example 3. The 8 in the numerator and the 12 in the denominator are each divided by their common factor 4. We obtain the answer by multiplying all the remaining numerators for the new numerator and multiplying all the remaining denominators for the new denominator.

■ Example 5

Multiply $2\frac{1}{2} \times 3\frac{1}{4}$.

Solution $2\frac{1}{2} \times 3\frac{1}{4} = \frac{5}{2} \times \frac{13}{4} = \frac{65}{8} = 8\frac{1}{8}$

Explanation   When the quantities to be multiplied are mixed numbers, change them to improper fractions and then proceed as in previous examples.

Practice check: Do exercise 9 on the left.

Do the indicated multiplication.

9. $4\frac{2}{5} \times 5\frac{1}{4}$

The answer is in the margin on page 39.

■ **Example 6**

Find $\frac{2}{3}$ of $\frac{4}{5}$.

Solution $\qquad \frac{2}{3} \times \frac{4}{5} = \frac{2 \times 4}{3 \times 5} = \frac{8}{15}$

Answer to exercise 7

$\frac{25}{2}$ or $12\frac{1}{2}$

Explanation The word *of* has the same meaning as the multiplication sign ($\times$); therefore, $\frac{2}{3}$ of $\frac{4}{5}$ means $\frac{2}{3} \times \frac{4}{5} = \frac{8}{15}$.

■ **Example 7**

Multiply $\frac{3}{4} \times \frac{5}{8} \times \frac{2}{7}$.

Solution $\qquad \dfrac{3}{\underset{2}{4}} \times \dfrac{5}{8} \times \dfrac{\overset{2}{2}}{7} = \dfrac{3 \times 5 \times 1}{2 \times 8 \times 7} = \dfrac{15}{112}$

Explanation The 2 in the numerator and the 4 in the denominator are each divided by their common factor 2. The product of the remaining numerators, $3 \times 5 \times 1 = 15$, is the numerator of the answer; the product of the remaining denominators, $2 \times 8 \times 7 = 112$, is the denominator of the answer. The answer is $\frac{15}{112}$.

Practice check: Do exercise 10 on the right.

Do the indicated multiplication.

10. $3\frac{1}{3} \times \frac{3}{5} \times 5\frac{7}{10}$

The answer is in the margin on page 39.

■ **Problems 2–6**

Do the indicated multiplications. Make sure your answers are reduced to lowest terms. Use the space provided below to work out the answers.

1. $\frac{1}{3} \times 5$

2. $\frac{1}{2} \times 6$

3. $\frac{3}{4} \times 8$

4. $\frac{5}{8} \times 16$

5. $\frac{1}{2} \times \frac{2}{3}$

6. $\frac{2}{5} \times \frac{10}{11}$

7. $\frac{3}{4} \times \frac{5}{16}$

8. $\frac{3}{8} \times \frac{16}{27}$

9. $1\frac{1}{2} \times 3$

10. $1\frac{2}{3} \times 1\frac{1}{2}$

11. $2\frac{1}{2} \times 1\frac{1}{4}$

12. $\frac{1}{2} \times 3\frac{1}{4} \times 4$

13. $\frac{16}{21} \times \frac{3}{4}$

14. $\frac{13}{51} \times \frac{17}{39}$

15. $\frac{1}{3} \times \frac{1}{3}$

16. $\frac{7}{12} \times \frac{36}{49}$

17. $1\frac{1}{3} \times 8\frac{3}{5}$ **18.** $\frac{2}{7} \times 28$ **19.** $2\frac{1}{16} \times 1\frac{1}{16}$ **20.** $12\frac{1}{2} \times 7\frac{1}{2} \times \frac{2}{3}$

21. $\frac{1}{2} \times \frac{1}{2} \times \frac{1}{2}$ **22.** $\frac{3}{4} \times \frac{1}{3}$ **23.** $\frac{5}{6} \times \frac{6}{15}$ **24.** $\frac{9}{16} \times \frac{2}{3} \times 10\frac{1}{2} \times 12\frac{7}{8}$

25. What is the total length of 8 pieces of steel, each $5\frac{1}{2}''$ long?

26. What is the total length of 25 pieces of drill rod, each $5\frac{1}{10}''$ long?

27. The volume of a rectangular block of wood is found by multiplying the length by the width by the height. What is the volume of a rectangular block of wood $15\frac{1}{2}''$ long, $4\frac{1}{4}''$ wide, and $3''$ high? (Your answer will be in cubic inches.)

28. To find the approximate circumference of a circle, we multiply the diameter by $3\frac{1}{7}$. Find the approximate circumference of a circle whose diameter is $2\frac{7}{8}''$.

29. What is the volume of a block of steel $8\frac{1}{2}''$ long, $2\frac{1}{8}''$ wide, and $1\frac{3}{4}''$ thick? (Your answer will be in cubic inches.)

30. What is $\frac{1}{2}$ the volume of the block of steel in Problem 29? (Your answer will be in cubic inches.)

2–7   DIVISION OF FRACTIONS

To divide one fraction by another, multiply the dividend (the number to be divided) by the *reciprocal* of the divisor. The reciprocal of a fraction is formed by using the numerator of the original fraction as the denominator and by using the denominator of the original fraction as the numerator (inverting the fraction). For instance, the reciprocal of $\frac{2}{9}$ is $\frac{9}{2}$.

■ **Example 1**

Divide $\frac{5}{7}$ by $\frac{3}{4}$. (The reciprocal of $\frac{3}{4}$ is $\frac{4}{3}$.)

Solution $\frac{5}{7} \div \frac{3}{4} = \frac{5}{7} \times \frac{4}{3} = \frac{20}{21}$

Answer to exercise 8

$\frac{3}{5}$

Answer to exercise 9

$23\frac{1}{10}$

Answer to exercise 10

$\frac{57}{5}$ or $11\frac{2}{5}$

Explanation   Change the division problem to a multiplication problem by changing the divisor to its reciprocal and proceeding as in multiplication. The reciprocal of a whole number is 1 written over the whole number, such as $\frac{1}{6}$, which is the reciprocal of 6. This procedure is commonly called ***inverting.*** Thus, for division involving fractions, invert the divisor and multiply.

Rule

To divide fractions, invert the divisor (the second fraction) and proceed as in multiplication.

Practice check: Do exercise 11 on the right.

Do the indicated division.

11. $\frac{9}{8} \div \frac{3}{4}$

The answer is in the margin on page 41.

■ **Example 2**

Find $\frac{2}{3} \div 5$.

Solution   Invert the 5 and multiply.

$$\frac{2}{3} \div \frac{5}{1} = \frac{2}{3} \times \frac{1}{5} = \frac{2}{15}$$

■ **Example 3**

Divide $4\frac{5}{8}$ by $2\frac{1}{3}$.

Solution   Change each of the mixed numbers to an improper fraction. Thus, $4\frac{5}{8} = \frac{37}{8}$ and $2\frac{1}{3} = \frac{7}{3}$. Invert the divisor and multiply.

$$4\frac{5}{8} \div 2\frac{1}{3} = \frac{37}{8} \div \frac{7}{3} = \frac{37}{8} \times \frac{3}{7} = \frac{111}{56} = 1\frac{55}{56}$$

■ **Example 4**

Divide 7 by $\frac{2}{5}$.

Solution   Invert the divisor, $\frac{2}{5}$, and multiply.

$$7 \div \frac{2}{5} = \frac{7}{1} \times \frac{5}{2} = \frac{35}{2} = 17\frac{1}{2}$$

Do the indicated division.

12. $2\frac{1}{2} \div 10$

13. $8\frac{3}{4} \div 5\frac{3}{5}$

The answers are in the margin on page 41.

Practice check: Do exercises 12 and 13 on the right.

■ **Problems 2–7**

Do the indicated divisions. Simplify your answers. Use the space provided below to work out the answers.

1. $\frac{18}{25} \div 3$ **2.** $\frac{12}{9} \div 4$ **3.** $\frac{7}{8} \div 5$ **4.** $\frac{3}{4} \div 7$

5. $\frac{9}{16} \div \frac{3}{8}$ **6.** $\frac{11}{12} \div \frac{1}{2}$ **7.** $7 \div \frac{1}{2}$ **8.** $4\frac{1}{2} \div \frac{1}{4}$

9. $3\frac{3}{4} \div \frac{1}{5}$ **10.** $8 \div \frac{2}{3}$ **11.** $15 \div \frac{3}{8}$ **12.** $18\frac{2}{3} \div 6$

13. $12\frac{3}{4} \div 8$ **14.** $10\frac{7}{16} \div \frac{3}{4}$ **15.** $4\frac{1}{3} \div \frac{2}{5}$ **16.** $5 \div 4\frac{2}{3}$

17. $7 \div 3\frac{1}{8}$ **18.** $6 \div 2\frac{1}{4}$ **19.** $2\frac{1}{3} \div 3\frac{1}{2}$ **20.** $4\frac{1}{16} \div 8\frac{1}{8}$

21. $12\frac{5}{6} \div 4\frac{5}{18}$ **22.** $3\frac{1}{10} \div 18\frac{1}{3}$ **23.** $8\frac{8}{9} \div 6\frac{2}{3}$ **24.** $20\frac{1}{2} \div 16\frac{2}{3}$

25. $6\frac{1}{5} \div 5\frac{1}{4}$ **26.** $3\frac{1}{16} \div 4\frac{7}{8}$ **27.** $7\frac{3}{10} \div 12\frac{3}{4}$ **28.** $6\frac{7}{8} \div 2\frac{1}{2}$

29. $6\frac{1}{2} \div 2\frac{1}{4}$ **30.** $15\frac{7}{10} \div 5\frac{1}{10}$ **31.** $11\frac{2}{3} \div 4\frac{5}{6}$ **32.** $7\frac{1}{8} \div \frac{1}{16}$

33. $10\frac{1}{6} \div 3\frac{1}{4}$ **34.** $9\frac{3}{16} \div 5\frac{1}{8}$ **35.** $28\frac{1}{10} \div 14\frac{1}{20}$ **36.** $40\frac{7}{8} \div 20\frac{7}{16}$

37. The approximate diameter of a circle is found by dividing the circumference by $3\frac{1}{7}$. What is the diameter of a circle whose circumference is $26\frac{3}{4}''$?

38. A space $50\frac{3}{4}''$ long is to be divided into 14 equal parts. What is the length of each part?

Answer to exercise 11

$\frac{3}{2}$ or $1\frac{1}{2}$

Answer to exercises 12 and 13

12. $\frac{1}{4}$

13. $\frac{25}{16}$ or $1\frac{9}{16}$

39. A cubic foot of liquid contains approximately $7\frac{1}{2}$ gallons. How many cubic feet are there in a 100-gallon tank?

40. The diagonal of a square is approximately the length of a side divided by $\frac{7}{10}$. How long is the diagonal of a square whose side is $1\frac{3}{4}''$?

2–8   PROBLEMS INVOLVING MULTIPLICATION AND DIVISION

Sometimes a problem involves more than two numbers and more than one operation. If there are parentheses in the problem, then do the operations inside the parentheses first. If the operations are multiplication and division, you can do the multiplication and division from left to right. If parentheses are part of the exercise, you can always reach the correct result by doing the operations inside the parentheses before doing the operations outside the parentheses.

■ **Example**

Find the value of $\left(\frac{2}{3} \times \frac{7}{8}\right) \div 3\frac{1}{2}$.

Solution $\left(\overset{1}{\cancel{\frac{2}{3}}} \times \frac{7}{\underset{4}{\cancel{8}}}\right) \div 3\frac{1}{2} = \frac{7}{12} \div 3\frac{1}{2} = \frac{7}{12} \div \frac{7}{2} = \frac{\overset{1}{\cancel{7}}}{\underset{6}{\cancel{12}}} \times \frac{2}{\underset{1}{\cancel{7}}} = \frac{1}{6}$

Explanation   Do the operation inside the parentheses first. Thus, $\left(\frac{2}{3} \times \frac{7}{8}\right) = \frac{7}{12}$. Then divide the $\frac{7}{12}$ by $3\frac{1}{2}$.

Practice check: Do exercises 14 and 15 on the right.

Find the following values.

14. $\left(\frac{3}{11} \div \frac{5}{6}\right) \times 5\frac{1}{2}$

15. $15 \times \left(2\frac{1}{2} \div 1\frac{1}{4}\right) \div \frac{3}{4}$

The answers are in the margin on page 43.

■ **Problems 2–8**

Do the following problems involving multiplication and division. Make sure you perform the operations inside the parentheses before any operations outside the parentheses. Reduce your answers to lowest terms.

1. $\left(\frac{2}{5} \times \frac{1}{3}\right) \div \frac{3}{4}$

2. $\left(\frac{7}{8} \times \frac{1}{2}\right) \div \frac{1}{7}$

3. $\left(2\frac{1}{3} \times \frac{5}{8}\right) \div 1\frac{1}{4}$

4. $(3\frac{1}{5} \times 1\frac{1}{2}) \times (2\frac{1}{2} \div 4\frac{2}{3})$

5. $(\frac{2}{3} \times \frac{15}{16}) \times (\frac{7}{8} \div \frac{1}{2})$

6. $(\frac{2}{7} \times \frac{5}{9}) \times (\frac{3}{10} \div 4)$

7. $(2 \times \frac{5}{6}) \times (6 \div \frac{7}{10})$

8. $(3\frac{1}{7} \times 2) \div 4$

9. $(4\frac{1}{2} \times 4\frac{1}{2} \times 16) \div 144$

10. $(18\frac{2}{3} \times 16\frac{1}{2}) \div 144$

11. $(5\frac{1}{3} \times 5\frac{1}{3} \times 4\frac{1}{2}) \div 1,728$

12. $(3\frac{3}{4} \times 3\frac{3}{4} \times 3\frac{1}{5}) \div 144$

13. $(\frac{7}{8} \times \frac{3}{8}) \div \frac{21}{64}$

14. $(\frac{3}{4} \times \frac{7}{8}) \div \frac{7}{8}$

15. $(1\frac{1}{2} \times \frac{5}{16}) \div 15$

16. $(2\frac{3}{4} \times 1\frac{2}{3}) \div \frac{5}{12}$

17. $(1\frac{5}{8} \times \frac{5}{8}) \div \frac{7}{8}$

18. $(2\frac{7}{16} \times \frac{7}{16}) \div \frac{1}{144}$

19. $(3\frac{1}{8} \times 1\frac{1}{2}) \div \frac{3}{16}$

20. $(1\frac{1}{2} \times 2\frac{3}{4}) \div \frac{3}{64}$

21. $(\frac{5}{8} \div \frac{7}{8}) \times 14$

22. $(\frac{7}{16} \div \frac{3}{4}) \times \frac{1}{2}$

23. $(\frac{1}{64} \div \frac{7}{8}) \times \frac{3}{4}$

24. $(2\frac{1}{64} \div \frac{1}{32}) \times (2\frac{1}{64} \div 1\frac{1}{2})$

2–9   CHANGING A FRACTION WITH A GIVEN DENOMINATOR TO ONE WITH A HIGHER DENOMINATOR

When adding, subtracting, or comparing fractions, you must often change the denominator of a fraction to some other denominator, usually a higher or greater denominator. This is done by multiplying both terms of the fraction by the same number, that is, multiplying the fraction by 1. Multiplying any number by 1 does not change the value of the number.

Rule

Multiplying both terms of a fraction by the same number gives a fraction equal in value to the original fraction.

■ **Example 1**

Change $\frac{3}{4}$ to a fraction having 20 for its denominator.

Solution $\dfrac{3}{4} = \dfrac{3 \times 5}{4 \times 5} = \dfrac{15}{20}$

Explanation   Since the required denominator of the new fraction is 20, multiply both numerator and denominator by a number that will make the new denominator 20 (5 is the required number). The original fraction and the fraction obtained by this process are called ***equivalent fractions.***

■ **Example 2**

Change $\frac{5}{16}$ to a fraction having 64 for its denominator.

Solution $\dfrac{5}{16} = \dfrac{5 \times 4}{16 \times 4} = \dfrac{20}{64}$

Explanation   Since the required denominator of the new fraction is 64, multiply both the numerator and denominator by 4. (The denominator is the key! You must ask yourself, "What number multiplies 16 to give an answer of 64?")

Practice check: Do exercises 16 and 17 on the right.

Answer to exercises 14 and 15

14. $\frac{9}{5}$ or $1\frac{4}{5}$

15. 40

Change the given fractions to equivalent fractions having the required higher denominators.

16. $\frac{3}{4}$ to 48ths

17. $\frac{5}{3}$ to 90ths

The answers are in the margin on page 46.

■ **Problems 2–9**

Change the given fractions to equivalent fractions having the required higher denominators. Use the space provided below to work out the answers.

1. $\frac{1}{2}$ to 4ths

2. $\frac{1}{2}$ to 10ths

3. $\frac{3}{8}$ to 32nds

4. $\frac{5}{8}$ to 64ths

5. $\frac{2}{3}$ to 12ths

6. $\frac{3}{5}$ to 10ths

7. $\frac{2}{3}$ to 15ths

8. $\frac{5}{7}$ to 35ths

9. $\frac{11}{12}$ to 180ths

10. $\frac{5}{13}$ to 52nds

11. $\frac{23}{24}$ to 144ths

12. $\frac{23}{24}$ to 360ths

13. $\frac{11}{18}$ to 360ths

14. $\frac{11}{18}$ to 144ths

15. $\frac{9}{16}$ to 64ths

16. $\frac{9}{16}$ to 32nds

17. $\frac{9}{16}$ to 144ths

18. $\frac{5}{6}$ to 60ths

19. $\frac{3}{4}$ to 60ths

20. $\frac{2}{5}$ to 60ths

21. $\frac{5}{4}$ to 16ths

22. $\frac{7}{10}$ to 100ths

23. $\frac{13}{8}$ to 32nds

24. $\frac{7}{5}$ to 60ths

2–10　　CHANGING TWO OR MORE FRACTIONS TO EQUIVALENT FRACTIONS HAVING A COMMON DENOMINATOR

When fractions have a ***common denominator,*** it is easy to compare them and determine which one is the greater by comparing their numerators—the greater numerator is the greater fraction. To add or subtract fractions with common denominators, add or subtract their numerators and place the result over their common denominator.

The lowest common denominator (***l.c.d.***) of the desired fractions will be the least common multiple (***l.c.m.***) of the given denominators. Very often the least common multiple can be obtained at sight. For example, the l.c.m. of 2, 4, 8, and 16 is 16; that is, 16 is the lowest number that 2, 4, 8, and 16 will divide evenly.

■ **Example 1**

Change $\frac{1}{2}$, $\frac{3}{4}$, $\frac{5}{8}$, and $\frac{5}{16}$ to equivalent fractions having the lowest common denominator.

Solution
$$\frac{1}{2} = \frac{1}{2} \times \frac{8}{8} = \frac{8}{16}$$
$$\frac{3}{4} = \frac{3}{4} \times \frac{4}{4} = \frac{12}{16}$$
$$\frac{5}{8} = \frac{5}{8} \times \frac{2}{2} = \frac{10}{16}$$
$$\frac{5}{16} = \frac{5}{16} \times \frac{1}{1} = \frac{5}{16}$$

Explanation　Beside each of the given fractions write a fraction line with the l.c.d. 16 below it. To obtain the new numerator, divide the l.c.d. by the given denominator and multiply the result by the given numerator. Thus,

16 divided by 2 is 8; 8 times 1 is 8, the new numerator.

16 divided by 4 is 4; 4 times 3 is 12, the new numerator.

16 divided by 8 is 2; 2 times 5 is 10, the new numerator.

16 divided by 16 is 1; 1 times 5 is 5, the new numerator.

When the least common multiple of two or more denominators cannot be obtained on sight, the following method may be used. Find all the prime factors of each denominator. (A prime factor is a factor that is divisible only by itself and 1.) Make a list of the prime factors according to the number of times the prime factor appears in any one denominator; that is, if the prime factor appears twice in a denominator, then list it twice, and so on. Every different prime factor must appear at least once on the list. The lowest common denominator (l.c.d.) is the product of all the prime factors on the list.

Practice check: Do exercise 18 on the right.

■ **Example 2**

Change $\frac{5}{8}$, $\frac{2}{3}$, and $\frac{1}{10}$ to equivalent fractions with the lowest common denominator.

Solution   Factor the denominators 8, 3, and 10 into prime factors.

$$8 = 2 \times 2 \times 2$$
$$3 = 3$$
$$10 = 2 \times 5$$

Every different prime factor must be used, and if a prime factor appears more than once in any number, then it is used as many times as it appears. The 8 in this problem has the prime number 2 appear three times, and therefore this factor must be used three times.

The prime factors 3 and 5 appear once. The lowest common denominator is therefore $2 \times 2 \times 2 \times 3 \times 5$, or 120.

$$\frac{5}{8} \times \frac{15}{15} = \frac{75}{120}$$
$$\frac{2}{3} \times \frac{40}{40} = \frac{80}{120}$$
$$\frac{1}{10} \times \frac{12}{12} = \frac{12}{120}$$

Practice check: Do exercise 19 on the right.

Change the following given fractions to equivalent fractions having the lowest common denominator.

18. $\frac{1}{3}$, $\frac{7}{10}$

The answer is in the margin on page 48.

Change the following given fractions to equivalent fractions having the lowest common denominator.

19. $\frac{2}{9}$, $\frac{5}{12}$, $\frac{1}{6}$

The answer is in the margin on page 48.

■ **Problems 2–10**

Change the following given fractions to equivalent fractions having the lowest common denominator. Use the space provided below to work out the answers.

1. $\frac{1}{2}$, $\frac{3}{4}$ **2.** $\frac{1}{2}$, $\frac{1}{3}$ **3.** $\frac{1}{2}$, $\frac{2}{3}$ **4.** $\frac{1}{2}$, $\frac{3}{8}$ **5.** $\frac{1}{3}$, $\frac{1}{4}$

6. $\frac{1}{3}$, $\frac{1}{5}$ **7.** $\frac{1}{3}$, $\frac{5}{6}$ **8.** $\frac{1}{2}$, $\frac{1}{3}$, $\frac{1}{4}$ **9.** $\frac{1}{2}$, $\frac{2}{3}$, $\frac{3}{4}$ **10.** $\frac{1}{2}$, $\frac{1}{3}$, $\frac{1}{6}$

11. $\frac{2}{3}, \frac{3}{4}, \frac{5}{6}$

12. $\frac{1}{2}, \frac{2}{3}, \frac{5}{6}, \frac{7}{8}$

13. $\frac{3}{8}, \frac{5}{16}$

14. $\frac{5}{8}, \frac{1}{4}, \frac{1}{16}$

15. $\frac{3}{4}, \frac{7}{8}, \frac{5}{16}$

16. $\frac{3}{10}, \frac{1}{2}, \frac{3}{4}$

17. $\frac{5}{12}, \frac{5}{24}$

18. $\frac{15}{16}, \frac{7}{8}, \frac{3}{4}, \frac{1}{2}$

19. $\frac{7}{8}, \frac{1}{10}, \frac{7}{12}$

20. $\frac{5}{32}, \frac{3}{64}, \frac{1}{2}$

21. $\frac{1}{13}, \frac{1}{11}$

22. $\frac{1}{5}, \frac{1}{7}, \frac{3}{17}$

23. $\frac{1}{7}, \frac{3}{35}, \frac{1}{2}$

24. $\frac{5}{64}, \frac{1}{100}, \frac{3}{10}$

25. Write the following fractions in ascending (smallest first) order: $\frac{2}{3}, \frac{5}{7}, \frac{5}{8}$ (*Hint*: Change the fractions having the same l.c.d.)

26. Write the following fractions in descending order: $\frac{3}{16}, \frac{5}{32}, \frac{11}{64}, \frac{1}{8}$

27. Which is greater: $\frac{3}{5}$ or $\frac{2}{3}$?

28. Which is greater: $\frac{7}{10}$ or $\frac{3}{4}$?

29. Is $\frac{13}{32}$ larger than $\frac{25}{64}$?

30. Is $\frac{2}{3}$ less than $\frac{43}{64}$?

2–11   ADDITION OF FRACTIONS

To add fractions having the same denominator, add their numerators and place the sum over the common denominator. Thus, $\frac{3}{8} + \frac{5}{8} + \frac{7}{8}$ means 3 eighths + 5 eights + 7 eights = 15 eights = $\frac{15}{8} = 1\frac{7}{8}$. To add fractions with different denominators, such as $\frac{3}{4} + \frac{7}{8} + \frac{5}{6}$, first change the given fractions to equivalent fractions having a common denominator. Then add the fractions by adding the numerators and placing the sum over the common denominator.

■ **Example**

Find the sum of $\frac{3}{4}$, $\frac{7}{8}$, and $\frac{5}{6}$.

Solution

$$\frac{3}{4} = \frac{18}{24}$$
$$\frac{7}{8} = \frac{21}{24}$$
$$\frac{5}{6} = \frac{20}{24}$$
$$\overline{\phantom{\frac{5}{6} = }\frac{59}{24} = 2\frac{11}{24}}$$

Explanation   The sum of 18 twenty-fourths and 21 twenty-fourths and 20 twenty-fourths is 59 twenty-fourths, written $\frac{59}{24}$ and changed to the mixed number $2\frac{11}{24}$.

Practice check: Do exercises 20 and 21 on the right.

Find the sums.

20. $\frac{3}{4} + \frac{1}{5}$

21. $\frac{1}{6} + \frac{7}{15} + \frac{2}{5}$

The answers are in the margin on page 50.

■ **Problems 2–11**

Find the sums of the following fractions. Your answers should be in lowest terms. Use the space provided below to work out your answers.

1. $\frac{1}{2}$, $\frac{3}{4}$

2. $\frac{1}{3}$, $\frac{1}{2}$

3. $\frac{2}{3}$, $\frac{3}{4}$

4. $\frac{1}{5}$, $\frac{3}{10}$

5. $\frac{2}{3}$, $\frac{5}{8}$

6. $\frac{5}{8}$, $\frac{3}{4}$

7. $\frac{7}{8}$, $\frac{7}{16}$

8. $\frac{5}{16}$, $\frac{5}{32}$

9. $\frac{3}{4}$, $\frac{7}{8}$

10. $\frac{3}{32}$, $\frac{15}{16}$

11. $\frac{15}{16}$, $\frac{31}{32}$

12. $\frac{29}{32}$, $\frac{63}{64}$

13. $\frac{1}{8} + \frac{3}{8} + \frac{5}{8} + \frac{7}{8}$

14. $\frac{29}{21} + \frac{13}{21} + \frac{5}{21} + \frac{38}{21}$

15. $\frac{3}{16} + \frac{7}{16} + \frac{5}{16} + \frac{9}{16} + \frac{11}{16}$

16. $\frac{1}{2} + \frac{3}{4} + \frac{5}{8} + \frac{7}{16}$

17. $\frac{7}{12} + \frac{3}{10} + \frac{8}{15}$

18. $\frac{3}{7} + \frac{5}{14} + \frac{7}{18} + \frac{1}{9}$

19. $\frac{7}{8} + \frac{2}{3} + \frac{1}{6} + \frac{1}{2}$

20. $\frac{5}{6} + \frac{3}{14} + \frac{3}{4} + \frac{11}{12}$

21. $\frac{15}{16} + \frac{8}{3} + \frac{2}{3}$

22. $\frac{9}{16} + \frac{7}{8} + \frac{15}{32}$

23. $\frac{1}{8} + \frac{5}{12} + \frac{11}{18} + \frac{23}{24}$

24. $\frac{2}{3} + \frac{3}{4} + \frac{5}{7} + \frac{7}{12}$

25. $\frac{2}{5} + \frac{5}{9} + \frac{1}{8} + \frac{7}{20}$ **26.** $\frac{3}{5} + \frac{3}{7} + \frac{3}{8} + \frac{3}{10}$

Answer to exercise 18

$\frac{10}{30}, \frac{21}{30}$

Answer to exercise 19

$\frac{8}{36}, \frac{15}{36}, \frac{6}{36}$

27. $\frac{9}{28} + \frac{31}{49} + \frac{16}{21} + \frac{32}{35}$ **28.** $\frac{1}{10} + \frac{1}{100} + \frac{1}{1000}$

29. $\frac{1}{10} + \frac{3}{100} + \frac{17}{1000}$ **30.** $\frac{7}{10} + \frac{21}{100} + \frac{7}{1000}$

2–12   ADDITION OF MIXED NUMBERS

When adding mixed numbers, you will often find it easier to add the mixed numbers if their fractional parts have a common denominator. Remember that the mixed number represents a fraction added to a whole number; that is, $1\frac{1}{2}$ means $\frac{1}{2}$ added to 1.

■ **Example**

Find the sum of $5\frac{3}{8}$, $6\frac{1}{4}$, $2\frac{1}{3}$, and $7\frac{1}{2}$.

Solution   The sum of the mixed numbers is equal to the sum of the whole numbers added to the sum of the fractions. The work may conveniently be arranged as shown.

$$
\begin{aligned}
5\tfrac{3}{8} &= 5\tfrac{9}{24} \\
6\tfrac{1}{4} &= 6\tfrac{6}{24} \\
2\tfrac{1}{3} &= 2\tfrac{8}{24} \\
7\tfrac{1}{2} &= 7\tfrac{12}{24} \\
\hline
& 20\tfrac{35}{24} = 21\tfrac{11}{24}
\end{aligned}
$$

Find the sums.

22. $2\frac{2}{3} + 3\frac{1}{4}$

23. $4\frac{2}{5} + 3 + 8\frac{7}{15} + 3\frac{3}{10}$

The answers are in the margin on page 50.

Explanation   The sum of the fractions is $\frac{35}{24}$, but $\frac{35}{24} = 1\frac{11}{24}$. The 1 is added to the 20 which is the sum of the whole number part. The complete answer is $21\frac{11}{24}$.

Practice check: Do exercises 22 and 23 on the left.

■ **Problems 2–12**

Find the sums of the following mixed numbers. Your answers should be in lowest terms. Use the space provided to work out the answers.

1. $2\frac{1}{2} + 4\frac{3}{4}$

2. $5\frac{1}{3} + 2\frac{1}{2}$

3. $1\frac{3}{4} + 3\frac{3}{8}$

4. $3\frac{1}{2} + 5\frac{2}{3}$

5. $6\frac{2}{3} + 2\frac{1}{6}$

6. $10\frac{1}{8} + 9\frac{15}{16}$

7. $3\frac{5}{8} + 1\frac{5}{16}$

8. $5\frac{7}{8} + 12\frac{7}{16}$

9. $7\frac{1}{10} + 5\frac{4}{5}$

10. $2\frac{5}{16} + 7\frac{3}{4}$

11. $11\frac{1}{3} + 15\frac{1}{8}$

12. $17\frac{1}{2} + 10\frac{9}{10}$

13. $2\frac{1}{2} + 5\frac{1}{2} + 3\frac{1}{2} + 7\frac{1}{2}$

14. $3\frac{7}{16} + 4\frac{5}{16} + 2\frac{11}{16} + 8\frac{9}{16}$

15. $5\frac{3}{4} + 2\frac{1}{8} + 3\frac{1}{16} + 3\frac{7}{32}$

16. $2\frac{1}{5} + 8\frac{2}{3} + 10\frac{3}{4} + 4\frac{3}{7}$

17. $6\frac{7}{16} + \frac{7}{8} + \frac{2}{3} + 3\frac{3}{4}$

18. $4 + 11\frac{3}{5} + \frac{7}{8} + 3\frac{1}{2}$

19. $14\frac{3}{10} + 8\frac{7}{15} + 6\frac{9}{20} + 2\frac{17}{30}$

20. $\frac{19}{32} + \frac{11}{16} + 18 + 12$

21. $1\frac{3}{8} + 5\frac{7}{8} + 3\frac{5}{8} + 6\frac{7}{8}$

22. $9\frac{3}{4} + 6\frac{1}{2} + 4\frac{3}{8}$

23. $5\frac{3}{8} + 4\frac{1}{4} + 8\frac{11}{12}$

24. $9\frac{5}{6} + 12\frac{3}{4} + 10\frac{15}{16} + 4\frac{11}{12}$

25. $8\frac{1}{2} + 6\frac{1}{3} + 5\frac{1}{4} + 7\frac{1}{6} + 12\frac{1}{18}$

26. $8\frac{1}{2} + 4\frac{5}{6} + 3\frac{2}{3} + 2\frac{1}{6}$

27. $1\frac{11}{16} + 15 + 12 + 3\frac{7}{8}$

28. $5\frac{2}{3} + \frac{49}{240} + 4\frac{7}{120} + 8\frac{21}{80}$

29. Add the following dimensions: $1\frac{7}{8}''$, $2\frac{1}{4}''$, $\frac{9}{64}''$, $\frac{11}{32}''$, $3\frac{1}{2}''$.

30. Add the following weights: $1\frac{1}{4}$ lb, $\frac{1}{2}$ lb, $\frac{7}{16}$ lb, $1\frac{5}{16}$ lb, $\frac{7}{8}$ lb.

31. Add the following dimensions: $6\frac{1}{4}''$, $1\frac{3}{64}''$, $3\frac{7}{8}''$, $2\frac{11}{32}''$, $\frac{1}{2}''$, $1\frac{9}{16}''$, $5\frac{5}{8}''$.

32. Add the following dimensions: $\frac{15}{64}''$, $\frac{13}{32}''$, $\frac{5}{8}''$, $\frac{53}{64}''$, $1\frac{7}{16}''$, $\frac{35}{64}''$, $1\frac{9}{16}''$.

33. Cheryl worked $5\frac{1}{2}$ hours on Monday, $6\frac{1}{4}$ hours on Tuesday, 8 hours on Wednesday, $6\frac{1}{2}$ hours on Thursday, and $7\frac{1}{4}$ hours on Friday. What is the total number of hours she worked?

34. John worked $4\frac{1}{2}$ hours on Monday and $8\frac{1}{2}$ hours on Tuesday. How many hours did he work?

35. Betsy worked $6\frac{1}{2}$ hours on Monday, $6\frac{1}{2}$ hours on Tuesday, $6\frac{1}{2}$ hours on Wednesday, $6\frac{1}{2}$ on Thursday, and 8 hours on Friday. How many hours did she work this week?

2–13 SUBTRACTION OF FRACTIONS

To subtract two fractions, you must have a common denominator. The procedure is much like adding fractions, except that the numerators are subtracted and the difference is placed over the common denominator.

■ **Example 1**

Take $\frac{3}{9}$ from $\frac{7}{9}$.

Solution

$$\begin{array}{ll} \frac{7}{9} & \text{(minuend)} \\ -\frac{3}{9} & \text{(subtrahend)} \\ \hline \frac{4}{9} & \text{(difference)} \end{array}$$

Explanation   Since the two fractions have a common denominator, subtract the subtrahend's numerator, 3, from the minuend's numerator, 7, and write the result over the common denominator. As shown, 3 ninths subtracted from 7 ninths is 4 ninths.

■ **Example 2**

Take $\frac{2}{3}$ from $\frac{7}{8}$.

Solution

$$\begin{array}{lll} \frac{7}{8} & = & \frac{21}{24} \\ -\frac{2}{3} & = & \frac{16}{24} \\ \hline & & \frac{5}{24} \end{array}$$

Explanation   Since the fractions have different denominators, change them to fractions having a common denominator. The lowest common denominator of the two fractions is the least common multiple of 8 and 3, namely, 24. The given fractions become $\frac{21}{24}$ and $\frac{16}{24}$. The problem is now similar to Example 1.

Practice check: Do exercises 24 and 25 on the right.

Do the indicated subtractions.

24. $\frac{5}{8} - \frac{1}{3}$

25. $\frac{5}{6} - \frac{7}{15}$

The answers are in the margin on page 55.

■ Problems 2–13

Do the indicated subtractions. Simplify your answers. Use the space provided below to work out the answers.

1. Take $\frac{5}{16}$ from $\frac{9}{16}$.

2. Take $\frac{3}{8}$ from $\frac{7}{8}$.

3. From $\frac{3}{4}$ take $\frac{1}{2}$.

4. From $\frac{29}{32}$ take $\frac{5}{8}$.

5. $\frac{1}{2} - \frac{1}{3}$

6. $\frac{4}{5} - \frac{3}{4}$

7. $\frac{1}{8} - \frac{5}{16}$

8. $\frac{13}{15} - \frac{9}{20}$

9. $\frac{14}{15} - \frac{7}{12}$

10. $\frac{5}{12} - \frac{5}{18}$

11. $\frac{7}{8} - \frac{3}{10}$

12. $\frac{1}{12} - \frac{1}{144}$

13. How much more is $\frac{5}{8}$ than $\frac{1}{2}$?

14. How much more is $\frac{3}{4}$ than $\frac{2}{3}$?

15. How much less is $\frac{1}{4}$ than $\frac{1}{3}$?

16. How much less is $\frac{3}{5}$ than $\frac{5}{7}$?

17. A planer takes a $\frac{3}{32}''$ cut from a piece of steel $\frac{3}{8}''$ thick. What is the remaining thickness?

18. By how much must the diameter of a $\frac{3}{4}''$ shaft be reduced to bring it down to a diameter of $\frac{41}{64}''$?

19. What is the difference in thickness between a $\frac{9}{16}''$-thick steel plate and a $\frac{7}{8}''$-thick steel plate?

20. A steel casting $\frac{7}{8}''$ thick is finished by taking a cut $\frac{3}{64}''$ deep. What is the final thickness?

2–14   SUBTRACTION OF MIXED NUMBERS

To subtract two mixed numbers, you must use common denominators. Subtract the fractional parts first and the whole number parts second. It may be necessary to borrow a 1 from the greater whole number and add it to the fractional part so the fractions can be subtracted. Usually, the **minuend** is the larger number in a subtraction problem and the **subtrahend** is the smaller number. The answer is called the **difference.**

■ **Example 1**

Take $7\frac{3}{4}$ from $15\frac{5}{6}$.

Solution
$$
\begin{array}{rl}
15\frac{5}{6} = & 15\frac{10}{12} \quad \text{(minuend)} \\
- \ 7\frac{3}{4} = - & 7\frac{9}{12} \quad \text{(subtrahend)} \\
\hline
& 8\frac{1}{12} \quad \text{(difference)}
\end{array}
$$

Explanation   This case is similar to that of adding mixed numbers. First find the difference between the fractions and then find the difference between the whole numbers. The complete answer consists of the two results.

■ **Example 2**

Take $5\frac{2}{3}$ from $8\frac{3}{8}$.

Solution
$$
\begin{array}{rl}
8\frac{3}{8} = & 8\frac{9}{24} = (7\frac{24}{24} + \frac{9}{24}) = & 7\frac{33}{24} \\
- \ 5\frac{2}{3} = - & 5\frac{16}{24} = & - 5\frac{16}{24} \\
\hline
& & 2\frac{17}{24}
\end{array}
$$

Do the indicated subtractions.

26. $7\frac{7}{12} - 2\frac{1}{4}$

27. $8\frac{2}{5} - 5$

The answers are in the margin on page 55.

Explanation   It is not possible to take $\frac{2}{3}$ from $\frac{3}{8}$, because $\frac{2}{3}$ is larger than $\frac{3}{8}$. When written with a common denominator, $\frac{3}{8} = \frac{9}{24}$, and $\frac{2}{3} = \frac{16}{24}$. Since $\frac{16}{24}$ is larger than $\frac{9}{24}$, borrow 1 from the whole number 8 and change this borrowed 1 to twenty-four twenty-fourths, $\frac{24}{24}$. Add $\frac{24}{24}$ to $\frac{9}{24}$, giving $(24 + 9)/24 = \frac{33}{24}$. Instead of writing $8\frac{9}{24}$ in the minuend, write its equivalent number $7\frac{33}{24}$. To take $5\frac{16}{24}$ from $7\frac{33}{24}$, take $\frac{16}{24}$ from $\frac{33}{24}$, giving $\frac{17}{24}$; and take 5 from 7, giving 2. The answer is $2\frac{17}{24}$.

Practice check: Do exercises 26 and 27 on the left.

■ **Example 3**

A steel bar $6'$ long has four pieces, each $4\frac{5}{8}''$ long, cut from it. The saw kerf (width of the saw cut) is $\frac{1}{8}''$. How long is the remainder of the bar?

Solution

$$4\tfrac{5}{8}'' + \tfrac{1}{8}'' = 4\tfrac{3}{4}'' \text{ (length of one piece plus the saw cut)}$$
$$4\tfrac{3}{4}'' + 4\tfrac{3}{4}'' + 4\tfrac{3}{4}'' + 4\tfrac{3}{4}'' = 19'' \text{ or } 1'7'' \text{ (length of four pieces}$$
$$\text{plus four saw cuts)}$$

$$6' - 1'7'' = 5'12'' - 1'7''$$
$$= 4'5'' \text{ (length of remaining bar)}$$

Explanation   The kerf must be added to the length of each cut. The sum of the lengths of all the pieces plus the width of the saw cuts must be subtracted from the length of the original bar to find the length of the bar that would be left in stock. Note that $12''$ is borrowed from the $6'$ to enable the subtraction of $1'7''$ from $6'0''$.

Practice check: Do exercise 28 on the right.

28. The total length of a piece of square bar stock is $15\tfrac{7}{8}$ in. If $7\tfrac{7}{16}$ in. of this bar stock is turned to a cylinder, what length will remain square?

The answer is in the margin on page 55.

■ **Problems 2–14**

Do the indicated subtractions. Simplify your answers. Use the space provided below to work out the answers.

1. Take 5 from $9\tfrac{7}{8}$.

2. Take 2 from $6\tfrac{3}{4}$.

3. Take $2\tfrac{13}{32}$ from $5\tfrac{19}{32}$.

4. From $18\tfrac{15}{16}$ take $4\tfrac{9}{16}$.

5. From $12\tfrac{5}{8}$ take $8\tfrac{1}{4}$.

6. Take $\tfrac{1}{3}$ from 2.

7. $4 - \tfrac{3}{8}$

8. $9 - \tfrac{7}{16}$

9. $6\tfrac{1}{2} - 2\tfrac{1}{8}$

10. $3\tfrac{2}{5} - 1\tfrac{1}{3}$

11. $10\tfrac{1}{3} - 7\tfrac{3}{4}$

12. $5\tfrac{2}{5} - 3\tfrac{7}{8}$

13. $14\tfrac{7}{18} - 8\tfrac{19}{24}$

14. $32\tfrac{5}{16} - 21\tfrac{11}{12}$

15. $5\tfrac{3}{8} - \tfrac{7}{8}$

16. $6\tfrac{3}{4} - \tfrac{15}{16}$

17. $8\tfrac{3}{7} - 6\tfrac{7}{9}$

18. $16\tfrac{25}{64} - 10\tfrac{7}{8}$

19. $4\tfrac{1}{4} - 2\tfrac{3}{5}$

20. $8\tfrac{1}{3} - 6\tfrac{5}{7}$

21. $15\tfrac{3}{4} - 12\tfrac{7}{8}$

22. $19\tfrac{1}{16} - 10\tfrac{17}{32}$

23. $12\frac{3}{4} - 7\frac{15}{16}$

24. $75\frac{1}{2} - 37\frac{63}{64}$

25. $16\frac{1}{8} - 7\frac{3}{16}$

26. $12\frac{2}{3} - 8\frac{1}{8}$

27. $9\frac{3}{4} - 2\frac{1}{2}$

28. $3\frac{5}{8} - 1\frac{5}{16}$

29. $7\frac{13}{32} - 3\frac{13}{32}$

30. $4\frac{17}{64} - 2\frac{5}{8}$

31. $2\frac{7}{8} - 2\frac{3}{4}$

32. $8\frac{1}{2} - 6$

33. $17\frac{15}{16} - 11\frac{7}{8}$

34. $23\frac{1}{16} - 15\frac{15}{16}$

35. $14\frac{7}{16} - 11\frac{23}{32}$

36. $29\frac{3}{4} - 21\frac{7}{8}$

37. $67\frac{1}{8} - 37\frac{1}{4}$

38. $43\frac{5}{16} - 23\frac{9}{16}$

39. $22\frac{13}{64} - 12\frac{1}{2}$

40. $104\frac{1}{2} - 75\frac{3}{4}$

Do the indicated subtractions and additions in the order that they appear in the problem.

41. Simplify $16\frac{3}{4} - 2\frac{1}{2} - 4\frac{3}{16} + 2$.

42. Find the value of $5\frac{3}{7} + 8\frac{7}{10} - 3\frac{1}{3} - 1\frac{1}{5}$.

43. Find the value of $8\frac{3}{4} - 1\frac{7}{16} + 4 - 2\frac{1}{3}$.

44. From a bar of brass $16\frac{1}{2}''$ long the following three pieces are cut: $1\frac{1}{8}''$, $3\frac{1}{2}''$, $3\frac{3}{32}''$. What is the final length of the bar, allowing $\frac{1}{16}''$ for each cut?

45. Allowing $\frac{1}{8}''$ for each cut, what must be the length of a bar of brass to provide the following four pieces: $1\frac{7}{8}''$, $3\frac{3}{16}''$, $\frac{11}{32}''$, $2\frac{1}{4}''$?

46. Add the following dimensions: $2\frac{7}{16}''$, $3\frac{1}{8}''$, $5\frac{9}{64}''$, $4\frac{1}{32}''$, $\frac{1}{2}''$.

47. A shaft required to be $\frac{11}{16}''$ in diameter measures $\frac{49}{64}''$. How much oversized is it?

48. A planer takes a $\frac{3}{32}''$ cut on a plate that is $1\frac{3}{4}''$ thick. What is the final thickness of the plate?

49. A piece of tapered stock has a diameter of $1\frac{1}{10}''$ at one end and $\frac{57}{64}''$ at the other end. Find the difference between the two diameters.

50. Sam could have worked 40 hours this week. He worked $7\frac{1}{2}$ hours on Monday, $5\frac{1}{4}$ hours on Tuesday, and $8\frac{1}{2}$ hours on Wednesday. He did not work the rest of the week. How many hours was he short of 40 hours?

Answer to exercises 24 and 25

24. $\frac{7}{24}$

25. $\frac{11}{30}$

Answer to exercises 26 and 27

26. $5\frac{4}{12}$ or $5\frac{1}{3}$

27. $3\frac{2}{5}$

Answer to exercise 28

$8\frac{7}{16}$ in.

2–15 PROBLEMS INVOLVING ADDITION, SUBTRACTION, MULTIPLICATION, AND DIVISION

Problems involving more than two numbers and more than one operation with a combination of multiplication and division were discussed in Section 2–8. Problems can involve all four operations with more than two numbers. When this is the case, operations within parentheses are done first. Operations with multiplication and division from left to right are done next, and operations with addition and subtraction from left to right are done last.

Rule _____

Order of Operations

1. *Operations within parentheses are done first.*
2. *Operations with multiplication and division from left to right are done second.*
3. *Operations with addition and subtraction from left to right are done last.*

■ **Example 1**

Find the value of $\frac{2}{3} \times (16 + 8) - 9 \div \frac{3}{4}$.

Solution　$\frac{2}{3} \times (24) - 9 \div \frac{3}{4} = 16 - 9 \div \frac{3}{4} = 16 - 12 = 4$

Explanation　First do the operation inside the parentheses; thus, $(16 + 8) = 24$. Second, multiply $\frac{2}{3}$ by 24. Third, divide 9 by $\frac{3}{4}$. Finally, subtract 12 from 16.

■ Example 2

Find the value of $3 + 2\frac{1}{2} \times (1\frac{1}{2} + \frac{1}{4}) \div \frac{3}{2}$.

Solution　$3 + 2\frac{1}{2} \times (1\frac{1}{2} + \frac{1}{4}) \div \frac{3}{2} = 3 + \frac{5}{2} \times (\frac{3}{2} + \frac{1}{4}) \div \frac{3}{2} = 3 + \frac{5}{2} \times (\frac{9}{4}) \div \frac{3}{2}$
$= 3 + \frac{45}{8} \div \frac{3}{2} = 3 + \frac{45}{8} \times \frac{2}{3} = 3 + \frac{15}{4}$
$= \frac{12}{4} + \frac{15}{4} = \frac{27}{4} \text{ or } 6\frac{3}{4}$

Find the value of the following.

29. $7\frac{1}{2} - (4\frac{1}{2} + \frac{1}{3} \div \frac{1}{3})$

The answer is in the margin on page 58.

Explanation　First, change the mixed numbers to improper fractions. Second, add $\frac{3}{2}$ and $\frac{1}{4}$. Third, multiply $\frac{5}{2}$ by $\frac{9}{4}$. Fourth, divide $\frac{45}{8}$ by $\frac{3}{2}$. Finally, add 3 and $\frac{15}{4}$.

Practice check: Do exercise 29 on the left.

■ Problems 2–15

Do the following problems involving multiplication, division, addition, and subtraction. Make sure you perform the operations inside the parentheses before performing any operations outside the parentheses. Reduce your answers to lowest terms. Use the space provided below to work out the answers.

1. $\frac{3}{4} + 33 \div 4\frac{1}{8}$

2. $9\frac{1}{6} - 1\frac{1}{4} \times 4$

3. $(\frac{1}{2} + \frac{5}{8}) \times \frac{2}{3}$

4. $\frac{7}{8} - \frac{5}{6} \div \frac{4}{3} + \frac{5}{6}$

5. $2\frac{1}{3} \div (1\frac{1}{4} + 3\frac{1}{6})$

6. $(2\frac{3}{4} - 1\frac{1}{5}) \div 3$

7. $\frac{3}{8} - (\frac{2}{3} \div \frac{4}{5} - \frac{1}{2})$

8. $\frac{3}{4} + (\frac{4}{5} \times \frac{5}{8} + \frac{2}{3})$

9. $2\frac{2}{3} \times (\frac{3}{7} + 2) \div 4\frac{6}{7}$

10. $3\frac{1}{3} \times (\frac{3}{5} + 3 \div 2\frac{2}{5})$

11. $(1\frac{5}{6} + \frac{2}{3} \times 2\frac{1}{2}) - 3$

12. $\frac{5}{8} + \frac{2}{3} \div (\frac{1}{4} + \frac{1}{6}) - \frac{1}{8}$

13. $(6 \times 3 + 4) \times \frac{2}{11} - 2\frac{1}{2}$

14. $6\frac{2}{3} \div 2 - \frac{1}{3} + 8 \times (\frac{3}{4} - \frac{5}{8})$

15. $1\frac{3}{7} \div 5 + (8 - 6) \times 2\frac{1}{4} - 4$

16. $5\frac{1}{4} + 11 \div (3\frac{1}{8} - 1\frac{3}{4} + \frac{1}{8}) - 3$

17. $(\frac{1}{2} \times 3\frac{1}{3} + \frac{2}{3}) \div (3\frac{2}{3} - \frac{4}{5} \times 2\frac{1}{2})$

18. $\left(\frac{2}{3} \times 3\frac{1}{2} - 1\frac{1}{4}\right) \div \left(2\frac{1}{4} + \frac{3}{4} \times 5\frac{1}{3}\right)$ **19.** $\left(\frac{3}{4} \times \frac{2}{3} - \frac{1}{2} \times \frac{1}{6}\right) \times 3\frac{1}{4} - 1$ **20.** $\left(\frac{2}{5} \times \frac{1}{2} + \frac{3}{5} \times \frac{1}{2}\right) \div 4\frac{1}{2} + 2$

2–16   COMPLEX FRACTIONS

Sometimes an arithmetic operation involves a fraction where a fraction or a mixed number is the numerator and a fraction or a mixed number is the denominator. These fractions are called *complex fractions.*

     A complex fraction is one in which the numerator, the denominator, or both are fractions or mixed numbers. It was pointed out that a fraction may be regarded as an indicated division in which the numerator is the dividend and the denominator is the divisor. To evaluate the given expression, perform the indicated division.

■ **Example 1**

Find the value of the expression $\dfrac{\frac{2}{3}}{\frac{5}{7}}$.

Solution $\dfrac{\frac{2}{3}}{\frac{5}{7}} = \frac{2}{3} \div \frac{5}{7} = \frac{2}{3} \times \frac{7}{5} = \frac{14}{15}$

Explanation   This complex fraction indicates that $\frac{2}{3}$ is to be divided by $\frac{5}{7}$. Proceed as in division of fractions by inverting the divisor ($\frac{5}{7}$) and then multiplying the two fractions.

■ **Example 2**

Find the value of $\dfrac{2\frac{3}{4}}{16}$.

Solution $2\frac{3}{4} \div \frac{16}{1} = \frac{11}{4} \times \frac{1}{16} = \frac{11}{64}$

Explanation   This complex fraction indicates that $2\frac{3}{4}$ is to be divided by 16. Change the $2\frac{3}{4}$ to an improper fraction and proceed as in the division of fractions by inverting the divisor to $\frac{1}{16}$ and then multiplying the two fractions.

     Practice check: Do exercise 30 on the right.

■ **Example 3**

Find the value of $\dfrac{3\frac{1}{4} - 2\frac{1}{8}}{4\frac{1}{8} - 1\frac{1}{2}}$.

Solution   Subtraction is involved in both the numerator and the denominator. To solve this problem, simplify the numerator and denominator separately before simplifying the resulting fraction. Thus, $3\frac{1}{4} - 2\frac{1}{8} = 1\frac{1}{8}$ and $4\frac{1}{8} - 1\frac{1}{2} = 2\frac{5}{8}$.

$$\frac{3\frac{1}{4} - 2\frac{1}{8}}{4\frac{1}{8} - 1\frac{1}{2}} = \frac{1\frac{1}{8}}{2\frac{5}{8}} = \frac{\overset{3}{\cancel{9}}}{\cancel{8}} \times \frac{\overset{1}{\cancel{8}}}{\cancel{21}_{7}} = \frac{3}{7}$$

Find the value of the following.

30. $\dfrac{4\frac{1}{4}}{2\frac{1}{16}}$

The answer is in the margin on page 60.

Answer to exercise 29

2

Explanation   First simplify the numerator by performing the indicated subtraction, giving the mixed number $1\frac{1}{8}$. Then simplify the denominator in the same way, obtaining $2\frac{5}{8}$. Finally, simplify the resulting fraction $\dfrac{1\frac{1}{8}}{2\frac{5}{8}}$ by carrying out the indicated division of $1\frac{1}{8}$ divided by $2\frac{5}{8}$.

■ **Example 4**

Find the value of $\dfrac{\frac{3}{4} \text{ of } 2\frac{3}{8}}{3\frac{1}{4} \div 14\frac{1}{3}}$.

Find the value of the following.

31. $\dfrac{5\frac{5}{9} + \frac{2}{3}}{7\frac{2}{3} - 6\frac{1}{2}}$

The answer is in the margin on page 60.

Solution $\frac{3}{4} \text{ of } 2\frac{3}{8} = \frac{3}{4} \times \frac{19}{8} = \frac{57}{32}$

$3\frac{1}{4} \div 14\frac{1}{3} = \frac{13}{4} \times \frac{3}{43} = \frac{39}{172}$

$\dfrac{\frac{3}{4} \text{ of } 2\frac{3}{8}}{3\frac{1}{4} \div 14\frac{1}{3}} = \frac{57}{32} \div \frac{39}{172}$

$\frac{57}{32} \div \frac{39}{172} = \dfrac{\overset{19}{\cancel{57}}}{\underset{8}{\cancel{32}}} \times \dfrac{\overset{43}{\cancel{172}}}{\underset{13}{\cancel{39}}} = \frac{817}{104} = 7\frac{89}{104}$

Practice check: Do exercise 31 on the left.

■ **Problems 2–16**

Simplify the following complex fractions. Reduce your answers to lowest terms. Use the space provided below to work out the answers.

1. $\dfrac{\frac{1}{2}}{\frac{1}{3}}$

2. $\dfrac{\frac{3}{5}}{4\frac{1}{8}}$

3. $\dfrac{6}{3\frac{2}{3}}$

4. $\dfrac{12\frac{1}{2}}{16\frac{2}{3}}$

5. $\dfrac{2\frac{5}{8}}{4}$

6. $\dfrac{4\frac{9}{16} - 3\frac{1}{8}}{2\frac{1}{4} + 1\frac{3}{16}}$

7. $\dfrac{5\frac{1}{2} - 4\frac{3}{4}}{3\frac{7}{8} - 2\frac{15}{16}}$

8. $\dfrac{3\frac{1}{2} + 5\frac{1}{4}}{4\frac{1}{4} + 2\frac{1}{2}}$

9. $\dfrac{\frac{2}{3} \text{ of } \frac{3}{4}}{\frac{1}{2} \text{ of } \frac{7}{8}}$

10. $\dfrac{2\frac{1}{3} \times 3\frac{1}{3}}{7 \times 6\frac{2}{3}}$

11. $\dfrac{8\frac{4}{5} \div 3\frac{1}{7}}{10\frac{1}{2} \times 6\frac{2}{3}}$

12. $\dfrac{\frac{7}{16} \text{ of } 4\frac{2}{5}}{\frac{2}{5} \text{ of } \frac{9}{14}}$

13. $\dfrac{4\frac{1}{4} - 2\frac{1}{2}}{3\frac{1}{2} + 2\frac{1}{2}}$

14. $\dfrac{3\frac{1}{8} + 4\frac{3}{8}}{5 - 1\frac{1}{2}}$

15. $\dfrac{\frac{7}{8} - \frac{1}{2}}{\frac{7}{8} - \frac{1}{2}}$

16. $\dfrac{\frac{5}{8} \times \frac{8}{9}}{\frac{3}{4} \div \frac{3}{8}}$

17. $\dfrac{\frac{5}{16} \div \frac{3}{16}}{4 \times \frac{1}{2}}$

18. $\dfrac{3\frac{1}{2} \times 4\frac{1}{2}}{\frac{5}{8} \times \frac{3}{8}}$

19. $\dfrac{\frac{2}{3} \div 1\frac{2}{3}}{\frac{2}{5} \div 3\frac{1}{5}}$

20. $\dfrac{4 + 1\frac{2}{3}}{5 + 3\frac{1}{3}}$

21. $\dfrac{\frac{4}{5} + \frac{5}{16}}{\frac{3}{20} + 1\frac{3}{4}}$

22. $\dfrac{1\frac{2}{5} + 3\frac{1}{4}}{5\frac{1}{8} - 1\frac{1}{10}}$

23. $\dfrac{\frac{2}{3} \times 3\frac{1}{16}}{\frac{5}{16} \div \frac{3}{4}}$

24. $\dfrac{3\frac{1}{2} \div 4\frac{1}{2}}{5\frac{1}{2} \times 6\frac{1}{2}}$

■ Miscellaneous Problems in Common Fractions

1. Find the total length of the shaft in Fig. 2–2.

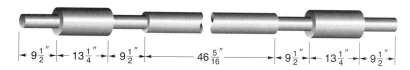

Figure 2–2

2. Find the total width of the boiler plate in Fig. 2–3.

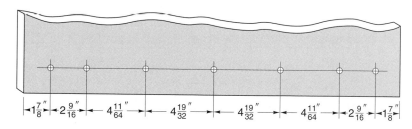

Figure 2–3

3. What is the outside diameter (O.D.) of the 1-in.-I.D. tubing in Fig. 2–4?

Figure 2–4

4. Find the distance between the end holes of the splice plate in Fig. 2–5.

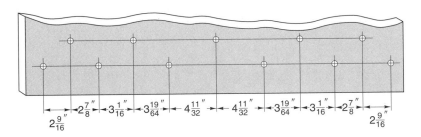

Figure 2–5

5. Find the total length of the bolt in Fig. 2–6.

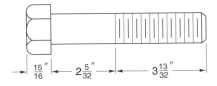

Figure 2–6

6. Find the total length of the spindle shown in Fig. 2–7.

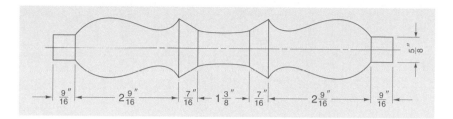

Figure 2–7

7. What is the total length of the piece needed to make the handle for the gavel shown in Fig. 2–8?

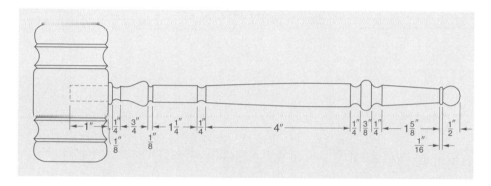

Figure 2–8

8. What is the length of stock needed to make the bracket shown in Fig. 2–9? (Add $\frac{1}{2}$ of radius, R, for each bend.)

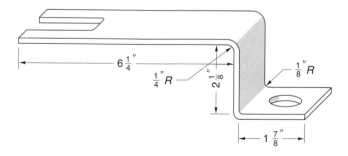

Figure 2–9

9. Compute the overall length of the bevel gear blank in Fig. 2–10.

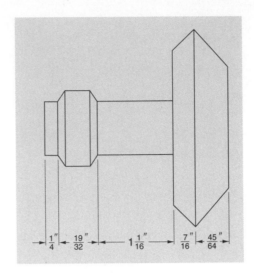

Figure 2–10

10. Compute the overall length of the gland in Fig. 2–11.

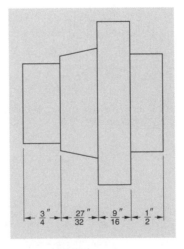

Figure 2–11

11. Compute the overall length of the wheel hub in Fig. 2–12.

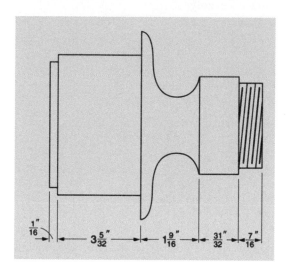

Figure 2–12

12. Find the difference between the diameters at the ends of the tapered piece in Fig. 2–13.

Figure 2–13

13. Supply the missing dimension in Fig. 2–14 if the overall length is $1\frac{5}{8}''$.

Figure 2–14

14. Find the center-to-center distance of holes in the crankcase cover in Fig. 2–15 if the holes are spaced evenly.

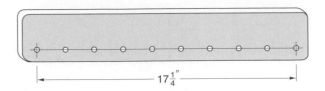

$\longmapsto$ $17\frac{1}{4}''$ $\longrightarrow$

Figure 2–15

15. Find the distance between end holes in the girder connection in Fig. 2–16.

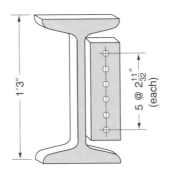

1'3"   5 @ $2\frac{11}{32}''$ (each)

Figure 2–16

16. How many binding posts $1\frac{7}{8}''$ long can be cut from a piece of brass rod $27\frac{1}{2}''$ long? Allow $\frac{1}{8}''$ waste for each cut.

17. How long a piece of drill rod is required to make 15 drills $3\frac{1}{16}''$ long? Allow $\frac{1}{8}''$ waste for each cut.

18. Four standard distance blocks are used to measure a certain space. The blocks are $\frac{5}{16}''$, $\frac{7}{64}''$, $\frac{3}{8}''$, and $\frac{1}{32}''$. How wide is the space?

19. How many holes spaced $1\frac{7}{16}''$ center to center can be drilled in an angle iron $22\frac{5}{8}''$ long, allowing $1\frac{1}{4}''$ distance on each end?

20. Find the length of the steel plate required to make the drill jig shown in Fig. 2–17.

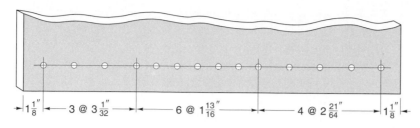

Figure 2–17

21. Seven pins, each $6\frac{7}{0}''$ long, were cut from a piece of drill rod 72″ long. Allowing $\frac{3}{32}''$ waste for each cut, what was the length of the piece left?

22. How many washers, each $\frac{3}{32}''$ thick, can be made from a piece of stock $25\frac{1}{2}''$ long, allowing $\frac{1}{16}''$ waste for each cut?

23. Allowing $1\frac{1}{4}$ minutes for putting away the finished pin and placing the stock in the lathe, how long will it take to machine 25 pins if each pin requires $10\frac{1}{2}$ minutes of machining time?

24. Find the length of the piece of stock required for eight taper keys, each $6\frac{1}{2}''$ long, allowing $\frac{1}{8}''$ waste for each cut.

25. A machinist spends $2\frac{1}{4}$ hours at the lathe, $4\frac{1}{2}$ hours at the planer, and the rest of the day at the shaper. If the working day is 8 hours, how much time is spent at the shaper? If the pay is $19.75 per hour, how much is paid for the time spent at each machine?

26. Find the two missing dimensions in Fig. 2–18.

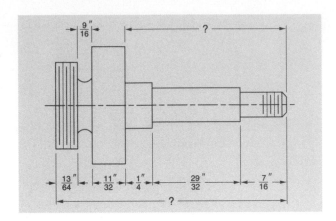

Figure 2–18

27. If the run of each step in the flight of stairs shown in Fig. 2–19 is $9\frac{3}{8}''$, what is the total run of the flight of stairs?

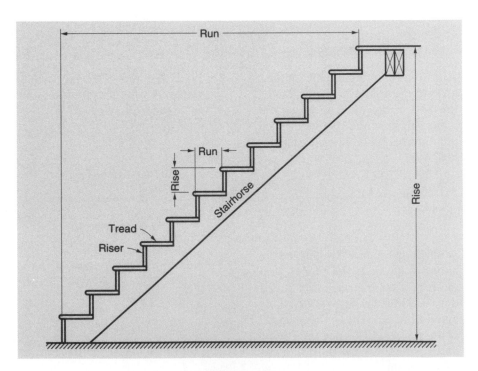

Figure 2–19

28. Find the inside diameter of the pipe shown in cross section in Fig. 2–20.

29. Find the missing dimension in Fig. 2–21.

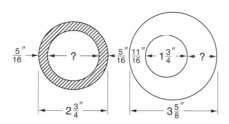

Figure 2–20   **Figure 2–21**

30. Compute the overall dimension in Fig. 2–22.

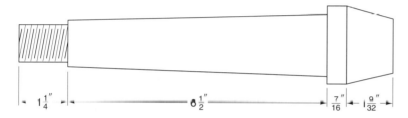

Figure 2–22

■ **Self-Test**

Do the following 10 problems. Use the space provided below to work out the answers. Check your answers with the answers in the back of the book.

1. $\frac{1}{8} + \frac{3}{4}$

2. $2\frac{1}{2} + 4\frac{3}{4}$

3. $\frac{3}{8} - \frac{1}{16}$

4. $3\frac{1}{4} - 1\frac{3}{8}$

5. $\frac{2}{3} \times \frac{4}{5}$

6. $1\frac{3}{4} \times 1\frac{1}{2}$

7. $\frac{1}{3} \div \frac{1}{2}$

8. $2\frac{1}{2} \div 1\frac{1}{2}$

9. $\dfrac{1\frac{1}{2} + \frac{1}{4}}{3\frac{1}{2} \times \frac{1}{2}}$

10. $\dfrac{1\frac{1}{3} - \frac{1}{2}}{4\frac{1}{2} \div 2}$

■ Chapter Test

Rewrite the following examples. Do the indicated operations, and reduce your answers to lowest terms. Use the space provided below to work out the answers.

Add.

1. $\frac{1}{2} + \frac{3}{4}$

2. $1\frac{2}{3} + \frac{2}{5}$

3. $\frac{3}{8} + 2\frac{3}{4}$

4. $1\frac{5}{8} + 2\frac{3}{4}$

5. $2\frac{1}{16} + \frac{3}{32}$

6. $5\frac{15}{16} + \frac{7}{8}$

7. $3\frac{7}{8} + \frac{31}{32}$

8. $2\frac{3}{4} + 5\frac{63}{64}$

Subtract.

9. $\frac{3}{4} - \frac{2}{3}$

10. $1\frac{3}{4} - \frac{7}{8}$

11. $4 - \frac{15}{16}$

12. $2\frac{1}{2} - 1\frac{7}{8}$

13. $12\frac{1}{2} - 3\frac{7}{16}$

14. $3\frac{1}{4} - 2\frac{15}{16}$

15. $1\frac{1}{2} - \frac{31}{32}$

16. $3\frac{1}{8} - 1\frac{21}{64}$

Multiply.

17. $\frac{1}{2} \times \frac{3}{4}$

18. $\frac{7}{8} \times 1\frac{1}{2}$

19. $2\frac{1}{2} \times \frac{3}{4}$

20. $2\frac{3}{8} \times 5$

21. $2\frac{1}{8} \times 3\frac{3}{4}$

22. $4\frac{1}{2} \times 3\frac{1}{8}$

23. $5\frac{1}{2} \times 6\frac{2}{3}$

24. $\frac{33}{77} \times \frac{7}{16}$

Divide.

25. $1\frac{1}{2} \div 4$

26. $2\frac{3}{4} \div \frac{7}{8}$

27. $3\frac{1}{2} \div \frac{1}{2}$

28. $5\frac{3}{8} \div \frac{1}{3}$

29. $2\frac{1}{2} \div 3\frac{1}{4}$

30. $\frac{1}{2} \div 2\frac{1}{2}$

31. $4\frac{1}{4} \div 2\frac{3}{8}$

32. $1\frac{1}{2} \div 1\frac{15}{16}$

Simplify.

33. $\dfrac{3\frac{1}{2} + \frac{7}{8}}{4\frac{1}{2} + 3\frac{3}{4}}$

34. $\dfrac{2\frac{1}{2} - 1\frac{1}{4}}{3\frac{1}{8} - \frac{7}{8}}$

35. $\dfrac{2\frac{1}{2} \times 3\frac{1}{4}}{2\frac{1}{2} \times 4\frac{1}{8}}$

36. $\dfrac{1\frac{3}{4} \div 4}{3\frac{1}{16} \div 2\frac{3}{16}}$

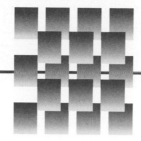

CHAPTER 3

Decimal Fractions

3–1   INTRODUCTION

Decimal fractions are frequently used by mechanics in their daily work. Look at any drawing or machine design and you will see that decimal fractions, or **decimals,** are commonly used to show the various dimensions.

There are still some occasions to use common fractions (halves, quarters, eighths, sixteenths, etc.), since dimensions involving drill sizes and so on are generally given in common fractions. It is easy for the mechanic to visualize these dimensions. The mechanic should be able to work with both kinds of fractions and know the decimal equivalents of halves, quarters, eighths, sixteenths, and so forth.

A **decimal fraction** is a fraction whose denominator is 10 or some power of 10, such as 100, 1,000, or 10,000. Thus, $\frac{3}{10}$, $\frac{25}{100}$, and $\frac{625}{1000}$ are decimal fractions. Since the denominator of a decimal fraction is always 10 or a power of 10 (that is, the denominator is always a 1 with zeros), we can write a decimal fraction more compactly, and therefore more conveniently, by omitting the denominator entirely. Thus, $\frac{3}{10}$ can be written 0.3; $\frac{25}{100}$, 0.25; and $\frac{625}{1000}$, 0.625.

The use of the period (.) in a number indicates that the number following the period is a fraction with a power of 10 as its denominator. This period is called the **decimal point.** Any number with a decimal point in front of it is a fraction whose numerator is the number after the point and whose denominator is a 1 with as many zeros after it as there are figures in the number to the right of this point. Thus, 0.3 means $\frac{3}{10}$; 0.35 means $\frac{35}{100}$; 0.03 means $\frac{3}{100}$; 0.003 means $\frac{3}{1000}$; and so on.

The relative values of the figures in any number are shown in the table at the top of the next page.

In reading a numeral that is written as a whole number plus a decimal fraction, we read first the whole number part and then the decimal fraction part according to the number of places to the right of the decimal point. The numeral 4.01 is read as *four and one one-hundredth.* The whole number part is 4, and the two places to the right of the decimal point indicate that the decimal fraction part of the numeral is hundredths. The word *and* indicates that one one-hundredth is to be added to the whole number 4. If the whole number part is zero, then only the decimal fraction part is read. For example, the numeral 0.132 is read as *one hundred thirty-two thousandths.* Notice that the word *and* does not appear in this reading. The word

Values of Figures	
100	= one hundred
10	= ten
1	= one
0.1	= one tenth
0.01	= one hundredth
0.001	= one thousandth
0.0001	= one ten-thousandth
0.00001	= one hundred-thousandth

1. Read the numeral 1.069.

2. Write the numeral three and five-hundredths.

The answers are in the margin on page 74.

and is used only to indicate the separation of the whole number part from the decimal fraction part.

Practice check: Do exercises 1 and 2 on the left.

■ **Problems 3–1**

Use the space provided below to work out the answers for the following problems.

Read the following numerals.

1. 0.1   **2.** 0.01   **3.** 0.001   **4.** 0.011

5. 0.15   **6.** 0.83   **7.** 0.083   **8.** 2.016

Write the following numerals.

9. Six tenths.   **10.** Two and five tenths.

11. Three hundredths.   **12.** Twenty-five hundredths.

13. One and thirty-three hundredths.   **14.** Three thousandths.

15. Ten and one one-hundredth.

16. One and one one-thousandth.

17. Three hundred two thousandths.

18. Two hundred five and sixty-four thousandths.

19. One thousand one ten-thousandths.

20. Four and one hundred twenty-five ten-thousandths.

21. Ten and one tenth.

22. One hundred one and one hundred one ten-thousandths.

23. Two and six hundred twenty-five ten-thousandths.

24. One and three thousand one hundred twenty-five hundred-thousandths.

25. Five hundred twelve and seventy-five ten-thousandths.

26. Sixty-five ten-thousandths.

27. Five and forty-two millionths.

28. One hundred twenty-five thousand and one hundred twenty-five thousandths.

Answer to exercises 1 and 2

1. One and sixty-nine thousandths

2. 3.05

3–2   CHANGING A DECIMAL FRACTION TO A COMMON FRACTION

It is sometimes convenient to change a decimal fraction to a common fraction. The numerator of the common fraction is all the numerals that are present in the decimal fraction, and the denominator of the common fraction is a 1 followed by as many zeros as there are places to the right of the decimal point. The fraction is then reduced to lowest terms following the procedures in Chapter 2.

■ **Example 1**

Change 0.4 to a common fraction.

Solution $0.4 = \dfrac{4}{10} = \dfrac{4 \div 2}{10 \div 2} = \dfrac{2}{5}$

Explanation   The number after the decimal point is the numerator of the fraction; the denominator is a 1 with one zero because there is one figure to the right of the decimal point. Thus, $0.4 = \frac{4}{10}$. This is changed to the lowest terms by dividing both numerator and denominator by 2.

Practice check: Do exercise 3 on the left.

3. Change 0.6 to a common fraction.

The answer is in the margin on page 76.

■ **Example 2**

Change 0.75 to a common fraction.

Solution $0.75 = \dfrac{75}{100} = \dfrac{75 \div 5}{100 \div 5} = \dfrac{15}{20}$ $\dfrac{15 \div 5}{20 \div 5} = \dfrac{3}{4}$

or

$$\frac{75}{100} = \frac{\overset{1}{\cancel{5}} \times \overset{1}{\cancel{5}} \times 3}{\underset{1}{\cancel{5}} \times \underset{1}{\cancel{5}} \times 4} = \frac{3}{4}$$

Explanation   The numerator of the common fraction is 75, and the denominator is 1 with two zeros because there are two figures after the decimal point. Thus, $0.75 = \frac{75}{100}$. To reduce $\frac{75}{100}$ to its lowest terms, divide both terms by 5 and get $\frac{15}{20}$; then divide each term of $\frac{15}{20}$ by 5 to obtain $\frac{3}{4}$. The desired result could be obtained in one step by dividing both terms of the fraction $\frac{75}{100}$ by 25. Thus, $\dfrac{75 \div 25}{100 \div 25} = \dfrac{3}{4}$.

Practice check: Do exercise 4 on the left.

4. Change 0.55 to a common fraction.

The answer is in the margin on page 76.

■ **Example 3**

Change 0.024 to a common fraction.

Solution $0.024 = \dfrac{24}{1,000} = \dfrac{24 \div 2}{1,000 \div 2} = \dfrac{12}{500}$

$\dfrac{12 \div 2}{500 \div 2} = \dfrac{6}{250}$ $\dfrac{6 \div 2}{250 \div 2} = \dfrac{3}{125}$

These steps can be condensed as follows.

$$\frac{\overset{3}{\cancel{6}}\overset{}{\cancel{12}}\overset{}{\cancel{24}}}{\underset{}{\cancel{1,000}}\,\underset{}{\cancel{500}}\,\underset{}{\cancel{250}}\,125} = \frac{3}{125}$$

Explanation To reduce $\frac{24}{1000}$ to lowest terms, proceed as shown in the previous chapter by finding all the common factors of 24 and 1,000 and canceling them. If you found 8 to be the highest common factor of both these numbers, you could find the common fraction by dividing 8 into both of these terms in one step. Thus, $\frac{24 \div 8}{1,000 \div 8} = \frac{3}{125}$.

Practice check: Do exercise 5 on the right.

5. Change 0.175 to a common fraction.

The answer is in the margin on page 77.

■ Problems 3–2

Change the following decimal fractions to common fractions and write each answer in its lowest terms. Use the space provided below to work out the answers for the problems.

1. 0.5   **2.** 0.15   **3.** 0.125   **4.** 0.25   **5.** 0.75

6. 0.375   **7.** 0.875   **8.** 0.16   **9.** 0.8   **10.** 0.06

11. 0.0625   **12.** 0.187   **13.** 0.042   **14.** 0.625   **15.** 0.03125

16. 0.015625   **17.** 0.0125   **18.** 0.0375   **19.** 0.9625   **20.** 0.005

21. 0.0075   **22.** 0.002   **23.** 0.0875   **24.** 0.982   **25.** 0.96875

26. 0.984375   **27.** 0.953125   **28.** 0.844   **29.** 0.9375   **30.** 0.6875

31. 0.8125   **32.** 0.7875   **33.** 0.8750   **34.** 0.328125   **35.** 0.84375

36. 0.609375   **37.** 0.140625   **38.** 0.921875   **39.** 0.453125   **40.** 0.21875

Answer to exercise 3

$$0.6 = \frac{6}{10} = \frac{6 \div 2}{10 \div 2}$$
$$= \frac{3}{5}$$

Answer to exercise 4

$$0.55 = \frac{55}{100} = \frac{55 \div 5}{100 \div 5}$$
$$= \frac{11}{20}$$

3–3   CHANGING A COMMON FRACTION TO A DECIMAL FRACTION

Round-off Rule

All common fractions can be written as repeating decimal fractions. A decimal approximation of the common fraction is often sufficiently accurate for solving a shop problem. It will be necessary to adopt a rule to use when expressing a decimal equivalent of common fractions. The mechanic will determine how many decimal places are needed to solve the problem. A carpenter would seldom use an accuracy greater than one-hundredth, while a machinist would sometimes want to work to the nearest ten-thousandth.

In this book the following rules will be used to round off decimal fractions.

1. Determine the number of places that are necessary for the particular problem.
2. Now look at the remaining places. If the remaining places start with a 5 or a number greater than 5, add 1 to the last place of the decimal that is to be kept. For example, if **7.36482** is to be rounded to **three** decimal places, then observe that the numeral after the third place is an 8, which is greater than 5. Therefore, 1 is added to the third place (4), and the number 7.36482, to three decimal places, is 7.365.

If **344.06250** is to be rounded to **three** decimal places, then observe that the numeral after the third place is a 5. Therefore, 1 is added to the third place (2), and the number 344.06250, to three decimal places, is 344.063.
3. Suppose a mechanic wants to round **6.28236** to **three** decimal places. Observe that the numeral directly after the third place is a 3, which is less than 5. Therefore, all the numerals after the third place are ignored, and the numeral 6.28236, rounded to three decimal places, is 6.282.

To change a common fraction, such as halves, quarters, eighths, sixteenths, and so on, to a decimal fraction, divide the denominator into the numerator as shown in the following examples.

■ **Example 1**

Change $\frac{7}{8}$ to a decimal fraction.

Solution

```
    0.875
  8)7.000
    6 4
    ─────
      60
      56
      ──
      40
      40
      ──
```

Explanation   Since a fraction is an indicated division, divide 7 by 8. Place a decimal point after the 7, and annex zeros. The division is carried out as shown. Thus, $\frac{7}{8} = 0.875$.

Practice check: Do exercise 6 on the right.

■ **Example 2**

Change $\frac{2}{3}$ to a decimal fraction accurate to the hundredths place.

Solution

```
    0.66 . . .
  3)2.000
    1 8
    ─────
      20
      18
      ──
       2
       .
       .
       .
```

Explanation   Proceeding as in Example 1, we find that no matter how many zeros are added after the decimal point, the quotient will not come out evenly. In such a case, you must decide how far the answer should be carried out. The answer to the nearest hundredth, which means two places to the right of the decimal point, is $\frac{2}{3} = 0.67$.

Practice check: Do exercise 7 on the right.

■ **Example 3**

Change $\frac{3}{16}$ to a decimal of three places.

Solution

```
       0.1875
   16)3.0000
      1 6
      ──────
      1 40
      1 28
      ──────
        120
        112
        ────
         80
         80
         ──
          0
```

Answer to exercise 5

$$\frac{175}{1,000} = \frac{175 \div 25}{1,000 \div 25}$$
$$= \frac{7}{40}$$

6. Change $\frac{13}{20}$ to a decimal fraction.

The answer is in the margin on page 80.

7. Change $\frac{5}{12}$ to a decimal fraction accurate to the thousandths place.

The answer is in the margin on page 80.

8. Change $\frac{1}{16}$ to a decimal of two places.

The answer is in the margin on page 80.

Explanation Carrying out the indicated division, we find that the digit after the third place is exactly 5. Add 1 to the third place, since the round-off rule says to add 1 if the remaining places start with a 5 or a numeral greater than 5. Thus, $\frac{3}{16}$ = 0.188 to three decimal places.

Practice check: Do exercise 8 on the left.

■ **Problems 3–3**

Change the following common fractions to decimal fractions of three places. Use the space provided below to work out the answers for the problems.

NOTE: *If the digit in the fourth place is a numeral equal to or greater than 5, add 1 to the third place. Thus, 8.7236 would be 8.724 to three decimal places, and 4.3425 would be 4.343 to three decimal places. If the numeral in the fourth place is less than 5, leave the third numeral unchanged. Thus, 6.1482, to three decimal places, would be 6.148.*

1. $\frac{1}{4}$ 2. $\frac{3}{8}$ 3. $\frac{7}{16}$ 4. $\frac{1}{3}$ 5. $\frac{2}{3}$

6. $\frac{3}{5}$ 7. $\frac{1}{80}$ 8. $\frac{3}{20}$ 9. $\frac{11}{64}$ 10. $\frac{3}{32}$

11. $\frac{7}{8}$ 12. $\frac{13}{16}$ 13. $\frac{5}{6}$ 14. $\frac{1}{11}$ 15. $\frac{1}{12}$

16. $\frac{1}{25}$ 17. $\frac{1}{250}$ 18. $\frac{1}{7}$ 19. $\frac{1}{9}$ 20. $\frac{9}{10}$

21. $\frac{19}{144}$ 22. $\frac{450}{1728}$ 23. $\frac{490}{1728}$ 24. $\frac{11}{17}$ 25. $\frac{631}{5280}$

26. $\frac{87}{144}$ 27. $\frac{625}{1728}$ 28. $\frac{26}{27}$ 29. $\frac{440}{5280}$ 30. $\frac{158}{314}$

31. $\frac{158}{3141}$ **32.** $\frac{158}{31416}$ **33.** $\frac{1}{144}$ **34.** $\frac{235}{350}$ **35.** $\frac{187}{250}$

36. What decimal part of a dollar is 42 cents?

37. What decimal fraction of 1 pound is 13 ounces?

38. What decimal fraction of 1 foot is 8″?

39. Change 100 sq in. into a decimal fraction of 1 square foot.

40. Change 500 cu in. into a decimal fraction of 1 cubic foot.

3–4   USING A TABLE OF DECIMAL EQUIVALENTS

The answer to a problem may often be a fraction such as $\frac{5}{7}$ in. or $\frac{3}{19}$ in. To lay off such a dimension with the ordinary scale, change the answer to the nearest eighth, sixteenth, thirty-second, or sixty-fourth. This is done most conveniently by means of a *table of decimal equivalents.* See Table 4 at the back of this book.

■ **Example 1**

On a scale divided into sixty-fourths of an inch, what is the dimension nearest to $\frac{13}{37}$ in.?

Solution $\frac{13}{37}$ in. = 0.35135 in. ≈ $\frac{22}{64}$ in. (The symbol ≈ means *is approximately equal to.*)

Explanation   First change the common fraction $\frac{13}{37}$ to the decimal 0.35135. Looking down the table of decimal equivalents, observe that 0.35135 lies between 0.34375 and 0.359375. By subtracting, find that 0.35135 − 0.34375 = 0.0076, whereas 0.359375 − 0.35135 = 0.008025. Therefore, 0.35135 is nearer to 0.34375, and its nearest equivalent in sixty-fourths of an inch is $\frac{22}{64}$ in.

     To find the equivalent of $\frac{13}{37}$ in sixteenths: $\frac{5}{16} = 0.3125$; $\frac{6}{16}$ or $\frac{3}{8} = 0.375$. Since 0.35135 is nearer to 0.375, the nearest equivalent of $\frac{13}{37}$ in. is $\frac{6}{16}$ in.

     Practice check: Do exercise 9 on the right.

     Where great accuracy is not required, it is sufficient to convert the given fraction into a decimal of three places and use only the first three figures of the table.

     If a table of decimal equivalents is not available, any fraction can be converted into any other fraction with the desired denominator, as shown in the following example.

9. Using Table 4, on a scale divided into thirty-secondths of an inch, find the dimension nearest to $\frac{7}{17}$.

The answer is in the margin on page 82.

Answer to exercise 6

0.65

Answer to exercise 7

0.417

Answer to exercise 8

0.06

10. Change $\frac{7}{17}$ to thirty-secondths.

The answer is in the margin on page 82.

■ **Example 2**

Change $\frac{13}{37}$ to sixty-fourths.

Solution $\frac{13}{37} \times 64 = \frac{832}{37} = 22\frac{18}{37} \approx 22$. Therefore, $\frac{13}{37} \approx \frac{22}{64}$ to the nearest sixty-fourth.

Explanation   Since in 1 unit there are 64 sixty-fourths, in $\frac{13}{37}$ of a unit there will be $\frac{13}{37}$ of 64 or $22\frac{18}{37}$ sixty-fourths. Since $\frac{18}{37}$ is less than 0.5, round off to $\frac{22}{64}$.

■ **Example 3**

Change $\frac{2}{7}$ to sixteenths.

Solution $\frac{2}{7} \times 16 = \frac{32}{7} = 4\frac{4}{7} \approx 5$. Thus, $\frac{2}{7} \approx \frac{5}{16}$ to the nearest sixteenth.

Explanation   Since in 1 unit there are 16 sixteenths, in $\frac{2}{7}$ of a unit there will be $\frac{2}{7}$ of 16 or $4\frac{4}{7}$ sixteenths. Rounded off, that is $\frac{5}{16}$.

Practice check: Do exercise 10 on the left.

■ **Problems 3–4**

Use Table 4 at the back of the book to find the decimal equivalent of each of the following fractions to the nearest sixteenth of an inch. Use the space provided below to work out the answers for the problems.

1. $\frac{1}{2}$ 2. $\frac{3}{4}$ 3. $\frac{5}{8}$ 4. $\frac{29}{32}$ 5. $\frac{8}{17}$ 6. $\frac{5}{21}$

7. $\frac{2}{3}$ 8. $\frac{2}{9}$ 9. $1\frac{9}{21}$ 10. $3\frac{5}{9}$ 11. $4\frac{7}{11}$ 12. $5\frac{5}{12}$

13. $2\frac{1}{3}$ 14. $6\frac{3}{11}$ 15. $8\frac{6}{19}$ 16. $7\frac{5}{14}$

Using Table 4, find the decimal equivalent of each of the following fractions to the nearest thirty-second of an inch.

17. $\frac{2}{3}$ 18. $\frac{3}{5}$ 19. $\frac{3}{10}$ 20. $\frac{15}{31}$ 21. $2\frac{8}{17}$ 22. $5\frac{1}{8}$

23. $3\frac{6}{19}$ **24.** $2\frac{2}{7}$ **25.** $6\frac{5}{6}$ **26.** $1\frac{11}{13}$ **27.** $8\frac{19}{144}$ **28.** $2\frac{4}{5}$

29. $4\frac{325}{1728}$ **30.** $12\frac{880}{5280}$ **31.** $10\frac{1}{144}$ **32.** $3\frac{1320}{1728}$

Using Table 4, find the decimal equivalent of each of the following fractions to the nearest sixty-fourth of an inch.

33. $\frac{2}{3}$ **34.** $\frac{3}{4}$ **35.** $\frac{3}{5}$ **36.** $\frac{7}{10}$ **37.** $2\frac{9}{17}$ **38.** $5\frac{5}{8}$

39. $3\frac{16}{19}$ **40.** $2\frac{6}{7}$ **41.** $6\frac{5}{6}$ **42.** $1\frac{11}{13}$ **43.** $8\frac{143}{144}$ **44.** $2\frac{4}{5}$

45. $4\frac{1000}{1728}$ **46.** $12\frac{880}{1760}$ **47.** $10\frac{440}{5280}$ **48.** $3\frac{121}{144}$

3-5   CONVERSION OF DIMENSIONS

It is sometimes necessary in computation to convert inches and fractions of an inch into feet and decimal fractions of a foot. The following examples illustrate the procedure in problems of this type. (The examples use the table of decimal equivalents, Table 4, in the back of the book.)

Rule _____

Dividing inches by 12 will change inches to feet.

_____

■ **Example 1**

Convert $3\frac{3}{4}$ inches into a decimal fraction of a foot. State the answer to three decimal places.

Solution $3\frac{3}{4}$ in. = 3.75 in. $\dfrac{3.75}{12} = 0.3125$ ft

           $3\frac{3}{4}$ in. = 0.313 ft   (to three decimal places)

Answer to exercise 9

$\frac{13}{32}$

Answer to exercise 10

$\frac{13}{32}$

11. Convert $25\frac{3}{16}$ inches into a decimal fraction of a foot. State the answer to three decimal places.

The answer is in the margin on page 84.

12. Convert 27.469 inches into feet and inches and the nearest thirty-secondths of an inch.

The answer is in the margin on page 84.

13. Convert 0.628 foot into inches and a fraction of an inch.

The answer is in the margin on page 84.

■ **Example 2**

Convert $46\frac{1}{8}$ inches into feet and a decimal fraction of a foot to three decimal places.

Solution $46\frac{1}{8}$ in. = 46.125 in. $\dfrac{46.125}{12} = 3.84375$ ft

             $46\frac{1}{8}$ in. = 3.844 ft   (to three decimal places)

Practice check: Do exercise 11 on the left.

■ **Example 3**

Convert 43.172 inches into feet and inches and the nearest sixty-fourth of an inch. (Use the table of decimal equivalents.)

Solution   43.172 in. = 43 in. + 0.172 in.

$$\dfrac{43\text{ in.}}{12} = 3\tfrac{7}{12} = \quad 3\text{ ft } 7 \quad \text{ in.}$$
$$0.172 \text{ in.} = + \qquad \tfrac{11}{64}\text{ in.}$$
$$\overline{3\text{ ft }7\tfrac{11}{64}\text{ in.}}$$

Explanation   For easier solving, break 43.172 into its two parts—a whole number (43) and a decimal fraction (0.172). Divide 43 by 12 to find the number of feet in 43 inches. Because there are 12 inches in 1 foot, $3\frac{7}{12}$ ft = 3 ft 7 in. To convert 0.172 to sixty-fourths of an inch, refer to the table, which shows that the decimal nearest in value is 0.171875, which is equivalent to $\frac{11}{64}$. Finally, add 3 ft 7 in. to $\frac{11}{64}$ in. for an answer of 3 ft $7\frac{11}{64}$ in.

Rule _____

Multiplying feet by 12 will change feet to inches.
_____

Practice check: Do exercise 12 on the left.

■ **Example 4**

Convert 0.319 foot into inches and a fraction of an inch. (Use the table of decimal equivalents.)

Solution   0.319 ft × 12 = 3.828 in. The figure in the table of decimal equivalents nearest to 0.828 is 0.828125, which is the decimal equivalent of $\frac{53}{64}$. Thus, 0.319 ft = $3\frac{53}{64}$ in.

Practice check: Do exercise 13 on the left.

■ **Problems 3–5**

Convert the following dimensions from inches into feet. State the answers in whole numbers and/or decimals to three decimal places. Use the space provided below to work out the answers for the problems.

1. $7\frac{1}{2}''$ **2.** $9\frac{3}{4}''$ **3.** $8\frac{7}{16}''$ **4.** $9\frac{5}{8}''$ **5.** $\frac{7}{8}''$ **6.** $6\frac{9}{16}''$

7. $71\frac{3}{4}''$ **8.** $5\frac{11}{16}''$ **9.** $2\frac{1}{4}''$ **10.** $53\frac{1}{2}''$ **11.** $75\frac{5}{8}''$ **12.** $96\frac{1}{8}''$

13. $27\frac{5}{8}''$ **14.** $33\frac{3}{4}''$ **15.** $25\frac{7}{16}''$ **16.** $16\frac{7}{8}''$ **17.** $21\frac{1}{4}''$ **18.** $42\frac{1}{2}''$

19. $37\frac{9}{16}''$ **20.** $120\frac{3}{8}''$

Convert the following dimensions into feet, inches, and the nearest sixty-fourth of an inch. (Use Table 4.)

21. $219.31''$ **22.** $28.19''$ **23.** $39.11''$ **24.** $165.3''$ **25.** $73.26''$

26. $326.83''$ **27.** $172.8''$ **28.** $91.3''$ **29.** $53.17''$ **30.** $220.94''$

31. $14.116''$ **32.** $19.1''$ **33.** $217.6''$ **34.** $32.31''$ **35.** $56.26''$

36. $181.33''$ **37.** $234.54''$ **38.** $121.116''$ **39.** $200.017''$ **40.** $147.32''$

Convert the following dimensions into inches and the nearest sixty-fourth of an inch. (Use Table 4.)

41. $0.91'$ **42.** $0.81'$ **43.** $0.864'$ **44.** $0.395'$ **45.** $0.217'$

46. 0.09′   **47.** 0.186′   **48.** 0.78′   **49.** 0.146′   **50.** 0.391′

51. 0.82′   **52.** 0.211′   **53.** 1.91′   **54.** 2.864′   **55.** 1.23′

56. 1.065′   **57.** 3.021′   **58.** 2.11′   **59.** 2.36′   **60.** 3.421′

Answer to exercise 11

$\dfrac{25.1875}{12} = 2.0989583$

Thus, $25\frac{3}{16}$ ft = 2.099 ft (to three decimal places)

Answer to exercise 12

27 + 0.469

$\dfrac{27 \text{ in.}}{12} = 2\frac{3}{12}$ ft = 2 ft 3   in.

0.469 in. $= \qquad \frac{15}{32}$ in.

                $= 2$ ft $3\frac{15}{32}$ in.

Answer to exercise 13

0.628 ft × 12 = 7.536 in.
0.536 ≃ 0.53125 = $\frac{17}{32}$
Thus, 0.628 ft = $7\frac{17}{32}$ in.

---

14. Add 65.987, 9.4703, 1.0003, 200.895, and 6744.02.

The answer is in the margin on page 86.

3–6   ADDITION OF DECIMALS

Addition of decimal fractions is much like addition of whole numbers. The difference is that when adding decimal fractions, you must place the decimal points one under the other. The decimal point appears in the answer in the same column as it is in the problem.

■ **Example 1**

Add 3.25, 72.004, 864.0725, 647, and 0.875.

Solution
$$
\begin{array}{r}
3.25 \\
72.004 \\
864.0725 \\
647. \\
+\quad 0.875 \\
\hline
1{,}587.2015
\end{array}
$$

$$
\begin{array}{r}
3.2500 \\
72.0040 \\
864.0725 \\
647.0000 \\
+\quad 0.8750 \\
\hline
1{,}587.2015
\end{array}
$$

Explanation   To add the given numbers, write them so that the decimal points are in the same column. Then proceed as in the addition of ordinary whole numbers, and place the decimal point in the answer in the same column as the other decimal points.

To facilitate the work and avoid errors, it is advisable to annex zeros to the numbers with the fewer decimal places so that all the numbers will have an equal number of places after the decimal point. The work then appears as shown in the second problem in the solution.

Practice check: Do exercise 14 on the left.

■ **Example 2**

Find the sum of $\frac{1}{2} + 0.662 + \frac{7}{8}$.

Solution $\frac{1}{2} = 0.5$ $\frac{7}{8} = 0.875$

$$
\begin{array}{r}
0.500 \\
0.662 \\
+\ 0.875 \\
\hline
2.037
\end{array}
$$

Explanation   First change $\frac{1}{2}$ and $\frac{7}{8}$ to decimals, obtaining 0.5 and 0.875, and then add the numbers as in Example 1.

Practice check: Do exercise 15 on the right.

15. Find the sum of $2.014 + \frac{3}{8} + 19.6 + \frac{13}{16}$.

The answer is in the margin on page 87.

■ **Problems 3–6**

Find the sums of the following numbers. Make sure that your answers are in decimal fraction form. Use the space provided below to work out the answers for the problems.

1. $3 + 5.6$

2. $46.2 + 37.5$

3. $3.1 + \frac{1}{2}$

4. $4.2 + \frac{1}{2} + 1.3$

5. $4.6 + 0.6 + \frac{7}{8}$

6. $5.31 + 4.21 + \frac{1}{4}$

7. $27.62 + 0.62 + \frac{3}{4}$

8. $\frac{3}{4} + \frac{1}{2} + 4.9$

9. $\frac{1}{2} + \frac{1}{4} + \frac{1}{8}$

10. $\frac{1}{2} + \frac{1}{4} + \frac{1}{8} + \frac{1}{16} + \frac{1}{32}$

Answer to exercise 14

7,021.3726

11. 235 + 49.2

12. 321.42 + 69.73

13. 1,624.08 + 12.236

14. 1.728 + 0.0084 + 6.52

15. 932.04 + 93.204 + 9.3204 + 0.93204

16. 0.732 + 4.896 + 0.153 + 18.654 + 2.404

17. $1.335 + \frac{5}{8} + 1\frac{1}{2}$

18. $2\frac{1}{3} + \frac{3}{4} + 0.625$

19. $3.482 + \frac{1}{6} + 0.025$

20. $4.603 + 2.136 + \frac{2}{3}$

21. $3.201 + \frac{1}{8} + \frac{1}{32}$

22. $5.206 + \frac{1}{64}$

23. $8.0123 + \frac{15}{16}$

24. $\frac{1}{2} + \frac{1}{4} + \frac{1}{8} + \frac{1}{16} + \frac{1}{32} + \frac{1}{64}$

25. $0.563 + 1.72 + 3.6131 + \frac{1}{8}$

26. $1.362 + 1.11 + 3.315 + \frac{7}{16}$

27. $\frac{3}{16}$ + 1.365 + $\frac{7}{8}$ + 2.812 + 3

28. 6.21 + 532.8 + 11.1 + 87.631

Answer to exercise 15

22.8015

29. 3.214 + 3.1416 + $\frac{7}{16}$ + 3.1182

30. 6.315 + $\frac{15}{16}$ + 5.312 + 4.117 + 8

31. 37.2 + 463.118 + 17.163 + 81.223

32.
```
    264.
      0.065
     30.832
    716.45
+     9.7
```

33.
```
  1.508
  4.391
  0.484
  0.745
+ 0.656
```

34.
```
  0.8435
  2.6264
  5.1706
  0.2035
+ 8.6302
```

35.
```
  0.5957
  0.4056
  2.9540
  3.3470
+ 0.4941
```

36. $4.35 + $6.75 + $3.00 + $5.75

37. $6.76 + $9.70 + $8.65 + $5.40

38. $2.05 + $18.99 + $36.18 + $45.17

39. $629.03 + $875.92 + $492.85 + $867.62

40. $103.00 + $4,621.01 + $2,317.05 + $6,417.07

3-7   SUBTRACTION OF DECIMALS

Subtraction of decimal fractions follows the same rules as subtraction of whole numbers. The decimal points in the problem must be in the same column. The decimal point in the answer appears in the same column as it is in the problem.

■ Example 1

Take 18.275 from 42.63.

Solution
$$\begin{array}{ll} 42.630 & \text{(minuend)} \\ -\ 18.275 & \text{(subtrahend)} \\ \hline 24.355 & \text{(difference)} \end{array}$$

Explanation   Write the numbers so that the decimal points are under each other. If the minuend contains fewer figures after the decimal point than the subtrahend, annex zeros and proceed as in subtraction of whole numbers. Place the decimal point in the difference under the other decimal points.

■ Example 2

From $2\frac{1}{3}$ take 0.675.

Solution $2\frac{1}{3} =$
$$\begin{array}{ll} 2.333 & \text{(minuend)} \\ -\ 0.675 & \text{(subtrahend)} \\ \hline 1.658 & \text{(difference)} \end{array}$$

Explanation   Change $\frac{1}{3}$ to a decimal fraction, and the number $2\frac{1}{3}$ becomes 2.333 (to three decimal places, to match the subtrahend). Perform the subtraction as in Example 1.

Practice check: Do exercises 16 and 17 on the left.

16. Take 0.0035 from 3.6.

17. From $5\frac{1}{8}$ take 0.598.

The answers are in the margin on page 90.

■ Problems 3-7

Do the following indicated subtractions. All answers are to be in decimal fraction form. Use the space provided below to work out the answers for the problems.

1. Take 234 from 625.75.

2. Take 148.35 from 436.62.

3. Take 82.875 from 125.

4. Take 139.25 from 218.08.

5. Take 0.4025 from 0.4929.

6. Take 0.6535 from 1.

7. Take 0.40875 from 2.856.

8. Take 0.545 from 5.9205.

9. Take 0.6346 from 0.8964.

10. Take 0.4058 from 0.4948.

11. From 3 take 1.0807.

12. From 18 take 2.198.

13. From 7.002 take 3.2.

14. From 8.078 take 7.706.

15. From 11.047 take 0.489.

16. From 60.0008 take 15.0655.

17. From $4\frac{5}{8}$ take 2.326.

18. From 5.604 take $1\frac{2}{3}$.

19. From 3.862 take $2\frac{1}{6}$.

20. From $5\frac{3}{8}$ take 3.078.

21. From $6\frac{3}{8}$ take 4.286.

22. From $5\frac{7}{16}$ take 3.416.

Answer to exercise 16

```
  3.6000
− 0.0035
  3.5965
```

Answer to exercise 17

$5\frac{1}{8} =$

```
    5.125
 − 0.598
    4.527
```

23. From 7.015 take $2\frac{1}{4}$.

24. From $11\frac{7}{8}$ take 8.109.

25. From 13.218 take $10\frac{9}{16}$.

26. From $9\frac{1}{2}$ take 3.614.

27. From 1.068 take $\frac{11}{32}$.

3–8   MULTIPLICATION OF DECIMALS

Multiplication of decimal fractions follows the same rules as multiplication of whole numbers. The product (answer) will have as many decimal places as the total number of decimal places in both the multiplicand and the multiplier. Study the example problem below. Notice that unlike in addition or subtraction, it is not necessary to align decimal points in multiplication. In fact, when multiplying decimal fractions, align the rightmost numerals, regardless of the position of the decimal points.

■   **Example 1**

Multiply 43.286 by 6.04.

Solution

```
      43.286   (multiplicand)
 ×     6.04    (multiplier)
     173144
    2597160
   261.44744   (product)
```

Explanation   Multiply as in whole numbers; then, beginning at the right, point off in the product as many decimal places as there are decimal places in both the multiplier and the multiplicand. Thus, in this example, the multiplicand has three places after the decimal point, and the multiplier has two, for a total of five places. Therefore, the product must have five decimal places.

The decimal part of this product denotes hundred-thousandths. But the machinist, the cabinetmaker, and the electrician are not interested in such small fractions of a unit. Therefore, rewrite the result so that it will be of practical value to the person in the shop or factory. The machinist may want measurements given to the nearest thousandth of an inch; therefore, the answer to a machinist's problem will be stated with three decimal places. The answer would then read 261.447. The figures after the third decimal place are dropped because their value is less than one-half of a thousandth.

The answer to two decimal places would be 261.45; to one decimal place, it would read 261.4.

Decide in advance the degree of accuracy required in the answer to any given problem. After performing the indicated operations and obtaining the exact answer, rewrite the answer with as many decimal places as are required to bring it to the degree of accuracy previously determined for the given problem. The degree of accuracy cannot be greater than the least accurate measure in the problem.

Practice check: Do exercise 18 on the right.

18. Multiply 6.204 by 18.32.

The answer is in the margin on page 94.

■ Example 2

Multiply 0.36 by 0.24.

Solution

$$
\begin{array}{r}
0.36 \\
\times\ 0.24 \\
\hline
144 \\
72 \\
\hline
0.0864
\end{array}
$$

Explanation Multiplying as in whole numbers, we find that there are only three figures in the product, although, according to the rule given previously, the product must have four places after the decimal point. The fourth place is supplied by a zero written in front of the three figures, thus, 0.0864.

Practice check: Do exercise 19 on the right.

19. Multiply 0.63 by 0.38.

The answer is in the margin on page 94.

■ Example 3

Multiply 0.85 by $1\frac{3}{4}$.

Solution

$$
\begin{array}{r}
1.75 \\
\times\ 0.85 \\
\hline
875 \\
1400 \\
\hline
1.4875
\end{array}
$$

Explanation Change the common fraction $\frac{3}{4}$ to a decimal fraction: $1\frac{3}{4} = 1.75$. Then multiply 0.85 by 1.75 as in Example 2.

Practice check: Do exercise 20 on the right.

20. Multiply $2\frac{7}{16}$ by 1.07.

The answer is in the margin on page 95.

Rule

To multiply a number by 10; 100; 1,000; 10,000; and so on, move the decimal point in the multiplicand as many places to the right as there are zeros in the multiplier. If necessary, annex zeros.

Thus, $10 \times 3.25 = 32.5$; $100 \times 3.25 = 325$; $10,000 \times 3.25 = 32,500$; and so forth.

■ Example 4

Multiply 1.0356 by 1,000.

Solution $1.0356 \times 1,000 = 1,035.6$

21. Multiply 3.16 by 10,000.

The answer is in the margin on page 95.

Explanation   Note that there are three zeros in 1,000. To multiply 1.0356 by 1,000, move the decimal point in 1.0356 three places to the right. The answer is 1,035.6

Practice check: Do exercise 21 on the left.

■ Problems 3–8

Find the following products. All answers are to be in decimal fraction form. Use the space provided below to work out the answers for the problems.

1. 0.25×10

2. 0.375×100

3. 0.5×100

4. $0.625 \times 1,000$

5. 0.3125×10

6. $0.4375 \times 10,000$

7. 1.875×10

8. $2.75 \times 1,000$

9. 15.25×100

10. $1.75 \times 1,000$

11. $6.375 \times 1,000$

12. 4.9375×100

13. $0.25 \times \frac{1}{2}$

14. $0.375 \times \frac{3}{4}$

15. 0.5×0.5

16. $0.5 \times \frac{1}{2}$

17. 0.5×1.5

18. $\frac{1}{2} \times \frac{1}{2}$

19. 0.8×0.9

20. 0.32×0.4

21. 0.57×0.61

22. 0.256×0.4

23. 0.25×0.375

24. 0.8125×4

25. 0.831×0.3

26. 4.26×3.7

27. 5.79×0.625

28. 0.0625×5

29. 0.031×0.062

30. 3.1416×5

31. 8×0.5346

32. 8.75×3.5

33. 8.94808×0.6

34. 86.4×0.0458

35. 1.125×3.375

36. $0.7854 \times 3.25 \times 3.25$

37. $5,280 \times 0.876$

38. 16×0.1875

39. 0.125×0.36

40. 0.008×0.072

41. 5.84×0.0059

42. 2.25×0.0125

43. $1,728 \times 0.621$

44. 231×3.125

45. 0.2465×12

46. 2.457×2.568

47. 0.457×12.36

48. 0.008×24.86

49. $3.1416 \times 2.5 \times 2.5$

50. 144×2.368

Answer to exercise 18

$$
\begin{array}{r}
6.204 \\
\times\ 18.32 \\
\hline
12408 \\
1\ 8612 \\
49\ 632 \\
62\ 04 \\
\hline
113.65728
\end{array}
$$

Answer to exercise 19

0.2394

51. Restate the answers to Problems 21–25 with three decimal places.

52. Restate the answers to Problems 26–30 with two decimal places.

53. Restate the answers to Problems 31–35 with four decimal places.

54. Restate the answers to Problems 36–40 with two decimal places.

55. $0.062 \times 2\frac{5}{8}$

56. $2.246 \times 4\frac{1}{3}$

57. $5\frac{2}{3} \times 0.425$

58. $4\frac{2}{9} \times 3.065$

59. $3\frac{5}{6} \times 2.083$

60. 72×10

61. 18×100

62. $92 \times 1,000$

63. 6.86×10

64. 7.2×100

65. $1.37 \times 1,000$

66. 0.721×10

67. 0.063×100

68. $0.375 \times 1,000$

3–9   DIVISION OF DECIMALS

Division of decimal fractions follows the same rules as division of whole numbers except for the placement of the decimal point in the quotient (answer). The divisor is converted to a whole number, and the dividend is changed correspondingly. Study the following example problems.

■ **Example 1**

Divide 76.125 by 24. State the answer to three decimal places.

Solution   (quotient) $3.171\frac{21}{24}$ or 3.172
          (divisor) $24\overline{)76.125}$ (dividend)

$$\begin{array}{r} 72 \\ \hline 4\,1 \\ 2\,4 \\ \hline 1\,72 \\ 1\,68 \\ \hline 45 \\ 24 \\ \hline 21 \end{array}$$

Explanation   Divide as in whole numbers and place the decimal point in the quotient above the point in the dividend. Since the remainder after the third place is $\frac{21}{24}$, or more than $\frac{1}{2}$, increase the third place by 1, giving the quotient 3.172.

Practice check: Do exercise 22 on the right.

■ **Example 2**

Divide 432 by 0.625.

Solution 691.2
          $_x625.\overline{)432_x000.0}$
          $375\,0$
          $\overline{57\,00}$
          $56\,25$
          $\overline{750}$
          625
          $\overline{1250}$
          1250

Explanation   When the dividend is a whole number and the divisor is a decimal fraction, convert the divisor into a whole number by moving the decimal point to the right of it, in this instance, three places to the right. Then move the decimal point in the dividend also three places to the right. To do this, annex three zeros after 432, giving the number 432,000, with the decimal point after the third zero. Proceed to divide as in ordinary division. Place the decimal point in the quotient directly above the decimal point in the dividend, annex zeros after the decimal point in the dividend, and continue the division to as many decimal places as is required. In this instance there is no remainder after the first decimal place.

Practice check: Do exercise 23 on the right.

Answer to exercise 20

$$\begin{array}{r} 2.4375 \\ \times\quad 1.07 \\ \hline 170625 \\ 2\,43750 \\ \hline 2.608125 \end{array}$$

Answer to exercise 21

31,600
There are four zeros in 10,000. Move the decimal point four places to the right. Annex two zeros.

22. Divide 201.825 by 32. State the answer to three decimal places.

The answer is in the margin on page 99.

23. Divide 972 by 0.108.

The answer is in the margin on page 99.

■ Example 3

Divide 0.78 by 0.964.

Solution

$$
\begin{array}{r}
0.809 \\
_x964.\overline{)_x780.000} \\
771\ 2 \\
\overline{8\ 800} \\
8\ 676 \\
\overline{124}
\end{array}
$$

24. Divide 452.4 by 0.052.

The answer is in the margin on page 99.

Explanation   First convert the divisor into a whole number by moving the decimal point to the right three places. Then move the decimal point in the dividend an equal number of places to the right. If, as in this instance, the dividend has fewer figures than the divisor, annex zeros. Then proceed to divide as explained in the preceding examples.

Practice check: Do execise 24 on the left.

Rule

To divide a number by 10; 100; 1,000; 10,000; and so on, move the decimal point in the dividend as many places to the left as there are zeros in the divisor. If necessary, prefix zeros.

Thus, $56.4 \div 10 = 5.64$; $56.4 \div 100 = 0.564$; $56.4 \div 1,000 = 0.0564$; and so forth.

■ Example 4

Divide 924.3 by 100.

Solution $924.3 \div 100 = 9.243$

25. Divide 40.73 by 1,000.

The answer is in the margin on page 99.

Explanation   Note that there are two zeros in 100. To divide 924.3 by 100, move the decimal point in 924.3 two places to the left. The answer is 9.243.

Practice check: Do exercise 25 on the left.

■ Problems 3–9

Give the answers to the following division problems to three places after the decimal point. Use the space provided below to work out the answers for the problems.

1. $25 \div 10$

2. $3.75 \div 10$

3. $4.375 \div 100$

4. $7.125 \div 10$

5. $625 \div 100$

6. $3.125 \div 1,000$

7. 93.75 ÷ 100

8. 312.5 ÷ 10

9. 25.625 ÷ 100

10. 35 ÷ 1,000

11. 6.25 ÷ 10

12. 0.0625 ÷ 1,000

13. 1 ÷ 16

14. 1 ÷ 32

15. 10 ÷ 16

16. 10 ÷ 32

17. 100 ÷ 16

18. 100 ÷ 32

19. 1,000 ÷ 32

20. 1 ÷ 3.14

21. 3.216 ÷ 0.9

22. 4.36 ÷ 0.07

23. 37 ÷ 0.063

24. 15.07 ÷ $\frac{2}{3}$

25. 6.21 ÷ $1\frac{1}{4}$

26. 3.21 ÷ 1.10

27. 5.623 ÷ 1.21

28. 1.23 ÷ 0.625

29. 124.625 ÷ 48

30. 0.2086 ÷ 32

31. 81.63 ÷ 0.09

32. 0.875 ÷ 0.634

33. 0.454 ÷ 6.05

34. 72 ÷ 0.625   **35.** 80 ÷ 4.05   **36.** 2.36 ÷ 5.48

37. 6.455 ÷ 0.008   **38.** 0.565 ÷ 8.673   **39.** 0.11 ÷ 1,848.43

40. 42 ÷ 0.0075   **41.** 0.762 ÷ 888   **42.** 1 ÷ 1,728

43. 7 ÷ 144   **44.** 1 ÷ 3.1416   **45.** 1 ÷ 0.7854

46. 450 ÷ 1,728   **47.** 233 ÷ 5,280   **48.** 0.0406 ÷ 40.08

49. $\frac{4}{7}$ ÷ 3.45   **50.** $\frac{5}{9}$ ÷ 2.083   **51.** $\frac{5}{12}$ ÷ 4.96

52. 6.38 ÷ $\frac{2}{3}$ **53.** 5.09 ÷ $\frac{3}{7}$ **54.** 766 ÷ 10

55. 13 ÷ 100   **56.** 7,216 ÷ 1,000   **57.** 1.28 ÷ 10

58. 42.6 ÷ 100   **59.** 11.9 ÷ 1,000   **60.** 0.153 ÷ 100

3–10 USING THE CALCULATOR TO SOLVE DECIMAL PROBLEMS

The calculator is a help in solving decimal problems quickly and accurately. In order not to become too dependent upon the calculator, the operator not only should understand the problem to be solved but also should analyze the result to judge whether the answer is reasonable for the problem at hand. A good habit is to check the answer after the problem is solved. The operator must make sure that the keys are pressed in the correct sequence. In using calculators and computers, the operator must remember *GIGO;* that is, *Garbage In = Garbage Out.*

The examples demonstrated here for use with the calculator were selected from the examples in the previous sections of this chapter. Compare how you solved the problems using pencil and paper with how you solved the problems using the calculator.

■ **Example 1**

Change $\frac{7}{8}$ to a decimal fraction.

Solution

The Display Shows

Turn on the calculator	0.
Enter 7	7.
Press ÷	7.
Enter 8	8.
Press =	0.875 (answer)

This is the decimal equivalent to $\frac{7}{8}$.

■ **Example 2**

Change $\frac{2}{3}$ to a decimal fraction accurate to the hundredths place.

Solution

The Display Shows

Turn on the calculator	$0.$
Enter 2	$2.$
Press ÷	$2.$
Enter 3	$3.$
Press =	0.666666666

Use the round-off rule and write 0.67 as your answer.

■ **Example 3**

Change $\frac{3}{16}$ to a decimal of three places.

Solution

The Display Shows

Turn on the calculator	$0.$
Enter 3	$3.$
Press ÷	$3.$
Enter 16	$16.$
Press =	0.1875

Use the round-off rule and write 0.188 as your answer.

■ **Example 4**

On a scale divided into sixty-fourths of an inch, what is the dimension nearest $\frac{13}{37}$ in.?

Solution

The Display Shows

Turn on the calculator	$0.$
Enter 13	$13.$
Press ÷	$13.$
Enter 37	$37.$
Press ×	0.351351351
Enter 64	$64.$
Press =	22.48648649

Round off to the nearest whole number (22), and use this whole number as the numerator and 64 as the denominator. This fraction reduces to $\frac{11}{32}$. Therefore, $\frac{13}{37}$ in. is approximately equal to $\frac{11}{32}$ in.

■ **Example 5**

Change $\frac{2}{7}$ to sixteenths.

Solution

The Display Shows

Turn on the calculator	0.
Enter 2	2.
Press ÷	2.
Enter 7	7.
Press ×	0.28714285
Enter 16	16.
Press =	4.571428571

Round off to the nearest whole number (5), and use this whole number as the numerator and 16 as the denominator. This fraction is $\frac{5}{16}$. Therefore, $\frac{2}{7}$ is approximately equal to $\frac{5}{16}$.

Practice check: Do exercise 26 on the right.

26. Change $\frac{5}{9}$ to thirty-secondths.

The answer is in the margin on page 103.

■ **Example 6**

Convert $3\frac{3}{4}$ in. into a decimal fraction of a foot. State the answer to three decimal places.

Solution

The Display Shows

Turn on the calculator	0.	⎤ The fraction part of
Enter numerator 3	3.	⎥ the mixed number
Press ÷	3.	⎥ is usually done first
Enter 4	4.	⎦ on calculators.
		⎤ The whole number
Press +	0.75	⎥ part (3) is now
Enter 3	3.	⎦ added.
Press =	3.75	
Press ÷	3.75	⎤ To convert inches to
Enter 12	12.	⎦ feet, divide by 12.
Press =	0.3125	

Using the round-off rule makes the answer 0.313 ft.

■ **Example 7**

Convert $46\frac{1}{8}$ in. into feet and a decimal fraction of a foot to three decimal places.

Solution

The Display Shows

Turn on the calculator	0.
Enter 1	1.
Press ÷	1.
Enter 8	8.
Press +	0.125

The Display Shows

Enter 46	46.
Press =	46.125
Press ÷	46.125
Enter 12	12.
Press =	3.84375 (answer)

---

27. Convert 29⅝ inches into feet and a decimal fraction of a foot to three decimal places.

The answer is in the margin on page 104.

Round this display answer off to three decimal places for an answer of 3.844 ft.

Practice check: Do exercise 27 on the left.

■ **Example 8**

Convert 43.172 in. into feet and inches and the nearest sixty-fourth of an inch.

Solution

The Display Shows

Turn on the calculator	0.
Enter 43.172	43.172
Press ÷	43.172
Enter 12	12.
Press =	3.597666667 (feet part)

The whole number part shows us that we have 3 ft in 43.172 in. Now write down the 3 ft and subtract 3 from our display to determine how many inches we have in the remainder of 0.597666667 ft.

Press −	3.597666667
Enter 3	3.
Press =	0.597666667
Press ×	0.597666667
Enter 12	12.
Press =	7.172 (inches part)

The whole number part shows us that we have 7 in. in 0.597666667 ft. Write down 7 in. and subtract 7 from our display to determine how many sixty-fourths are in the remainder 0.172 in.

Press −	7.172
Enter 7	7.
Press =	0.172
Press ×	0.172
Enter 64	64.
Press =	11.008 (64ths part)

---

28. Convert 32.312 inches into feet and inches and the nearest sixteenth of an inch.

The answer is in the margin on page 104.

The whole number part shows us that we have approximately 11 sixty-fourths in 0.172 in. Now we combine all the subtotals that we have and write 43.172 in. = 3 ft 7$\frac{11}{64}$ in.

Practice check: Do exercise 28 on the left.

■ **Example 9**

Convert 0.319 ft into inches and a fraction of an inch (to the nearest sixty-fourth inch).

Answer to exercise 26

$\frac{18}{32}$ or $\frac{9}{16}$

Solution

	The Display Shows	
Turn on the calculator	0.	
Enter 0.319	0.319	
Press ×	0.319	
Enter 12	12.	
Press =	3.828	(inches part)

This number tells us that we have 3 in. and a fraction of an inch remaining. Write down the 3 and subtract 3 from our display to work on the fractional part.

Press −	3.828	
Enter 3	3.	
Press =	0.828	
Press ×	0.828	
Enter 64	64.	
Press =	52.992	(64ths part)

This number rounds off to 53, the approximate number of sixty-fourths that are remaining. The answer is $3\frac{53}{64}$ in.

Practice check: Do exercise 29 on the right.

29. Convert 0.523 foot into inches and a fraction of an inch (to the nearest thirty-secondth inch).

The answer is in the margin on page 105.

■ **Example 10**

Add 3.25, 72.004, 864.0725, 647, and 0.875.

Solution

	The Display Shows	
Turn on the calculator	0.	
Enter 3.25	3.25	
Press +	3.25	
Enter 72.004	72.004	
Press +	75.254	
Enter 864.0725	864.0725	
Press +	939.3265	
Enter 647	647.	
Press +	1586.3265	
Enter 0.875	0.875	
Press =	1587.2015	(answer)

■ **Example 11**

Find the sum of $\frac{1}{2}$ + 0.662 + $\frac{7}{8}$.

Solution

	The Display Shows
Turn on the calculator	0.
Enter 1	1.
Press ÷	1.
Enter 2	2.
Press =	0.5
Press +	0.5
Enter 0.662	0.662
Press +	1.162
Enter 7	7.
Press ÷	7.
Enter 8	8.
Press =	2.037 (answer)

■ **Example 12**

Take 18.275 from 42.63.

Solution

	The Display Shows
Turn on the calculator	0.
Enter 42.63	42.63
Press −	42.63
Enter 18.275	18.275
Press =	24.355 (answer)

■ **Example 13**

From $2\frac{1}{3}$ take 0.675. Round off the answer to three decimal places.

Solution

	The Display Shows
Turn on the calculator	0.
Enter 2	2.
Press +	2.
Enter 1	1.
Press ÷	1.
Enter 3	3.
Press −	2.333333333
Enter 0.675	0.675
Press =	1.658333333 (answer)

Round off your answer to 1.658.

■ Example 14

Multiply 43.286 by 6.04.

Answer to exercise 29

$6\frac{9}{32}$

Solution

	The Display Shows
Turn on the calculator	0.
Enter 43.286	43.286
Press ×	43.286
Enter 6.04	6.04
Press =	261.44744 (answer)

■ Example 15

Multiply 0.85 by $1\frac{3}{4}$. (It is usually easier to enter the mixed number first. Special care is needed when entering common fractions into a calculator.)

Solution

	The Display Shows
Turn on the calculator	0.
Enter 1	1.
Press +	1.
Enter 3	3.
Press ÷	3
Enter 4	4.
Press =	1.75
Press ×	1.75
Enter 0.85	0.85
Press =	1.4875 (answer)

■ Example 16

Divide 78.125 by 24. State the answer to three decimal places.

Solution

	The Display Shows
Turn on the calculator	0.
Enter 76.125	76.125
Press ÷	76.125
Enter 24	24.
Press =	3.1718 75 (answer)

This answer is 3.172 when rounded to three decimal places.

Practice check: Do exercises 30 and 31 on right.

30. Multiply 102.68 by 7.12

31. Divide 62.5 by 0.723. State the answer to three decimal places.

The answers are in the margin on page 107.

■ Problems 3–10

Fill out the following equivalent chart using the calculator. Write the decimal equivalent of the fraction in inches to five decimal places and the millimeter equivalent to four places. One inch is equal to 25.4 millimeters.

	Common Fraction (in.)	Decimal Equiva-lent	mm Equiva-lent		Common Fraction (in.)	Decimal Equiva-lent	mm Equiva-lent
1.	$\frac{1}{32}$	_____	_____	2.	$\frac{1}{16}$	_____	_____
3.	$\frac{3}{32}$	_____	_____	4.	$\frac{1}{8}$	_____	_____
5.	$\frac{5}{32}$	_____	_____	6.	$\frac{3}{16}$	_____	_____
7.	$\frac{7}{32}$	_____	_____	8.	$\frac{1}{4}$	_____	_____
9.	$\frac{9}{32}$	_____	_____	10.	$\frac{5}{16}$	_____	_____
11.	$\frac{11}{32}$	_____	_____	12.	$\frac{3}{8}$	_____	_____
13.	$\frac{13}{32}$	_____	_____	14.	$\frac{7}{16}$	_____	_____
15.	$\frac{15}{32}$	_____	_____	16.	$\frac{1}{2}$	_____	_____
17.	$\frac{17}{32}$	_____	_____	18.	$\frac{9}{16}$	_____	_____
19.	$\frac{19}{32}$	_____	_____	20.	$\frac{5}{8}$	_____	_____
21.	$\frac{21}{32}$	_____	_____	22.	$\frac{11}{16}$	_____	_____
23.	$\frac{23}{32}$	_____	_____	24.	$\frac{3}{4}$	_____	_____
25.	$\frac{25}{32}$	_____	_____	26.	$\frac{13}{16}$	_____	_____
27.	$\frac{27}{32}$	_____	_____	28.	$\frac{7}{8}$	_____	_____
29.	$\frac{29}{32}$	_____	_____	30.	$\frac{15}{16}$	_____	_____
31.	$\frac{31}{32}$	_____	_____	32.	1	1.00000	25.4000

Use the calculator to change each of the following decimal fractions to the nearest sixteenth, thirty-second, and sixty-fourth. Round each answer to the nearest whole number.

	Decimal Fraction	Nearest 16th	Nearest 32nd	Nearest 64th
33.	0.9684	_____	_____	_____
34.	0.7211	_____	_____	_____
35.	0.5925	_____	_____	_____
36.	0.2459	_____	_____	_____
37.	0.1183	_____	_____	_____
38.	0.6720	_____	_____	_____
39.	0.0477	_____	_____	_____
40.	0.0656	_____	_____	_____
41.	0.4239	_____	_____	_____
42.	0.6874	_____	_____	_____

Use the calculator to change the following millimeters to decimal parts of an inch and to the nearest thirty-second of an inch.

Answer to exercise 30

731.0816

Answer to exercise 31

86.445

	mm	in.	Nearest 32nd
43.	1	0.03937	$\frac{1}{32}$
44.	2	_____	_____
45.	3	_____	_____
46.	4	_____	_____
47.	5	_____	_____
48.	6	_____	_____
49.	7	_____	_____
50.	8	_____	_____
51.	9	_____	_____
52.	10	_____	_____
53.	11	_____	_____
54.	12	_____	_____
55.	13	_____	_____
56.	14	_____	_____
57.	15	_____	_____
58.	16	_____	_____
59.	17	_____	_____
60.	18	_____	_____
61.	19	_____	_____
62.	20	_____	_____

Solve the following problems with a calculator. Round off the answers to three places.

63. What is the product of 46.887 and 63.921?

64. What is the sum of 65.34, 56.38, 788.45, and 28.447?

65. Subtract 345.967 from 9876.455.

66. Divide 78.431 by 7.43.

67. Divide 0.056 by 0.543333.

68. Multiply 458.998 by 397.596.

69. Add 56.45, 365.65, 387.54, 2987.87, 9832.77, and 45.75.

70. Subtract 85.357 from 3578.911.

71. Multiply 5678.9 by 387.56.

72. Divide 65.929 by 196.45.

73. Gold weights 1,206 lb per cubic foot. Find the weight of 1 cubic inch of gold if there are 1,728 cubic inches in 1 cubic foot.

74. How many cubic feet are there in 1 ton of aluminum if 1 cubic foot of aluminum weights 160 lb? (There are 2,000 lb in a ton.)

75. There is 0.4536 kilogram in 1 pound. How many kilograms are there in 10 tons?

76. There are 231 cubic inches in 1 gallon. How many cubic inches are there in 20 gallons?

77. How many seconds are there in a 24-hour day?

78. How many seconds have you lived?

79. Light travels 186,000 miles per second. How many seconds does it take for light to reach the Earth from our sun if the Earth is 93,000,000 miles from the sun?

80. How many dollars would you have if you took a penny the first day, two cents the second day, four cents the third day, and kept doubling in this manner for 30 days?

81. The volume of a sphere (ball) is found by multiplying, using this formula: $\frac{4}{3} \times 3.1416 \times$ radius $\times$ radius $\times$ radius. What is the volume of the Earth in cubic miles if the radius of the Earth is 4,000 miles?

82. A car travels 394 miles on one tankful of gasoline. If the gasoline tank holds 21 gallons, how many miles per gallon does this car get?

■ Miscellaneous Problems in Decimal Fractions

Use the space provided below to work out the answers for the following problems.

NOTE: *Where the number of decimal places desired in the answer is not stated, decide what degree of accuracy is required in the answer to any given problem.*

1. Cast iron weighs 450 lb per cubic foot; wrought iron, 480 lb per cubic foot; and steel, 490 lb per cubic foot. Find the weight of each per cubic inch (1 cu ft = 1728 cu in.). State the answers as decimal fractions of three places.

2. What will be the weight of a wrought-iron rod that contains 62.25 cu in.?

3. What part of a cubic foot is contained in a steel shaft weighing 215.75 lb? State the answer as a decimal fraction to three places.

4. What will be the weight of 312 iron castings if each contains 86.5 cu in.?

5. The actual inside diameter of a 1-inch pipe is 1.04″; the actual outside diameter is 1.315″. Find the thickness of the pipe (Fig. 3–1).

Figure 3–1

6. A $\frac{7}{8}''$ bolt (without nut) weighs 0.664 lb. How many such bolts are there in a keg of 250 lb?

7. $3 \times 3 \times \frac{3}{8}$ angle iron weighs 7.2 lb per foot. Find the weight of a piece 12'3" long.

8. Find the difference in diameters of the piece of tapered work in Fig. 3–2.

$1\frac{7}{8}''$ 1.194"

Figure 3–2

9. The circumference of a circle is found by multiplying its diameter by 3.1416. Find the circumference of a circle whose diameter is 2.35". State the answer to two decimal places.

10. To find the diameter of a circle when the circumference is known, divide the circumference by 3.1416. Find the diameter of a circle having a circumference of 8.7865". State the answer to one decimal place.

11. Find the whole depth of the gear tooth shown in Fig. 3–3. State the answer to the nearest ten-thousandths of an inch.

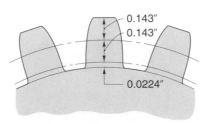

0.143"
0.143"
0.0224"

Figure 3–3

12. How many pounds of water are there in a tank that contains 904.7 cu ft? (One cubic foot of water weighs 62.5 lb.)

13. How many gallons of water will the tank in Problem 12 contain if there are 231 cu in. in a gallon?

14. Find (*a*) the weight of a cubic inch of water and (*b*) the weight of a gallon of water.

15. With gasoline at $1.419 per gallon, what will it cost to fill a tank whose capacity is 1.75 cu ft? (One cu ft is approximately $7\frac{1}{2}$ gal.)

16. Find the diameter at *A* of the tapered shank in Fig. 3–4 if the difference between the small diameter and the diameter at *A* is 0.392″.

Figure 3–4

17. A machinist's helper gets $12.26 per hour. How much will he earn in a week of 40 hours?

18. A welder gets $22.67 per hour for a 40-hour week. During the week she works 6 hours overtime at time-and-a-half. How much does she earn?

19. A sheet metal worker works a 40-hour week at $21.75 per hour. How much does he earn?

20. If the basic work week is 40 hours at $16.56 per hour and a woman works 49 hours, how much does she earn during the week, allowing time-and-a-half for overtime?

21. The rim or cutting speed on cylindrical work is found by multiplying the circumference of the work by the number of revolutions per minute. Find the cutting speed on a piece of work whose circumference is 4.273″ if the work is making 120 revolutions per minute. Express the answer in feet per minute.

22. At 97 cents per pound, find the cost of 125 castings, each weighing 8.68 lb.

23. Compute the depth of tooth in Fig. 3–5.

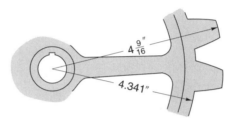

Figure 3–5

24. Compute the root diameter of the thread in Fig. 3–6.

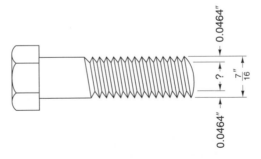

Figure 3–6

25. Find the corner measurement for laying out an octagonal end on the square bar in Fig. 3–7.

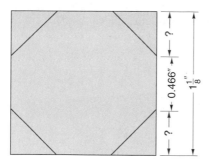

Figure 3–7

26. Find the depth of cut required to mill a square end on the round shaft in Fig. 3–8.

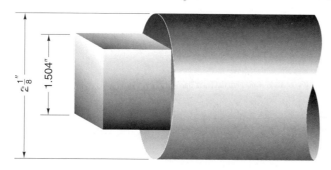

Figure 3–8

27. What is the thickness of metal in the cross section of pipe in Fig. 3–9?

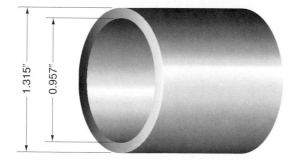

Figure 3–9

28. (a) Find the diameter at the large end of the tapered portion in Fig. 3–10.
  (b) Find the difference between the diameters at the large and small ends of the tapered part.
  (c) Compute the missing dimension.

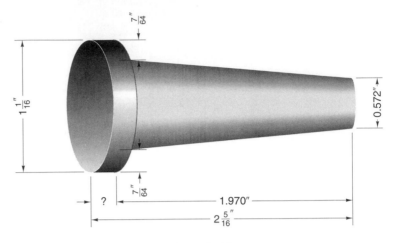

Figure 3–10

29. A shaft that should be $1\frac{3}{16}''$ in diameter measures 1.317″. How much is it oversize?

30. A round piece of work is $1\frac{5}{8}''$ in diameter. How deep a cut should be taken to bring the diameter down to 1.594″?

31. Figure 3–11 shows an American Standard thread of $\frac{1}{9}''$ pitch. Find the flat at top and bottom if the flat is $\frac{1}{8}$ of the pitch. State the answer to the nearest ten-thousandth of an inch.

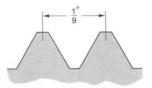

Figure 3–11

32. Find the value of the following expressions. State the answers as decimals of three places.

(a) $2.689 + 1\frac{7}{12}$

(b) $1\frac{2}{7} - 0.875$

(c) $3.256 \times 4\frac{1}{8}$

(d) $4.586 \div \frac{9}{16}$

(e) $\dfrac{\frac{1}{3} \text{ of } \frac{7}{8}}{\frac{1}{2} \text{ of } \frac{3}{4}}$

(f) $\dfrac{\frac{3}{4} \div 2\frac{1}{2}}{\frac{5}{8} \times 4\frac{1}{4}}$

33. Fourteen holes, equally spaced, are to be drilled as shown in Fig. 3–12. Find the distance center to center of adjacent holes. Give the answer to the nearest sixty-fourth of an inch.

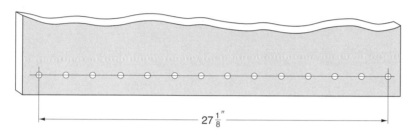

$27\frac{1}{8}''$

Figure 3–12

34. A No. 10 wire has a diameter of 0.1019 in., and a No. 12 wire has a diameter of 0.0808 in. (American Wire Gage sizes). What is the difference between their diameters?

35. One 100-W bulb draws 0.87 A of current. How many amperes will 17 such bulbs draw?

36. If roofing shingles cost $7.91 per bundle and three bundles make a square (100 sq ft), how much would 20 such squares cost?

37. The yearly rainfall in Cherrapunji, India, is 449.8 in. What is the average rainfall per month?

38. Suppose the population of the United States is 241,078,000 and the annual increase in population is estimated at 1.2 times the present population. What will the population be next year?

39. Mt. McKinley in Alaska is 20,320 ft high and is the highest mountain in the United States. How many miles high is Mt. McKinley (5,280 ft = 1 mi)? (Source: U.S. Geological Survey.)

40. The distance from Albuquerque, New Mexico, to Philadelphia, Pennsylvania, is 2,124 miles. If you want to make a trip in three days, how many miles per day must you travel?

■ Self-Test

Do the following problems. Use the space provided below to work out the answers. All answers are to be in decimal form. Check your answers with the answers in the back of the book.

1. $0.377 + \frac{1}{4}$

2. $1.427 + 0.675$

3. $5.329 - \frac{5}{8}$

4. $7.62 - 1.81$

5. $3.2 \times \frac{3}{4}$

6. 4.7×4.3

7. $5.3 \div \frac{1}{2}$

8. $4.32 \div 2.35$

9. $\dfrac{3.2 + 1.7}{4.6}$

10. $\dfrac{7.35 \times 6.25}{\frac{5}{8}}$

■ Chapter Test

Do the following problems. Use the space provided below to work out the answers. All answers are to be in decimal form. Round off all answers to three decimal places.

Add.

1. $0.513 + \frac{1}{8}$

2. $\frac{3}{4} + 1.007$

3. $1.06 + 0.391$

4. $6\frac{1}{2} + 0.762$

5. $0.83 + 4.205$

6. $1.07 + 3.6121$

7. $0.006 + 7$

8. $3.217 + 15.3928$

9. $8 + 0.312$

10. $4.37 + 5$

Subtract.

11. $1\frac{5}{16} - 0.3125$

12. $0.362 - \frac{1}{8}$

13. $1.71 - 0.007$

14. $8.62 - 1.317$

15. $15.73 - 9.635$

16. $8.6 - 0.082$

17. $97 - 9.708$

18. $0.438 - 0.099$

19. $14.293 - 8$

20. $63 - 12.62$

Multiply.

21. $6.2 \times \frac{1}{16}$

22. $1\frac{1}{2} \times 0.079$

23. 0.03×0.07

24. 8.6×0.43

25. $1\frac{2}{3} \times 6.21$

26. 1.27×6.33

27. 4.81×0.0162

28. $0.625 \times \frac{1}{16}$

29. 4×6.357

30. $\$46.20 \times 6$

Divide.

31. $1 \div 16$

32. $3.2 \div \frac{1}{8}$

33. $0.01 \div 0.7$

34. $8.6 \div 3.21$

35. $1.2 \div 0.63$

36. $10 \div 1.63$

37. $100 \div 0.51$

38. $0.6 \div 0.006$

39. $\$5.00 \div 4$

40. $\$655.60 \div 8$

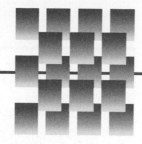

CHAPTER 4

Percentage

4-1   DEFINITIONS

One *percent* of a number means one-*hundredth* of a number, six percent means six-*hundredths,* and so on,

The symbol % stands for the word *percent.* Thus, 4 percent is written 4%, and 6 percent is written 6%. The symbol % does the work of two decimal places. For example, 6% = 0.06; 25% = 0.25; 110% = 1.10; 0.5% = 0.005; $2\frac{1}{4}\%$ = 0.0225 or $0.02\frac{1}{4}$; and 5.7% = 0.057.

Since percent means hundredths, the whole of any number contains 100% of itself. Thus, 100% of 75 is 75.

Every problem in percentage contains three elements: the **base,** the **rate,** and the **percentage.** When we say that 6% of $100 is $6, the base is $100, the rate is 6% or 0.06, and the percentage is $6. If any two of these three elements are known, the third may be found.

The following form will help you solve percentage problems.

$$\text{Percentage} \searrow \qquad \swarrow \text{Rate}$$
$$\frac{5}{8} = 0.625 = 62.5\%$$
$$\text{Base} \nearrow \qquad \nwarrow \text{Decimal equivalent}$$

Memorize the following table, which gives the percentage equivalents of the common fractions most frequently used in practice.

Percentage Equivalents of Common Fractions			
$\frac{1}{8} = 12\frac{1}{2}\%$	$\frac{5}{8} = 62\frac{1}{2}\%$	$\frac{1}{6} = 16\frac{2}{3}\%$	$\frac{1}{5} = 20\%$
$\frac{1}{4} = 25\%$	$\frac{3}{4} = 75\%$	$\frac{1}{3} = 33\frac{1}{3}\%$	$\frac{2}{5} = 40\%$
$\frac{3}{8} = 37\frac{1}{2}\%$	$\frac{7}{8} = 87\frac{1}{2}\%$	$\frac{2}{3} = 66\frac{2}{3}\%$	$\frac{3}{5} = 60\%$
$\frac{1}{2} = 50\%$		$\frac{5}{6} = 83\frac{1}{3}\%$	$\frac{4}{5} = 80\%$

4–2   FINDING THE PERCENTAGE, GIVEN THE BASE AND THE RATE

To find the percentage when the base and the rate are known, multiply the rate by the base. Remember to change the rate to its decimal equivalent; for example, 6% = 0.06, and so on.

$$\text{Decimal equivalent of rate} \times \text{Base} = \text{Percentage}$$

■ Example

Find 4% of 683.

Solution   4% = 0.04   0.04 × 683 = 27.32

1. Find 6% of 598.

The answer is in the margin on page 122.

Explanation   Since percent means hundredths, 4% of a number means 0.04 of a number. To obtain 4% of 683, multiply 683 by 0.04 and get the percentage 27.32.

Practice check: Do exercise 1 on the left.

Rule

To find the percentage, multiply the base by the rate.

■ Problems 4–2

Find the following percentages. Use the space provided below to work out the answers for the problems.

1. 10% of 50

2. 10% of 500

3. 10% of 5000

4. 1% of 500

5. 1% of 5000

6. 20% of 50

7. 20% of 500

8. 20% of 5000

9. 50% of 15

10. 50% of 500

11. 50% of 5000

12. 12% of 30

13. 15% of 75

14. 18% of 28

15. 5% of 87

16. $12\frac{1}{2}$% of 88

17. 17% of 300   **18.** $83\frac{1}{3}$% of 660   **19.** $\frac{1}{2}$% of 100   **20.** $\frac{1}{2}$% of 50

21. White metal is made up of 3.7% copper, 88.8% tin, and 7.5% antimony. How many pounds of each metal are there in 465 lb of the alloy?

22. An apprentice earning $405 per week got a wage increase of 15%. What did the apprentice earn after the increase?

23. If an individual is allowed a discount of 2% from a bill amounting to $242.85, what must the person pay?

24. A dealer in electrical supplies allows discounts of 15% and 5% from the list price, and a 2% discount for cash. How much cash must be paid for a bill of goods whose list price amounts to $162.50? (Deduct 15% from the original amount, then 5% from the remainder, and 2% from the second remainder.)

25. Muntz metal is 59.5% copper, 39.9% zinc, and 0.6% lead. How many pounds of each are there in 432 lb of the alloy?

26. U.S. Navy specifications for phosphor bronze call for 85% copper, 7% tin, 0.06% iron, 0.2% lead, 0.3% phosphorus, and the remainder of zinc. How many pounds of each element are required to make 500 lb of phosphor bronze?

27. On a $15,000 contract, 72% was paid for labor and materials, 6% for supervision, and the remainder for profit. How much was paid for each item?

28. The manufacturer of a certain automobile estimates the following costs: materials, $38\frac{1}{2}$%; labor, $41\frac{1}{4}$%; overhead, $6\frac{1}{2}$%; and profit, $13\frac{3}{4}$%. Find the cost of each item in an automobile that sells for $17,589.

29. A standard formula for type metal calls for $77\frac{1}{2}$% lead, $6\frac{1}{2}$% tin, and 16% antimony. How many pounds of each metal are there in 250 lb of the type metal?

30. A certain iron ore yields $4\frac{1}{4}$% iron. How many pounds of iron are there in a ton of this ore?

31. A man's weekly pay is $650.85. At the rate of 6.13%, how much does he pay in Social Security taxes per week?

32. A woman works a 40-hour week at the rate of $16.75 per hour. How much should she find in her pay envelope at the end of the week if 18% is deducted for income tax and 6.13% for Social Security tax?

33. If sales tax is 6%, how much was the tax on a used car that sold for $21,495?

34. A customer buys an electric drill for $8.98, a disc sander for $36.75, and an electric chain saw for $39.99. How much sales tax must be paid if the sales tax is 6%?

35. Three suits are purchased at $299.99 each. The sales tax is 6% and is added to the bill. How much is the total bill?

36. Shirts are advertised at $14.75 each, less 25%. The sales tax is 6%. What will the bill be for five shirts?

37. What is $5\frac{1}{4}$% of $1,500?

38. What is $5\frac{1}{2}$% of $100?

39. If a certain bill is paid within 20 days of the billing date, a discount of 2% is allowed. How much must be paid if the original bill was for $596.50 and it is paid within the 20-day limit?

40. The interest charge per month on a charge card is $1\frac{1}{2}$% of the balance due. What is the month's interest charge on a $2,067.58 balance?

4–3   FINDING THE RATE, GIVEN THE BASE AND THE PERCENTAGE

Suppose a company makes 5,000 articles and 150 are found to be defective. This relationship between articles made and articles rejected is often expressed as a rate; that is, 3% of the articles made were rejected. The problem can be stated as: What percent of 5,000 is 150? The solution follows.

$$150 \text{ parts out of } 5{,}000 \text{ is } \tfrac{150}{5000} = 0.03 \quad \text{and} \quad 0.03 \times 100 = 3\%$$

To find the rate use the following rule.

Rule _____

To find the rate, write a fraction with the percentage as the numerator and the base as the denominator, and change this fraction to a decimal. Multiply the quotient by 100 to get the final percent.

$$\frac{\text{Percentage}}{\text{Base}} = \text{Decimal equivalent of rate}$$

■ **Example 1**

What percent of 8 is 7?

Solution   7 is $\frac{7}{8}$ of 8, and $\frac{7}{8} = 0.875 = 87.5\%$.

■ **Example 2**

What percent of 64.8 is 16.5?

Solution $\dfrac{16.5}{64.8} = 0.2546 = 25.46\%$

Practice check: Do exercises 2 and 3 on the right.

2. What percent of 16 is 13?

3. What percent of 32.4 is 14.8?

The answers are in the margin on page 125.

■ Problems 4–3

Find the following rates. Use the space provided below to work out the answers for the problems.

1. What percent of 10 is 5?

2. 12 is what percent of 100?

3. What percent of 80 is 20?

4. What percent of 30 is $7\frac{1}{2}$?

5. $12\frac{1}{2}$ is what percent of 50?

6. What percent of 75 is 25?

7. What percent of 20 is 4?

8. 3 is what percent of 15?

9. What percent of 36 is 12?

10. 25 is what percent of 125?

11. What percent of 58 is 29?

12. What percent of 208 is 52?

13. 10 is what percent of 61?

14. What percent of 51 is 17?

15. $3\frac{1}{4}$ is what percent of $5\frac{1}{2}$?

16. What percent of $2\frac{1}{2}$ is $\frac{1}{2}$?

17. What percent of 1.32 is 0.6?

18. 0.25 is what percent of 8.75?

19. What percent of 0.625 is 0.25?

20. What percent of 0.025 is 0.005?

21. A motor receiving 8 hp delivers 6.8 hp of work. What percent of the input is the output?

22. One ton of ore yields 80 lb of iron. What percent of the ore is iron?

23. A man pays $5.75 for an article and sells it for $6.50. What percent profit does he make?

24. The usual allowance made for shrinkage when casting iron pipe is $\frac{1}{8}''$ per foot. What percent is this?

25. In making 95 lb of solder, 38 lb of lead and 57 lb of tin were used. What percent of each was used?

26. The indicated horsepower of a steam engine is 9.4 and the effective horsepower is 8.1. What percent of the indicated horsepower is the effective horsepower?

27. What percent of a foot is $\frac{5}{8}$ of an inch?

28. What percent of a mile is 100 feet?

29. Out of a total production of 2,715 ball bearings manufactured during a day, 107 were rejected by the inspectors as imperfect. What percent of the total was rejected?

30. An alloy of common yellow brass is made of the following ingredients: copper, 170.5 lb; lead, 7.7 lb; tin, 0.55 lb; and zinc, 96.25 lb. What percent of the entire alloy does each of the metals represent?

31. A machine shop job required 42 hours on the lathe, $7\frac{1}{2}$ hours on the milling machine, and $11\frac{1}{4}$ hours on the planer. What percent of the total time should be charged to each machine?

32. A company made 1,000 automobile doors, and 20 were found to be defective. What percent of the output was found to be defective?

33. A company projects that if no more than 1% of its output is defective, it will make a profit. If the company averages 98 rejected articles out of every 10,000 articles made, does it meet the projection for making a profit?

Fill out the following weekly production chart.

	Day	Articles Made	Articles Rejected	Rate of Rejections
34.	Monday	4,020	43	_____
35.	Tuesday	4,070	37	_____
36.	Wednesday	4,100	40	_____
37.	Thursday	4,075	29	_____
38.	Friday	4,125	32	_____
39.	Saturday	3,760	30	_____
40.	Totals	_____	_____	_____

4–4 FINDING THE BASE, GIVEN THE PERCENTAGE AND THE RATE

If a certain ore contained 5% iron and a company required a production schedule of 100 tons of iron per week, then it would have to process a certain amount of ore to produce the necessary 100 tons of iron. This is an example of knowing the percentage and the rate and using them to find the base.

Rule _____

To find the base, divide the percentage by the rate expressed as hundredths.

$$\frac{\text{Percentage}}{\text{Decimal equivalent of rate}} = \text{Base}$$

■ Example 1

1,022 is 28% of what number?

Solution

$$
\begin{array}{r}
36\ 50 \\
{}_{x}28.\overline{)1022_{x}00.} \\
\underline{84} \\
182 \\
\underline{168} \\
140 \\
\underline{140}
\end{array}
$$

Explanation   Dividing by 28 gives 1% of the number. To get 100%, that is, the whole number, multiply by 100. To obtain the result in one step, divide 1,022 by 0.28.

Practice check: Do exercise 4 on the right.

■ Example 2

300 is 20% more than what number?

Solution

$$
\begin{array}{r}
2\ 50 \\
1_{x}20.\overline{)300_{x}00.} \\
\underline{240} \\
600 \\
\underline{600}
\end{array}
$$

Explanation   The problem states that 300 is more than the original number by 20%. Therefore, part of the 300 is 100% of the original number, and part of the 300 is 20% of the original number. The 300 then represents 120% of the original number. Since 300 is 120% of the number, divide by 120 to get 1% of the number. To get 100%, or the whole number, multiply by 100. To obtain the result in one step, divide 300 by 1.20.

Practice check: Do exercise 5 on the right.

4. 3,024 is 56% of what number?

The answer is in the margin on page 129.

5. 819 is 30% more than what number?

The answer is in the margin on page 129.

■ **Example 3**

210 is 30% less than what number?

Solution

$$\begin{array}{r} 3\ 00 \\ {}_x\overline{70.)210_x00.} \\ \underline{210} \end{array}$$

Explanation   The problem states that 210 is really 70% of the original number, since 100% would equal the original number, and 30% less than 100% is 70%. Since 210 is 70% of the number, divide by 70 to get 1% of the number. To get 100%, or the whole number, multiply by 100. To obtain the result in one step, divide 210 by 0.70.

Practice check: Do exercise 6 on the left.

6. 880 is 60% less than what number?

The answer is in the margin on page 130.

■ **Problems 4–4**

In each of the following problems find the base. Remember the rule:

Rule _____

To find the base, divide the percentage by the rate expressed as hundredths.
_____

Use the space provided below to work out the answers for the problems.

1. 50% of what number is 8?

2. 70 is 10% of what number?

3. 20 is 100% of what number?

4. 116 is $16\frac{2}{3}$% of what number?

5. $83\frac{1}{3}$% of what number is 125?

6. 20% of what number is 67?

7. $28\frac{1}{2}$ is 40% of what number?

8. 49 is 70% of what number?

9. 198 is 90% of what number?

10. 2% of what number is 5?

11. 5 is 25% of what number?

12. $\frac{2}{3}$ is 75% of what number?

13. 80% of what number is $8\frac{1}{2}$?

14. 30 is 10% more than what number?

15. 10% more than what number is 15?

16. 10% more than what number is 5.5?

17. $6\frac{7}{8}$ is 1% more than what number?

18. 48 is 20% less than what number?

19. 15% less than what number is 5.8?

20. $3\frac{1}{3}$ is 6% less than what number?

21. 75% of what number is 3?

22. 22.4 is 65% of what number?

23. 25% of what number is $\frac{2}{3}$?

24. 30 is $16\frac{2}{3}\%$ more than what number?

25. 18.2% less than what number is 48?

26. After a man's wages were reduced 15%, he got $684.20. How much did he get before the reduction?

27. A dealer sells coal at $159.50 per ton. If her profit is 12%, what does the coal cost her?

28. The efficiency of a motor is 90%; that is, the output is 90% of the input. If the motor delivers 8 hp, what is the input?

29. A person sells an article for $3.60, thereby losing 10%. What did the article originally cost?

30. A certain ore yields 5% of iron. How many tons of ore are required to produce $2\frac{1}{2}$ tons of iron?

31. An article loses $3\frac{1}{2}$% of its weight by drying. If it weighs $8\frac{1}{2}$ lb when dry, what was its weight before drying?

32. The inspectors in a factory rejected 33 pieces as imperfect. This represented $1\frac{1}{2}$% of the daily production. How many pieces were produced?

33. A motor whose efficiency is 86% delivers $10\frac{3}{4}$ hp. What horsepower does it receive?

34. What must be the length of a pattern for an $18\frac{1}{2}$″-long casting if the allowance for shrinkage is $\frac{3}{4}$ of 1%?

35. A mechanic gets an increase of $27.20 per week, which represents an increase of 4%. What is the mechanic's new weekly salary?

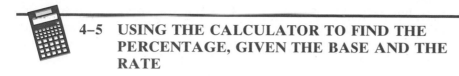

4–5 USING THE CALCULATOR TO FIND THE PERCENTAGE, GIVEN THE BASE AND THE RATE

The calculator problems in this chapter may be worked on a calculator like the one shown here. To find the percentage when the base and rate are known, multiply the base by the rate expressed in hundredths.

■ **Example**

What is 6% of $35.65?

Solution

	The Display Shows
Turn on the calculator	0.
Enter 6	6.
Press ÷	6.
Enter 100	100.
Press ×	0.06
Enter 35.65	35.65
Press =	2.139

Since this is a problem involving money, round the answer to $2.14.

Practice check: Do exercise 7 on the right.

7. What is $2\frac{1}{2}$% of $155.22?

The answer is in the margin on page 133.

■ **Problems 4–5**

Use the calculator to find the following percentages. Round your answers to fit the problems; that is, a money problem should be rounded to the nearest cent, and so on.

1. A car was bought for $14,095, and the state sales tax was 6%. What was the cost of the car including the sales tax?

2. A company making angle brackets expects a 1.5% reject rate daily. If 5,973 angle brackets are made daily, how many rejected brackets can be expected each day?

3. 15% of a class of 35 students got a grade of A. How many students got an A in this class?

4. A firm gives successive discounts of 10% and 2% for their preferred customers who pay their bills promptly. What amount should a preferred customer pay if the original bill was $1,258.00?

5. A baseball player is hitting at 0.247, which means that he has averaged a 24.7% chance of hitting the ball for each time he is up to bat. If he was up to bat 490 times, how many hits did he get?

6. What is the total Social Security tax on a $20,000 yearly income if the Social Security tax is 6.13%?

7. A cottage is sold for $56,500. If 72% of the selling price is for labor and materials and 8% is for supervision, what is the profit made on the cottage?

8. A certain gold ore contains 0.02% gold. How many ounces of gold will there be in 1 ton of this ore?

9. The surface area of a sphere (ball) is found by using this formula: $4 \times 3.1416 \times$ radius $\times$ radius. If the radius of the Earth is 4,000 miles and 70% of the Earth's surface is covered by water, then how many square miles of water are there on the Earth's surface?

10. In 1994 there were 6,613,970 passenger cars sold in the United States, and 9.7% of these were luxury cars. Consumers bought 53.6% of these luxury cars. How many luxury cars were bought by consumers? (Source: American Automobile Manufacturers Association.)

4–6   USING THE CALCULATOR TO FIND THE RATE, GIVEN THE BASE AND THE PERCENTAGE

Answer to exercise 7

$3.88

To find the rate, divide the percentage by the base and multiply the quotient by 100.

■ **Example**

What was the rate of death by heart disease in 1977 if 723,878 people in the United States died from heart disease that year? The population at that time was 216,332,000.

Solution

The Display Shows

Turn on the calculator	0 .
Enter 723878	723878 .
Press ÷	723878 .
Enter 216332000	216332000 .
Press ×	0.003346143
Enter 100	100 .
Press =	0.334614389

Round this answer to 0.33% as the rate of death by heart disease in 1977. (Source: U.S. Health Service.)

Practice check: Do exercise 8 on the right.

■ **Problems 4–6**

Use the calculator to solve the following problems. Round the answer according to the nature of problem.

1. In a recent year there were 134,036,000 licensed drivers in the United States and 6,938,000 licensed drivers in Pennsylvania. What percentage of the licensed drivers lived in Pennsylvania? (Source: Federal Highway Administration.)

2. One ton of ore yields 36 lb of copper. What percent of the ore is copper?

3. What percent of the Earth's surface is Asia if Asia's area is 17,129,000 square miles? (Refer to Problem 9 of Problem Set 4–5.)

4. The diameter of the sun is 864,900 miles and the diameter of the Earth is 8,000 miles. What percent of the sun's diameter is the Earth's diameter?

8. The sales tax is $2,103.30 on an automobile purchase price of $37,558.92. What is the sales tax rate?

The answer is in the margin on page 135.

5. Out of 1,399,000 college graduates, 322,000 received master's degrees. What percent of these college graduates received master's degrees?

6. Out of a total output of 380 castings per day, 27 were rejected. What is the percentage of bad castings?

7. If a person borrows $5,000 and repays $5,650, what is the rate of interest expressed in percent?

8. By installing new machinery, a manufacturer can make 10,000 special bolts where only 7,500 bolts could be made before. What is the gain in output of the new process over the old expressed in percent?

9. In 1968 gasoline was 54.9 cents per gallon, and in 1996 gasoline was $1.419 per gallon. What was the increase in cost per gallon expressed in percent?

10. In 1994 the U.S. Department of Labor estimated that the number of workers in the United States was 123,060,000 and that about 3,629,000 workers were in farm occupations. What percentage of the U.S. workers were in farm occupations?

 4–7   USING THE CALCULATOR TO FIND THE BASE, GIVEN THE PERCENTAGE AND THE RATE

To find the base, divide the percentage by the rate expressed in hundredths.

■ **Example**

A certain iron ore yields 4.7% of iron. How many tons are required to produce 150 tons of iron? Round your answer to hundreds.

Solution

The Display Shows

Turn on the calculator	0.
Enter 150	150.
Press ÷	150.
Enter 4.7	4.7
Press ×	31.91489362.
Enter 100	100.
Press =	3191.489362.

Round this answer to 3,200 tons.

Practice check: Do exercise 9 on the right.

9. A small gas engine pump delivers 34 hp. This represents 85% of its total capable horsepower. How much horsepower can it deliver?

The answer is in the margin on page 137.

■ **Problems 4–7**

Use the calculator to solve the following problems. Round the answer according to the nature of the problem.

1. A certain iron ore yields 3.6% of iron. How many tons are required to produce 100 tons of iron?

2. Inspectors rejected 137 parts as not usable. This represented 1.2% of the total output. What was the total output?

3. A casting is to be 1 foot $6\frac{1}{2}$ inches long. What must be the length of the pattern if the shrinkage allowance is 0.0075%?

4. 65 is 11.5% of what number?

5. 60 is 12% more than what number?

6. 2.75 is 5.5% less than what number?

7. A motor has an efficiency rating of 87.5% and delivers 17.5 hp. What horsepower does it receive?

8. In 1995 there were 463,701 U.S. Navy personnel on active duty. Of this total, 13.2% were commissioned officers. How many commissioned officers were on active duty in 1995? (Source: U.S. Department of the Navy.)

9. In 1890 there were 43,731 graduates from high school. This represents 3.4% of those who could have graduated based on age. How many could have graduated in 1890 if school attendance was mandatory?

10. If approximately 11.97% of the population of the United States lives in California and the population of California is 29,780,061, then what is the population of the United States? (Source: U.S. Bureau of Statistics.)

4–8   THE *P–R–B* TRIANGLE

Sometimes the device shown in Fig. 4–1 is used to help remember the rules involving percentage problems. Draw a triangle and place *P* (meaning *percentage*), *R* (meaning *rate*), and *B* (meaning *base*) as shown in the figure.

When the percentage is required, cover the *P* with your finger, and the device shows that the base (*B*) must be multiplied by the rate (*R*). If the rate is required, cover the *R* with your finger, and the device shows that the percentage (*P*) must be divided by the base (*B*). If the base is required, cover the *B* with your finger, and the device shows that the percentage (*P*) must be divided by the rate (*R*). You might want to try the *P–R–B triangle* to solve the percentage problems in the following tests.

Figure 4–1

■ Self-Test

Do the following problems. Use the space provided below to work out the answers for the problems. Check your answers with the answers in the back of the book.

Answer to exercise 9

40 hp

1. $\frac{1}{2}$ = _____%

2. $\frac{1}{3}$ = _____%

3. 10% of 75 is _____.

4. What percent of 25 is 5?

5. 40 is 20% of what number?

6. $8\frac{1}{4}$% of \$32,000 is _____.

7. 1% of 80 is _____.

8. $\frac{1}{2}$% of 200 is _____.

9. Inspectors rejected 15 pieces of a 500-piece lot. What was the percentage of rejections?

10. If an article was priced at \$230.50 and is marked down 20%, what is its new price?

■ Chapter Test

Do the following problems involving percentages. Make sure that your answer is reasonable by estimating the answer first. Then use the space provided below to work out the answers for the problems.

1. $\frac{1}{8}$ = _____%

2. $\frac{1}{4}$ = _____%

3. $\frac{3}{4}$ = _____%

4. $\frac{2}{3}$ = _____%

5. $\frac{1}{6} =$ _____%   **6.** $\frac{5}{8} =$ _____%

7. 10% of 250 is _____.   **8.** 18% of 508 is _____.

9. 45% of 127 is _____.   **10.** 0.75% of 30 is _____.

11. What percent of 15 is 5?   **12.** What percent of 35 is $17\frac{1}{2}$?

13. What percent of 40 is 6?   **14.** What percent of 75 is 5?

15. What percent of 1 is 16?   **16.** 4 is 75% of what number?

17. 36 is 20% of what number?   **18.** 72 is 20% less than what number?

19. 57 is $16\frac{2}{3}$% of what number?   **20.** 100% of 3.1416 is _____.

21. Out of 1,200 persons who work for Stetson and Ross Machine and Foundry, 75 are supervisors. What percent of the persons are supervisors?

22. A company that builds airplanes has 12,000 employees. It can predict that approximately 1.7% of its employees will be absent from work on any particular day. How many would be absent on a Thursday?

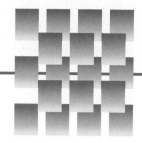

CHAPTER 5

Ratio and Proportion

5–1   RATIO

A *ratio* is a comparison of two like quantities. It is expressed by the quotient obtained when the first quantity is divided by the second. For example, the relation between 10 and 5 can be stated in one of four ways: the ratio of 10 to 5, or $10 \div 5$, or $10:5$, or $\frac{10}{5}$. In each instance the expression represents the ratio of 10 to 5, and the value of the ratio is obtained by dividing 10 by 5, giving 2.

Every common fraction may be regarded as a ratio. The fraction $\frac{2}{3}$ is the ratio of 2 to 3. The two numbers compared are the *terms* of the ratio. The *inverse ratio* is the reciprocal of the original ratio. The ratio of 3 to 2 is the inverse of the ratio of 2 to 3, and vice versa.

5–2   REDUCTION OF RATIOS TO LOWEST TERMS

Working with ratios is much like working with common fractions, and the same rules apply. The ratio of 6 to 30 is the same as the ratio of 2 to 10 and the same as the ratio of 1 to 5. To reduce a ratio to lowest terms, proceed as in reducing a common fraction to lowest terms; that is, find the highest common factor of both terms and divide both terms using the highest common factor as their divisor.

■ **Example 1**
Express in lowest terms the ratio of 6 to 30.

Solution $\dfrac{6}{30} = \dfrac{6 \div 6}{30 \div 6} = \dfrac{1}{5}$ or   1 to 5

Explanation   Express the ratio as a fraction $\frac{6}{30}$ and divide both terms by the highest common divisor, 6, giving the fraction $\frac{1}{5}$. The relation between 6 and 30 is the same as between 1 and 5. The ratio of 6 to 30 is in lowest terms when expressed as the ratio of 1 to 5.

■ Example 2

Express in lowest terms the ratio $\frac{2}{3}$ to $\frac{4}{5}$.

Solution   Ratio of $\frac{2}{3}$ to $\frac{4}{5} = \frac{2}{3} \div \frac{4}{5} = \frac{2}{3} \times \frac{5}{4} = \frac{5}{6}$

Explanation   Since the value of a ratio is the quotient obtained by dividing the first term by the second, divide $\frac{2}{3}$ by $\frac{4}{5}$, obtaining the quotient $\frac{5}{6}$. That is, the ratio of $\frac{2}{3}$ to $\frac{4}{5}$ is the same as the ratio of 5 to 6.

Practice check: Do exercises 1 and 2 on the left.

■ Example 3

Divide $28 between two people in the ratio of 2 to 5.

Solution $\frac{2}{7}$ of $28 = $8
$\frac{5}{7}$ of $28 = $20

Explanation   To divide in the ratio of 2 to 5 means that for every $2 given to one person, $5 must be given to the other person. In other words, out of every $7, one person gets $\frac{2}{7}$ or $2 and the other gets $\frac{5}{7}$ or $5. In general, one person gets $\frac{2}{7}$ of the total amount and the other gets $\frac{5}{7}$.

Practice check: Do exercise 3 on the left.

1. Express in lowest terms the ratio of 15 to 25.

2. Express in lowest terms $\frac{3}{5}$ to $\frac{9}{10}$.

The answers are in the margin on page 144.

3. Divide a piece of cable 30 in. long in the ratio of 3 to 7.

The answer is in the margin on page 144.

■ Problems 5–2

Use the space provided below to work out the answers for the following problems.

Express the following ratios in lowest terms.

1. 12 to 3   **2.** 3 to 12   **3.** 6 to 5   **4.** $\frac{3}{4}$ to $\frac{9}{16}$

5. $\frac{5}{8}$ to $\frac{7}{8}$ **6.** $1\frac{1}{2}$ to 3   **7.** $2\frac{1}{2}$ to $3\frac{1}{4}$ **8.** $3\frac{1}{2}$ to $16\frac{2}{3}$

Write the inverse of each of the following ratios and express it in lowest terms.

9. 3 to 7   **10.** 5 to 8   **11.** $1\frac{1}{2}$ to 3   **12.** $2\frac{1}{2}$ to $3\frac{1}{2}$

13. $5\frac{1}{8}$ to 6   **14.** 3 to $2\frac{1}{2}$ **15.** 5.5 to 0.5   **16.** 0.25 to 0.75

17. Express the following ratios in lowest terms: 15 to 3; 8 to 16; 12 to 18; $\frac{7}{8}$ to $\frac{9}{16}$; $\frac{1}{2}$ to $\frac{1}{3}$; $2\frac{1}{2}$ to 10; $3\frac{1}{3}$ to $16\frac{2}{3}$.

18. Find the inverse of the following ratios: $5:3$; $7:8$; $4\frac{1}{2}:15$; $2\frac{2}{3}:3\frac{1}{4}$; $\frac{6}{7}:\frac{9}{10}$.

19. The smaller of two belted pulleys makes 240 revolutions per minute, and the larger one makes 80. What is the ratio of their speeds?

20. Of two gears in mesh, the smaller gear makes 75 rpm and the larger gear makes 50 rpm. What is the ratio of their speeds?

21. A train runs at the rate of 60 mph, and an airplane flies at the rate of 640 mph. Find the ratio of their speeds.

22. Tool steel may be worked at a cutting speed of 20′ per minute in a lathe, and cast iron may be worked at a cutting speed of 45′ per minute. Find the ratio of the cutting speeds.

23. A high-speed drill $\frac{3}{4}''$ in diameter drills through 240 castings in the same time a carbon steel drill of the same diameter drills only 65 castings. Find the ratio of their speeds.

24. One foot of copper wire 0.001″ in diameter has a resistance of 10.4 ohms, whereas a foot of aluminum wire of the same diameter has a resistance of 18.7 ohms. What is the ratio of the two resistances?

25. One teenager earns $72 per week and another earns $64. What is the ratio of their earnings?

26. Bell metal is made of 4 parts of copper and 1 part of tin. Find the amount of each in a bell weighing 8.5 lb.

Answer to exercise 1

$$\frac{15}{25} = \frac{15 \div 5}{25 \div 5} = \frac{3}{5}$$

or 3 to 5

Answer to exercise 2

$\frac{3}{5} \div \frac{9}{10} = \frac{3}{5} \times \frac{10}{9} = \frac{2}{3}$

or 2 to 3

Answer to exercise 3

$\frac{3}{10}$ of 30 in. = 9 in.

$\frac{7}{10}$ of 30 in. = 21 in.

27. Divide $40 between two persons in the ratio of 3 to 5.

28. Britannia metal consists of 2 parts antimony, 1 part bismuth, and 1 part tin. How many pounds of each are there in a casting of Britannia metal weighing 24 lb?

29. White pine weighs 25 lb per cubic foot; steel, 490 lb per cubic foot. Find the ratio of their weights.

30. The circumference of a $2\frac{3}{4}''$ circle is $8.64''$. Find the ratio between the circumference and the diameter. State the answer as a decimal of two places.

31. Refer to Table 7 (Weights of Materials) in the back of this book, and find the ratio of the weight of aluminum to the weight of brass.

32. Refer to Table 7 and the column headed *Average Weight in Grams per CM*^{3}, and find the ratio of the weight of gold to the weight of silver.

33. Compare the weight of the lightest metal in Table 7 to the weight of the heaviest metal in Table 7. What metals are they?

34. Compare the weight of the lightest wood in Table 7 to the weight of the heaviest wood in Table 7. What woods are they?

5–3   PROPORTION

A ***proportion*** is an equality between two ratios. Study the following problems carefully and see what a powerful idea proportion is! When it is shown that one ratio is equal to another ratio, the expression is a proportion; for example, the ratio of 2 to 5 is equal to the ratio of 4 to 10. This proportion can be written $2 : 5 = 4 : 10$, or as $\frac{2}{5} = \frac{4}{10}$. Read this proportion, "Two is to five as four is to ten."

In the proportion $2 : 5 = 4 : 10$, the *outside* terms 2 and 10 are called the ***extremes,*** and the *inside* terms 4 and 5 are called the ***means.*** In the proportion $2:5 = 4:10$, notice the following:

$$5 \times 4 = 2 \times 10 \qquad\qquad \textbf{[1]}$$

The product of the means is equal to the product of the extremes.

$$2 = \frac{5 \times 4}{10} \quad \text{and} \quad 10 = \frac{5 \times 4}{2} \qquad\qquad \textbf{[2]}$$

To find either extreme, multiply the means and divide the product by the other extreme.

$$5 = \frac{2 \times 10}{4} \quad \text{and} \quad 4 = \frac{2 \times 10}{5} \qquad\qquad \textbf{[3]}$$

To find either mean, multiply the extremes and divide by the other mean.

■ Example 1

In 15 minutes a worker can machine 12 studs. How long will it take to machine 250 studs?

Solution   Let x be the number of minutes it will take to machine 250 studs.

$$15:12 = x:250 \quad (\text{min}:\text{studs} = \text{min}:\text{studs})$$

$$\therefore x = \frac{15 \times 250}{12} = 312\tfrac{1}{2}\,\text{min} = 5\,\text{hours}\,12\tfrac{1}{2}\,\text{min}$$

(The symbol $\therefore$ means *therefore*.)

Explanation   Let x stand for the number of minutes it will take to machine 250 studs. The ratio of the 15 min to the 12 studs machined in those 15 min is the same as the ratio of the time (x min) to the 250 studs to be machined in x min. The proportion then would be

$$15\,\text{min}:12\,\text{studs} = x\,\text{min}:250\,\text{studs}$$

Then we can find the mean (x min) by multiplying the extremes (15 and 250) and dividing the product by the other mean (12); we get $x = 5$ hours $12\tfrac{1}{2}$ min.

   The proportion could be read, "15 minutes compares to 12 studs as x minutes compares to 250 studs." This type of problem can be written as two equal fractions; that is,

$$\frac{15}{12} = \frac{x}{250}$$

By cross-multiplying, we get

$$12x = 15 \times 250$$

$$12x = 3{,}750$$

$$x = \frac{3{,}750}{12}$$

$$x = 312.5\,\text{min} \quad \text{or} \quad 5\,\text{hours}\,12\tfrac{1}{2}\,\text{min}$$

■ **Example 2**

An 18-in. gear meshes with a 6-in. gear. If the large gear has 72 teeth, how many teeth will the small gear have?

Solution $\qquad 18:6 = 72:x \qquad \therefore x = \dfrac{72 \times 6}{18} = 24$ teeth

Explanation   Let x represent the number of teeth on the smaller gear. The ratio of the size of the larger gear to the size of the smaller gear (18 in. to 6 in.) is the same as the ratio of the number of teeth on the larger gear to the number of teeth on the smaller gear (72 teeth to x teeth). The proportion would be

$$18 \text{ in.}:6 \text{ in.} = 72 \text{ teeth}:x \text{ teeth}$$

The extreme (x teeth) can be found by multiplying the means (6 and 72) and dividing by the other extreme (18); we get $x = 24$ teeth.

Practice check: Do exercises 4 and 5 on the left.

■ **Example 3**

If a young man earns $285 per week, how long must he work to earn $3,420?

Solution $\qquad 285:3{,}420 = 1:x \qquad \therefore x = \dfrac{3{,}420 \times 1}{285} = 12$ weeks

Explanation   Since the same relation exists between the lengths of time as between the amounts earned, the ratio $285:3{,}420$ is equal to the ratio $1:x$, where x is the length of time he must work to earn $3,420.

Practice check: Do exercise 6 on the left.

If two pulleys are connected together by a belt and are rotating, the speed at the rim of each pulley must be the same. A larger pulley must rotate more slowly than a smaller pulley for their rim speeds to be equal. The ratio of the sizes of the pulleys is inversely proportional to the ratio of their revolutions per minute (rpm).

■ **Example 4**

A 2-in. pulley on a 3,450-rpm motor drives a 3-in. pulley. What are the revolutions per minute of the larger pulley?

Solution $\qquad 2 \text{ in.}:3 \text{ in.} = x \text{ rpm}:3{,}450 \text{ rpm}$

$$\therefore x = \frac{2 \times 3{,}450}{3} = 2{,}300 \text{ rpm}$$

Explanation   The larger pulley must rotate more slowly than the smaller pulley, so that the sizes are inversely proportional to the speeds. Let x be the rpm of the larger pulley, and compare their sizes inversely as in the example.

Practice check: Do exercises 7 and 8 on the left.

---

4. In a metal alloy, the ratio of iron to copper is 5 to 17. If there are 650 lb of iron, how much copper is there?

5. A certain airplane engine has a power-to-weight ratio of 5 to 9. If the engine weighs 450 lb, how much power does it produce?

The answers are in the margin on page 148.

---

6. A clothing store advertised athletic socks at 5 pairs for $9.50. At this rate, what would it cost for 2 pairs?

The answer is in the margin on page 148.

---

7. It takes 6 persons 8 hours to complete an assembly job. If 8 persons worked at the same rate, how long would it take them to do the same job?

8. A 32-tooth gear is set on a shaft of a motor that runs at 1,800 rpm. It meshes with a 72-tooth gear. What is the speed of the second gear?

The answers are in the margin on page 148.

■ Problems 5–3

Use the space provided below to work out the answers for the following problems.

Solve for x in the following proportions.

1. $3:4 = 9:x$

2. $4:7 = 8:x$

3. $15:9 = x:3$

4. $3:6 = x:24$

5. $24:x = 8:12$

6. $7:x = 28:84$

7. $x:6 = 4:12$

8. $x:18 = 24:6$

9. $4.5:10 = 9:x$

10. $2.6:x = 9.1:1.75$

11. $3:10.5 = x:5.25$

12. $x:6.5 = 5.52:15$

Solve the following problems involving proportion.

13. If a 14-lb casting costs $1.04, what would a 30-lb casting cost?

14. A pattern made of white pine weighs 3.74 lb. What will a brass casting made from this pattern weigh if white pine weighs 25 lb per cubic foot and brass weighs 520 lb per cubic foot?

15. The lengths of the two rectangles shown in Fig. 5–1 are proportional to their widths. What is the length of the smaller rectangle?

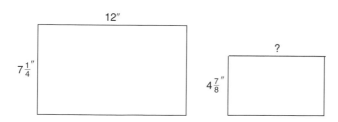

Figure 5–1

16. A copper wire 400′ long has a resistance of 1.084 ohms. What is the resistance of 2500′ of copper wire?

17. Cast iron weighs 450 lb per cubic foot; white pine, 25 lb per cubic foot. If a certain pattern made of white pine weighs $2\frac{1}{4}$ lb, what will the casting weigh?

18. A 24″ pulley running at 180 rpm drives a 14″ pulley. How many revolutions per minute will the smaller pulley make?
NOTE: *The ratio of their speeds is the inverse of the ratio of their sizes.*

19. A 14″ pulley makes 240 rpm and drives a larger pulley making 210 rpm. What is the diameter of the larger pulley?

20. A 16″ grinding wheel makes 1,000 rpm, and the driving pulley is 10″ in diameter. If the driven pulley is 6″ in diameter, how many revolutions per minute does the driving pulley make?

21. A 15″ shaft is found to have a taper of 0.204″ in 4″. Find the taper in the entire length of the shaft.

22. If the corresponding sides of the two triangles in Fig. 5–2 are to be proportional, find the missing dimensions of the small triangle.

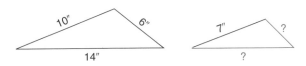

Figure 5–2

23. In the two triangles shown in Fig. 5–3, the heights are proportional to the bases. Find the height of the smaller triangle.

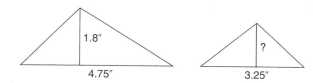

Figure 5–3

24. The circumferences of circles are proportional to their diameters. Find the circumference of a 7″ circle if the circumference of a 3″ circle is 9.42″.

25. The areas of circles are proportional to the squares of their diameters. Find the area of a 9″ circle if the area of a 5″ circle is 19.635 sq in.

26. The pressure of water increases with the depth. If the pressure is 6.51 lb per square inch at a depth of 15′, what is the pressure at a depth of 80′?

27. The power of a gas engine increases with the area of the piston. If an engine with a piston area of 8.30 sq in. develops 25.5 hp, how many horsepower will be developed by an engine with a piston area of 7.07 sq in.?

5–4 USING THE CALCULATOR TO SOLVE PROPORTION PROBLEMS

Proportion problems are one ratio (fraction) equated to another ratio (fraction). The solution to a proportion problem is that of finding the missing term of one of the ratios. To find the missing term, multiplication and division are usually required. A calculator, such as the one shown here, is an excellent tool to solve these problems. The operator must be careful to multiply those parts that should be multiplied and divide those parts that should be divided.

Rule

The product of the extremes is equal to the product of the means.

■ **Example**

If a worker can stamp out 36 receptacle plates in 15 minutes, how many can he stamp out in an 8-hour shift?

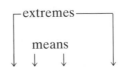

Solution   The problem becomes $36:15 = x:(8 \times 60)$. Stampings are compared to minutes as unknown stampings (x) are compared to minutes (i.e., 8 hours $\times$ 60 minutes = 480 minutes).

The Display Shows

Turn on the calculator	0 .
Enter 36	36 .
Press ×	36 .
Enter 480	480 .
Press ÷	17280 .
Enter 15	15 .
Press =	1152 . (answer)

The answer is the number of stampings the worker could make in an 8-hour shift by working at the rate of 36 stampings for each 15 minutes.

Practice check: Do exercise 9 on the right.

9. Use the calculator. The maintenance on 21 machines for 1 year is $150. What is the maintenance on 50 machines for 1 year (round to the nearest dollar)?

The answer is in the margin on page 153.

■ **Problems 5–4**

Use the calculator to solve the following proportions. Analyze each answer to determine if the answer is reasonable for the problem that was solved.

1. Copper wire has an average weight of 550 lb per cubic foot. What will 1 cubic inch of copper weigh? There are 1,728 cubic inches in 1 cubic foot.

2. There are 16.387 cubic centimeters in 1 cubic inch. How many cubic centimeters are there in 1 cubic foot?

3. One cubic foot of water weighs 62.5 lb. A gallon contains 231 cubic inches. What would 1,000 gallons of water weigh?

4. There are 3.58267 inches in 91 millimeters. How many inches are there in 25 millimeters?

5. Black walnut wood weighs 38 lb per cubic foot when dry. Refer to Problem 2 and give the weight of dry black walnut in grams per cubic centimeter (1 lb = 452.6 g).

6. A 1.25″ disc has a small hole in its center as shown in Fig. 5–4. To measure the hole, you project the disc through a filmstrip projector; the shadow of the disc measures $6\frac{7}{16}$″, and the image of the hole measures 2.5 mm. How large is the hole in the disc, to the nearest 0.1 mm?

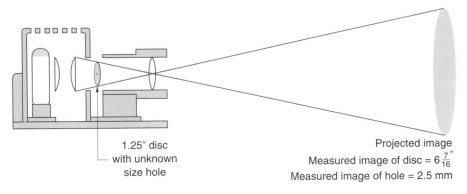

1.25″ disc with unknown size hole

Projected image
Measured image of disc = $6\frac{7}{16}$″
Measured image of hole = 2.5 mm

Figure 5–4

7. The material in a 15.75-lb casting costs $4.76. What would a 67.5-lb casting of the same material cost?

8. Silver weighs 10.51 grams per cubic centimeter. If a silver charm weighs 15.5 grams, then how many cubic centimeters does it contain?

9. Spruce weighs 25 lb per cubic foot. What would be the weight of a brass casting (brass weighs 512 lb per cu ft) made from a spruce pattern weighing 16.8 lb?

10. A 2″ pulley running at 3,400 rpm drives a 2.5″ pulley. What is the rate at which the driven pulley is turning? (*Hint:* Does the larger pulley turn faster or more slowly?)

11. Fill out the following chart. The water pressure is proportional to the depth.

Depth (ft)	10	20	30	40	50	100	200	300
Water pressure (lb)	4.34							

12. The areas of circles are proportional to the square of their diameters. Find the area of an 8.5-inch circle if a 10-inch circle has an area of 78.53975 square inches.

Answer to exercise 9

$$\frac{21}{150} = \frac{50}{x}$$

$$x = \$357$$

5–5   AVERAGES

To find the **average,** or **mean,** grade for five tests whose scores are 90, 85, 75, 80, and 100, add the five scores and divide the sum by 5. In this example, 90 + 85 + 75 + 80 + 100 = 430. Dividing 430 by 5 yields 86, which is the average grade for the five tests. This is the procedure to follow for all average problems: Add together all the items and divide by the number of items.

Rule

To find the average of two or more quantities, divide their sum by the number of quantities.

■ **Example**

Find the average length of four rods whose lengths are 5 ft, 7 ft, 12 ft, and 16 ft.

Solution   Total length = 5 + 7 + 12 + 16 = 40 ft
                 Average length = $\frac{40}{4}$ = 10 ft

Explanation   Divide the total length of the rods by the number of rods.

Practice check: Do exercise 10 on the right.

■ **Problems 5–5**

Answer the following questions involving averages. Use the space provided below to work out the answers for the problems.

1. Find the average weight of the following five castings: 8 lb, 10 lb, 14 lb, 9 lb, 19 lb.

2. Find the average length of the following rods: 6′8″, 5′4″, 3′10″, 4′6″, 5′2″, 4′9″.

10. A supervisor needed to find the average working time for a certain machine. Five machinists were given the job, and their completion times were as follows:

2 hr 34 min
2 hr 44 min
2 hr 24 min
2 hr 22 min
2 hr 36 min

What was the average working time?

The answer is in the margin on page 155.

3. One of three apprentices turns out 184 cotter pins, and the other two turn out 206 and 195. What is the average output per apprentice?

4. A machinist works $8\frac{1}{2}$ hours on Monday, $8\frac{1}{4}$ hours on Tuesday, 8 hours on Wednesday, $9\frac{1}{4}$ hours on Thursday, 9 hours on Friday, and $4\frac{1}{2}$ hours on Saturday. What is the average number of hours worked per day?

5. What would you consider the correct diameter of a ball bearing if you measure it three times with a vernier caliper and find a different reading each time as follows: 0.214″, 0.212″, 0.213″?

6. The measurements of the diameter of a piece of work are found to be as follows: 0.4206″, 0.4203″, 0.4209″, 0.4204″. What is the most probable diameter?

7. Find the mean of the following dimensions: 1.6435″, 1.6440″, 1.6438″, 1.6429″, 1.6432″, 1.6426″.

8. Which of the dimensions in Problem 7 differed most from the mean, and by how much?

9. The average weight of 24 castings is 16.5 lb. What does the lot weigh?

10. Four mechanics (A, B, C, and D) in a shop require different amounts of time for a certain job. A can do it in 3 hours 20 minutes, B in 2 hours 50 minutes, C in 2 hours 45 minutes, and D in 3 hours 35 minutes. What is the average amount of time required in the shop for that job?

11. In a shop where a careful inspection is maintained, the following numbers of pieces were rejected during a certain week: Monday 140, Tuesday 166, Wednesday 161, Thursday 171, Friday 155, and Saturday 93. What was the average number of rejects per day?

12. Several samples of steel wire taken from the same lot are tested for tensile strength with the following results: 278 lb, 276 lb, 285 lb, 270 lb, 281 lb, 275 lb. What is the average strength of the wire?

Answer to exercise 10

$$
\begin{array}{rl}
& 2 \text{ hr} \quad 34 \text{ min} \\
& 2 \text{ hr} \quad 44 \text{ min} \\
& 2 \text{ hr} \quad 24 \text{ min} \\
& 2 \text{ hr} \quad 22 \text{ min} \\
+ & 2 \text{ hr} \quad 36 \text{ min} \\
\hline
& 10 \text{ hr} \ 160 \text{ min}
\end{array}
$$

$$
\text{Average} = \frac{10 \text{ hr}}{5} \ \frac{160 \text{ min}}{5}
$$
$$
= 2 \text{ hr } 32 \text{ min}
$$

13. Here is a trick question you can use on your friends. (It is impossible to do.) A man drives up a hill at 30 mph. How fast must he drive down the hill to average 60 mph?

5–6   USING THE CALCULATOR TO FIND THE AVERAGE OF A SET OF NUMBERS

You can use a calculator to solve average problems by adding the individual items and dividing by the number of items. For example, suppose that you wanted to know the average daily temperature at noon in Philadelphia during January. It would be necessary to know the temperature for each day. Suppose the temperature at noon was as follows for each day of January.

Date	1	2	3	4	5	6	7	8	9	10	11	12	13	14
Temp. at noon	21°	25°	35°	40°	42°	41°	37°	35°	20°	15°	12°	10°	17°	19°

Date	15	16	17	18	19	20	21	22	23	24	25	26	27	28
Temp. at noon	15°	17°	21°	25°	35°	40°	45°	50°	52°	50°	40°	35°	30°	21°

Date	29	30	31
Temp. at noon	25°	27°	12°

To use the calculator to solve this problem, enter each temperature and add it to the previous sum of temperatures; when the last temperature is added, divide the total sum by 31 (the number of days in January). For repetitive operations, a flow chart (Fig. 5–5) is often helpful to show the order of operations.

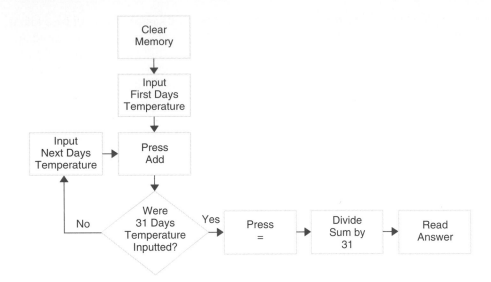

Figure 5–5

■ Problems 5–6

Use the calculator to find the following averages.

1. Use the temperature chart for January and the flow chart in Fig. 5–5 to find the average temperature in January.

2. The linemen of the football team of Eureca Prep School weighed as follows: right end, 165 lb; left end, 185 lb; right guard, 210 lb; left guard, 192 lb; center, 217 lb. What was the average weight of the Eureca Prep line?

3. Using a ten-thousandth micrometer, a mechanic made the following five measurements of a steel ball: 2.0156″, 2.0155″, 2.0157″, 2.0156″, 2.0155″. What measurement should be accepted as probably correct?

4. In one year 14,033 ships used the Panama Canal and transported 147,907,000 metric tons of cargo. What was the average tonnage (metric tons) per ship?

5. John, Kenny, and Malcolm went fishing on 11 successive days last June. A tally of their total catch follows.

Date	11	12	13	14	15	16	17	18	19	20	21
Fish caught	15	22	5	33	8	17	12	8	17	6	10
Weight in pounds of catch	39	55	11	75	19	43	29	14	36	14	24

What was the average number of fish caught per person? What was the average weight per fish?

■ **Self-Test**

Do the following problems. Use the space provided below to work out the answers. Check your answers with the answers in the back of the book.

1. Express the ratio 8 to 6 in lowest terms.

2. Express the ratio 4 to 16 in lowest terms.

3. Write the inverse of the ratio 3 to 4.

4. Write the inverse of the ratio 6 to 5.

5. Express the ratio $4\frac{1}{4}$ to $8\frac{1}{2}$ in lowest terms.

6. If 100 ft of copper wire has a resistance of 0.54 ohm, what is the resistance of 200 ft of copper wire?

7. Two gears in mesh have 36 teeth and 18 teeth, respectively. What is the ratio of their speeds?

8. Divide $100 between two people in the ratio of 3 to 2.

9. If five castings weigh 8 lb, 7 lb, 8 lb, 6 lb, and 6 lb, what is their average weight?

10. If your average earnings are $150 a week, what is your yearly income?

■ **Chapter Test**

Do the following examples involving ratios, proportions, and averages. Use the space provided below to work out the answers. Make sure that all your answers are in lowest terms.

Express the following ratios in lowest terms.

1. 12 to 4

2. $\frac{3}{4}$ to $\frac{5}{8}$

3. $1\frac{1}{2}$ to 6

4. $2\frac{1}{2}$ to $5\frac{1}{2}$

Write the inverse of the following ratios.

5. 3 to 8

6. $\frac{1}{4}$ to $\frac{3}{8}$

7. $3\frac{1}{2}$ to 7

8. $5\frac{1}{4}$ to $6\frac{1}{8}$

Find the following ratios.

9. Of two gears in mesh, the smaller gear makes 66 rpm, and the larger gear makes 22 rpm. What is the ratio of their speeds?

10. Divide $60 between two people in the ratio of 3 to 2.

11. One worker earns $60 per day, and another earns $50. What is the ratio of their earnings?

12. What is the ratio of one week to one year?

Solve the following proportions.

13. A copper wire 200' long has a resistance of 1.084 ohms. What is the resistance of 1000'?

14. An 18" shaft is found to have a taper of 0.375" in 6". Find the taper in the entire length of the shaft.

15 A 10" pulley makes 150 rpm and drives a larger pulley 75 rpm. What is the diameter of the larger pulley?

Find the following averages.

16. Find the average weight of the following five castings: 10 lb, 12 lb, 16 lb, 20 lb, 22 lb.

17. Find the average length of the following rods: 5'6", 3'2", 6'5", 2'7".

18. The average weight of 19 castings is 12 lb. What is the weight of the entire lot?

19. If you earn $300 a week for three weeks and $280 a week for the following five weeks, what are your average weekly earnings?

20. If your average earnings are $305.50 a week, what is your yearly income?

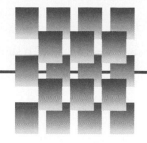

CHAPTER 6

Practical Algebra

6–1   USE OF LETTERS

Algebra uses letters to represent numbers. This is of great advantage because rules and formulas can be expressed in concise form. For example, to find the area of a rectangle, multiply the length by the width. This rule can be expressed in algebraic form.

$$\text{If:} \quad A = \text{area} \qquad l = \text{length} \qquad w = \text{width}$$

$$\text{Then:} \quad A = l \times w \quad \text{or} \quad A = lw$$

The times sign ($\times$) is understood when letters are written next to each other, with no sign or space between them.

Thus,

$$lw = l \times w$$

6–2   NEGATIVE NUMBERS

Negative numbers are numbers less than zero. Numbers greater than zero are ***positive numbers.*** We are familiar with the negative numbers used in winter weather reports when the temperature is below zero degrees. On a February morning in Maine, it might be minus 10 degrees ($^-10°$) with a wind chill factor of minus 30 degrees ($^-30°$). You may also have heard of a business having a negative cash flow, which means that there is more money being spent than being earned.

Negative numbers are distinguished from positive numbers by a minus sign. Thus, $^-3$ means 3 below zero, while $^+3$ or 3 means 3 above zero. In Fig. 6–1, positive numbers are represented by spaces *above* zero, and negative numbers by spaces *below* zero. To add $^+3$ to $^+1$, start at $^+1$, go *up* three spaces, and arrive at $^+4$; that is, $^+1 + {}^+3 = {}^+4$. To add $^+5$ to $^-2$, start at $^-2$, go up five spaces, and arrive at $^+3$; that is, $^-2 + {}^+5 = {}^+3$.

To subtract $^+3$ from $^+5$, start at $^+5$, go down three spaces, and arrive at $^+2$; that

Figure 6–1

1. Add ⁻3 and ⁺6.

2. Subtract ⁻2 and ⁺4.

The answers are in the margin on page 164.

is, $^+5 - {}^+3 = {}^+2$. To subtract $^+5$ from $^+2$, start at $^+2$, go down five spaces, and arrive at $^-3$; that is, $^+2 - {}^+5 = {}^-3$. To subtract $^+3$ from $^-2$, start at $^-2$, go down three spaces, and arrive at $^-5$; that is, $^-2 - {}^+3 = {}^-5$.

Practice check: Do exercises 1 and 2 on the left.

■ **Problems 6–2**

Fill in the following charts using the diagram of positive and negative numbers (Fig. 6–1). (Note that a positive number does not need a plus sign to indicate that it is positive.)

	Adding		
	Starting Number	**Add n (Go up n)**	**Arrive at**
1.	1	3	_____
2.	3	2	_____
3.	5	3	_____
4.	⁻2	4	_____
5.	⁻5	3	_____

	Subtracting		
	Starting Number	**Subtract n (Go down n)**	**Arrive at**
6.	5	3	_____
7.	3	1	_____
8.	2	5	_____
9.	⁻1	2	_____
10.	⁻3	3	_____

Use the space provided below to work out the answers to the following problems.

11. How many degrees of latitude are there between ⁻5° and 3°?

12. On a certain morning the tempera ture in Chitna, Alaska, was ⁻5°. Three hours later the temperature rose 3°. What was the temperature then?

13. A construction elevator starts at the sixth floor, goes up 3, then down 5. At what floor does it end up?

14. A gambler loses $3 playing roulette one evening and then wins $5 the next night. How much did the gambler win?

15. At 1:00 A.M. the temperature was ⁻7°, and at 11:00 A.M. it was 6°. What was the change in temperature?

6–3 DEFINITIONS

Absolute Value

The *absolute value* of a number is the distance between the number and zero without reference to its sign. Thus, the absolute value of ⁺6 is 6; the absolute value of ⁻6 is also 6. In algebra, the symbol used to indicate that the absolute value of a number is to be used rather than the number itself is two vertical bars, one on either side of the number, such as $|^-12| = 12$. When distance is discussed, the signs can be used to indicate direction; that is, away from a particular point can be considered positive, and toward the point can be considered negative. Distance, however, can be only positive. You couldn't drive ⁻20 miles, for instance.

■ **Example**

State the value of $|9|$.

Solution $|9| = 9$

Explanation The distance between 0 and 9 is 9 units.

Practice check: Do exercise 3 on the right.

3. State the value of $|^-19|$.

The answer is in the margin on page 165.

Factors

Numbers that can be multiplied together to produce a given number are *factors* of the given number. When the given number is a prime number, only the prime number and 1 are considered to be the factors. Prime numbers are numbers greater than 1. The first 10 prime numbers are 2, 3, 5, 7, 11, 13, 17, 19, 23, and 27. If the given number is non-prime, then look for all the prime numbers that can be factors. Thus, the factors of 5 are 5 and 1, because 5 is a prime number. The factors of 6 are 2 and 3. Since 6 is a non-prime number, we are interested in the prime numbers that are factors of 6, but we are not interested in using 1 as a factor.

Answer to exercise 1

⁺3

Answer to exercise 2

⁻6

4. Factor 30*m* into prime factors.

The answer is in the margin on page 167.

Algebra uses numbers and letters together. An algebraic expression such as 10*xy* means 10 times *x* times *y*. Since 10, *x*, and *y* are multiplied together, the 10 and the letters are known as factors of 10*xy*. We may not know what values *x* and *y* represent, so we cannot factor them other than claiming $x = x$ times 1 and $y = y$ times 1. We can, however, factor the expression 10*xy* as 5 times 2 times *x* times *y*. The factor 1 is generally not necessary for operations such as this.

■ **Example**

Factor 18 into prime factors.

Solution $18 = 2 \times 9 = 2 \times 3 \times 3$

Explanation   The product of $2 \times 3 \times 3$ is 18, and all the factors of 18 are prime.

Practice check: Do exercise 4 on the left.

Power

A *power* is the product of two or more identical factors. Thus, *aa*, which means *a* times *a*, is the second power of *a* and is written a^2; *aaa*, which means *a* times *a* times *a*, is the third power of *a* and is written a^3; *aaaa*, which means *a* times *a* times *a* times *a*, is the fourth power of *a* and is written a^4; and so on.

5. State the power to which x^8 is to be raised.

The answer is in the margin on page 167.

■ **Example**

State the power to which 5^3 is to be raised.

Solution   Third.

Practice check: Do exercise 5 on the left.

Exponent

An *exponent* is a number written above and to the right of a number or letter; it shows how many times the number or letter is to be used as a factor. In the expression a^4, 4 is the exponent of *a* and shows that *a* is to be used as a factor four times; that is, $a^4 = aaaa$. When no exponent is written over a number or letter, the exonent 1 is understood; that is, $a^1 = a$; $2^1 = 2$; and so forth.

6. Write 7*xyyyzzzz* using exponents.

The answer is in the margin on page 167.

■ **Example**

Write 3*yy* using an exponent.

Solution $3yy = 3y^2$

Practice check: Do exercise 6 on the left.

Coefficients

In a product any factor or the product of two or more factors is the *coefficient* of the remaining factor or of the product of the remaining factors. Thus, in the expression 3*a*, 3 is the coefficient of *a*; in the expression 5*abx*, 5 is the coefficient of *abx*, 5*a* is the coefficient of *bx*, and 5*ab* is the coefficient of *x*. In elementary algebra, the coefficient is the numerical part of the term. Where no coefficient is shown, the coefficient 1 is understood; thus, *a* means 1*a*; *xy* means 1*xy*.

You should learn to distinguish between the significance of exponents and that of coefficients. For example, $5^3 - 5 \times 5 \times 5 - 125$, but $3 \times 5 = 15$; likewise, $a^3 = aaa$, but $3a = 3$ times a.

Answer to exercise 3

$|{}^-19| = 19$

■ **Example**

In the expression $9x^2y$, what is the coefficient of x^2y?

Solution   9 is the coefficient of x^2y.

Practice check: Do exercise 7 on the right.

7. What is the coefficient of n^3 in the expression $31mn^3$?

The answer is in the margin on page 169.

Roots

A *root* is one of the equal factors of a number. Since $5 \times 5 = 25$, 5 is the square root of 25; since $2 \times 2 \times 2 = 8$, 2 is the cube root of 8. The root of a number is indicated by the symbol $\sqrt{}$. Thus, the square root of 25 is written $\sqrt{25}$; the cube root of 8 is written $\sqrt[3]{8}$.

The *index* of the root is the number that tells what root is to be taken. It is written to the left and above the root sign. When no index appears, the square root is understood. Thus, $\sqrt{16}$ denotes the square root of 16, which is 4 since $4 \times 4 = 16$; but $\sqrt[4]{16}$ denotes the fourth root of 16, which is 2, since $2 \times 2 \times 2 \times 2 = 16$.

■ **Example**

Give the root of $\sqrt{36}$.

Solution $\sqrt{36} = 6$

Explanation $\sqrt{36} = 6$ since $6 \times 6 = 36$.

Practice check: Do exercise 8 on the right.

8. Give the root of $\sqrt[3]{8a^3}$.

The answer is in the margin on page 169.

■ **Problems 6–3**

Use the space below to work out the answers to the following problems.

State the value of the following numbers.

1. $|+7|$ **2.** $|-7|$ **3.** $|85|$ **4.** $|-2.6|$ **5.** $|-1010|$

Fill in the blanks with factors for the number or expression as indicated.

6. $10 = \underline{\quad} \times \underline{\quad}$ **7.** $27 = \underline{\quad} \times \underline{\quad} \times \underline{\quad}$ **8.** $6x = \underline{\quad} \times \underline{\quad} \times \underline{\quad}$

9. $17xyz = \underline{\quad} \times \underline{\quad} \times \underline{\quad} \times \underline{\quad}$ **10.** $39mm = \underline{\quad} \times \underline{\quad} \times \underline{\quad} \times \underline{\quad}$

State the power to which each of the following numbers is to be raised.

11. 2^2 **12.** a^4 **13.** 7^9 **14.** b^3 **15.** y^5

Write the following expressions using exponents.

16. $aaaa$ **17.** $2xx$ **18.** $xxyyy$ **19.** $mmmnn$ **20.** $aaabbbcccddd$

Factor the following expressions without exponents.

21. a^5 **22.** a^2b^3 **23.** $7x^2y^2$ **24.** $23m^3n$ **25.** $3x^5y^2z^3$

Answer the following questions involving coefficients.

26. In the expression $6x$ the coefficient of the x is what?

27. In the expression $5ax$ the coefficient of the x is what?

28. The coefficient of xy^2z in the expression $3xy^2z$ is what?

29. The coefficient of bc in the expression $7abc$ is what?

30. In the expression $25a^3xy^2$ the coefficient of a^3xy^2 is what?

Give a root of the following.

31. $\sqrt{25}$ **32.** $\sqrt{225}$ **33.** $\sqrt{x^2}$ **34.** $\sqrt[3]{27}$ **35.** $\sqrt[4]{625}$

6–4   SUBSTITUTION

Study the examples shown below. If the numbers that the letters represent are known, then you can find the numerical value of the expression by following the rules of arithmetic. Remember that an expression such as *4a* means *4 times the value of a*. Parentheses indicate that the expression inside the parentheses is to be treated as a unit. Example 4 shows how to solve a problem that includes parentheses.

Evaluate the expression inside the parentheses first. Roots and powers are done before multiplication and division, which is done before addition and subtraction.

$$\left\{ \begin{array}{c} \text{Parentheses} \\ \text{First} \end{array} \right\} \rightarrow \left\{ \begin{array}{c} \text{Roots} \\ \text{or} \\ \text{Powers} \end{array} \right\} \rightarrow \left\{ \begin{array}{c} \text{Multiplication} \\ \text{or} \\ \text{Division} \end{array} \right\} \rightarrow \left\{ \begin{array}{c} \text{Addition} \\ \text{or} \\ \text{Subtraction} \end{array} \right\}$$

Answer to exercise 4

$30m = 2 \times 3 \times 5 \times m$

Answer to exercise 5

Eighth

Answer to exercise 6

$7xy^3z^4$

■ **Example 1**

If $a = 3$ and $x = 5$, find the value of $a + x$.

Solution $a + x = 3 + 5 = 8$

■ **Example 2**

If $a = 3$ and $x = 5$, find the value of $4a + 7x$.

Solution $4a + 7x = 4 \times 3 + 7 \times 5$
$= 12 + 35 = 47$

■ **Example 3**

If $a = 3$ and $x = 5$, find the value of $6a^2 + 2x^3$.

Solution $6a^2 + 2x^3 = 6 \times 3^2 + 2 \times 5^3$
$= 6 \times 9 + 2 \times 125 = 54 + 250$
$= 304$

■ **Example 4**

If $a = 4$, $b = 6$, and $y = 9$, find the value of $5(3a + 2b^2)\sqrt{y}$.

Solution $5(3a + 2b^2)\sqrt{y} = 5(3 \times 4 + 2 \times 6^2)\sqrt{9}$
$= 5(12 + 72)3 = 5 \times 84 \times 3$
$= 1,260$

Practice check: Do exercises 9 and 10 on the right.

9. If $a = 7$, $b = 4$, and $c = 3$, find the value of $a(3b - 4c) + 5a$.

10. If $m = 4$, $n = 2$, and $x = 5$, find the value of $6x^2 + 5n^2 - 3\sqrt{m}$.

The answers are in the margin on page 169.

■ **Problems 6–4**

Find the numerical values of the following expressions if $a = 2$, $b = 5$, $c = 4$, $x = 3$, and $y = \frac{1}{2}$. Use the space provided below to work out the answers.

1. $5c$ **2.** $4x$ **3.** $2by$ **4.** $4a^2$

5. $5b^2$

6. $6abcx$

7. $3a^2b^2c^2$

8. $4(b + x^2)$

9. $2a(cy + bc)$

10. $4y + 3b$

11. $8ac + 5ax$

12. $2(a^2 + b^2 + x)$

13. $\sqrt{c}$

14. $\sqrt{x^2}$

15. $\sqrt[3]{2c}$

16. $\sqrt[3]{b^3}$

17. $a\sqrt{b^2}$

18. $(a + c)\sqrt{10ab}$

19. $c(b - a)$

20. $(c + x)^2$

21. $a^2 - y^2$

22. $\dfrac{c}{a} + \dfrac{b}{y}$

23. $\dfrac{1}{a} + \dfrac{1}{c}$

24. $\dfrac{1}{x} + \dfrac{1}{y}$

25. $4bc - 3ax$

26. $2c^2 - x^2$

27. $x^2 + y^2$

28. $(x + y)^2$

29. $(x + y)^3$

30. $(x - y)^2$

6–5   ADDITION

If an expression is written as the sum of several quantities, each of these quantities is called a **_term_** of the expression. Thus, x, $5cb$, $3ab^2 \times 5cd$, $\dfrac{2a^3}{3c}$ are one term each.

The expression $2a - 3bc + d$ consists of the three terms $2a$, ^-3bc, and d.

    Similar or **_like terms_** are terms having the same letters raised to the same powers; thus, $3a^2bc$ and $^-5a^2bc$ are like terms.

    Dissimilar or **_unlike terms_** are terms that are not similar; $3a^2bc$ and $4ab^2c^2$ are unlike terms because the same letters are not raised to the same power; $2ax$ and $3dy$ are also dissimilar terms because they have different letters.

Rule

To add similar terms having the same sign, add the absolute values of the numerical coefficients, annex the common letters, and prefix the common sign to the result.

■ **Example 1**

Add $3a$, $5a$, and $4a$.

Solution
$$
\begin{array}{r}
3a \\
5a \\
+\ \underline{4a} \\
12a
\end{array}
$$

Explanation   Adding the numerical coefficients, we get $3 + 5 + 4 = 12$. Annex to this sum the common letter a; and since all terms are positive, their sum is positive. Hence, the answer is $12a$.

■ **Example 2**

Add ^-6ab, ^-2ab, and ^-9ab.

Solution
$$
\begin{array}{r}
^-6ab \\
^-2ab \\
+\ \underline{^-9ab} \\
^-17ab
\end{array}
$$

Explanation   The sum of the absolute values of the numerical coefficients is $6 + 2 + 9 = 17$. Annexing the common letters ab gives $17ab$; and since the common sign is negative, the sign of the sum is negative. Hence, the answer is ^-17ab.

Rule

If the terms to be added have unlike signs, proceed as follows:

1. Add the positive coefficients.

2. Add the negative coefficients.

3. Evaluate the absolute value of the coefficients of both sums, and subtract the smaller from the larger. Use the sign of the larger as the sign of the result.

4. Annex the common letters.

■ **Example 3**

Add $4x$, ^-7x, ^-3x, $6x$, and ^-5x.

Solution	**Add Positive Coefficients (Step 1)**	**Add Negative Coefficients (Step 2)**	**Follow Steps 3 and 4**
		^-7x	
	$4x$	^-3x	^-15x
	$+\ \underline{6x}$	$+\ \underline{^-5x}$	$+\ \underline{10x}$
	$10x$	^-15x	^-5x

Answer to exercise 7

$31m$ is the coefficient of n^3.

Answer to exercise 8

$\sqrt[3]{8a^3} = 2a$ since $2a \times 2a \times 2a = 8a^3$.

Answer to exercise 9

$a(3b - 4c) + 5a$
$= 7(3 \times 4 - 4 \times 3) + 5 \times 7$
$= 7(12 - 12) + 35$
$= 7(0) + 35 = 35$

Answer to exercise 10

$6x^2 + 5n^2 - 3\sqrt{m}$
$= 6 \times 5^2 + 5 \times 2^2 - 3\sqrt{4}$
$= 6 \times 25 + 5 \times 4 - 3 \times 2$
$= 150 + 20 - 6$
$= 170 - 6 = 164$

Explanation   Evaluate the absolute value of both sums.

Absolute value of $^-15 = 15$

Absolute value of $10 = 10$

Subtract the smaller from the larger, use the sign of the larger, and annex the common letters.

■ **Example 4**

Add $7a - 3b$, $4a + 6b$, and $^-5a + 4b$.

_____

11. Add ^-4x, $3x$, $7x$, and ^-5x.

12. Add $10r + 9s^2$, $^-r + 5s^2$, and $^-12r - 7s^2$.

The answers are in the margin on page 173.

_____

Solution $7a - 3b$
$\quad\quad\ \ 4a + 6b$
$\ \underline{+\ ^-5a + 4b}$
$\quad\quad\ \ 6a + 7b$

Explanation   Write like terms under one another; then add each column separately as in Example 3.

Practice check: Do exercises 11 and 12 on the left.

■ **Problems 6–5**

Perform the following additions. Use the space provided below to work out the answers.

1. $7a$
$\ \ 5a$
$\ \ 4a$
$\underline{+\ 3a}$

2. $16bx$
$\ \ 12bx$
$\underline{+\ \ \ 8bx}$

3. ^-5abc
$\ \ ^-8abc$
$\underline{+\ ^-10abc}$

4. $^-2a^2x$
$\ \ ^-9a^2x$
$\underline{+\ ^-4a^2x}$

5. ^+25by
$\ \ ^+18by$
$\underline{+\ ^-12by}$

6. ^-16xy
$\ \ ^-14xy$
$\underline{+\ ^+15xy}$

7. $3b$
$\ \ ^-5b$
$\underline{+\ \ 8b}$

8. ^-mn
$\ \ 5mn$
$\underline{+\ ^-12mn}$

9. $^-15ax^2$
$\ \ ^-18ax^2$
$\ \ 22ax^2$
$\underline{+\ \ 12ax^2}$

10. $11b^2y$
$\ \ ^-6b^2y$
$\ \ ^-7b^2y$
$\underline{+\ \ 4b^2y}$

11. $6ab$
$\ \ 5ab$
$\ \ ^-8ab$
$\underline{+\ ^-13ab}$

12. $^-14bc^2$
$\ \ 6bc^2$
$\ \ ^-18bc^2$
$\underline{+\ \ 9bc^2}$

13.
$$
\begin{array}{r}
4a - 6c \\
{}^-5a - 5c \\
+ \ {}^-8a + 4c \\
\hline
\end{array}
$$

14.
$$
\begin{array}{r}
{}^-7x + 4y \\
{}^-9x - 6y \\
+ \ \ 5x + 7y \\
\hline
\end{array}
$$

15.
$$
\begin{array}{r}
8b + \ \ 3d \\
{}^-3b - \ \ 5d \\
+ \ {}^-12b - 10d \\
\hline
\end{array}
$$

16.
$$
\begin{array}{r}
{}^-5x^2 - 3y \\
{}^-4x^2 + \ \ y \\
+ \ {}^-x^2 + \ \ y \\
\hline
\end{array}
$$

17.
$$
\begin{array}{r}
9r^2 - 8st \\
5r^2 + 2st \\
+ \ {}^-7r^2 - 3st \\
\hline
\end{array}
$$

18.
$$
\begin{array}{r}
{}^-13x^2 - 12x \\
{}^-24x^2 + 30x \\
+ \ 12x^2 - 11x \\
\hline
\end{array}
$$

6-6 SUBTRACTION

Rule

To subtract, change the sign of the number to be subtracted (subtrahend), and then combine the two numbers as in addition.

■ **Example 1**

Subtract $5x$ from $12x$.

Solution
$$
\begin{array}{r}
12x \\
+ \ {}^-5x \\
\hline
7x
\end{array}
$$

Explanation   Changing the sign of $5x$ makes it ${}^-5x$. Adding $12x$ and ${}^-5x$ gives $7x$.

■ **Example 2**

Subtract $5x$ from ${}^-12x$.

Solution
$$
\begin{array}{r}
{}^-12x \\
+ \ {}^-5x \\
\hline
{}^-17x
\end{array}
$$

Explanation   Changing the sign of $5x$ makes it ${}^-5x$. Adding ${}^-12x$ and ${}^-5x$ gives ${}^-17x$.

Practice check: Do exercise 13 on the right.

13. Subtract $7x$ from ${}^-15x$.

The answer is in the margin on page 173.

■ **Example 3**

Subtract ${}^-5x$ from $12x$.

Solution
$$
\begin{array}{r}
12x \\
+ \ \ 5x \\
\hline
17x
\end{array}
$$

Explanation   Changing the sign of ${}^-5x$ makes it $5x$. Adding $12x$ and $5x$ gives $17x$.

■ **Example 4**

Subtract ⁻5x from ⁻12x.

Solution   12x
              + ⁻5x
              ────────
                  ⁻7x

14. Subtract ⁻8x from ⁻3x.

The answer is in the margin on page 176.

Explanation   Changing the sign of ⁻5x makes it 5x. Adding ⁻12x and 5x gives ⁻7x.

Practice check: Do exercise 14 on the left.

■ **Example 5**

From 5x − 2y take 8x + 12y.

Solution   5x + ⁻2y
              −   8x + 12y
              ──────────────────
                  ⁻3x + ⁻14y = ⁻3x − 14y

15. From 9a − 4b take ⁻3a + 7b.

The answer is in the margin on page 176.

Explanation   Write like terms under each other, and proceed with each pair of like terms as explained in the preceding examples.

Practice check: Do exercise 15 on the left.

■ **Problems 6–6**

Do the following subtraction problems as indicated. Use the space provided below to work out the answers.

1. Subtract 12ax from 18ax.   **2.** Subtract 14b^2y^2 from ⁻5b^2y^2.

3. Subtract ⁻16a^2b^2 from ⁻22a^2b^2.   **4.** Subtract ⁻10mm from 8mn.

5. Subtract ⁻36x^2y^2 from ⁻24x^2y^2.   **6.** Subtract 21ac from ⁻9ac.

7. Subtract 7ay from 3ay.   **8.** Subtract ⁻15abc from ⁻18abc.

9. Subtract ^-24xy from $32xy$.

10. Subtract $9ac^2$ from $^-6ac^2$.

11. Subtract $48z + 14y$ from $36z - 12y$.

12. Subtract $^-30abc - 10bx$ from $20abc - 18bx$.

13. Subtract $8a + 9b$ from $6a + 4b$.

14. From $^-21x - 25y$ take $33x + 28y$.

15. From $14x + 16y$ take $^-7x - 8y$.

16. From $5a^2 - 6a$ take $^-21a^2 - 13a$.

17. From $8x^2y - 4xy^2$ take $^-5x^2y + xy^2$.

18. From $9m^2 + 7n^2$ take $3m^2 - 3n^2$.

19. From $^-10a^2 + 3a$ take $7a^2 + 4a$.

20. From $^-21a^3b^2 + 18a^2b^3$ take $^-51a^3b^2 + 32a^2b^3$.

Answer to exercise 11

$$\begin{array}{rr} 3x & ^-4x \\ + \quad 7x & + \quad ^-5x \\ \hline 10x & ^-9x \end{array}$$

$$\begin{array}{r} 10x \\ + \quad ^-9x \\ \hline x \end{array}$$

Answer to exercise 12

$$\begin{array}{r} 10r + 9s^2 \\ ^-r + 5s^2 \\ + \quad ^-12r - 7s^2 \\ \hline ^-3r + 7s^2 \end{array}$$

Answer to exercise 13

$$\begin{array}{r} ^-15x \\ + \quad ^-7x \\ \hline ^-21x \end{array}$$

6–7 SYMBOLS OF GROUPING

The symbols of grouping are parentheses ( ), brackets [ ], and braces { }. The method of removing signs of grouping and expressing the resulting quantity in its simplest form will be clear from a study of the following examples.

■ **Example 1**

$5 + (3 + 4)$

Solution $5 + (3 + 4) = 5 + 7 = 12$

The same result would be obtained if the parentheses were dropped first and then the numbers added; thus, $5 + (3 + 4) = 5 + 3 + 4 = 12$.

Rule

Parentheses preceded by a positive sign (+) may be dropped without changing the sign of any term.

■ Example 2

$15 - (6 + 2)$

Solution $15 - (6 + 2) = 15 - 8 = 7$

16. Simplify $8 - (7 + 3)$.

The answer is in the margin on page 176.

The negative sign before the parentheses means that the value of the numbers within the parentheses is to be subtracted from 15. The rule for subtraction is as follows. Change the sign in the subtrahend, and combine with the minuend as in addition.

Practice check: Do exercise 16 on the left.

Rule

Parentheses preceded by a negative sign (−) may be dropped if the sign of every term within the parentheses is changed. When one pair of parentheses appears within another, remove one pair at a time, beginning with the innermost and proceeding outward.

■ Example 3

$6a - [7a - (3a + 2b)]$

Solution	$6a - [7a - (3a + 2b)]$	[1]
	$= 6a - [7a - 3a - 2b]$	[2]
	$= 6a - 7a + 3a + 2b$	[3]
	$= 2a + 2b$	[4]

17. Simplify the following.

$^-5x - [^-5x - (x - 2y)] + 6y$

The answer is in the margin on page 176.

Explanation   Removing the parentheses around $3a + 2b$ gives line (2); next remove the brackets and obtain line (3). Combining like terms gives line (4), the answer.

Practice check: Do exercise 17 on the left.

■ Problems 6–7

Simplify the following expressions by removing the parentheses and combining like terms. Use the space provided below to work out the answers.

1. $12 + (5 + 2)$

2. $12 + (5 - 2)$

3. $12 - (5 + 2)$

4. $12 - (5 - 2)$

5. $a + (b + a)$

6. $a + (6 - 2a)$

7. $a - (b - 2a)$

8. $4a + (b - a)$

9. $5a - (a - b)$

10. $9 - (2 - 4)$

11. $10 - (6 + 3)$

12. $(3a - 3b) - (b + c)$

13. $8ab - (2ab - 2a^2)$

14. $a - [^-(^-a)]$

15. $a + [b - (a - b)]$

16. $[2x - (3x - y)] - (x - 3y)$

17. $3x^2 - (5x^2 + 3y) + [(x^2 + 7y) - 5y]$

18. $5a - [(^-7a + b^2) + 7a] - 5b^2$

19. $^-m - (3m - 11n) - [6m - (?m + n)] - (^-7m + 9n)$

20. $(x^2 - 6y^2) - 6x^2 - [7x^2 + (^-2x^2 - 19y^2) + 8y^2]$

6-8   MULTIPLICATION

In the multiplication of algebraic quantities, two important laws must be followed.

The Law of Signs

The product of two numbers having like signs is positive; the product of two numbers having unlike signs is negative.

Thus,

$$(^+3) \times (^+5) = {^+15}$$
$$(^-3) \times (^-5) = {^+15}$$
$$(^+3) \times (^-5) = {^-15}$$
$$(^-3) \times (^+5) = {^-15}$$

Answer to exercise 14

$$\begin{array}{r} {}^-3x \\ +\ 8x \\ \hline 5x \end{array}$$

Answer to exercise 15

$$\begin{array}{l} 9a\ -\ 4b \\ \underline{3a\ -\ 7b} \quad \text{Change signs.} \\ 12a\ -\ 11b \quad \text{Add.} \end{array}$$

Answer to exercise 16

$8 - (7 + 3) = 8 - 10 = {}^-2$

Answer to exercise 17

$$\begin{array}{l} {}^-5x - [{}^-5x - (x - 2y)] + 6y \\ = {}^-5x - [{}^-5x - x + 2y] + 6y \\ = {}^-5x + 5x + x - 2y + 6y \\ = x + 4y \end{array}$$

NOTE: ${}^-5x + 5x = 0$

18. Multiply ${}^-5xy^3$ by ${}^-2x^2y^2$.

The answer is in the margin on page 179.

19. Simplify the following.

$\quad {}^-5a({}^-3a^2 + 7ab^2)$

The answer is in the margin on page 179.

The Law of Exponents

To find the exponent of a letter in the product, add the exponents of that letter in the factors.

Thus,

$$a^2 \times a^3 = a^{2+3} = a^5 \qquad ab^2c \times a^2b^3c^2 = a^3b^5c^3$$

Rule

To multiply one term by another, find the product of their numerical coefficients; prefix the sign according to the law of signs; annex the letters that appear in the factors; and obtain the exponent for each letter in the product by adding the exponents of that letter in the factors.

■ **Example 1**

Multiply $6a^2x$ by ${}^-4ax^2y$.

Solution
$$\begin{array}{r} 6a^2x \\ \times\ {}^-4ax^2y \\ \hline {}^-24a^3x^3y \end{array}$$

Explanation   Multiplying 6 by 4 gives 24. Since the signs are unlike, prefix the negative sign to the product. Annex all the letters that appear in both factors, writing each letter but once. Since the exponent of a in the multiplicand is 2 and in the multiplier 1, the exponent of a in the product is $2 + 1$, or 3. Similarly the exponent of x is $1 + 2$, or 3. The exponent of y is 1 because it appears in the multiplier to the first power and does not appear at all in the multiplicand.

Practice check: Do exercise 18 on the left.

■ **Example 2**

Simplify ${}^-3a(2x - 7y)$.

Solution ${}^-3a(2x - 7y) = {}^-6ax - {}^-21ay = {}^-6ax + 21ay$

Explanation   To simplify the given expression means to perform the indicated operations. Each term within the parentheses is multiplied by the multiplier ${}^-3a$. To obtain the first term in the product, we find that ${}^-3a \times 2x = {}^-6ax$. The second term is equal to ${}^-3a \times {}^-7y = 21ay$. The complete answer is ${}^-6ax + 21ay$.

Practice check: Do exercise 19 on the left.

■ **Example 3**

Multiply $8a - 3b$ by $2a - 5b$.

Solution
$$\begin{array}{r} 8a\ -\ 3b \\ \times\ \ 2a\ -\ 5b \\ \hline 16a^2\ -\ \ 6ab \\ -\ 40ab\ +\ 15b^2 \\ \hline 16a^2\ -\ 46ab\ +\ 15b^2 \end{array}$$

Explanation   First multiply each term of the multiplicand by the first term of the multiplier ($2a$), giving the partial product $16a^2 - 6ab$. Next multiply each term of the multiplicand by the second term of the multiplier (^-5b), giving the partial product, $^-40ab + 15b^2$. The term ^-40ab is written under the similar term ^-6ab in the first partial product. Adding the two partial products gives the final result, $16a^2 - 46ab + 15b^2$.

Practice check: Do exercise 20 on the right.

20. Multiply $3a - 5b$ by $3a + 5b$.

The answer is in the margin on page 179.

■ **Problems 6–8**

Perform the following multiplications as indicated. Use the space provided below to work out the answers.

1. $a \times a$

2. $c^2 \times c$

3. $a^2 \times a^4$

4. $2a^2 \times 3a^3$

5. $^-5b \times ^-3b^2$

6. $^-b^2 \times ^-b^3$

7. $4a \times ^-3a^2$

8. $^-3c \times 8c^3$

9. $2a \times 3a^2 \times 5a$

10. $6a^2 \times ^-4a \times 3a$

11. $^-x^2 \times ^-3x \times ^-2x^3$

12. $2ab(a + b + c)$

13. $3a^2c(2a - 3b - c)$

14. $^-5a(2x^2 - 2y^2)$

15. $(3a - 4b)(5a - 6b)$

16. $(a + b)(a - b)$

17. $(2a + 3b)(4a + 5b)$

18. $(8a - 2b)(2b + 4a)$

19. $(2x + 3y)(2x - 3y)$ **20.** $(2x + 3y)(2x + 2y)$ **21.** $(7a - 3b)(7a - 3b)$

22. $(x + 2y)(3x - y)$ **23.** $(3x + 4y)(3x + 4y)$ **24.** $(2ab - 3ac)(5ab + 2ac)$

6–9   DIVISION

To perform a division, reverse the process of multiplication.

Rule _____

To divide one term by another, divide the numerical coefficient of the dividend by the numerical coefficient of the divisor; the result is the numerical coefficient of the quotient. Then follow the law of signs and the law of exponents to determine the sign and exponent of the quotient.

_____

The Law of Signs

The quotient is positive if the signs are the same; the quotient is negative if the signs are opposite.

The Law of Exponents

Subtract the exponent of a letter in the divisor from the exponent of the same letter in the dividend; the result is the exponent of the same letter in the quotient.

■ **Example 1**

Divide $40a^2x^4$ by $5ax^2$.

Solution $\dfrac{40a^2x^4}{5ax^2} = \dfrac{40}{5} \times \dfrac{a^2}{a} \times \dfrac{x^4}{x^2} = 8 \times a^{2-1} \times x^{4-2} = 8ax^2$

Explanation   Dividing 40 by 5 gives 8. Since both signs are alike, the sign of the quotient is positive. Write after the numerical quotient 8 the letters that appear in both dividend and divisor, namely, a and x. To obtain the exponent of a in the quotient, subtract the exponent of a in the divisor from the exponent of a in the dividend; thus, $2 - 1 = 1$, which is not written after the a in the quotient. Similarly we find the exponent of x in the quotient by subtracting the exponent of x in the divisor from the exponent of x in the dividend; thus, $4 - 2 = 2$.

■ **Example 2**

Divide $5a^2x^4$ by $5a^2x$.

Solution $\dfrac{\cancel{5}\cancel{a^2}x^4}{\cancel{5}\cancel{a^2}x} = x^3$

Answer to exercise 18

$10x^3y^5$

Answer to exercise 19

$15a^3 - 35a^2b^2$

Answer to exercise 20

$9a^2 - 25b^2$

Explanation   The 5 in the dividend contains the 5 in the divisor once; the a^2 in the dividend contains the a^2 in the divisor once. Hence, the given example is equivalent to $1 \times 1 \times \dfrac{x^4}{x} = 1 \times 1 \times x^3 = x^3$.

■ **Example 3**

Divide $^-12a^3x^2y^5$ by $3ay^2$.

Solution $\dfrac{^-12a^3x^2y^5}{3ay^2} = {}^-4a^2x^2y^3$

Explanation $12 \div 3 = 4$; the signs being unlike, the sign of the quotient is negative. Write in the quotient the letters that appear in both dividend and divisor. The exponent of a is $3 - 1 = 2$; the exponent of x is $2 - 0 = 2$; the exponent of y is $5 - 2 = 3$. The answer is $^-4a^2x^2y^3$.

Practice check: Do exercises 21 and 22 on the right.

21. Divide $72m^5s^2$ by $^-8m^2s^2$.

22. Divide $^-36x^8y^{10}z^5$ by $^-12x^2y^{10}z^3$.

The answers are in the margin on page 182.

■ **Example 4**

Divide $20a^2b^2 - 16a^3b$ by $4ab$.

Solution $\dfrac{20a^2b^2 - 16a^3b}{4ab} = \dfrac{20a^2b^2}{4ab} - \dfrac{16a^3b}{4ab} = 5ab - 4a^2$

Explanation   Dividing $4ab$ into the first term of the dividend $20a^2b^2$ gives $5ab$; dividing $4ab$ into the second term of the dividend gives $^-4a^2$. The quotient is $5ab - 4a^2$.

Practice check: Do exercise 23 on the right.

23. Divide $40x^3 - 88x^2$ by $4x^2$.

The answer is in the margin on page 182.

■ **Example 5**

Divide $a^2 - 8ab + 15b^2$ by $a - 3b$.

Solution
$$
\begin{array}{r}
a - 5b \\
a - 3b{\overline{\smash{)}a^2 - 8ab + 15b^2}} \\
\underline{a^2 - 3ab\phantom{{}+ 15b^2}} \\
-5ab + 15b^2 \\
\underline{-5ab + 15b^2}
\end{array}
$$

Explanation   Divide the first term of the divisor into the first term of the dividend to obtain the first term of the quotient; thus, $a^2 \div a = a$. Multiply each term of the divisor by the first term of the quotient; thus, $a(a - 3b) = a^2 - 3ab$. Subtract this product from the dividend, leaving $^-5ab + 15b^2$. Divide the first term of the divisor into the first term of the remainder to obtain the second term of the quotient; thus, $^-5ab \div a = {}^-5b$. Multiply the divisor by the second term of the quotient; thus,

24. Divide $42a^2 + 23ab - 10b^2$ by $7a - 2b$.

The answer is in the margin on page 182.

$^-5b(a - 3b) = ^-5ab + 15b^2$. Subtracting this product from the remainder previously obtained leaves 0. Since the remainder is exactly 0, the quotient is exactly $a - 5b$.

Practice check: Do exercise 24 on the left.

■ Problems 6–9

Perform the following divisions as indicated. Use the space provided below to work out the answers.

1. $18x^3 \div 6x$

2. $24x^4y^3 \div 3x^2y^2$

3. $^-16x^3y^4 \div ^-8xy$

4. $^-12abc \div ^-2ab$

5. $32a^2x^3 \div ^-4a^2x$

6. $27b^5c^4 \div ^-9bc$

7. $^-36x^3y^3 \div 4xy^2$

8. $^-40a^3d^3 \div 5a^2d$

9. $(24a^2b^2 + 18a^3b^3) \div 2ab$

10. $(16a^3x^2 - 12a^2x^3) \div 4ax$

11. $(21b^4y^5 + 18b^5y^6) \div ^-3b^2y^2$

12. $(14a^4y^4 - 7a^2y^2) \div ^-7ay$

13. $(x^2 + 5x + 6) \div (x + 2)$

14. $(x^2 - 8x + 15) \div (x - 3)$

15. $(x^2 - y^2) \div (x + y)$

16. $(a^2 + 2ab + b^2) \div (a + b)$

17. $(x^2 - 2xy + y^2) \div (x - y)$

18. $\dfrac{48m^7n^7p}{-6mp}$

19. $\dfrac{-14a^2b^3c^4}{-14a^2b^3c^4}$

20. $\dfrac{-12xy^3}{12y^3}$

21. $(12b^2 + 13b - 14) \div (3b - 2)$

22. $(6x^2 - 13xy + 6y^2) \div (2x - 3y)$

6–10   EQUATIONS

An **equation** is a statement that two quantities are equal. For example, the equation $3x = 15$ states that $3x$ and 15 are equal to each other.

Members

The part of an equation that is to the left of the equality sign ($=$) is called the *first member, left side,* or *left-handed member.* The part of an equation that is on the right of the equality sign is called the *second member, right side,* or *right-hand member.* In the equation $3x = 15$, $3x$ is the first or left-hand member, and 15 is the second or right-hand member.

An equation is used to find the value of an unknown number from its relation to known numbers. The unknown number is usually represented by some letter, such as x, y, or z. The first letters of the alphabet, a, b, or c, usually represent known numbers.

The solution of equations depends upon the following axioms (statements or principles that are accepted as true).

Axioms

1. *Equal numbers added to equal numbers give equal sums.*
   If $x = y$, then $x + 2 = y + 2$.

2. *Equal numbers subtracted from equal numbers give equal remainders.*
   If $x = y$, then $x - 2 = y - 2$.

3. *Equal numbers multiplied by equal numbers give equal products.*
   If $x = y$, then $2x = 2y$.

4. *Equal numbers divided by equal numbers give equal quotients.*
   If $x = y$, then $\dfrac{x}{2} = \dfrac{y}{2}$.

Transposition of Terms

To **transpose** a term is to transfer it from one side of the equation to the other. A term may be transposed from one member of an equation to the other, provided the sign of the term is changed.

Proof 1

Given the equation

$$x - a = b$$

Add a to each member.

$$x - a + a = b + a$$

But $^-a + a = 0$; therefore, $x = b + a$.

The result is the same as if the ^-a from the left side were transposed to the right and its sign changed.

Proof 2

Given the equation

$$x + a = b$$

Subtract a from each member.

$$x + a - a = b - a$$

But $a - a = 0$; therefore, $x = b - a$.

In this case the same result is obtained by transposing a from the left side to the right side and changing the sign.

The sign of every term of an equation may be changed without affecting the equality.

Proof 3

If in the equation

$$a - x = b + c \qquad\qquad \textbf{[1]}$$

each term is multiplied by $^-1$ (Axiom 3), the result is

$$^-a + x = {}^-b - c \qquad\qquad \textbf{[2]}$$

Equation (2) is the same as Equation (1) with the sign of each term changed.

Solution of Equations

To solve an equation, it is usual to transpose all the unknown values to the left side of the equation and all the known values to the right side of the equation. Then by combining and simplifying, it is possible to solve for the unknown value. Study the examples that follow.

■ Example 1

Solve the equation $6x - 4 = 4x + 6$.

Solution   Transposing $4x$ to the left side and $^-4$ to the right side, we have

$$6x - 4x = 6 + 4$$

Combining like terms yields

$$2x = 10$$

Dividing both members by the coefficient of x, we have

$$x = 5$$

Check Substitute 5 for x in both members of the equation.

$$6 \times 5 - 4 = 4 \times 5 + 6$$
$$26 = 26$$

Therefore, the answer $x = 5$ is correct.

Practice check: Do exercise 25 on the right.

■ **Example 2**

Solve the equation $\frac{2}{3}(x + 3) = \frac{1}{2}(x + 8)$.

Solution Multiply by 6 to clear fractions.

Simplify	$4(x + 3) = 3(x + 8)$
Simplify	$4x + 12 = 3x + 24$
Transpose	$4x - 3x = 24 - 12$
Combine	$x = 12$

Check Substitute 12 for x in the given equation.

$$\frac{2}{3}(12 + 3) = \frac{1}{2}(12 + 8)$$
$$\frac{2}{3} \times 15 = \frac{1}{2} \times 20$$
$$10 = 10$$

Practice check: Do exercise 26 on the right.

■ **Example 3**

Solve the equation for x.

$$\frac{2x}{a} = b + c$$

Solution Multiply both members by a.

$$2x = a(b + c)$$

Divide by the coefficient of x.

$$x = \frac{a(b + c)}{2}$$

Practice check: Do exercise 27 on the right.

25. Solve the following equation.

$$8 - 3x = 2x - 2$$

The answer is in the margin on page 186.

26. Solve the following equation.

$$\frac{1}{4}(2x - 2) = x + 4$$

The answer is in the margin on page 186.

27. Solve for x.

$$\frac{3x + 2}{a} = b - 2c$$

The answer is in the margin on page 186.

■ **Example 4**

Solve the equation for x.

$$ax + 2b = bx + c$$

Solution   Transposing, we have

$$ax - bx = c - 2b$$

Factoring the left side gives

$$(a - b)x = c - 2b$$

28. Solve the equation for x.

$$b - ax = 3c - bx$$

The answer is in the margin on page 186.

Dividing by the coefficient of x, we obtain

$$x = \frac{c - 2b}{a - b}$$

Practice check: Do exercise 28 on the left.

From the preceding examples, we formulate the following rule for solving a simple equation.

Rule

Transpose all unknown terms to the left side and all known terms to the right side. Combine like terms and divide both members by the coefficient of the unknown term.

■ **Example 5**

Solve the equation for x.

$$4x^2 - 6 = 2x^2 + 12$$

Solution   Transpose.

$$4x^2 - 2x^2 = 12 + 6$$

Combine like terms.

$$2x^2 = 18$$

Divide by the coefficient of x^2.

$$x^2 = 9$$

Take the square root of both sides.

$$\sqrt{x^2} = \sqrt{9}$$

$$x = 3 \quad \text{or} \quad x = {}^-3$$

29. Solve the equation for x.

$$6x^2 + 7 = 10x^2 - 9$$

The answer is in the margin on page 186.

Since ${}^+3^2 = 3 \times 3 = 9$, and ${}^-3^2 = {}^-3 \times {}^-3 = 9$, the square root of 9 is either 3 or ${}^-3$. From the conditions of the problem, it would be impossible to tell which of the two values of x is the correct one.

Practice check: Do exercise 29 on the left.

■ **Example 6**

Solve for P.

$$D = \frac{N+2}{P}$$

Solution   Multiplying by P yields

$$PD = N + 2$$

Dividing by D, we have

$$P = \frac{N+2}{D}$$

Practice check: Do exercise 30 on the right.

30. Solve for x.

$$m = \frac{n+2}{2x}$$

The answer is in the margin on page 187.

■ **Problems 6–10**

Solve the following equations for x or as indicated. Use the space provided below to work out the answers.

1. $5x = 10 + 4x$

2. $12x - 4 = 7x + 6$

3. $3x + 3 = 18$

4. $4x - 5 = 3x + 2$

5. $ax = b$

6. $ax + b = c$

7. $ax + b = cx + d$

8. $6(x + 2) = 5(x + 4)$

9. $4(x - 2) = 2(x + 6)$

10. $\dfrac{3x}{b} = a - c$

Answer to exercise 25

$8 - 3x = 2x - 2$
$^-3x - 2x = ^-2 - 8$
$^-5x = ^-10$
$x = 2$ (Divide by $^-5$.)

Answer to exercise 26

$\frac{1}{4}(2x - 2) = x + 4$
$2x - 2 = 4x + 16$
$^-2x = 18$
$x = ^-9$

Answer to exercise 27

$\frac{3x + 2}{a} = b - 2c$
$3x + 2 = a(b - 2c)$
$3x + 2 = ab - 2ac$
$3x = ab - 2ac - 2$
$x = \frac{ab - 2ac - 2}{3}$

Answer to exercise 28

$b - ax = 3c - bx$
$^-ax + bx = 3c - b$
$bx - ax = 3c - b$
$x(b - a) = 3c - b$
$x = \frac{3c - b}{b - a}$

Answer to exercise 29

$6x^2 + 7 = 10x^2 - 9$
$^-4x^2 = ^-16$
$x^2 = 4$
$\sqrt{x^2} = \sqrt{4}$
$x = 2 \text{ or } ^-2$

11. $\frac{5x}{a} = b + c$

12. $3x^2 + 3 = 15$

13. $2x^2 - 12 = x^2 + 24$

14. If $C = 2\pi r$, solve for r.

15. $V = lwh$. Solve for h.

16. $N = PD$. Solve for D.

17. $E = IR$. Solve for I and R.

18. Horsepower $= \frac{PLAN}{33,000}$. Solve for P.

19. Solve the formula in Problem 18 for L.

20. Solve the formula in Problem 18 for A.

21. $R = \frac{10.4 \times L}{CM}$. Solve for L and CM.

22. $D = \frac{N + 2}{P}$. Solve for N.

23. $A = \pi r^2$. Solve for r.

24. $D = M - 3W + 1.732P$. Solve for P.

25. $\frac{a}{b} = \frac{c}{d}$. Solve for b and c.

26. $V^2 = 2gh$. Solve for g and h.

27. $C = \frac{5}{9}(F - 32)$. Solve for F.

28. $a = \frac{V - v}{t}$. Solve for V, v, and t.

Answer to exercise 30

$$m = \frac{n + 2}{2x}$$
$$2mx = n + 2$$
$$x = \frac{n + 2}{2m}$$

29. $M = \frac{1}{6}bh^2$. Solve for b and h.

30. $A = \pi ab$. Solve for a and b.

31. $S = 4\pi r^2$. Solve for r.

32. Horsepower $= 0.4D^2N$. Solve for D.

33. $OD = \frac{(N + 2)P_c}{\pi}$. Solve for N.

34. $A = lw$. Solve for l and w.

35. $V = \frac{4}{3}\pi r^3$. Solve for r.

36. $P = \frac{\pi}{P_c}$. Solve for P_c.

37. $A = 0.7854d^2$. Solve for d.

38. $N = \frac{\pi d}{P_c}$. Solve for d and P_c.

39. Power factor $= \frac{\text{Effective power}}{\text{Apparent power}}$. Solve for effective power and apparent power.

40. Efficiency $= \frac{\text{Output}}{\text{Input}}$. Solve for output and input.

41. $BHP = \dfrac{2\pi LNW}{33{,}000}$. Solve for L, N, and W.

42. Cutting speed = Circumference × rpm. Solve for circumference and rpm.

43. Watts = Amperes × Volts. Solve for amperes and volts.

44. Area $= \dfrac{ab}{2}$. Solve for a and b.

45. $A = 0.866s$. Solve for s.

46. $s = 0.707d$. Solve for d.

47. $f = 1.732s$. Solve for s.

48. $C = 1.083f$. Solve for f.

49. $A = 0.433s^2$. Solve for s.

50. $A = 0.866f^2$. Solve for f.

■ **Self-Test**

Use the space provided below to work out the answers to the following problems. Check your answers with the answers in the back of the book.

1. Add 3 to $^-2$.

2. What is the absolute value of $^-11$?

3. What is the exponent of the letter x in the term $5x^3y^2$?

4. What is the coefficient in $15x^2yz^3$?

5. If $a = 3$, $b = {}^-2$, and $c = 7$, what is the value of $4ab^2c$?

6. Add $5ab^2$, ${}^-2ab^2$, $10ab^2$, ${}^-7ab^2$, and $3ab^2$.

7. Subtract ${}^-3x^2y$ from $5x^2y$.

8. Multiply: $(3x + 2y)(2x - 5y)$.

9. Divide $x^2 + 2xy + y^2$ by $x + y$.

10. Solve for a if $V = \frac{1}{3}\pi r^2 a$.

■ **Chapter Test**

Use the space provided below to work out the answers to the following problems.

1. Add ${}^-4$, ${}^+7$, and ${}^-2$.

2. What is the absolute value of ${}^-7x + 3x$?

3. If $x = 7$, $y = 3$, and $z = {}^-3$, what is the value of $2x^2yz^3 - 4xy^2$?

4. What is the sum of $3mn^2$, ${}^-2mn^2$, mn^2, and $12mn^2$?

5. Subtract ${}^-2xy^3$ from ${}^-5xy^3$.

6. Multiply: $(4a^2 - b)(4a^2 + b)$.

7. Divide $8x^3y^2z$ by $2x^2yz$.

8. Solve for r if $A = \pi r^2$.

9. Solve for b if $A = \frac{1}{2}ab$.

10. $I = \dfrac{E}{R}$. Solve for R.

11. $\dfrac{2}{x} = \dfrac{x}{8}$. Solve for x.

12. $\frac{9}{5}(C) + 32 = F$. Solve for C.

13. $E = mc^2$. Solve for c.

14. $V = \dfrac{\pi d^3}{6}$. Solve for d.

15. $P = I^2R$. Solve for R and I.

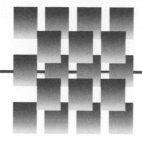

CHAPTER 7

Rectangles and Triangles

7–1 AREA OF SURFACES AND UNITS OF AREA

Common units of area are the square inch, the square foot, the square yard, the acre, and the square mile. It is interesting that acre is automatically a square measure; there is no need to use the word *square* when referring to land measure in acres. In metric measure the common units are the square millimeter, the square centimeter, the square meter, the are, and the square kilometer. The metric *are* is automatically a square measure and is used for land measure (see Chapter 10). An are is much smaller than an acre.

A *square inch* is the area contained in a square 1 in. on each side (Fig. 7–1).

A *square foot* is the area contained in a square 1 ft on each side (Fig. 7–2).

A *square yard* is the area contained in a square 1 yd on each side.

Since there are 12 in. in a foot, a square foot contains 12 in. $\times$ 12 in., or 144 sq in.

Similarly, since there are 3 ft in a yard, a square yard contains 3 ft $\times$ 3 ft, or 9 sq ft (Fig. 7–3).

A figure frequently encountered is the rectangle. A *rectangle* is a four-sided figure whose angles are right angles (90°) and whose opposite sides are equal in length (Fig. 7–4).

Rule

To find the area of a rectangle, multiply the length by the width.

If l = length and w = width, the rule may be expressed by the formula

$$\text{Area} = l \times w$$

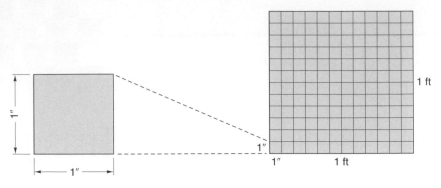

Figure 7–1   Figure 7–2

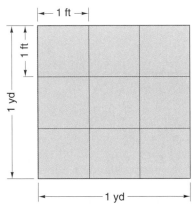

Figure 7–3   Figure 7–4

■ **Example 1**

Find the area of a rectangle whose length is 5 in. and whose width is 3 in.

Solution   Area = $l \times w$
                = 5 in. $\times$ 3 in. = 15 sq in.

Explanation   Since the length of the rectangle is 5 in. and the width 3 in., to get the area, multiply the length 5 in. by the width 3 in., obtaining the area 15 sq in.

■ **Example 2**

Find the area in square feet of a floor that is 18 ft 5 in. long and 12 ft 9 in. wide.

Solution   Change 18 ft 5 in. and 12 ft 9 in. to inches.

$$18 \text{ ft } 5 \text{ in.} = (18 \times 12) + 5 = 216 \text{ in.} + 5 \text{ in.} = 221 \text{ in.}$$

$$12 \text{ ft } 9 \text{ in.} = (12 \times 12) + 9 = 144 \text{ in.} + 9 \text{ in.} = 153 \text{ in.}$$

$$\text{Area} = l \times w = 221 \times 153 = 33{,}813 \text{ sq in.}$$

$$33{,}813 \div 144 = 234.8125 \text{ sq ft} \approx 234.8 \text{ sq ft}$$

Alternate Solution   Change 18 ft 5 in. and 12 ft 9 in. to feet before applying the formula for area.

$$18 \text{ ft } 5 \text{ in.} = 18\tfrac{5}{12} \text{ ft} = 18.417 \text{ ft}$$

$$12 \text{ ft } 9 \text{ in.} = 12\tfrac{9}{12} \text{ ft} = 12.75 \text{ ft}$$

$$A = l \times w = 18.417 \text{ ft} \times 12.75 \text{ ft}$$

$$= 234.81675 \text{ sq ft} \approx 234.8 \text{ sq ft}$$

Explanation   Floor area is generally expressed in square feet. Divide the 33,813 sq in. by 144 (the number of sq in. in a sq ft). Since one decimal place is sufficiently accurate for this type of work, state the answer as 234.8 sq ft. In the alternate solution, both dimensions are changed to feet before the formula for area is used. It is not necessary to divide by 144 because the units are already in feet.

Practice check: Do exercise 1 on the right.

1. Find the area in square feet of a flat roof that is 52 ft 7 in. long and 25 ft 3 in. wide.

The answer is in the margin on page 195.

The wise mechanic estimates the answer before solving a problem. This practice avoids errors in computation. To estimate the answer of Example 2, round the two dimensions to the nearest foot. If both dimensions were in feet, then it would be a matter of simply multiplying the two; 18 ft 5 in. is a little over 18 ft, and 12 ft 9 in. is almost 13 ft. An approximate answer to the original problem would therefore be the product of 18 ft and 13 ft. Therefore, the estimated area would be

$$\text{Area}_{est} = 18 \text{ ft} \times 13 \text{ ft} = 234 \text{ sq ft} \quad \text{(estimated answer)}$$

A comparison of the estimated answer with the actual answer shows that the computed answer is probably correct. In general, the estimated answer will be fairly accurate if one number can be rounded up and the other number can be rounded down. Common sense must be used when comparing the computed answer with the estimated answer. If there is a great difference between the two, then the processes used to find the answers must be reviewed to discover where a mistake might lie. Mistakes are costly and should always be avoided.

2. Check exercise 1 above by estimation.

The answer is in the margin on page 195.

Practice check: Do exercise 2 on the right.

■ Problems 7–1

Fill in the areas (in square feet) in the following chart. Estimate the answer before doing the necessary computation in the space provided in the margin. Use one decimal place in the answer.

Rectangles				
	Length	Width	Estimated Area	Computed Area
1.	5′	4′	20 sq ft	_____
2.	5′3″	4′0″	20 sq ft	_____
3.	5′3″	4′9″	25 sq ft	_____
4.	6′9″	4′4″	_____	_____
5.	8′9″	6′6″	_____	_____

Rectangles				
	Length	Width	Estimated Area	Computed Area
6.	12'3"	10'9"	_____	_____
7.	7'4"	5'3"	_____	_____
8.	9'6"	7'9"	_____	_____
9.	5'8"	4'4"	_____	_____
10.	3'4"	2'6"	_____	_____
11.	6'9"	3'3"	_____	_____
12.	12'10"	8'2"	_____	_____

Use the space provided below to find the following areas. In your answers use units that would be reasonable for each particular problem. Estimate the answers before doing the problems.

13. Find the area of a page of a book $7\frac{1}{2}$" long and 5" wide.

14. Find the area of the floor of a room 16'4" long and 10'8" wide. Give the answer in square feet to one decimal place.

15. How many square feet of pavement are there in a sidewalk 198' long and 18' wide?

16. Find the area of the walls of a room 9'2" high if the room is 19'4" long and 12'3" wide. (Make no allowance for doors or windows.)

17. How many square feet are there in the top of a desk 52" long and 28" wide? (144 sq in. = 1 sq ft.)

18. How many square inches are there in a pane of glass $15\frac{1}{2}$" $\times$ $19\frac{1}{2}$"?

19. A parking lot is 186' $\times$ 91'. How many square yards does it contain?

20. How many square feet are there in a rectangular lot 88'4" deep with a frontage of 74'2"?

21. Allowing 25 sq ft per pupil, how many children may safely play in a playground 200′ long and 140′ wide?

22. How many square inches are there in the surface of a closed cubical box that measures 8″ on a side?

23. How many square feet are there in the six sides of a closed box $8\frac{1}{2}″ \times 6″ \times 3\frac{1}{2}″$?

24. How many square feet of corrugated iron are required for the walls of a garage 15′6″ × 8′9″ and 8′3″ high? (Subtract 64 sq ft as the door and window allowance.)

Answer to exercise 1

52 ft 7 in. = $52\frac{7}{12}$ ft
           = 52.583 ft

25 ft 3 in. = $25\frac{3}{12}$ ft
           = 25.25 ft

A = 52.583 ft × 25.25 ft
   = 1,327.72 sq ft
   ≃ 1,327.7 sq ft

Answer to exercise 2

Area$_{est}$ = 53 ft × 25 ft
        = 1,325 sq ft
          (Estimated answer)

Use the space provided below to find the dimensions of the parts marked (?) in the following composite rectangular figures. Remember that a property of the rectangle is that its opposite sides are equal.

25.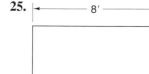

26.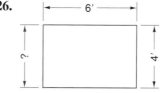

27.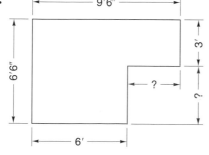

28.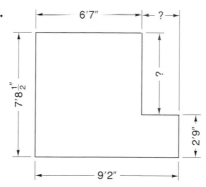

29.

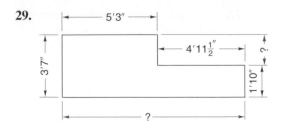

Use the space provided below to find the areas of the following figures, which are composite rectangles. Divide each figure into rectangles, and the area of the figure will be the sum of the areas of the rectangles into which the figure can be divided. Try estimating the areas before doing the actual computation. Area = length times width.

30.

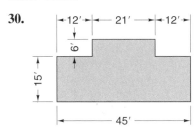

31.

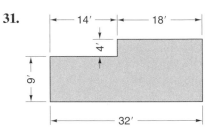

32.

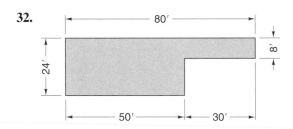

33.

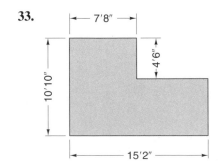

34.

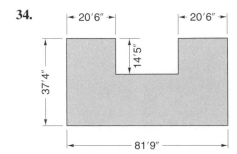

35.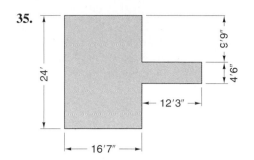

36.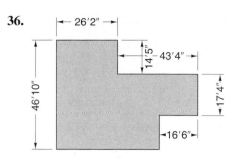

37. Find the cost of covering the floor of a room 13'9" × 21'6" with carpet costing $12.95 a square yard.

38. How many 9"-by-9" vinyl floor tiles will be needed to cover a floor of a room 15'9" by 18'9"?

39. If the tiles in Problem 38 cost 64 cents each, what would be the cost of covering the floor in that problem?

40. If carpeting costs $13.50 a square yard and 9"-by-9" cork tiles cost 85 cents each, which floor covering would be less expensive to buy for the floor in Problem 38, and what would be the difference in cost between the two floor coverings?

7-2   THE PERIMETER OF A RECTANGLE

The **perimeter** of a rectangle is the distance around the outside of the rectangle. Since a property of the rectangle is that its opposite sides are equal, the dimensions given for a particular rectangle need be only a given length and a given width. The perimeter of a rectangle then would be twice the sum of the width and length.

$$p = 2(l + w)$$

■ **Example**

Find the perimeter of a rectangle 5 ft long and 3 ft wide.

Solution $p = 2(l + w)$
$= 2(5 \text{ ft} + 3 \text{ ft})$
$= 2(8 \text{ ft})$
$= 16 \text{ ft}$

NOTE: *Perimeter is length or linear measure.*

Practice check: Do exercise 3 on the right.

3. Find the perimeter of a rectangular-shaped shop floor 30 ft long and 21 ft wide.

The answer is in the margin on page 199.

■   **Problems 7–2**

Fill in the perimeters in the following chart. Determine the units that best fit the information given. Use the space provided in the margin to work out the answers.

	Rectangles		
	Length	**Width**	**Perimeter**
1.	15′	10′	_____
2.	12′	6′6″	_____
3.	10′	6′8″	_____
4.	12′6″	5′6″	_____
5.	9′9″	5′4″	_____
6.	3′2″	1′8″	_____
7.	2′6½″	3′7″	_____
8.	5′3¼″	3′2¾″	_____

Use the space provided below to work out the answers to the following problems.

9. A rectangular tent is 3.5 yd long and 2.7 yd wide. Find the perimeter.

10. A contractor has to replace the baseboard in a bathroom. If the bathroom measures 10 ft 4 in. by 5 ft 3 in., how many feet of baseboard should he buy?

11. In a baseball game, Cal hit a homerun with the bases loaded. If there are 90 ft between the bases, how many total feet did the baserunners travel?

7–3   FINDING THE WIDTH OR LENGTH OF A RECTANGLE

If the area and the length of a rectangle are known, divide the area by the length to find the width.

$$\text{Width} = \frac{\text{Area}}{\text{Length}}$$

■ **Example 1**

The area of a rectangle is 72 sq in. and the length is 12 in. Find the width.

Solution   Width $= \dfrac{\text{Area}}{\text{Length}}$

$= \dfrac{72 \text{ sq. in.}}{12 \text{ in.}}$

$= 6$ in.

If the area and the width of a rectangle are known, divide the area by the width to find the length.

$$\text{Length} = \dfrac{\text{Area}}{\text{Width}}$$

Practice check: Do exercise 4 on the right.

■ **Example 2**

Find the length of a rectangle whose area is 135 sq in. and whose width is 7.5 in.

Solution   Length $= \dfrac{\text{Area}}{\text{Width}}$

$= \dfrac{135 \text{ sq in.}}{7.5 \text{ in.}}$

$= 18$ in.

Practice check: Do exercise 5 on the right.

Answer to exercise 3

$p = 2(l + w)$
$= 2(30 \text{ ft} + 21 \text{ ft})$
$= 2(51 \text{ ft})$
$= 102 \text{ ft}$

4. The area of a picture frame is 88 sq in. and the length is 11 in. Find the width.

The answer is in the margin on page 201.

5. Find the length of a rectangular-shaped lot whose area is 5,292 sq ft and whose width is 50.4 ft.

The answer is in the margin on page 201.

■ **Problems 7–3**

Use the space provided in the margin to find the missing dimensions indicated in the following chart. Include the units of measure in your answers.

	Rectangles		
	Width	**Length**	**Area**
1.	5′	_____	15 sq ft
2.	_____	10′6″	126 sq ft
3.	4′8″	_____	$30\frac{1}{3}$ sq ft
4.	5′6″	_____	$47\frac{2}{3}$ sq ft
5.	_____	3′4″	15 sq ft
6.	2′6″	_____	$9\frac{1}{6}$ sq ft
7.	_____	2′9″	$9\frac{5}{8}$ sq ft
8.	3′6″	_____	$14\frac{7}{8}$ sq ft
9.	_____	12′8″	$107\frac{2}{3}$ sq ft
10.	11′6″	_____	$140\frac{7}{8}$ sq ft

Use the space provided below to solve the following problems. Make sure that the units of measure are reasonable for each problem.

11. A floor contains 360 sq ft and is 24′ long. How wide is it?

12. What is the length of a parking space 29,100 sq ft in area if it is 145′6″ wide?

13. A building lot has an area of 3,125 sq ft. If it is 25′0″ wide, how deep is the lot?

14. The floor of a room has an area of 308 sq ft. If the room is 22′ long, how wide is it?

15. A rug 14 ft 6 in. long has an area of 152.25 sq ft. How wide is the rug?

16. A playground covers an area of 11,250 sq ft. If it is 62 ft 6 in. wide, how long is it?

17. Find the length of a factory floor 49 ft 6 in. wide if its area is 6,162.75 sq ft.

18. What is the width of a loft 82 ft 9 in. long if its area is 2,027.38 sq ft?

19. A rectangle 2 ft 4 in. wide has an area of 1,456 sq in. How long is it?

20. The floor area of a garage 65 ft by 85 ft is to be increased by 1,000 sq ft. If it is made 6 ft longer, how much will the width have to be increased?

21. The area of a square with 6-in. sides is how many times greater than the area of a square with 3-in. sides?

22. The area of a square with 15-in. sides is how many times greater than the area of a square with 3-in. sides?

23. If you double the length of the sides of a square, then how many times is the area increased?

24. The area of a square with 8-in. sides is how many times greater than the area of a square with 2-in. sides?

25. The area of a square with 7-in. sides is how many times greater than the area of a square with 3-in. sides?

26. The area of a rectangle 8 ft by 4 ft is how many times greater than the area of a rectangle 4 ft by 2 ft?

27. The area of a tile 9 in. by 9 in. is how much smaller than the area of a tile 1 ft by 1 ft?

28. The area of a rectangle whose length is 3 ft 6 in. and whose width is 1 ft 6 in. is how many times smaller than the area of a rectangle whose length is 4 ft 6 in. and whose width is 2 ft 3 in.?

29. If the length and width of a rectangle are doubled, then its area is _____.

30. If only the length or only the width of a rectangle is doubled, then its area is _____.

7–4   SQUARES AND SQUARE ROOTS

In Fig. 7–5 the 1-in. square *a* contains 1 sq in.; the 2-in. square *b* contains 4 sq in.; the 5-in. square *c* contains 25 sq in. In each instance the area of the square is obtained by multiplying the length of a side by itself.

(*a*)   area of 1-in. square = 1 in. × 1 in. = 1 sq in.

(*b*)   area of 2-in. square = 2 in. × 2 in. = 4 sq in.

(*c*)   area of 5-in. square = 5 in. × 5 in. = 25 sq in.

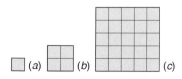

(*a*)   (*b*)   (*c*)

Figure 7–5

There are short ways of saying and writing certain expressions in mathematics. For example, to show that 3 multiplied by itself (3 × 3) equals 9, you may write $3^2 = 9$. The small number 2 used in this way is called an **exponent** and indicates how many times the number is used as a factor. When the exponent of a number is 2, the resulting product is called the **square** of the number.

The square of 3 is 9, or $3^2 = 9$.

The square of 7 is 49, or $7^2 = 49$.

The square of 5 is 25, or $5^2 = 25$.

To find the length of the side of a square that contains a certain number of square inches, or the number that must be multiplied by itself to obtain a certain number, it is necessary to reverse the operation. For instance, what number multiplied by itself equals 9? Since 3 × 3 = 9, 3 is the **square root** of 9. The sign $\sqrt{}$ takes the place of the four words *the square root of,* so that $\sqrt{9} = 3$ means "the square root of 9 equals 3."

You should memorize the following table of squares and square roots.

Squares	Square Roots
$1^2 = 1$	$\sqrt{1} = 1$
$2^2 = 4$	$\sqrt{4} = 2$
$3^2 = 9$	$\sqrt{9} = 3$
$4^2 = 16$	$\sqrt{16} = 4$
$5^2 = 25$	$\sqrt{25} = 5$
$6^2 = 36$	$\sqrt{36} = 6$
$7^2 = 49$	$\sqrt{49} = 7$
$8^2 = 64$	$\sqrt{64} = 8$
$9^2 = 81$	$\sqrt{81} = 9$
$10^2 = 100$	$\sqrt{100} = 10$

Find the following.

6. 15^2

7. $\sqrt{400}$

The answers are in the margin on page 205.

Practice check: Do exercises 6 and 7 on the left.

■ **Problems 7–4**

Use the space provided below to perform the following indicated operations.

1. 6^2 **2.** 11^2 **3.** 12^2 **4.** 45^2 **5.** 61^2

6. 100^2 **7.** 125^2 **8.** 50^2 **9.** 60^2 **10.** 80^2

11. $\sqrt{49}$ **12.** $\sqrt{64}$ **13.** $\sqrt{81}$ **14.** $\sqrt{121}$ **15.** $\sqrt{144}$

16. $\sqrt{225}$ **17.** $\sqrt{10,000}$ **18.** $\sqrt{3,600}$ **19.** $\sqrt{4,900}$ **20.** $\sqrt{1,000,000}$

21. Find the area of a square if one of its sides is 6 in.

22. A square room has an area of 625 sq ft. Find the length of a side of the room.

23. A machine shop floor is a square 20 ft on a side. Find the width of the square floor if the area is increased by four times.

7–5   FINDING THE SQUARE ROOT OF A WHOLE NUMBER

When a number is an exact square, there is generally no difficulty in finding its square root. The mechanic usually has a reference book or a calculator available to find the square root of a number. The following method may be used to find the square root of a number when other methods are not available.

■ **Example**

Find the square root of 5,625.

Solution

$$
\begin{array}{r}
7\ \ 5 \\
\sqrt{56\ 25} \\
49 \\
\hline
145)\ \ 7\ 25 \\
7\ 25 \\
\hline
0
\end{array}
$$

Explanation   Write the number and group it into pairs of figures from right to left, $\overline{56}\ \overline{25}$. The first pair of figures is 56. The largest square equal to or less than 56 is 49. The square root of 49 is 7. Write the 49 under the 56 and the 7 over the 56. Subtracting 49 from 56 leaves 7. Bring down the next pair of figures, 25, with the 7, making 725. Now double the answer already obtained: $2 \times 7 = 14$. Put this before the remainder 725 as a *trial divisor*. Disregard the last figure of the 725 and divide 14 into 72. This gives 5 for the second figure of the answer. Write the new figure 5 above the second pair, 25, and after the trial divisor, 14, making the complete divisor 145. Multiply the divisor 145 by the new figure 5; $5 \times 145 = 725$. Subtracting 725 from the divisor leaves no remainder. This shows that the exact square root of 5,625 is 75.

8. Find the square root of 4,489.

The answer is in the margin on page 206.

Check $75 \times 75 = 5,625$

Practice check: Do exercise 8 on the left.

■ Problems 7–5

Use the space provided below to find the square root of the following whole numbers.

1. $\sqrt{7225}$ **2.** $\sqrt{6400}$ **3.** $\sqrt{8836}$ **4.** $\sqrt{1849}$ **5.** $\sqrt{2025}$

6. $\sqrt{576}$ **7.** $\sqrt{4225}$ **8.** $\sqrt{784}$ **9.** $\sqrt{3969}$ **10.** $\sqrt{1024}$

11. $\sqrt{1225}$ **12.** $\sqrt{2809}$ **13.** $\sqrt{1369}$ **14.** $\sqrt{2116}$ **15.** $\sqrt{5329}$

16. $\sqrt{6561}$ **17.** $\sqrt{1156}$ **18.** $\sqrt{8649}$ **19.** $\sqrt{529}$ **20.** $\sqrt{3721}$

7–6   THE SQUARE ROOT OF MIXED NUMBERS

Numbers like 16 and 81 are called perfect squares because their square roots have no remainder. Numbers that occur in the trades usually are not perfect squares, and their square roots do have a remainder. Before using the procedure to find the square root of a number, the mechanic must decide how accurate the square root of the number should be.

To obtain the square root of 56.25, or $56\frac{1}{4}$, proceed in the same manner. Group the number 56.25 into pairs of figures *to the right and left of the decimal point,* and place the decimal point of the square root directly above the decimal point of the square.

$$\frac{7\,.\,5}{\sqrt{56.25}} \qquad \sqrt{56.25} = 7.5 \quad \text{or} \quad 7\frac{1}{2}$$

The square root of 731, for example, has a remainder after getting the second figure. Suppose the square root of 731 is to be accurate to two decimal places. Four figures after the decimal point are needed in the square to give two decimal places in the square root.

Answers to exercises 6 and 7

6. $15 \times 15 = 225$

7. 20 since $20 \times 20 = 400$

Rule _____

Two decimal places are required in the square for each decimal place in the root.

Therefore, to find the square root of 731 to two decimal places, write 731.0000 and mark off the figures into pairs to the right and left of the decimal point: $\overline{7}\,\overline{31}.\overline{00}\,\overline{00}$.

■ **Example**

Find the square root of 731 to two decimal places.

Solution

$$
\begin{array}{r}
2\ 7\ .\ 0\ \ 3 \\
\sqrt{7\,\overline{31}.\overline{00}\,\overline{00}} \\
4 \\
\hline
47)\overline{3\ 31} \\
3\ 29 \\
\hline
5403)\ \ \ 2\ 00\ 00 \\
1\ 62\ 09 \\
\hline
37\ 91
\end{array}
$$

Explanation   The largest square less than 7 is 4, of which the square root is 2. Write the 2 over the 7 and the 4 under the 7. Subtracting 4 from 7 leaves a remainder of 3. Bring down the next pair, 31. Double the root already obtained; $2 \times 2 = 4$, the trial divisor. 4 divides into 33 eight times, but if we try 8, we will find it too large; so we try a smaller number, 7. Write the 7 over the 31 and after the trial divisor, making the complete divisor 47. Multiply the divisor 47 by the new root figure 7; $7 \times 47 = 329$. Subtract 329 from 331, leaving the remainder 2.

Bring down the next pair of figures, 00. Double the root already obtained, 27; $2 \times 27 = 54$, the new trial divisor. 54 divides into 20 zero times. Write the 0 over the 00 and after the 54. Bring down the next pair of figures, 00, making the dividend 20,000. Dividing the trial divisor, 540, into 2,000 gives the figure 3 as the fourth figure of the root. Multiply the complete divisor 5403 by the new root figure 3; $3 \times 5,403 = 16,209$. Subtracting 16,209 from 20,000 leaves a remainder of 3,791. Because the fraction $\frac{3791}{5403}$ is more than $\frac{1}{2}$, we add 1 to the last figure of the root, making the answer 27.04. The square root of 731, accurate to two decimal places, equals 27.04. We see that after the operation of obtaining the first figure of the root, all the other figures are obtained by exactly the same process.

The important things to bear in mind are as follows.

1. Begin at the decimal point and group the number into pairs of figures on each side of the decimal point.

2. Decide the number of decimal places the work requires.

3. Arrange your work for that number of decimal places.

4. Place the decimal point in the root directly over the decimal point in the square.

Practice check: Do exercise 9 on the right.

9. Find the square root of 371 to two decimal places.

The answer is in the margin on page 207.

■ Problems 7–6

The following examples illustrate how to arrange the work for finding the square roots of various numbers. Use the space provided below to work out the answers.

1. Find $\sqrt{19}$ to two decimal places. $\sqrt{19.00\ 00}$

2. Find $\sqrt{43}$ to two decimal places. $\sqrt{43.00\ 00}$

3. Find $\sqrt{3649}$ to two decimal places. $\sqrt{36\ 49.00\ 00}$

4. Find $\sqrt{3859}$ to two decimal places. $\sqrt{38\ 59.00\ 00}$

5. Find $\sqrt{2}$ to three decimal places. $\sqrt{2.00\ 00\ 00}$

6. Find $\sqrt{343}$ to two decimal places. $\sqrt{3\ 43.00\ 00}$

7. Find $\sqrt{846}$ to two decimal places. $\sqrt{8\ 46.00\ 00}$

8. Find $\sqrt{6154}$ to three decimal places. $\sqrt{61\ 54.00\ 00\ 00}$

9. Find $\sqrt{28143}$ to two decimal places. $\sqrt{2\ 81\ 43.00\ 00}$

Answer to exercise 9

$$
\begin{array}{r}
1\ 9\ .\ 2\ \ 6 \\
\sqrt{3\ 71.00\ 00} \\
\underline{1} \\
29)\overline{2\ 71} \\
\underline{2\ 61} \\
382)\overline{\ \ 10\ 00} \\
\underline{7\ 64} \\
3846)\overline{23\ 6\ 00} \\
\underline{23\ 0\ 76} \\
5\ 24
\end{array}
$$

10. Find $\sqrt{2733816}$ to no decimal places. $\sqrt{2\ 73\ 38\ 16.}$

11. Find $\sqrt{713.801}$ to three decimal places. $\sqrt{7\ 13.80\ 10\ 00}$

12. Find $\sqrt{8.2733}$ to three decimal places. $\sqrt{8.27\ 33\ 00}$

13. Find $\sqrt{0.273}$ to three decimal places. $\sqrt{0\ 27\ 30\ 00}$

14. Find $\sqrt{0.08601}$ to three decimal places. $\sqrt{0.08\ 60\ 10}$

15. Find $\sqrt{29.68}$ to three decimal places. $\sqrt{29.68\ 00\ 00}$

16. Find $\sqrt{0.0351}$ to three decimal places. $\sqrt{0.03\ 51\ 00}$

17. Find $\sqrt{3\frac{5}{8}}$ to three decimal places. $\sqrt{3.62\ 50\ 00}$

18. Find $\sqrt{5\frac{2}{3}}$ to two decimal places. $\sqrt{5.\overset{\cdot}{66}\ \overline{67}}$

19. Find $\sqrt{6\frac{1}{3}}$ to two decimal places. $\sqrt{6.\overset{\cdot}{33}\ \overline{33}}$

20. Find $\sqrt{22\frac{3}{8}}$ to two decimal places. $\sqrt{22.\overset{\cdot}{37}\ \overline{50}\ \overline{00}}$

7–7   FINDING THE SQUARE ROOT OF A FRACTION

The simplest method of finding the square root of a fraction is to convert the fraction into a decimal of twice as many places as are required in the root. For instance, to get the square root of $\frac{2}{3}$ to three decimal places, convert the fraction $\frac{2}{3}$ to the decimal 0.666667. The square root of 0.666667 is 0.816; hence, $\sqrt{\frac{2}{3}} = 0.816$.

10. Find the square root of $\frac{29}{4.6}$ to two decimal places.

The answer is in the margin on page 210.

To obtain the square root of $\frac{191}{3.14}$ to two decimal places, divide 191 by 3.14, carrying work to four decimal places: $\frac{191}{3.14} = 60.8280$. The square root of 60.8280 is 7.80; hence, $\sqrt{\frac{191}{3.14}} = 7.80$.

Practice check: Do exercise 10 on the left.

■ Problems 7–7

Use the space provided below to find the square root of the following numbers.

1. $\sqrt{2\ \overline{25}}$ **2.** $\sqrt{6\ \overline{25}}$ **3.** $\sqrt{12\ \overline{25}}$ **4.** $\sqrt{121}$ **5.** $\sqrt{256}$

6. $\sqrt{169}$ **7.** $\sqrt{289}$ **8.** $\sqrt{361}$ **9.** $\sqrt{324}$ **10.** $\sqrt{400}$

11. $\sqrt{1.44}$ **12.** $\sqrt{72.25}$

Find the square root of the following numbers to two decimal places.

13. $\dfrac{51}{2.6}$ **14.** $\dfrac{73}{2.19}$ **15.** $\dfrac{123}{3.21}$ **16.** $\dfrac{318}{5.32}$

Try estimating the answers to the following problems before performing the operation in the space provided.

17. Find the square root of 4,624.

18. Find the square root of 54,756.

19. Find the square root of 161.29.

20. Find the square root of 2,316 to two decimal places.

21. Find the square root of 8.609 to three decimal places.

22. Find the square root of 75.3 to three decimal places.

23. Find the length of the side of a square that contains 1,000 sq in.

24. Find the length of the side of a square that contains 2 sq in.

25. Find the length of the side of a square that contains 5 sq in.

26. Find the length of the side of a square that contains 10 sq ft.

Answer to exercise 10

2.51

27. Find the length of the side of a square that contains 35 sq ft.

28. Find the length of the side of a square that contains 50 sq ft.

 7–8   USING THE CALCULATOR TO SOLVE SQUARE ROOT PROBLEMS

Many calculators, like the one shown here, have a square root function key. On these calculators, you can compute the square root of the number in the display by pressing the square root function key.

■ **Example**

Find the square root of 367.

Solution   Use the calculator in the following manner.

The Display Shows

Turn on the calculator	0.
Enter 367	367.
Press √	19.15724406

Round off the solution to the number of decimal places desired. The square root of 367 to two decimal places is 19.16.

Practice check: Do exercise 11 on the left.

11. Find the square root of 3.67 to two decimal places.

The answer is in the margin on page 212.

■ **Problems 7–8**

Use the calculator to solve the following problems.

1. $\sqrt{37}$ to two decimal places

2. $\sqrt{58}$ to two decimal places

3. $\sqrt{189}$ to three decimal places

4. $\sqrt{232}$ to four decimal places

5. $\sqrt{458}$ to three decimal places

6. $\sqrt{6104}$ to four decimal places

7. $\sqrt{4005}$ to two decimal places

8. $\sqrt{8.1}$ to two decimal places

9. $\sqrt{2.76}$ to three decimal places

10. $\sqrt{39.4}$ to three decimal places

11. $\sqrt{81.62}$ to three decimal places

12. $\sqrt{560.1}$ to four decimal places

13. $\sqrt{109.31}$ to three decimal places

14. $\sqrt{10.04}$ to three decimal places

15. $\sqrt{0.426}$ to three decimal places

16. $\sqrt{0.4891}$ to four decimal places

Answer to exercise 11

1.92

17. $\sqrt{0.0893}$ to three decimal places

18. $\sqrt{0.00524}$ to four decimal places

19. $\sqrt{0.000618}$ to four decimal places

20. $\sqrt{\frac{1}{3}}$ to three decimal places

21. $\sqrt{9\frac{3}{4}}$ to three decimal places

22. $\sqrt{2\frac{7}{16}}$ to three decimal places

23. $\sqrt{1.02}$ to three decimal places

24. $\sqrt{\frac{2}{7}}$ to three decimal places

25. $\sqrt{\frac{5}{12}}$ to three decimal places

26. $\sqrt{\frac{23}{32}}$ to three decimal places

27. $\sqrt{\dfrac{1.33}{3.14}}$ to four decimal places

28. $\sqrt{\dfrac{2763}{0.83}}$ to one decimal place

29. $\sqrt{\dfrac{11\frac{1}{8}}{96\frac{3}{4}}}$ to four decimal places

30. $\sqrt{0.4}$ to one decimal place

7–9   APPLICATIONS OF SQUARE ROOT

A common and most useful application of square root is in finding the third side of a right triangle when the lengths of two of the sides are known. A ***right triangle*** is a triangle that has a right angle (90°). Of the six triangles shown in Fig. 7–6, only (*a*) and (*b*) are right triangles; that is, they each have a right angle.

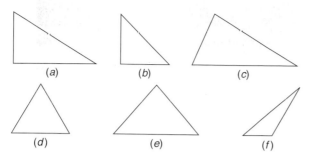

Figure 7-6

Shown in Fig. 7-7 are the names given to the three sides of a right triangle: **base, altitude,** and **hypotenuse.**

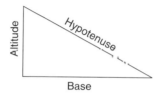

Figure 7-7

To find the relation between the three sides, construct a right triangle with a 4-in. base and a 3-in. altitude (Fig. 7-8). Measure the hypotenuse. It should be 5 in. The area of the square built on the hypotenuse is equal to the sum of the areas of the squares built on the base and the altitude.

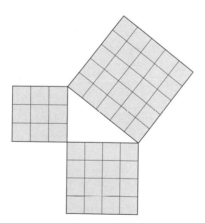

Figure 7-8

$$\text{Base}^2 = (4 \text{ in.})^2 = 16 \text{ sq in.}$$

$$\text{Altitude}^2 = (3 \text{ in.})^2 = \underline{ 9 \text{ sq in.}}$$

$$\text{Sum of squares} = 25 \text{ sq in.}$$

The square root of 25 sq in. is 5 in., the length of the hypotenuse. From this is derived the rule for the relation between the three sides of a right triangle.

Rule

The square of the hypotenuse is equal to the sum of the squares of the base and the altitude.

If we call the base b, the altitude a, and the hypotenuse c, we can express the rule thus: $c^2 = a^2 + b^2$. To find c, the hypotenuse, we must square a, square b, add the squares, and get the square root of the sum. This may be expressed by the formula

$$c = \sqrt{a^2 + b^2}$$

which says that the hypotenuse c is equal to the square root of the sum of a^2 and b^2. This rule is called the ***Pythagorean theorem.***

■ **Example 1**

Find the hypotenuse of the right triangle in Fig. 7–9.

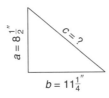

Figure 7–9

Solution $c = \sqrt{a^2 + b^2}$
$= \sqrt{(8\frac{1}{2})^2 + (11\frac{1}{4})^2}$
$= \sqrt{72.25 + 126.5625}$
$= \sqrt{198.8125}$
$= 14.10$

The length of the hypotenuse is 14.10″.

If the hypotenuse and one of the sides are known, then you can find the other side by subtracting the squares instead of adding them.

To find the altitude, use the formula

$$a = \sqrt{c^2 - b^2}$$

To find the base, use the formula

$$b = \sqrt{c^2 - a^2}$$

Practice check: Do exercise 12 on the left.

12. Find the hypotenuse of a right triangle whose altitude is 10″ and base is 24″.

The answer is in the margin on page 216.

■ **Example 2**

Find the altitude of the right triangle in Fig. 7–10.

Figure 7–10

Solution
$$a = \sqrt{c^2 - b^2}$$
$$= \sqrt{20^2 - 16^2}$$
$$= \sqrt{400 - 256}$$
$$= \sqrt{144} = 12$$

The altitude is 12″.

Practice check: Do exercise 13 on the right.

13. Find the altitude of a right triangle whose base is 9 ft and hypotenuse is 15 ft.

The answer is in the margin on page 217.

■ **Example 3**

Find the base of the right triangle in Fig. 7–11.

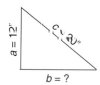

$a = 12″$

$c = 20″$

$b = ?$

Figure 7–11

Solution
$$b = \sqrt{c^2 - a^2}$$
$$= \sqrt{20^2 - 12^2}$$
$$= \sqrt{400 - 144}$$
$$= \sqrt{256} = 16$$

The base is 16″.

Practice check: Do exercise 14 on the right.

14. Find the base of a right triangle whose hypotenuse is 13 ft and altitude is 12 ft.

The answer is in the margin on page 217.

Figures 7–10 and 7–11 in Examples 2 and 3 are right triangles that have whole numbers for their sides. Note that the angle between the two shorter sides is a right angle. Carpenters frequently use the right triangle whose sides are 3–4–5 to check that the corner of a building is a right angle. The carpenter makes a mark 3 units along one side of the corner to be checked and then makes a mark 4 units along the other side. If the measure from mark to mark is 5 units, then the corner is a right angle.

Summary

To find the hypotenuse of a right triangle: Square both sides, add the squares, and take the square root of the sum.

To find the base or the altitude of a right triangle: Square the hypotenuse and the given side, subtract the square of the side from the square of the hypotenuse, and take the square root of the difference.

Answer to exercise 12

$c^2 = a^2 + b^2$
$\quad = 10^2 + 24^2$
$\quad = 100 + 576$
$\quad = 676$
$c = 26$

The length of the hypotenuse is 26″.

■ Problems 7–9

Use the space provided in the margin to find the missing sides of the right triangles in the following chart. The letter a indicates the altitude, b indicates the base, and c indicates the hypotenuse.

	Right Triangles		
	a	b	c
1.	5″	12″	_____
2.	8″	15″	_____
3.	7″	24″	_____
4.	2′6″	6′0″	_____
5.	1′4″	_____	2′10″
6.	6″	_____	10″
7.	1′2″	_____	4′2″
8.	9″	_____	15″
9.	_____	2′0″	2′2″
10.	_____	3′9″	4′3″
11.	_____	3′0″	3′1$\frac{1}{2}$″
12.	_____	1′10$\frac{1}{2}$″	2′1$\frac{1}{2}$″

Use the space provided below to solve the following problems.

13. A common problem is that of computing the diagonal of a square. How long is the diagonal of a square with 1-in. sides? A square with 2-in. sides? A square with 3-in. sides?

14. Compute the center-to-center distance between the bolt holes in Fig. 7–12.

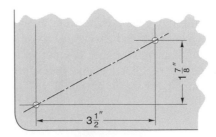

Figure 7–12

15. In printing, good proportion calls for a page whose diagonal is twice the width. What should be the length of a page that is $3\frac{7}{8}$ in. wide?

16. Compute the overall length along the center line of the piece of work shown in Fig. 7–13.

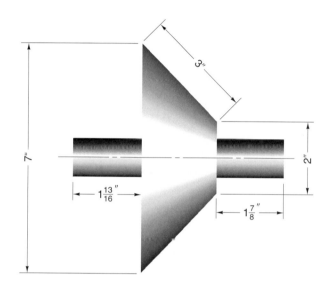

Figure 7–13

17. Compute the length of the stair stringer shown in Fig. 7–14.

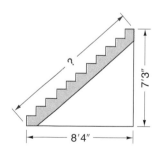

Figure 7–14

18. Figure 7–15 shows a roof truss. Would a 20′ rafter be long enough to build this truss? (The overhang is a part of the rafter.)

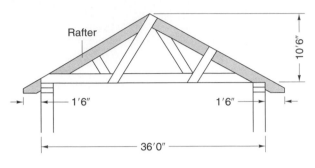

Figure 7–15

19. In Fig. 7–16, pulley *A* is 18′4″ above and 20′8″ to the right of pulley *B*. Find the distance center to center of the pulleys.

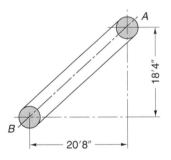

Figure 7–16

20. In building a pier, workers fastened the stringers to the piles with square-headed bolts 2″ on a side. If the heads are to be countersunk, what size holes should be bored in the stringers, allowing $\frac{1}{2}$″ clearance around the bolt heads? See Fig. 7–17.

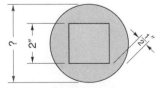

Figure 7–17

7–10   TRIANGLES

We shall deal with the following types of triangles: right, isosceles, equilateral, and scalene.

A *right triangle* is a triangle that has one right angle (Fig. 7–18).

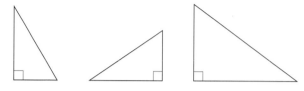

Figure 7–18

An *isosceles triangle* is a triangle that has two equal sides and two equal angles (Fig. 7–19).

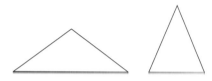

Figure 7–19

An *equilateral triangle* is a triangle in which the three sides are equal and the three angles are equal (Fig. 7–20).

Figure 7–20

A *scalene triangle* is a triangle that has no two sides equal (Fig. 7–21).

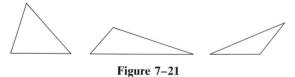

Figure 7–21

Areas of Triangles

To find the area of the rectangle in Fig. 7–22(*a*), multiply the altitude by the base; Area = 10 in. × 16 in. = 160 sq in. Drawing the diagonal *CF* divides the rectangle into two equal right triangles, *CEF* and *CDF*.

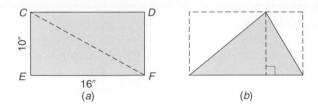

Figure 7–22

The area of each right triangle is $\frac{1}{2}$ of 160 sq in., or 80 sq in.; $\frac{1}{2} \times 16$ in. $\times$ 10 in. = 80 sq in. The area of the right triangle is therefore equal to $\frac{1}{2}$ the base times the altitude. Calling the altitude a and the base b, we find

$$\text{Area of right triangle} = \tfrac{1}{2} \times a \times b$$

A study of Fig. 7–22(b) shows that the same rule applies to that triangle. In fact, the following is a general formula for finding the area of any triangle when the base and the altitude are known or can be computed.

$$\text{Area} = \tfrac{1}{2} \times a \times b$$

It is customary to omit the sign of multiplication in mathematical formulas such as the preceding. (The expression ab always means a *times* b.) The formula is then written

$$\text{Area} = \tfrac{1}{2}ab$$

■ **Example 1**

Find the area of the right triangle shown in Fig. 7–23.

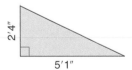

Figure 7–23

Solution $2'4'' = 28''$, the altitude
              $5'1'' = 61''$, the base

$$\text{Area} = \tfrac{1}{2}ab$$
$$= \tfrac{1}{2} \times 28 \text{ in.} \times 61 \text{ in.}$$
$$= 854 \text{ sq in.}$$
$$= 5.93 \text{ sq ft}$$

Practice check: Do exercise 15 on the left.

15. Find the area of a right triangle whose altitude is $6'8''$ and base is $4'6''$.

The answer is in the margin on page 222.

■ **Example 2**

Find the area of the right triangle shown in Fig. 7–24.

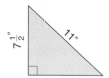

Figure 7–24

Solution   First find the length of the base; then proceed as in Example 1.

$$b = \sqrt{11^2 - (7\tfrac{1}{2})^2} = \sqrt{121 - 56.25} = \sqrt{64.75} = 8.05 \text{ in.}$$

$$\text{Area} = \tfrac{1}{2}ab$$

$$= \tfrac{1}{2} \times 7\tfrac{1}{2} \text{ in.} \times 8.05 \text{ in.}$$

$$= 30.19 \text{ sq in.}$$

Practice check: Do exercise 16 on the right.

16. Find the area of a right triangle whose altitude is $9\tfrac{3}{8}$ in. and hypotenuse is 13 in.

The answer is in the margin on page 223.

■ **Problems 7–10**

Use the space provided in the margin to find the areas of the right triangles in the following chart. Use the appropriate formula for the relation between the sides of a right triangle to determine the missing side: $c = \sqrt{a^2 + b^2}$; $a = \sqrt{c^2 - b^2}$; $b = \sqrt{c^2 - a^2}$.

	Right Triangles			
	Side a, Altitude	**Side b, Base**	**Side c, Hypotenuse**	**Area**
1.	5″	11″	_____	_____
2.	2′0″	2′6″	_____	_____
3.	$3\tfrac{1}{2}″$	$5\tfrac{1}{2}″$	_____	_____
4.	1′0″	_____	15″	_____
5.	8″	_____	1′5″	_____
6.	$7\tfrac{1}{2}″$	_____	$1′7\tfrac{1}{2}″$	_____
7.	$1′4\tfrac{1}{2}″$	$2′1\tfrac{1}{4}″$	_____	_____
8.	_____	2′0″	2′1″	_____
9.	_____	$10\tfrac{1}{2}″$	$3′1\tfrac{1}{2}″$	_____
10.	2′6″	2′6″	_____	_____
11.	1.732″	_____	2.000″	_____
12.	$3\tfrac{1}{2}″$	_____	7″	_____
13.	1″	_____	1.7″	_____
14.	_____	16″	34″	_____
15.	_____	8.66″	10″	_____
16.	6′	$2\tfrac{1}{2}′$	_____	_____
17.	9″	_____	$9\tfrac{3}{4}″$	_____
18.	_____	1′6″	3.00′	_____
19.	3′	1′3″	_____	_____
20.	_____	2″	$4\tfrac{1}{4}″$	_____

Answer to exercise 15

$6'8'' = 80''$, the altitude
$4'6'' = 54''$, the base

Area = $\frac{1}{2}ab$
   = $\frac{1}{2} \times 80'' \times 54''$
   = 2,160 sq in.
   = 15 sq ft

7–11   AREAS OF ISOSCELES TRIANGLES

In isosceles and equilateral triangles, computations are easily made because a perpendicular dropped from the **vertex,** that is, the point opposite the base, divides the triangles into two equal right triangles. In Fig. 7–25 the perpendicular divides the isosceles triangle into two equal right triangles; the base of each is 16 in., and the hypotenuse of each is 20 in. The altitude is found by the usual method for finding the altitude of a right triangle.

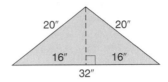

Figure 7–25

$$\text{Altitude} = \sqrt{20^2 - 16^2} = \sqrt{400 - 256} = \sqrt{144} = 12 \text{ in.}$$

The area of the isosceles triangle is found by the formula

17. Find the area of an isosceles triangle whose base is 18″ and sides are 15″.

The answer is in the margin on page 224.

$$\text{Area} = \frac{1}{2}ab$$
$$= \frac{1}{2} \times 12 \text{ in.} \times 32 \text{ in.}$$
$$= 192 \text{ sq in.}$$

Practice check: Do exercise 17 on the left.

■ Problems 7–11

Use the space provided in the margin to find the areas of the isosceles triangles in the following chart. The base and one of the equal sides are given. Use your knowledge of the relationship between the sides of a right triangle to find the altitude.

	Base	One of the Equal Sides	Altitude to the Base	Area
Isosceles Triangles				
1.	8″	5″	————	————
2.	2′6″	1′5″	————	————
3.	1′2″	2′1″	————	————
4.	3′0″	2′6″	————	————
5.	2′0″	2′0″	————	————
6.	1′4″	10″	————	————
7.	2′1″	2′6½″	————	————
8.	8′0″	5′0″	————	————

Isosceles Triangles

	Base	One of the Equal Sides	Altitude to the Base	Area
9.	19.2″	2′0″	_____	_____
10.	7.8″	7.8″	_____	_____
11.	38.4″	20.8″	_____	_____
12.	5′6″	5′6″	_____	_____
13.	5′3″	7′6″	_____	_____
14.	10.4 yd	6.3 yd	_____	_____
15.	$22\frac{1}{2}′$	$13\frac{3}{4}′$	_____	_____
16.	2′9″	8′	_____	_____
17.	44′0″	28′0″	_____	_____
18.	17.2″	17.2″	_____	_____
19.	13.2 yd	14.6 yd	_____	_____
20.	$56\frac{3}{8}″$	$31\frac{5}{16}″$	_____	_____

Answer to exercise 16

$$b = \sqrt{13^2 - (9\tfrac{3}{8})^2}$$
$$= \sqrt{169 - 87.89}$$
$$= \sqrt{81.11}$$
$$= 9 \text{ in.}$$

$$\text{Area} = \tfrac{1}{2}ab$$
$$= \tfrac{1}{2} \times 9\tfrac{3}{8} \text{ in.} \times 9 \text{ in.}$$
$$= 42.19 \text{ sq in.}$$

7-12 AREAS OF SCALENE TRIANGLES USING HERO'S FORMULA

In a scalene triangle, the area is found by the same formula

$$\text{Area} = \tfrac{1}{2}ab$$

provided the base and altitude are known. If the altitude is not known but the three sides are known, then the area can be found by using a special method called **Hero's formula,** shown here.

$$\text{Area of a triangle} = \sqrt{s(s - a)(s - b)(s - c)}$$

Denote the three sides by the letters a, b, and c; s denotes half the sum of the sides (semi-perimeter). The formula reads as follows: Area = the square root of the product of s, $(s - a)$, $(s - b)$, and $(s - c)$. In Fig. 7-26 call the 20-in. side a, the 14-in. side b, and the 12-in side c. Since the triangle is not known to be a right triangle, the formula $c^2 = a^2 + b^2$ cannot be used.

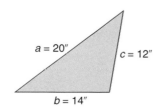

Figure 7-26

Answer to exercise 17

Altitude $= \sqrt{15^2 - 9^2}$
$\quad\quad\quad = \sqrt{225 - 81}$
$\quad\quad\quad = \sqrt{144}$
$\quad\quad\quad = 12$ in.

Area $= \frac{1}{2}ab$
$\quad\quad = \frac{1}{2} \times 12$ in. $\times 18$ in.
$\quad\quad = 108$ sq in.

■ Example

Find the area of the scalene triangle in Fig. 7–26.

Solution $20 + 14 + 12 = 46$ $s = \frac{1}{2}$ of $46 = 23$

Now subtract each side in turn from this half sum (*s*).

$$s - a = 23 - 20 = 3$$
$$s - b = 23 - 14 = 9$$
$$s - c = 23 - 12 = 11$$

Then multiply together the half sum (*s*) and the three quantities obtained by taking each side in turn from the half sum: $23 \times 3 \times 9 \times 11 = 6{,}831$. The square root of this product is the area of the triangle: $\sqrt{6{,}831} = 82.6$. The area is 82.6 sq in.

$$s = \tfrac{1}{2}\text{ sum of sides} = \frac{20 + 14 + 12}{2} = \frac{46}{2} = 23$$

$$\begin{aligned}
\text{Area} &= \sqrt{s(s - a)(s - b)(s - c)} \\
&= \sqrt{23(23 - 20)(23 - 14)(23 - 12)} \\
&= \sqrt{23 \times 3 \times 9 \times 11} \\
&= \sqrt{6{,}831} \\
&= 82.6
\end{aligned}$$

18. Find the area of a scalene triangle whose sides are 7′, 9′, and 10′.

The answer is in the margin on page 226.

Practice check: Do exercise 18 on the left.

■ Problems 7–12

Use the space provided below to find the area of the following triangles using Hero's formula.

1.

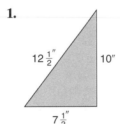

2.

3.

10″ 7″

5″

4.

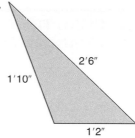

2′6″

1′10″

1′2″

5.

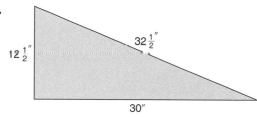

$32\frac{1}{2}''$

$12\frac{1}{2}''$

30″

6.

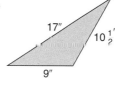

17″ $10\frac{1}{2}''$

9″

7.

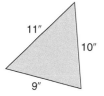

11″

10″

9″

8.

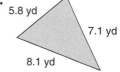

5.8 yd

7.1 yd

8.1 yd

9.

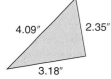

4.09″ 2.35″

3.18″

10.

$1\frac{1}{4}'$

$3\frac{1}{4}'$

3.00′

Answer to exercise 18

$$s = \frac{7 + 9 + 10}{2} = 13$$

$$
\begin{aligned}
A &= \sqrt{13(13-7)(13-9)(13-10)} \\
&= \sqrt{13(6)(4)(3)} \\
&= \sqrt{936} \\
&= 30.59 \text{ sq ft}
\end{aligned}
$$

7–13   USING THE CALCULATOR TO SOLVE TRIANGLE PROBLEMS

The calculator demonstration given here depends on your calculator having a square key (x^2), a square root key ($\sqrt{\ }$), a memory key ($x \rightarrow M$), a memory addition key ($M+$), and a memory recall key (RM). If your calculator does not have these keys, read your instruction manual, and adapt the examples accordingly.

■ Example 1

Using a calculator, find the hypotenuse of the right triangle in Fig. 7–27. Round the answer to two decimal places.

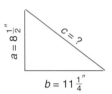

Figure 7–27

Solution   To solve this problem, use the Pythagorean theorem and solve for the hypotenuse c; that is, $c = \sqrt{a^2 + b^2}$.

The Display Shows

	Display	
Turn on the calculator	$0.$	
Enter 8	$8.$	
Press +	$8.$	
Enter 1	$1.$	
Press ÷	$1.$	
Enter 2	$2.$	
Press =	8.5	
Press x^2	72.25	
Press $x \rightarrow M$	72.25	The 72.25 is now in the calculator's memory.
Enter 11	$11.$	
Press +	$11.$	
Enter 1	$1.$	
Press ÷	$1.$	
Enter 4	$4.$	
Press =	11.25	
Press x^2	126.5625	
Press $M+$	126.5625	This has now been added to the 72.25 in the memory.
Press RM	198.8125	The new amount is in the memory.
Press $\sqrt{\ }$	14.10008865	(answer)

Round off the length of the hypotenuse to 14.10 in.

Practice check: Do exercise 19 on the left.

19. Using a calculator, find the hypotenuse of a right triangle whose sides are 12.4″ and 15.6″.

The answer is in the margin on page 228.

■ **Example 2**

Using a calculator, find the altitude of the right triangle in Fig. 7–28.

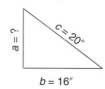

Figure 7–28

Solution   To solve this problem, use the Pythagorean theorem and solve for the altitude a; that is, $a = \sqrt{c^2 - b^2}$.

The Display Shows

Turn on the calculator	0.
Enter 20	20.
Press x^2	400.
Press $x \rightarrow$ M	400.
Enter 16	16.
Press x^2	256.
Press +/−	−256.
Press M+	−256.
Press RM	144.
Press $\sqrt{}$	12. (answer)

The length of the altitude is 12 in.

Practice check: Do exercise 20 on the right.

20. Using a calculator, find the altitude of a right triangle whose hypotenuse is 10.2′ and side is 9.1′.

The answer is in the margin on page 229.

■ **Example 3**

Using a calculator, find the area of the right triangle shown in Fig. 7–29. Round the answer to two decimal places.

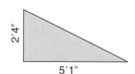

Figure 7–29

Solution   In solving for the area of a triangle when the base and altitude are known or can be found, use the formula $A = 0.5 \times a \times b$. Convert 2 ft 4 in. to 28 in. and 5 ft 1 in. to 61 in. before solving this problem.

The Display Shows

Turn on the calculator	0.
Enter 0.5	0.5
Press ×	0.5
Enter 28	28.

Answer to exercise 19

19.93″

The Display Shows

Press ×	14.
Enter 61	61.
Press =	854.
Press ÷	854.
Enter 144	144.
Press =	5.930555556 (answer)

Round off the area to 5.93 sq ft.

21. Using a calculator, find the area of a right triangle whose sides are 17′6″ and 20′4″.

The answer is in the margin on page 230.

Explanation   Because the numbers entered were converted to inches, the 854 means 854 sq in. It usually makes sense to reduce numbers to more manageable amounts whenever possible, so converting the answer to square feet is a good idea here. In every square foot there are 12 in. by 12 in., or 144 sq in. Dividing 854 by 144 will give the proper answer in square feet.

Practice check: Do exercise 21 on the left.

■ **Example 4**

Using a calculator, find the area of the right triangle shown in Fig. 7–30. Round the answer to two decimal places.

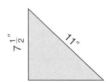

Figure 7–30

Solution   Use the formula for the area of a triangle; that is, $A = 0.5 \times a \times b$. To first solve for the unknown base, use the Pythagorean theorem, $b = \sqrt{c^2 - a^2}$.

The Display Shows

Turn on the calculator	0.
Enter 11	11.
Press x^2	121.
Press −	121.
Enter 7.5	7.5
Press x^2	56.25
Press =	64.75
Press √	8.04673847
Press $x \rightarrow$ M	8.04673847
Enter 0.5	0.5
Press ×	0.5
Enter 7.5	7.5
Press ×	3.75
Press RM	8.04673847
Press =	30.17526926 (answer)

22. Using a calculator, find the area of a right triangle whose hypotenuse is 6.25′ and side is 4.3′.

The answer is in the margin on page 230.

Round off the area to 30.18 in.

Practice check: Do exercise 22 on the left.

■ **Example 5**

Using a calculator, find the area of the scalene triangle in Fig. 7–31. Round the answer to one decimal place.

Answer to exercise 20

4.6′

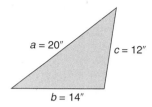

Figure 7–31

Solution Use Hero's formula to solve this problem.

$$\text{Area} = \sqrt{s(s-a)(s-b)(s-c)}$$

$$s = 0.5(a + b + c)$$

This problem requires a lot of keying in on the calculator, and you should use the memory to hold subtotals while the problem progresses. It is not difficult to do if the problem solution is thoroughly understood. (You may find that the work is easier to key in if some of the problem is worked out on paper and the calculator is used to do the more difficult operations.)

The Display Shows

Turn on the calculator	0.	
Enter 20	20.	a
Press +	20.	
Enter 14	14.	b
Press +	34.	
Enter 12	12.	c
Press =	46.	$a + b + c$
Press ×	46.	
Enter 0.5	0.5	
Press =	23.	$0.5(a + b + c) = s$
Press $x \rightarrow$ M	23.	s is now in memory.
Press −	23.	
Enter 20	20.	
Press =	3.	$s - a$
Press ×	3.	
Press MR	23.	
Press =	69.	$s(s - a)$
Press $x \rightarrow$ M	69.	$s(s - a)$ is now in memory, remembered as the value of s.
Enter 23	23.	
Press −	23.	
Enter 14	14.	
Press =	9.	$s - b$
Press ×	9.	
Press MR	69.	

Answer to exercise 21

177.89 sq in.

Answer to exercise 22

9.75 sq ft

---

23. Using a calculator, find the area of a scalene triangle whose sides are 43″, 15″, and 30″.

The answer is in the margin on page 232.

The Display Shows

Press =	621.	$s(s - a)(s - b)$
Press $x \rightarrow$ M	621.	$s(s - a)(s - b)$ is now in memory, remembered as the value of s.
Enter 23	23.	
Press −	23.	
Enter 12	12.	
Press =	11.	$s - c$
Press ×	11.	
Press MR	621.	
Press =	6831.	$s(s - a)(s - b)(s - c)$
Press √	82.64986388	(answer)

Round off the area to 82.6 sq in.

Practice check: Do exercise 23 on the left.

■ Test on Areas of Triangles

Find the areas of the following triangles. Make sure that you include the units of measure in your answer. Try to estimate the area of the triangle before doing the actual computation in the space provided below.

1.

1′9″

2′11″

2.

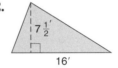

$7\frac{1}{2}'$

16′

3.

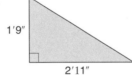

3.32′

45′

4.

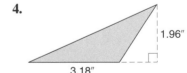

1.96″

3.18″

5.

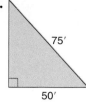

75'

50'

6.

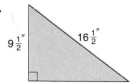

$9\frac{1}{2}''$ $16\frac{1}{2}''$

7.

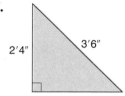

2'4"   3'6"

8.

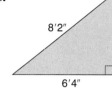

8'2"

6'4"

9.

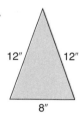

12"   12"

8"

10.

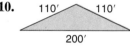

110'   110'

200'

11. $2\frac{1}{2}''$ $2\frac{1}{2}''$

$3\frac{1}{8}''$

12.

18"   18"

18"

7–14   ANGLES IN TRIANGLES

Angles are defined as follows.

A **right angle** is a 90-degree angle [Fig. 7–32(*a*)].

An **acute angle** is an angle that is less than 90 degrees [Fig. 7–32(*b*)].

An **obtuse angle** is an angle that is more than 90 degrees but less than 180 degrees [Fig. 7–32(*c*)].

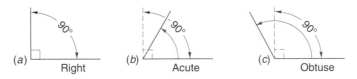

Figure 7–32

Rule _____

The sum of the three angles in any triangle is equal to 180 degrees.

In a right triangle, one angle is 90 degrees. The sum of the two acute angles is $180 - 90 = 90$ degrees.

In an isosceles triangle, the angles opposite the equal sides are equal.

In an equilateral triangle, since all the sides are equal, the angles are all equal; each angle $= \frac{180}{3} = 60$ degrees.

Angles are commonly measured in **degrees, minutes,** and **seconds.**

$$1 \text{ degree} = 60 \text{ minutes}$$

$$1 \text{ minute} = 60 \text{ seconds}$$

The symbols commonly used for indicating degrees, minutes, and seconds are degrees (°), minutes ('), seconds ("). Write 27 degrees, 18 minutes, and 30 seconds as 27°18'30".

Angles in degrees, minutes, and seconds can be converted to degrees and decimal parts of a degree for use in CNC applications. To convert, divide the minute part of the given angle by 60 and add it to the degree part. Then divide the seconds part by 3,600 and add it to the total.

The 27 degree 18 minute 30 second angle (27°18'30") can be converted to degrees and decimal parts of a degree by dividing the 18' by 60 (which is 0.3), dividing the 30" by 3,600 (which is 0.008$\overline{3}$), and adding these decimal parts to the 27 degrees, that is, $27 + 0.3 + 0.008\overline{3}$ which equals 27.308$\overline{3}$ degrees. The 27°18'30" is the same angle as 27.308$\overline{3}$°.

If your interest in angle measure is in degrees and decimal parts of a degree, then do the following examples in decimal form rather than in D.MS form as presented.

It is customary to label the vertices (the plural of *vertex*) of a triangle *A*, *B*, and *C*. The angle at each vertex is then called angle *A*, angle *B*, and angle *C*, respectively.

■ Example 1

In the right triangle of Fig. 7–33, find the value of angle *B*.

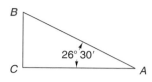

Figure 7–33

Solution
$$A + B = 90° = 89°60'$$
$$\underline{A \qquad\qquad\quad = 26°30'}$$
$$B \qquad\qquad = 63°30'$$

Explanation   Since angles *A* and *B* together are 90°, subtract 26°30′ from 90°, leaving 63°30′ for angle *B*. In order to subtract, change 90° to 89°60′.

Practice check: Do exercise 24 on the right.

■ Example 2

In the isosceles triangle of Fig. 7–34, compute the base angles *B* and *C*.

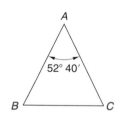

Figure 7–34

Solution
$$A + B + C = 180° = 179°60'$$
$$\underline{A \qquad\qquad\qquad\quad = \;\; 52°40'}$$
$$B + C \qquad\qquad = 127°20'$$

$$\text{Each base angle} = \frac{127°20'}{2} = \frac{126°80'}{2} = 63°40'$$

Explanation   Since the sum of the three angles is 180°, subtract 52°40′ from 180° (changed to 179°60′), leaving 127°20′ for the sum of the two base angles *B* and *C*. Each base angle is one-half of 127°20′. Changing 127°20′ to 126°80′ simplifies the division.

Practice check: Do exercise 25 on the right.

24. In the following right triangle, find the value of angle *A*.

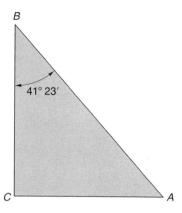

The answer is in the margin on page 235.

25. In the following isoceles triangle, compute the base angles *B* and *C*.

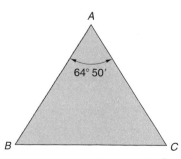

The answer is in the margin on page 235.

■ **Example 3**

In the scalene triangle of Fig. 7–35, angle $A = 52°40'30''$ and angle $B = 27°30'50''$. Find angle C.

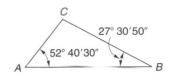

Figure 7–35

---

26. In the following scalene triangle, angle $C = 85°40'30''$ and angle $B = 24°10'40''$. Find angle A.

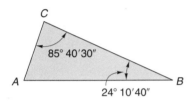

The answer is in the margin on page 238.

Solution

$$A = 52°40'30''$$
$$B = 27°30'50''$$
$$\overline{A + B = 79°70'80''}$$
$$= 80°11'20''$$

Sum of three angles $= A + B + C = 180° = 179°59'60''$
$$\underline{A + B \qquad\qquad = \ \ 80°11'20''}$$
$$C \qquad = \ \ 99°48'40''$$

Explanation Since the sum of the three angles is 180°, add the two given angles and subtract their sum from 180°.

Practice check: Do exercise 26 on the left.

■ **Example 4**

In the isosceles triangle of Fig. 7–36, the base angles are $47°39'51''$ each. Find the vertex angle, that is, the angle opposite the base.

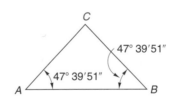

Figure 7–36

---

27. In the following isoceles triangle, the base angles are $34°56'8''$ each. Find the vertex angle.

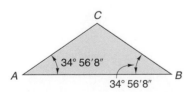

The answer is in the margin on page 239.

Solution

$$\text{Base angle} = \ 47°39' \ 51''$$
$$\text{Base angle} = \ 47°39' \ 51''$$
$$\overline{\text{Sum of base angles} = \ 94°78'102''}$$
$$= \ 95°19' \ 42''$$

Sum of three angles $= 180° = 179°59' \ 60''$
$$\underline{\text{Sum of base angles} = \ 95°19' \ 42''}$$
$$\text{Vertex angle} = \ 84°40' \ 18''$$

Explanation The base angles in an isosceles triangle are equal. To obtain the vertex angle, subtract the sum of the base angles from 180°.

Practice check: Do exercise 27 on the left.

■ Problems 7–14

Use the space provided below and in the margin to find the missing angles of the triangles in the following chart.

Triangles		
Angle A	**Angle B**	**Angle C**
1. 30°	60°	_____
2. 45°	45°	_____
3. 20°	_____	80°
4. _____	50°	40°
5. 27°10′	52°20′	_____
6. _____	70°27′	25°15′
7. 40°45′	_____	40°45′
8. 120°30′	15°30′	_____
9. 75°50′	25°28′	_____
10. 10°37′	_____	47°29′
11. 40°27′10″	_____	56°12′20″
12. _____	36°47′35″	72°25′42″
13. 33°50′09″	_____	16°20′33″
14. _____	66°30′09″	22°42′20°
15. 8°29′40″	_____	70°30′45″
16. _____	25°6′56″	2°46′06″

Answer to exercise 24

$$89°60'$$
$$- 41°23'$$
$$\overline{48°37'}$$

Answer to exercise 25

$$A + B + C = 179°60'$$
$$A \qquad\quad = \quad 64°50'$$
$$\overline{B + C = 115°10'}$$

$$B \text{ or } C = \frac{115°10'}{2} = \frac{114°70'}{2}$$
$$= 57°35'$$

7–15   USING THE CALCULATOR FOR ANGLE MEASURE IN D.MS

To do problems involving degrees, minutes, and seconds on the calculator, it is necessary on many calculators to do the problems in degrees and decimal parts of a degree and then convert back to degrees, minutes, and seconds by using the → D.MS function on the calculator. There may be some slight round-off errors with the calculator, because many calculators have only 10 decimal places.

To change minutes to a decimal part of a degree, divide minutes by 60.

To change seconds to a decimal part of a degree, divide seconds by 3,600.

■ Example 1

In the right triangle of Fig. 7–37, find the value of angle B.

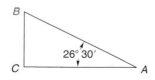

Figure 7–37

Solution   Angle *B* is the complement of angle *A*; that is, the sum of the two acute angles of a right triangle is 90° ($A + B = 90°$).

28. In the following right triangle, find the value of angle *A*.

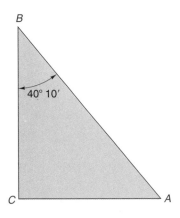

B

40° 10′

C   A

The answer is in the margin on page 239.

The Display Shows

	The Display Shows
Turn on the calculator	0 .
Enter 30	30 . (minutes)
Press ÷	30 . ⎫ Change minutes to deci-
Enter 60	60 . ⎭ mal parts of a degree.
Press =	0 .5
Press +	0 .5
Enter 26	26 .
Press =	26 .5
Press +/−	−26 .5
Press +	−26 .5
Enter 90	90 .
Press =	63 .5
Press 2nd F	63 .5
Press → D.MS	63 .300000

Degrees, minutes, and seconds are read from the calculator display thus: The numbers to the left of the decimal point indicate degrees, the first two numbers to the right of the decimal point indicate minutes, the next two numbers represent seconds, and the last two numbers indicate hundredths of a second.

The answer to this problem is 63°30′.

Practice check: Do exercise 28 on the left.

■ Example 2

In the isosceles triangle of Fig. 7–38, compute the base angles *B* and *C*.

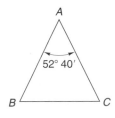

A

52° 40′

B   C

Figure 7–38

Solution   To solve this problem, we must remember that the base angles of an isosceles triangle are equal and that the sum of the three angles of a triangle is 180°. If we subtract the vertex angle from 180°, we then have the sum of the base angles. By dividing this sum by 2, we have the value of each base angle.

The Display Shows

	The Display Shows
Turn on the calculator	0 .
Enter 40	40 .
Press ÷	40 .
Enter 60	60 .
Press =	0 .666666666
Press +	0 .666666666

The Display Shows	
Enter 52	52 .
Press =	52 .66666667
Press +/−	−52 .66666667
Press +	−52 .66666667
Enter 180	180 .
Press =	127 .3333333
Press ÷	127 .3333333
Enter 2	2 .
Press =	63 .66666667
Press 2nd F	63 .66666667
Press → D.MS	63 .400000

We read this answer as 63°40′.

Practice check: Do exercise 29 on the right.

■ **Example 3**

In the scalene triangle of Fig. 7–39, angle $A = 52°40′30″$ and angle $B = 27°30′50″$. Find angle C.

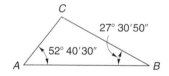

Figure 7–39

Solution   In this example, we have two angles in minutes and seconds that must be converted to decimal parts of a degree. We must use the memory of the calculator to hold one of these angles while we work on the other. To change the seconds to a decimal part of a degree, we must divide by 3,600. (60 parts of a degree is a minute, and 60 parts of a minute is a second.)

The sum of the three angles of the triangle is 180°. We are going to add the two angles we know and subtract their sum from 180° to solve our problem.

The Display Shows	
Turn on the calculator	0 .
Enter 50	50 .
Press ÷	50 .
Enter 3600	3600 .
Press =	0 .013888888
Press x → M	0 .013888888
Enter 30	30 .
Press ÷	30 .
Enter 60	60 .
Press =	0 .5
Press M+	0 .5
Enter 27	27 .
Press M+	27 .

29. In the following isosceles triangle, compute the base angles B and C.

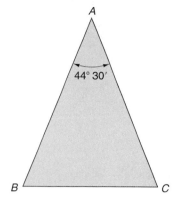

The answer is in the margin on page 239.

$$C = 85°40'30''$$
$$B = 24°10'40''$$
$$C + B = 109°50''70''$$
$$= 109°51'10''$$

$$A + B + C = 179°59'60''$$
$$C + B = 109°51'10''$$

$$\text{Angle } A = 70°\ 8'50''$$

The angle 27 degrees 30 minutes 50 seconds is now in decimal form and is saved in memory. We will now change $52°40'30''$ to decimal form and add it to the memory also.

	The Display Shows
Enter 30	30.
Press ÷	30.
Enter 3600	3600.
Press =	0.008333333
Press M+	0.008333333
Enter 40	40.
Press ÷	40.
Enter 60	60.
Press =	0.666666666
Press M+	0.666666666
Enter 52	52.
Press M+	52.

30. In the following scalene triangle, angle $A = 22°30'10''$ and angle $C = 47°20'50''$. Find angle B.

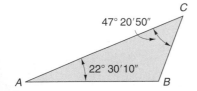

The answer is in the margin on page 240.

The memory now has the sum of the two angles in decimal form. We are going to subtract the contents of memory from 180 to get the decimal form of the answer.

	The Display Shows
Enter 180	180.
Press −	180.
Press MR	80.18888889
Press =	99.81111111
Press 2nd F	99.81111111
Press → D.MS	99.484000

Read this number as $99°48'40''$.

Practice check: Do exercise 30 on the left.

■ Problems 7–15

Use your calculator to do Problem Set 7–14, Problems 5–16. Check the answers you get with the D.MS key on the calculator against the results you wrote on page 235.

■ Self-Test

Use the space provided below to do the following problems. Check your answers with the answers in the back of the book.

1. Find the area of a rectangle whose width is 8″ and whose length is 10″.

2. Find the perimeter of the rectangle in Problem 1.

3. Find the missing dimension of the rectangle of Fig. 7–40.

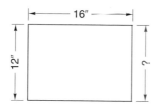

Figure 7–40

4. Find the width of a rectangle whose area is 234 sq in. and whose length is 18 in.

5. What is the square of 40?

6. What is the length of each side of a square whose area is 1,089 sq in.?

7. The legs of a right triangle are 6 in. and 8 in. What is the length of the hypotenuse?

8. One of the acute angles of a right triangle is 39°. What is the size of the other acute angle?

9. The hypotenuse of a right triangle is 12.5 in. and the altitude is 12 in. What is its area?

10. Find the area of an equilateral triangle whose sides are each 10 in.

Answer to exercise 30

Angle $B = 110°9'$

■ Chapter Test

Use the space provided below to do the following problems involving rectangles and triangles. Try to estimate the answer before doing the actual computation. Include the correct units of measure in your answer.

1. Find the area of a rectangle whose length is 15 in. and whose width is $10\frac{1}{2}$ in.

2. Find the area of a rectangle whose length is 2 ft 6 in. and whose width is 1 ft 8 in.

3. Find the area of a rectangle whose length is $30\frac{1}{2}$ in. and whose width is $25\frac{3}{4}$ in.

4. Find the missing dimension in the rectangle of Fig. 7–41.

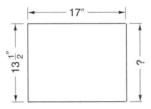

Figure 7–41

5. Find the missing dimension in the composite rectangle of Fig. 7–42.

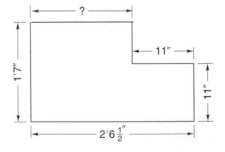

Figure 7–42

6. Find the perimeter of a rectangle whose length is 28 in. and whose width is 15 in.

7. Find the perimeter of a rectangle whose length is 2 ft 5½ in. and whose width is 1 ft 3¾ in.

8. Find the width of a rectangle whose area is 144 sq in. and whose length is 16 in.

9. Find the length of a rectangle whose area is 256 sq in. and whose width is 2 ft 8 in.

10. Find the square of 27.

11. Find the square of 32½.

12. Find the square root of 225.

13. Find the square root of 8,649.

14. Find the diagonal of a square 10 in. on a side.

15. Find the diagonal of a rectangle whose length is 3 ft 9 in. and whose width is 2 ft 0 in.

16. Find the hypotenuse of a right triangle whose altitude is 1 ft 4 in. and whose base is 1 ft 0 in. What is its perimeter?

17. Find the base of a right triangle whose altitude is 10 in. and whose hypotenuse is 26 in. What is its perimeter?

18. Find the altitude of a right triangle whose base is 7 in. and whose hypotenuse is 25 in.

19. Find the area of a triangle whose base is 33 in. and whose altitude is 31 in.

20. Find the area of a right triangle whose altitude is $10\frac{1}{2}$ in. and whose hypotenuse is $17\frac{1}{2}$ in.

21. Find the area of a square whose edge is $15\frac{7}{8}$ in.

22. Find the area of an isosceles triangle whose base is 40 in. and where one of the equal sides is 25 in.

23. Find the area of an equilateral triangle all of whose sides are 25 in.

24. If two angles of a triangle are 45° and 55°, how large is the remaining angle?

25. If two angles of a triangle are 50°27′ and 44°52′, how large is the third angle?

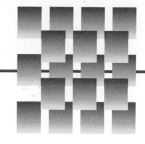

CHAPTER 8

Regular Polygons and Circles

8–1   DEFINITIONS

A *polygon* is a closed figure in a plane. All polygons have at least three sides. The type of polygon that occurs most often in practical work is the *regular polygon.* In a regular polygon all the sides are equal in length, and all the angles formed by any two adjacent sides have the same measure. The nine shapes in Fig. 8–1 depict regular polygons.

The area of a regular polygon, its perimeter, the length of its diagonal, and the perpendicular distance between any two sides are all related to one another in length. If one is known, the others can be computed by the application of the laws for right triangles. Considerable labor is involved in this computation, but the work may be greatly shortened by means of certain ratios, called factors or constants, that have been derived for the purpose. These ratios usually appear in a *Table of Constants* such as the one that follows on page 244.

The formulas for the regular polygons that appear in the Table of Constants

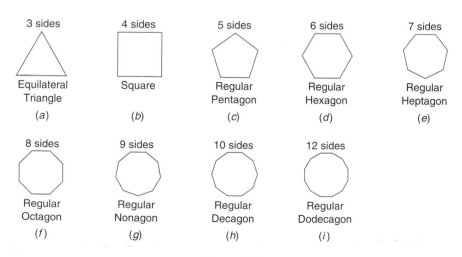

Figure 8–1

Table of Constants

Equilateral Triangle

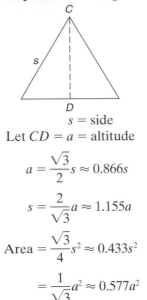

s = side

Let $CD = a$ = altitude

$$a = \frac{\sqrt{3}}{2}s \approx 0.866s$$

$$s = \frac{2}{\sqrt{3}}a \approx 1.155a$$

$$\text{Area} = \frac{\sqrt{3}}{4}s^2 \approx 0.433s^2$$

$$= \frac{1}{\sqrt{3}}a^2 \approx 0.577a^2$$

Square

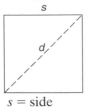

s = side

d = diagonal

$$d = \sqrt{2}s \approx 1.414s$$

$$s = \frac{\sqrt{2}}{2}d \approx 0.707d$$

$$\text{Area} = s^2$$
$$= 0.5d^2$$

Regular Hexagon

Let $AB = f$ = distance across flats

Let $DE = d$ = distance across corners (diagonal)

s = side

$$f = \frac{\sqrt{3}}{2}d \approx 0.866d \qquad f = \sqrt{3}s \approx 1.732s$$

$$d = 2s \qquad\qquad d = \frac{2}{\sqrt{3}}f \approx 1.155f$$

$$s = 0.5d \qquad\qquad s = \frac{1}{\sqrt{3}}f \approx 0.577f$$

$$\text{Area} = \frac{\sqrt{3}}{2}f^2 \approx 0.866f^2$$

$$= \tfrac{3}{8}\sqrt{3}d^2 \approx 0.650d^2$$

$$= \tfrac{3}{2}\sqrt{3}s^2 \approx 2.598s^2$$

Regular Octagon

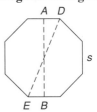

Let $AB = f$ = distance across flats

Let $DE = d$ = distance across corners (diagonal)

s = side

$$f \approx 0.924d \qquad f \approx 2.414s$$
$$d \approx 1.083f \qquad d \approx 2.613s$$
$$s \approx 0.414f \qquad s \approx 0.383d$$

$$\text{Area} \approx 0.828f^2$$

$$= \frac{\sqrt{2}}{2}d^2 \approx 0.707d^2$$

$$\approx 4.828s^2$$

NOTE: The symbol $\approx$ means "approximately equal to."

are accurate to three decimal places. Greater accuracy requires a knowledge of trigonometry.

8–2   EQUILATERAL TRIANGLES

An *equilateral triangle* is a regular polygon of three sides. All the sides are equal
to one another. The essential parts of the equilateral triangle are the side, the
altitude, the perimeter, and the area, If one of these is known, the other parts can
be computed by referring to the equilateral triangle shown in the Table of Constants.

■ **Example 1**

Find the altitude a of the equilateral triangle shown in Fig. 8–2.

Figure 8–2

Solution $a \approx 0.866s \approx 0.866 \times 6 \approx 5.196$ in.

Explanation   From the Table of Constants for the equilateral triangle, select the
formula that gives the altitude, $a = 0.866s$. Substituting the value of s in this formula
gives the altitude.

Practice check: Do exercise 1 on the right.

Note: Even though the equal sign (=) is used in place of the approximately
equal sign ($\approx$) during the remainder of the chapter, the assumption is that the
values are approximate when not exact.

■ **Example 2**

Find the side of the equilateral triangle shown in Fig. 8–3.

Figure 8–3

Solution $s = 1.155a = 1.155 \times 8 = 9.24$ in.

Explanation   From the Table of Constants for the equilateral triangle, select the
formula that gives the side in terms of the altitude, $s = 1.155a$. Substituting the
value of a (8 in.) in this formula gives the side.

Practice check: Do exercise 2 on the right.

1. Find the altitude of the equilateral triangle whose side is
25 in.

The answer is in the margin on
page 247.

2. Find the side of the equilateral triangle whose altitude is
31 in.

The answer is in the margin on
page 247.

■ Example 3

Find the area of the equilateral triangle shown in Fig. 8–4.

Figure 8–4

---

3. Find the area of the equilateral triangle whose side is 25 cm.

The answer is in the margin on page 249.

Solution   Area $= 0.433s^2 = 0.433 \times 12 \times 12 = 62.352$ sq in.

Explanation   From the Table of Constants for the equilateral triangle, select the formula that gives the area in terms of the side, Area $= 0.433s^2$. Substituting the value of s (12 in.) in this formula gives the area.

Practice check: Do exercise 3 on the left.

■ Example 4

Find the area of the equilateral triangle shown in Fig. 8–5.

Figure 8–5

---

4. Find the area of the equilateral triangle whose altitude is 15 ft.

The answer is in the margin on page 249.

Solution   Area $= 0.577a^2 = 0.577 \times 10 \times 10 = 57.7$ cm^2

Explanation   From the Table of Constants for the equilateral triangle, select the formula that gives the area in terms of the altitude, Area $= 0.577a^2$. Substituting the value of a (10 cm) in this formula gives the area.

Practice check: Do exercise 4 on the left.

■ Problems 8–2

Use the Table of Constants to find the missing information about equilateral triangles in the following chart. Give the answer to the nearest hundredth. Try estimating the answer before solving the problem.

Equilateral Triangles			
	Side	**Altitude**	**Area**
1.	10″	————	————
2.	8″	————	————
3.	$4\frac{1}{2}″$	————	————
4.	5.5 cm	————	————
5.	6.25 cm	————	————

Equilateral Triangles			
	Side	**Altitude**	**Area**
6.	$20\frac{3}{4}$ ft	————	————
7.	9.75 ft	————	————
8.	0.86 m	————	————
9.	————	5″	————
10.	————	11 cm	————
11.	————	$3\frac{1}{2}$″	————
12.	————	4.7 m	————
13.	————	7.75″	————
14.	————	13.1 ft	————
15.	————	$8\frac{1}{4}$ cm	————
16.	————	100 m	————

Answer to exercise 1

$a \approx 0.866s \approx 0.866 \times 25$
≈ 21.65 in.

Answer to exercise 2

$s = 1.155a = 1.155 \times 31$
$= 35.805$ in.

8–3 SQUARES

A square is a regular polygon of four sides. All the sides are equal to one another, and all the angles are 90°. The essential parts of a square are its side, its perimeter, its diagonal, and its area. If one of these is known, the other parts can be computed by referring to the square shown in the Table of Constants.

■ Example 1

Find the diagonal of the square shown in Fig. 8–6.

Figure 8–6

Solution $d = 1.414s = 1.414 \times 5 = 7.070$ in.

Explanation   From the Table of Constants for the square, select the formula that gives the diagonal in terms of the side, $d = 1.414s$. Substituting the value of s (5 in.) in this formula gives the diagonal.

Practice check: Do exercise 5 on the right.

5. Find the diagonal of the square whose side is 15 in.

The answer is in the margin on page 249.

■ Example 2

Find the side of the square shown in Fig. 8–7.

Figure 8–7

Solution $s = 0.707d = 0.707 \times 8 = 5.656$ in.

6. Find the side of the square whose diagonal is 10 ft.

The answer is in the margin on page 252.

Explanation   From the Table of Constants for the square, select the formula that gives the side in terms of the diagonal, $s = 0.707d$. Substituting the value of d (8 in.) in this formula gives the side.

Practice check: Do exercise 6 on the left.

■ **Example 3**

Find the area of the square shown in Fig. 8–8.

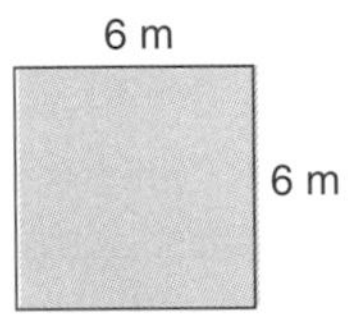

6 m
6 m

Figure 8–8

Solution   Area $= s^2 = 6 \times 6 = 36$ m^2

7. Find the area of the square whose side is 30 cm.

The answer is in the margin on page 252.

Explanation   From the Table of Constants for the square, select the formula that gives the area in terms of the side, Area $= s^2$. Substituting the value of s (6 m) in this formula gives the area.

Practice check: Do exercise 7 on the left.

■ **Example 4**

Find the area of the square shown in Fig. 8–9.

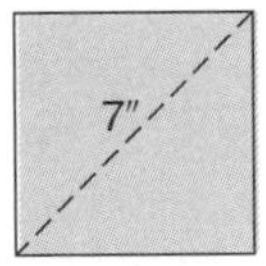

7"

Figure 8–9

Solution   Area $= 0.5d^2 = 0.5 \times 7 \times 7 = 24.5$ sq in.

8. Find the area of the square whose diagonal is 11 yd.

The answer is in the margin on page 252.

Explanation   From the Table of Constants for the square, select the formula that gives the area in terms of the diagonal, Area $= 0.5d^2$. Substituting the value of d (7 in.) in this formula gives the area.

Practice check: Do exercise 8 on the left.

■ **Problems 8–3**

Use the Table of Constants to find the missing information about squares in the following chart. Give your answer to the nearest hundredth. Try estimating the answer before solving the problem.

Squares			
	Side	**Diagonal**	**Area**
1.	7 m	————	————
2.	15″	————	————
3.	$2\frac{1}{2}″$	————	————
4.	$3\frac{1}{4}″$	————	————
5.	3.82 cm	————	————
6.	10 yd	————	————
7.	20.1 m	————	————
8.	9.25 ft	————	————
9.	————	9 cm	————
10.	————	$3\frac{1}{2}″$	————
11.	————	$4\frac{1}{4}″$	————
12.	————	6.5 cm	————
13.	————	3.05″	————
14.	————	2.65 m	————
15.	————	100 ft	————
16.	————	0.15 yd	————

Answer to exercise 3

Area = 0.433 × 25 × 25
     = 270.625 cm²

Answer to exercise 4

Area = 0.577 × 15 × 15
     = 129.825 sq ft

Answer to exercise 5

d = 1.414 × 15
   = 21.21 in.

8–4   THE REGULAR HEXAGON

The regular hexagon has six equal sides, and all the angles formed by any two adjacent sides are equal in measure. Probably the most common use of the hexagonal shape is for bolt heads and nuts. The essential parts of a regular hexagon are its side, its diagonal, the distance across its flats, and its area. (The distance across the flats is what determines the size of an open-end wrench.)

■ **Example 1**

Find the distance across the flats (line *GH*, or *f*) of the regular hexagon whose side is 5 in., as shown in Fig. 8–10.

Figure 8–10

Solution $f = 1.732s = 1.732 \times 5 = 8.66$ in.

Explanation   From the Table of Constants for the regular hexagon, select the formula that gives the distance across the flats (*f*) in terms of the side (*s*), *f* =

9. Find the distance across the flats of the regular hexagon whose side is 21 in.

The answer is in the margin on page 252.

1.732*s*. Substituting the value of *s* (5 in.) in this formula gives the distance across the flats.

Practice check: Do exercise 9 on the left.

■ **Example 2**

Find the diagonal (line *DA,* or *d*) of the regular hexagon whose side is 3 in., as shown in Fig. 8–11.

Figure 8–11

10. Find the diagonal of the regular hexagon whose side is 8 ft.

The answer is in the margin on page 252.

Solution $d = 2s = 2 \times 3 = 6$ in.

Explanation   From the Table of Constants for the regular hexagon, select the formula that gives the diagonal (*d*) in terms of the side (*s*), $d = 2s$. Substituting the value of *s* (3 in.) in this formula gives the length of the diagonal.

Practice check: Do exercise 10 on the left.

■ **Example 3**

Find the area of the regular hexagon whose side is 4 in., as shown in Fig. 8–12.

Figure 8–12

11. Find the area of the regular hexagon whose side is 10 cm.

The answer is in the margin on page 252.

Solution   Area $= 2.598s^2 = 2.598 \times 4 \times 4 = 41.568$ sq in.

Explanation   From the Table of Constants for the regular hexagon, select the formula that gives the area in terms of the side (*s*), Area $= 2.598s^2$. Substituting the value of *s* (4 in.) in this formula gives the area of the hexagon.

Practice check: Do exercise 11 on the left.

■ **Example 4**

Find the diagonal (line *DA*) of the regular hexagon whose distance across the flats (line *GH*) is 3 in., as shown in Fig. 8–13.

Figure 8–13

Solution $d = 1.155f = 1.155 \times 3 = 3.465$ in.

12. Find the diagonal of the regular hexagon whose distance across the flats is 11 yd.

The answer is in the margin on page 253.

Explanation   From the Table of Constants for the regular hexagon, select the formula that gives the diagonal (d) in terms of the distance across the flats (f), $d = 1.155f$. Substituting the value of f (3 in.) in this formula gives the diagonal.

Practice check: Do exercise 12 on the right.

■ **Example 5**

Find the area of the regular hexagon whose distance across the flats (line GH) is 6 cm, as shown in Fig. 8–14.

Figure 8–14

Solution   Area $= 0.866f^2 = 0.866 \times 6 \times 6 = 31.176$ cm²

13. Find the area of the regular hexagon whose distance across the flats is 13 cm.

The answer is in the margin on page 253.

Explanation   From the Table of Constants for the regular hexagon, select the formula that gives the area in terms of the distance across the flats (f), Area $= 0.866f^2$. Substituting the value of f (6 cm) in this formula gives the area.

Practice check: Do exercise 13 on the right.

■ **Example 6**

Find the area of the regular hexagon whose diagonal (line DA) is 4.5 in., as shown in Fig. 8–15.

Figure 8–15

Solution   Area $= 0.650d^2 = 0.650 \times 4.5 \times 4.5 = 13.1625$ sq in.

14. Find the area of the regular hexagon whose diagonal is 3.25 m.

The answer is in the margin on page 253.

Explanation   From the Table of Constants for the regular hexagon, select the formula that gives the area in terms of the diagonal (d), Area $= 0.650d^2$. Substituting the value of d (4.5 in.) in this formula gives the area.

Practice check: Do exercise 14 on the right.

■ **Problems 8–4**

Use the Table of Constants to find the missing information about regular hexagons in the following chart. Give your answer to the nearest hundredth. Try estimating the answer before solving the problem.

Answer to exercise 6

$s = 0.707 \times 10$
  $= 7.07$ ft

Answer to exercise 7

Area $= 30 \times 30$
  $= 900$ cm^2

Answer to exercise 8

Area $= 0.5 \times 11 \times 11$
  $= 60.5$ sq yd

Answer to exercise 9

$f = 1.732 \times 21$
  $= 36.372$ in.

Answer to exercise 10

$d = 2 \times 8$
  $= 16$ ft

Answer to exercise 11

Area $= 2.598 \times 10 \times 10$
  $= 259.8$ cm^2

	Regular Hexagons			
	Side	**Distance across Flats**	**Diagonal**	**Area**
1.	1″	———	———	———
2.	2 cm	———	———	———
3.	$2\frac{1}{2}$″	———	———	———
4.	3.75″	———	———	———
5.	7.25 m	———	———	———
6.	———	4 m	———	———
7.	———	$5\frac{1}{2}$″	———	———
8.	———	5.85″	———	———
9.	———	10.6 cm	———	———
10.	———	$2\frac{5}{8}$ in.	———	———
11.	———	———	6 cm	———
12.	———	———	$7\frac{1}{2}$″	———
13.	———	———	5.25″	———
14.	———	———	$8\frac{3}{4}$″	———
15.	———	———	2.275 m	———

8–5   THE REGULAR OCTAGON

A regular octagon has eight equal sides, and all the angles formed by any two adjacent sides are equal in measure. The essential parts of a regular octagon are its side, its diagonal, the distance across its flats, and its area.

■ Example 1

Find the distance across the flats (line *IJ*) of the regular octagon whose side is 5 in., as shown in Fig. 8–16.

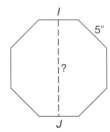

Figure 8–16

15. Find the distance across the flats of the regular octagon whose side is 7 m.

The answer is in the margin on page 255.

Solution $f = 2.414s = 2.414 \times 5 = 12.07$ in.

Explanation   From the Table of Constants for the regular octagon, select the formula that gives the distance across the flats (f) in terms of a side, $f = 2.424s$. Substituting the value of s (5 in.) in this formula gives the distance across the flats.

Practice check: Do exercise 15 on the left.

■ **Example 2**

Find the length of the diagonal of a regular octagon whose side is 3 in., as shown in Fig. 8–17.

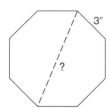

Figure 8–17

Solution $d = 2.613s = 2.613 \times 3 = 7.839$ in.

Explanation   From the Table of Constants for the regular octagon, select the formula that gives the diagonal (d) in terms of the side (s), $d = 2.613s$. Substituting the value of s (3 in.) in this formula gives the length of the diagonal.

Practice check: Do exercise 16 on the right.

16. Find the length of the diagonal of a regular octagon whose side is 10 in.

The answer is in the margin on page 255.

■ **Example 3**

Find the area of a regular octagon whose side is 4 in., as shown in Fig. 8–18.

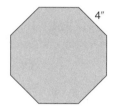

Figure 8–18

Solution   Area $= 4.828s^2 = 4.828 \times 4 \times 4 = 77.248$ sq. in.

Explanation   From the Table of Constants for the regular octagon, select the formula that gives the area in terms of the side (s), Area $= 4.828s^2$. Substituting the value of s (4 in.) in this formula gives the area.

Practice check: Do exercise 17 on the right.

17. Find the area of a regular octagon whose side is 12 ft.

The answer is in the margin on page 255.

■ **Example 4**

Find the length of the diagonal (line *EH*) of a regular octagon whose distance across the flats (line *IJ*) is 10 cm, as shown in Fig. 8–19.

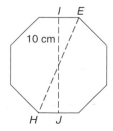

Figure 8–19

18. Find the length of the diagonal of a regular octagon whose distance across the flats is 15 cm.

The answer is in the margin on page 256.

Solution $d = 1.083f = 1.083 \times 10 = 10.83$ cm

Explanation   From the Table of Constants for the regular octagon, select the formula that gives the diagonal (d) in terms of the distance across the flats (f), $d = 1.083f$. Substituting the value of f (10 cm) in this formula gives the diagonal.

Practice check: Do exercise 18 on the left.

■ Example 5

Find the area of a regular octagon whose distance across the flats (line IJ) is 8 in., as shown in Fig. 8–20.

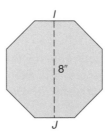

Figure 8–20

19. Find the area of the regular octagon whose distance across the flats is 4 m.

The answer is in the margin on page 256.

Solution   Area $= 0.828f^2 = 0.828 \times 8 \times 8 = 52.992$ sq in.

Explanation   From the Table of Constants for the regular octagon, select the formula that gives the area in terms of the distance across the flats (f), Area $= 0.828f^2$. Substituting the value of f (8 in.) in this formula gives the area.

Practice check: Do exercise 19 on the left.

■ Example 6

Find the area of a regular octagon whose diagonal (line EA) is 6 in., as shown in Fig. 8–21.

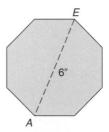

Figure 8–21

20. Find the area of a regular octagon whose diagonal is 15 yd.

The answer is in the margin on page 256.

Solution   Area $= 0.707d^2 = 0.707 \times 6 \times 6 = 25.452$ sq in.

Explanation   From the Table of Constants for the regular octagon, select the formula that gives the area in terms of the diagonal (d), Area $= 0.707d^2$. Substituting the value of d (6 in.) in this formula gives the area.

Practice check: Do exercise 20 on the left.

■ **Problems 8-5**

Use the space provided below to find the missing parts of the regular octagons as indicated. Use the Table of Constants. Try to estimate the answer before solving each problem.

1. Find the area of a regular octagon that is 4.78″ across the flats.

2. Find the area of a regular octagon with $5\frac{1}{2}$″ sides.

3. Find the flats of a regular octagon that has 6″ sides.

4. Find the side of a regular octagon that measures 6.3″ across the flats.

5. Find the area of a regular octagon 9″ across corners.

Find the missing parts of the regular octagons in the following chart. Use the given information to find the missing parts. Try to estimate the answer before solving each problem.

	Side	**Distance across Flats**	**Diagonal**	**Area**
		Regular Octagons		
6.	7″	_____	_____	_____
7.	$5\frac{1}{2}$″	_____	_____	_____
8.	_____	8 cm	_____	_____
9.	_____	$2\frac{1}{2}$″	_____	_____
10.	_____	_____	8 cm	_____
11.	_____	_____	$5\frac{1}{4}$″	_____
12.	3.02″	_____	_____	_____
13.	4.75″	_____	_____	_____

■ **Practice Problems on the Use of the Table of Constants**

Use the Table of Constants to find the missing parts of the regular figures as indicated in the following problems. Make a sketch of the figure in the space provided, and show the given part. Organize your computations and label all parts of the problem. Try to estimate the answer to each problem before solving it.

14. Find the area of an equilateral triangle with $4\frac{1}{2}''$ sides.

15. Find the area of an equilateral triangle whose altitude is $4\frac{3}{4}''$.

16. Find the diagonal of a 1.75″ square.

17. Find the side of a square whose diagonal is 3.45″.

18. Find the area of a regular hexagon that is 10″ across flats.

19. Find the area of a regular hexagon with $1\frac{3}{4}''$ sides.

20. Find the area of a regular hexagon that measures $\frac{7}{8}''$ across corners.

21. Find the side of an equilateral triangle whose altitude is $8\frac{1}{4}''$.

22. Find the altitude of an equilateral triangle with $5\frac{1}{2}''$ sides.

23. Find the area of a square whose diagonal is $7\frac{1}{2}''$.

24. Find the area of a regular hexagon that is $8\frac{1}{4}''$ across flats.

25. Find the diagonal of a regular octagon that is 6.03″ across flats.

26. Find the area of an equilateral triangle with 3.16″ sides.

27. Find the area of a regular hexagon with 1.09″ sides.

28. A sharp V-thread (Fig. 8–22) has the cross section of an equilateral triangle. If the pitch or distance AB is $\frac{1}{12}''$, find the depth CD.

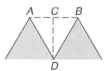

Figure 8–22

29. What should be the diameter of a piece of round stock (Fig. 8–23) if a $1\frac{3}{4}''$ square is to be milled on the end of it?

Figure 8–23

30. Eight bolt holes are to be drilled evenly spaced on the circumference of a 9″ circle, as shown in Fig. 8–24. What will be the distance center to center of the holes?

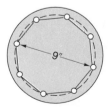

Figure 8–24

31. A wrench (Fig. 8–25) is required to be made that will fit a hexagon nut which measures $2\frac{1}{4}''$ across corners. What must be the width of the opening between the jaws of the wrench? Allow $0.02''$ for clearance.

Figure 8–25

32. How many feet of railing are required around a regular octagonal platform that measures $38'6''$ across flats?

33. Find the corner measurement (Fig. 8–26) for laying out a regular octagon on a piece of $2\frac{1}{2}''$ square stock.

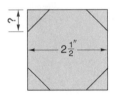

$2\frac{1}{2}''$

Figure 8–26

34. A piece of $1\frac{3}{4}''$ round stock is to have a regular hexagonal end milled on it, as shown in Fig. 8–27. Find the distance across flats of the hexagon.

$1\frac{3}{4}''$

Figure 8–27

35. Figure 8–28 represents a room with a bay window in the shape of one-half of a regular hexagon. The distance across *AB* is 2.6 m. Find the distance *CD*.

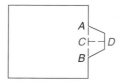

Figure 8–28

36. What is the largest square that can be cut from a piece of $\frac{7''}{8}$ round stock (Fig. 8–29)?

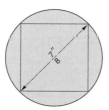

Figure 8–29

8–6　REGULAR POLYGONS AND THE CALCULATOR

To solve problems involving regular polygons with the calculator, we will be using the square root key ($\sqrt{\ }$) and the reciprocal key ($1/x$). On some calculators, the reciprocal key is a second function, and you must press the 2nd F key to activate the reciprocal function key. Read your manual to see what your sequence of keys must be to use these functions.

The problems done here will have the number of decimal places to match the example problems in order to show that slight differences will occur because of the accuracy of the calculator.

■   **Example 1**

Using a calculator, find the altitude of the equilateral triangle shown in Fig. 8–30. Round off your answer to three decimal places.

Figure 8–30

Solution   Use the formula $a = \dfrac{\sqrt{3}}{2}s$.

_____

Use the space below for notes.

_____

The Display Shows

Turn on the calculator	0.
Enter 3	3.
Press √	1.732050808
Press ÷	1.732050808
Enter 2	2.
Press ×	0.866025403
Enter 6	6.
Press =	5.196152423 (answer)

Round off this answer to 5.196 in.

■   **Example 2**

Using a calculator, find the side of the equilateral triangle shown in Fig. 8–31. Round off your answer to two decimal places.

Figure 8–31

Solution   Use the formula $s = \dfrac{2}{\sqrt{3}}a$.

The Display Shows

Turn on the calculator	0.
Enter 2	2.
Press ÷	2.
Enter 3	3.
Press √	1.732050808
Press ×	1.154700538
Enter 8	8.
Press =	9.237604307 (answer)

Round off this answer to 9.24 in.

■ **Example 3**

Using a calculator, find the area of the equilateral triangle shown in Fig. 8–32. Round off your answer to three decimal places.

Figure 8–32

Solution   Use the formula $A = \dfrac{\sqrt{3}}{4}s^2$.

The Display Shows

Turn on the calculator	0.
Enter 3	3.
Press $\sqrt{}$	1.732050808
Press ÷	1.732050808
Enter 4	4.
Press ×	0.433012701
Enter 12	12.
Press x^2	144.
Press =	62.35382907 (answer)

Round off this answer to 62.354 sq. in.

■ **Example 4**

Using a calculator, find the area of the equilateral triangle shown in Fig. 8–33. Round off your answer to one decimal place.

Figure 8–33

Solution   Use the formula $A = \dfrac{1}{\sqrt{3}}a^2$.

The Display Shows

Turn on the calculator	0.
Enter 3	3.
Press $\sqrt{}$	1.732050808
Press 2nd F	1.732050808
Press $1/x$	0.577350269
Press ×	0.577350269

	The Display Shows
Enter 10	10.
Press x^2	100.
Press =	57.73502692 (answer)

Round off this answer to 57.7 cm².

Problems involving squares, hexagons, and octagons use similar procedures.

■ **Problems 8–6**

Do Problem Sets 8–3, 8–4, and 8–5 again using the calculator. Compare the two sets of answers.

8–7   QUADRILATERALS

Figures with four sides are called ***quadrilaterals.*** The various types of quadrilaterals are shown in Fig. 8–34.

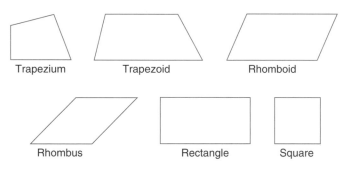

Figure 8–34

A ***parallelogram*** is a quadrilateral that has two pairs of parallel sides. The types of parallelograms are as follows.

The ***rectangle,*** in which all the angles are right angles.
The ***square,*** in which all the angles are right angles and the four sides are equal.
The ***rhomboid,*** which has two pairs of parallel sides but no right angles.
The ***rhombus,*** which has four equal sides but no right angles.

The ***trapezoid*** is a quadrilateral in which only two of the sides are parallel. The ***isosceles trapezoid*** is a trapezoid in which the two nonparallel sides are equal.
The ***trapezium*** is a quadrilateral that has no two sides parallel; in other words, it is an irregular four-sided figure.

8–8   AREA OF A TRAPEZOID

The area of the trapezoid *ABCD* in Fig. 8–35 can be found by the following method. Draw a center line parallel to and midway between the upper base *AB* and the lower base *CD*. Where this center line *EF* meets the sides, draw perpendiculars to it, forming the rectangle *KLHG*. The rectangle *KLHG* is equal in area to the trapezoid *ABCD* because the two triangles that are cut off the trapezoid in forming the rectangle are equal to the triangles that are gained. Triangle *KAE* is equal to triangle *EGD*, and triangle *BLF* is equal to triangle *FCH*.

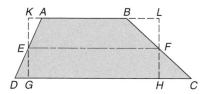

Figure 8–35

To find the area of a rectangle, multiply the base by the altitude. To find the area of the rectangle *KLHG* in Fig. 8–35, multiply the height *LH* by the width *EF*. But the area of the trapezoid *ABCD* is equal to the area of the rectangle *KLHG*. The area of the trapezoid can be found by multiplying the height by the width along the center line *EF*. Since *EF* is halfway between the upper and the lower bases, it is equal to their average, or half their sum.

$$\text{Area of trapezoid} = \text{Average width} \times \text{Altitude} = \left(\frac{b_1 + b_2}{2}\right)a$$

■ **Example 1**

Find the area of the trapezoid in Fig. 8–36.

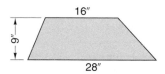

Figure 8–36

Solution

$$\text{Area} = \text{Average width} \times \text{Altitude} = \left(\frac{b_1 + b_2}{2}\right)a$$

$$\text{Average width} = \frac{16 + 28}{2} = \frac{44}{2} = 22 \text{ in.}$$

$$\begin{aligned}\text{Area} &= 22 \times 9 \\ &= 198 \text{ sq in.}\end{aligned}$$

Practice check: Do exercise 21 on the right.

21. Find the area of the trapezoid whose altitude is 2.5 in. and bases are 17 in. and 33 in.

The answer is in the margin on page 265.

■ Example 2

Find the altitude of the isosceles trapezoid in Fig. 8–37.

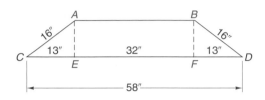

Figure 8–37

Solution $CE + FD = 58 - 32 = 26$ in.

$$CE = \tfrac{26}{2} = 13 \text{ in.}$$

$$AE = \sqrt{\text{Hypotenuse}^2 - \text{Base}^2}$$

$$= \sqrt{16^2 - 13^2}$$

$$= \sqrt{256 - 169} = \sqrt{87} = 9.33 \text{ in.}$$

22. Find the lower base of the following isosceles trapezoid.

The answer is in the margin on page 266.

Explanation   The perpendiculars AE and BF cut off equal distances CE and FD on either side of EF. But since $EF = AB = 32$ in., then

$$CE = FD = \frac{58 - 32}{2} = \frac{26}{2} = 13 \text{ in.}$$

In triangle ACE of Fig. 8–37, the base is 13 in. and the hypotenuse is 16 in. Hence, Altitude $= \sqrt{16^2 - 13^2} = 9.33$ in.

Practice check: Do exercise 22 on the left.

■ Problems 8–8

Find the average width and the area of each trapezoid in the following chart.

Trapezoids					
	Upper Base	**Lower Base**	**Average Width**	**Altitude**	**Area**
1.	10 cm	14 cm	_____	8 cm	_____
2.	8″	$10\tfrac{1}{2}″$	_____	10″	_____
3.	$6\tfrac{1}{2}″$	8″	_____	$7\tfrac{1}{2}″$	_____
4.	$5\tfrac{1}{2}″$	$6\tfrac{1}{2}″$	_____	4″	_____
5.	5.5″	7.75″	_____	6.2″	_____
6.	1′2″	1′8″	_____	8″	_____
7.	$1′6\tfrac{1}{2}″$	1′10″	_____	1′3″	_____
8.	$2′6\tfrac{1}{2}″$	$3′2\tfrac{1}{2}″$	_____	$2′7\tfrac{1}{2}″$	_____
9.	4.02″	8.07″	_____	6.02″	_____
10.	5.61″	7.22″	_____	5.05″	_____

Use the space provided below to find the areas of the following trapezoids.

11.

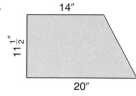

14″

$11\frac{1}{2}''$

20″

12.

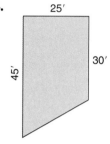

9″

7″

18″

Answer to exercise 21

$$\text{Area} = \left(\frac{17 + 33}{2}\right) \times 2.5$$
$$= \left(\frac{50}{2}\right) \times 2.5$$
$$= 25 \times 2.5$$
$$= 62.5 \text{ sq in.}$$

13.

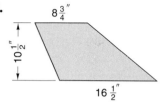

$8\frac{3}{4}''$

$10\frac{1}{2}''$

$16\frac{1}{2}''$

14.

25′

45′

30′

15.

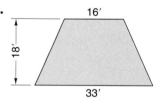

16′

18′

33′

16.

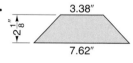

3.38″

$2\frac{1}{8}''$

7.62″

17.

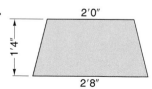

2′0″

1′4″

2′8″

18.

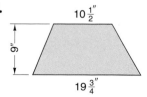

$10\frac{1}{2}''$

9″

$19\frac{3}{4}''$

Challenge: Find the altitudes and areas of the following isosceles trapezoids.

19.   **20.**

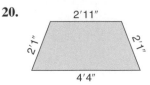

21.   **22.**

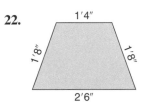

8–9   AREAS OF COMPOSITE FIGURES

It is sometimes necessary to find the area of a figure that does not belong to any of the classes of figures studied. If the figure can be divided into other shapes—the areas of which can be found by the rules already learned—then the area of the figure can be found.

■ Example 1

Find the area of the figure in Fig. 8–38.

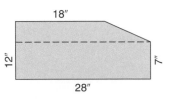

Figure 8–38

Solution   This figure can be divided into a trapezoid and a rectangle by the dotted construction line as shown.

$$\text{Altitude of trapezoid} = 12 - 7 = 5 \text{ in.}$$

$$\text{Average width of trapezoid} = \frac{18 + 28}{2} = \frac{46}{2} = 23 \text{ in.}$$

Area of trapezoid = 23 × 5 = 115 sq in.
Area of rectangle = 28 × 7 = <u>196 sq in.</u>
Total area = 311 sq in.

Practice check: Do exercise 23 on the right.

■ **Example 2**

Find the area of the figure in Fig. 8–39.

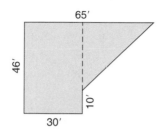

Figure 8–39

Solution   The dotted construction line divides the figure into a rectangle and a triangle.

Altitude of triangle = 46 − 10 = 36 ft

Base of triangle = 65 − 30 = 35 ft

Area of triangle = $\frac{1}{2}$ × 35 × 36 =   630 sq ft
Area of rectangle = 30 × 46   = <u>1,380 sq ft</u>
Total area = 2,010 sq ft

Practice check: Do exercise 24 on the right.

23. Find the area of the follow-ing figure.

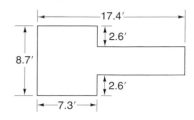

The answer is in the margin on page 269.

24. Find the area of the follow-ing figure.

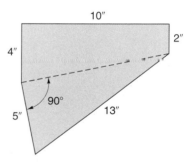

The answer is in the margin on page 269.

■ **Problems 8–9**

Find the area of the following composite figures. Using the space provided below, make a sketch of each figure and show how it can be divided into figures whose areas can be found. Label all parts of your computations.

1.

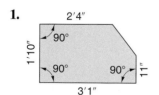

2.

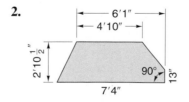

3.

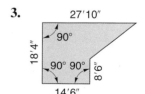

4.

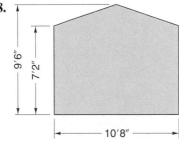

5.

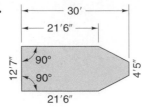

6.

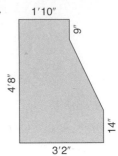

7.

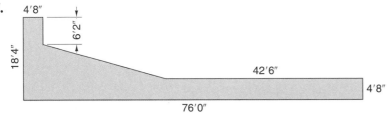

8.

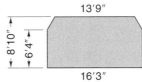

9.

10.

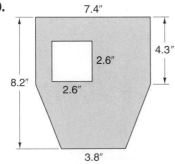

8–10   SCALE

Diagrams of objects are frequently drawn smaller than the objects they represent. For example, the top of an end table 3 ft long and 1 ft 3 in. wide may be represented by a diagram 3 in. long and $1\frac{1}{4}$ in. wide (Fig. 8–40). Every inch in the diagram would therefore represent a foot length in the table. Since the ratio of 1 in. to 1 ft is 1 to 12, the diagram is drawn to a scale of $\frac{1}{12}$, 1:12, or 1 in. = 1 ft. In general, scale means the ratio of any dimension on the drawing to the corresponding dimension on the object represented.

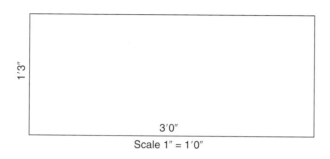

Scale 1″ = 1′0″

Figure 8–40

8–11   FINDING THE DRAWING MEASURE

If a drawing is to be made to a scale of $\frac{1}{4}$, it means that every distance on the object is to be represented by a line one-fourth as long. An object 8 ft $4\frac{1}{2}$ in. long drawn to a $\frac{1}{4}$ scale would be represented by a diagram 2 ft $1\frac{1}{8}$ in. in length, or $\frac{1}{4}$ of 8 ft $4\frac{1}{2}$ in.

Rule _____

To find the length of any line on a diagram, multiply the dimension on the object by the scale ratio.

_____

Practice check: Do exercise 25 on the right.

25. A draftsman needs to draw a 6-ft wall to a scale of $\frac{1}{12}$. What is the length of the wall in the drawing?

The answer is in the margin on page 271.

■ Problems 8–11

Use the space provided below to find the lengths of the following drawing measures.

1. A screwdriver 10″ long is drawn to a scale of $\frac{1}{2}$. What will be the length of the line in the drawing that represents the length of the screwdriver?

2. A wrench $10\frac{1}{2}$″ long is drawn to $\frac{1}{4}$ scale. What will be the length of the line in the drawing that represents the length of the wrench?

3. A box 5'4" long, 4'8" wide, and 3'0" high is drawn to a scale of $\frac{1}{8}$. Compute the lengths of the lines in the diagram that will represent the length, width, and height of the box.

4. Using a scale of $\frac{1}{4}$, find the length of a steel rod on a drawing if the rod is 18" tall.

5. The length of a metal plate is $5\frac{1}{2}$". What is the length on a scale drawing if the scale is $\frac{1}{4}$?

6. The length of an object is 2'4". Find the length of the corresponding line in a diagram drawn to **(a)** $\frac{1}{2}$ size, **(b)** $\frac{1}{4}$ size, **(c)** a scale of $\frac{3}{8}$, **(d)** a scale of $\frac{1}{8}$" = 1', **(e)** a scale of $\frac{1}{8}$.

7. A drawing of an object is made to a scale of 3:4. Find the lengths of the lines in the drawing that will represent the following measurements on the object: **(a)** 1'6", **(b)** $6\frac{1}{2}$", **(c)** $3'5\frac{1}{2}$", **(d)** $11\frac{3}{4}$", **(e)** 14".

8. A drawing is to be made to a scale of 1" = 1'. Compute the dimensions in the drawing that will correspond to the following dimensions on the object: **(a)** 3'9", **(b)** 2'3", **(c)** 2'6", **(d)** 18", **(e)** 4'0".

9. A house is 22' by 34'. A draftsman desires to draw floor plans on an $8\frac{1}{2}$"-by-11" piece of paper. Should he use a scale of $\frac{1}{4}$" = 1' or a scale of $\frac{1}{2}$" = 1'?

10. A drawing of a house roof gable is to be drawn at a scale of 1" = 4'. What are the dimensions on the drawing if two of the gable's measurements are $26\frac{1}{2}$' and $7\frac{1}{2}$'?

8–12   FINDING THE ACTUAL DIMENSION ON AN OBJECT FROM THE MEASURED DIMENSION ON THE DIAGRAM

If the scale of a diagram is $\frac{1}{8}$, it means that the length of a line on the object is 8 times as long as the corresponding line in the diagram. Thus, if a line in a diagram measures $1\frac{3}{4}$ in., the corresponding dimension on the object will be $8 \times 1\frac{3}{4} = 14$ in.; or $1\frac{3}{4}$ in. ÷ scale ratio $= 1\frac{3}{4} \div \frac{1}{8} = 1\frac{3}{4} \times \frac{8}{1} = 14$ in.

Rule _____

To find the measure on an object corresponding to the length of a line on the diagram, divide the diagram measure by the scale ratio.

NOTE: *Great care and good judgment must be used when measuring a drawing to determine the actual measure of an object. Temperature, humidity, the reproduction process, the type of paper, and so on, all have an effect on the actual lengths of the lines on a drawing. It is best to consider lengths found this way to be a good approximation only.*

■ **Example**

A plant floor plan measures $16\frac{1}{4}$ in. by $12\frac{1}{2}$ in. What is the actual measurement in feet if the scale is $\frac{1}{4}$ in. = 1 ft 0 in.?

Solution $16\frac{1}{4} \div \frac{1}{4} = \frac{65}{4} \times \frac{4}{1} = 65$ ft
              $12\frac{1}{2} \div \frac{1}{4} = \frac{25}{2} \times \frac{4}{1} = 50$ ft

The plant floor measures 65 ft by 50 ft.

Practice check: Do exercise 26 on the right.

■ **Problems 8–12**

Using the space provided below, find the actual dimensions of the following objects drawn to scale by measuring the diagram and dividing the diagram measure by the scale ratio.

1. The diagram in Fig. 8–41 is drawn to a scale of $\frac{3}{4}$. The length of the diagram measures $2\frac{15}{32}''$ and the width measures $1\frac{5}{16}''$. Find the corresponding dimensions of the object.

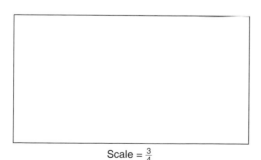

Scale = $\frac{3}{4}$

Figure 8–41

Answer to exercise 25

$\frac{1}{2}$ ft or 6 in.

26. On a blueprint, a vent measures $2\frac{1}{8}$ in. from the floor. What will be the actual distance if the scale is $\frac{1}{4}$ in. = 1.0 in.?

The answer is in the margin on page 273.

2. Measure the lines in the diagram in Fig. 8–42 and compute the actual dimensions of the object.

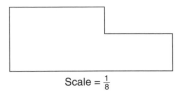

Scale = $\frac{1}{8}$

Figure 8–42

3. Measure the lines in the diagram in Fig. 8–43 and compute the actual dimensions of the object.

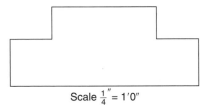

Scale $\frac{1}{4}'' = 1'0''$

Figure 8–43

4. Measure the lines in the diagram in Fig. 8–44 and compute the actual dimensions of the object.

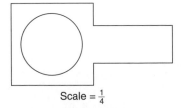

Scale = $\frac{1}{4}$

Figure 8–44

5. Measure the long and short diameters of the ellipse in Fig. 8–45 and compute the actual diameters of the object represented.

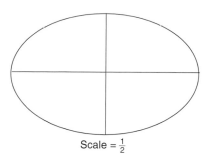

Scale = $\frac{1}{2}$

Figure 8–45

8–13 FINDING THE AREA OF A FIGURE DRAWN TO A CERTAIN SCALE

Measure each line in the diagram and compute the actual lengths of the corresponding lines in the object. Then find the area represented according to the rules given for the various figures.

■ **Example**

Find the area of the surface represented by the rectangle in Fig. 8–46.

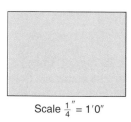

Scale $\frac{1}{4}'' = 1'0''$

Figure 8–46

Solution   Measure the length and width of the rectangle in the figure. The length measures $1\frac{1}{4}$ inches, and the width measures $\frac{7}{8}$ inch. Since the scale is $\frac{1}{4}$ in. = 1 ft, the scale ratio is $\dfrac{\frac{1}{4}\,\text{in.}}{12\,\text{in.}} = \frac{1}{4} \times \frac{1}{12} = \frac{1}{48}$. In other words, the length of a line on the object is 48 times the corresponding length on the diagram. Hence, the actual length of the object = $48 \times 1\frac{1}{4}$ in. = $48 \times \frac{5}{4}$ in. = 60 in., and the actual width of the object is $48 \times \frac{7}{8}$ in. = 42 in.

Actual area of the surface represented = $60 \times 42 = 2,520$ sq in.
$\qquad\qquad\qquad\qquad\qquad\qquad\qquad = 17.5$ sq ft

Practice check: Do exercise 27 on the right.

27. Find the area of the surface represented by the following rectangle.

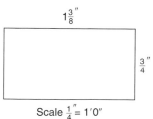

$1\frac{3}{8}''$

$\frac{3}{4}''$

Scale $\frac{1}{4}'' = 1'0''$

The answer is in the margin on page 275.

■ Problems 8–13

Using the space provided below, find the areas of the various surfaces represented in the following diagrams.

1. Find the area represented by the rectangle in Fig. 8–47 if $AB = 1\frac{3}{8}''$ and $BC = \frac{3}{4}''$.

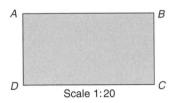

Scale 1:20

Figure 8–47

2. Find the actual diameter of a sanding disc represented by a circle whose diameter is $1\frac{1}{4}''$ when drawn to a scale of $\frac{1}{8}$.

3. The base of the right triangle in Fig. 8–48 is $1\frac{1}{8}''$ and the altitude is $\frac{5}{8}''$. Find the actual area represented by the diagram.

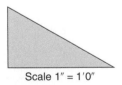

Scale 1" = 1'0"

Figure 8–48

4. Measure the base and altitude of the diagram in Fig. 8–49 and compute the actual area represented by the triangle.

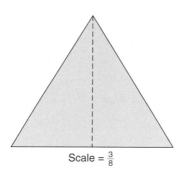

Scale = $\frac{3}{8}$

Figure 8–49

5. Measure the altitude and bases of the trapezoid in Fig. 8–50 and compute the area represented by the diagram.

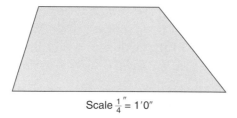

Scale $\frac{1}{4}'' = 1'0''$

Figure 8–50

8–14   CIRCLES

The perimeter of a circle is called the **circumference.** The distance from the center to any point on the circle is called the **radius.**

The distance across the circle through the center is called the **diameter.** The diameter is equal to two radii (see Fig. 8–51).

A 12-in. pulley means a pulley 12 in. in diameter. Similarly, a piece of $\frac{1}{2}$-in. round stock means a piece of stock $\frac{1}{2}$ in. in diameter. A 10-in. circle is a circle whose diameter is 10 in.

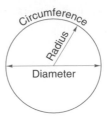

Figure 8–51

8–15   CIRCUMFERENCE

It has been found by measurement and computation that the ratio of the circumference of a circle to its diameter is a constant represented by the Greek letter π (pronounced *pie*). The value of π is nearly equal to 3.1416, and this is the approximation usually used in shop work involving π. If you press the π key on a 10-digit display calculator, you will see that π is given as 3.141592654. Use 3.1416 as a value for π.

The relation between the diameter and the circumference of a circle is expressed by the formula

$$C = \pi d$$

where C is the circumference of the circle and d is its diameter.

■ Example

Find the circumference of a 16-in. circle.

28. Find the circumference of a 22-in. circle.

The answer is in the margin on page 278.

Solution $C = \pi d$
$= 3.1416 \times 16$
$= 50.2656$

Practice check: Do exercise 28 on the left.

■ Problems 8–15

Using the space provided below, find the circumference of the following circles. (Use 3.1416 as an approximation of π.)

1. Find the circumference of a 14″ circle.

2. Find the circumference of a 5-cm circle.

3. Find the circumference of a 2″ circle.

4. Find the circumference of a circle whose radius is 0.5 in.

5. Find the circumference of a $2\frac{1}{8}''$ circle.

6. Find the circumference of a $1\frac{7}{8}''$ circle.

7. Find the circumference of a circle whose radius is 1.625 in.

8. Find the circumference of a circle 4′9″ in diameter.

9. Find the circumference of a circle 135′ in diameter.

10. What is the circumference of a 32″ pulley?

11. Find the circumference of a gear that is to be cut from a gear blank whose diameter is 20 cm to the nearest tenth.

12. A rectangular piece of insulation is to be wrapped around a water pipe whose diameter is $\frac{1}{2}''$. How wide does the piece of insulation need to be?

13. What is the distance the tip of a blade of a fan travels in one revolution if the length of the single fan blade from tip to tip to top is $3\frac{1}{4}'$?

14. What must be the length of a label to fit around a can 3″ in diameter, allowing $\frac{1}{4}''$ for pasting?

Answer to exercise 28

$C = \pi d$
   $= 3.1416 \times 22$
   $= 69.1152$ in.

15. What must be the length of a bar of steel to make a rim for a carriage wheel 4′2″ in diameter, allowing 1″ for weld?

16. An emery wheel is 8″ in diameter. How many inches will a point on the circumference travel while the wheel makes one revolution?

17. What distance does an automobile travel with each complete turn of the wheels if the LR78-15 tires measure 28 in. in diameter?

18. A circular running track in a gymnasium has an average diameter of 70 ft. Find the circumference.

19. A water tank is 10′6″ in diameter. Find the length of the tie rods around it, allowing 6″ for fastenings.

20. What must be the length of a belt connecting two 16″ pulleys that are 3′8″ center to center?

8–16   USING THE CALCULATOR TO FIND CIRCUMFERENCE

The π key on the 10-digit calculator shows π to be 3.141592654. When you use the calculator to solve practical problems, you will have to determine what accuracy is needed and when to round off your answers.

■ **Example**

Using a calculator, find the circumference of a 16-in. circle. Round off your answer to two decimal places; $C = \pi d$.

Solution

	The Display Shows
Turn on the calculator	0.
Enter 16	16.
Press $\times$	16.
Press π	3.141592654
Press $=$	50.26548246 (answer)

Round off this answer to 50.27 in.

■ **Problems 8–16**

Using a calculator, do all of the problems in Problem Set 8–15 again. Notice the difference in the answers that results from using a more accurate value for π.

8–17   FINDING THE DIAMETER OF A CIRCLE

When the circumference is given, the diameter is found by dividing the circumference by π. The formula is

$$d = \frac{C}{\pi}$$

■ **Example**

Find the diameter of a circle whose circumference is 42.5″.

Solution $d = \dfrac{C}{\pi}$

$$= \frac{42.5}{3.1416} = 13.5281 \text{ in.}$$

Practice check: Do exercise 29 on the right.

29. Find the diameter of a circle whose circumference is 18.25 ft.

The answer is in the margin on page 281.

■ **Problems 8–17**

Use the space provided below to find the diameters of the following circles (Use 3.1416 as an approximation of π.)

1. Find the diameter of a circle whose circumference is 12 m.

2. Find the diameter of a circle whose circumference is 280′.

3. Find the diameter of a circle whose circumference is 3.73″.

4. Find the diameter of a circle whose circumference is 3′9″.

5. Find the diameter of a circle whose circumference is 11″.

6. What is the diameter of a shaft if the distance around the shaft is $23\frac{1}{2}$″?

7. What is the pitch diameter of a gear if the circumference of the pitch circle is 20.42″?

8. The circumference of a pulley is 75.4″. Find its diameter.

9. What is the diameter of a piece of round stock whose circumference is 8.53″?

10. What is the diameter of a cylindrical tank if the distance around the tank is 62.5′?

11. One lap around a circular running track is 220 yd. What is the diameter of the track?

12. What must be the diameter of an emery wheel if a point on the circumference travels 20.41″ for each revolution?

13. The circumference of the trunk of a tree measures 7′4″. What is its diameter?

14. The circumference of a wheel is 88 cm. What is the diameter?

15. Find the diameter of a circular silo whose circumference is 52 ft.

8–18   FINDING THE AREA OF A CIRCLE: METHOD 1

To find the area of a circle, you can multiply the square of the radius by π. The formula is

$$\text{Area} = \pi r^2$$

■ **Example 1**

Find the area of a circle with a radius of 8 in.

Solution $A = \pi r^2$ $r = 8$ in.
$$= 3.1416 \times 8^2$$
$$= 3.1416 \times 64 = 201.1 \text{ sq in.}$$

■ Example 2

Find the area of an 11-in. circle.

Solution   Area $= \pi r^2$ $r = \frac{11}{2} = 5.5$ in.
$$= 3.1416 \times 5.5^2$$
$$= 3.1416 \times 30.25 = 95.0334 \text{ sq in.}$$

■ Example 3

Find the area of a circular steel shim 5.00 in. in diameter.

Solution $A = \pi r^2$ $r = \frac{5}{2} = 2.5$ in.
$$= 3.1416 \times 2.5^2$$
$$= 3.1416 \times 6.25 = 19.6 \text{ sq in.}$$

Practice check: Do exercise 30 on the right.

■ Problems 8–18

Fill in the following chart. Use 3.1416 for π if you are not using a calculator. In either case, estimate the area before solving each problem.

Circles

	Diameter	Radius	Estimated Area	Computed Area
1.	2″	____	____	____
2.	14″	____	____	____
3.	65″	____	____	____
4.	$4\frac{1}{2}″$	____	____	____
5.	$3\frac{1}{8}″$	____	____	____
6.	$1\frac{1}{4}″$	____	____	____
7.	1.85″	____	____	____
8.	3.6″	____	____	____
9.	5′9″	____	____	____
10.	108′	____	____	____

Use the space provided below to work out the answers to the following problems.

11. A sheet of steel $\frac{1}{8}″$ thick and 1′ square weighs 5.08 lb. What is its weight after an 8″ hole is cut in it?

12. The top of a gas tank 72′8″ in diameter is to be painted at a cost of $1.75 per square yard. Find the cost.

Answer to exercise 29

$$d = \frac{C}{\pi} = \frac{18.25}{3.1416}$$
$$= 5.809 \text{ ft}$$

---

30. Find the area of a circular region with a radius of 3.25 ft.

The answer is in the margin on page 283.

13. Find the area of a piston head $3\frac{3}{4}''$ in diameter.

14. Find the total pressure on the bottom of a cylindrical tank $8'4''$ in diameter if the pressure is 920 lb per square foot.

15. A splice plate $18'' \times 6''$ has 10 rivet holes, each $\frac{7}{8}''$ in diameter, punched in it. What is the net area of the plate?

8–19   FINDING THE AREA OF A CIRCLE: METHOD 2

To find the area of a circle, multiply the square of the diameter by $\frac{\pi}{4}$. The formula is

$$\text{Area} = \frac{\pi d^2}{4}$$

Since $\frac{\pi}{4} = \frac{3.1416}{4} = 0.7854$, write the formula thus:

$$\text{Area} = 0.7854 d^2$$

■ **Example 1**

Find the area of a 15-cm circle.

Solution $\text{Area} = 0.7854 d^2 = 0.7854 \times 15^2$
$= 0.7854 \times 225 = 176.715 \text{ cm}^2$

■ **Example 2**

Find the area of a circle whose radius is 8 in.

Solution $A = 0.7854 d^2 \qquad d = 2r$
$= 0.7854 \times 16^2 \qquad = 2 \times 8 = 16 \text{ in.}$
$= 0.7854 \times 256 = 201.06 \text{ sq in.}$

■ **Example 3**

Find the area of a circular window 3 ft in diameter.

31. Find the area of a circle whose radius is 5.4 cm.

The answer is in the margin on page 284.

Solution $A = 0.7854 d^2$
$= 0.7854 \times 3^2$
$= 0.7854 \times 9 = 7.07 \text{ sq ft}$

Practice check: Do exercise 31 on the left.

■   **Problems 8–19**

Find the area of the following circles using Method 2. Use the space provided to work out the answers.

1. Find the area of a 5″ circle.

2. Find the area of a 16-cm circle.

3. Find the area of a 72′ circle.

4. Find the area of a $5\frac{1}{2}$″ circle.

5. Find the area of a 9″ circle.

6. Find the area of a $2\frac{1}{2}$″ circle.

7. Find the area of a circle 2.64″ in diameter.

8. Find the area of a circle 5.8″ in diameter.

9. Find the area of a circle 6′3″ in diameter.

10. Find the net area of a manhole cover 26″ in diameter with 13 vent holes, each 1″ in diameter.

11. Find the area of the doorway with the semicircular top represented in Fig. 8–52.

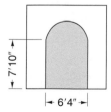

7′10″   6′4″

Figure 8–52

12. Find the area of the bottom of the clothes boiler with semicircular ends illustrated in Fig. 8–53.

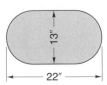

Figure 8–53

8–20   USING THE CALCULATOR TO FIND THE AREA OF A CIRCLE

■ **Example 1**

Using a calculator, find the area of an 11-in. circle. Round off your answer to two decimal places.

$$\text{Area} = \pi r^2 \qquad r = \tfrac{11}{2} = 5.5 \text{ in.}$$

Solution

	The Display Shows
Turn on the calculator	$0.$
Press π	3.141592654
Press $\times$	3.141592654
Enter 5.5	5.5
Press x^2	30.25
Press $=$	95.03317777 (answer)

Round off to 95.03 sq. in.

■ **Example 2**

Using a calculator, find the area of a 15-cm circle. Round off your answer to one decimal place.

$$\text{Area} = \frac{\pi}{4} d^2$$

Solution

	The Display Shows
Turn on the calculator	$0.$
Press π	3.141592654
Press $\div$	3.141592654
Enter 4	$4.$
Press $\times$	0.785398163

The Display Shows

Enter 15	15.
Press x^2	225.
Press =	176.7145868 (answer)

Round off to 176.7 cm².

■ **Problems 8–20**

Use either $A = \pi r^2$ or $A = \dfrac{\pi}{4} d^2$ to find the areas of the circles in Problem Sets 8–18 and 8–19. Compare the two sets of answers and notice the difference when using a more accurate value of π.

**8–21   FINDING THE DIAMETER OF A CIRCLE
WHEN THE AREA IS GIVEN: METHOD 1**

To find the diameter of a circle when the area A is known, you can use the formula

$$d = 2\sqrt{\frac{A}{\pi}}$$

This formula indicates the following operations.

1. Divide the area by 3.1416.
2. Find the square root of the quotient.
3. Multiply the square root of the quotient by 2.

■ **Example 1**

Find the diameter of a circle whose area is 36.8 sq in.

Solution $d = 2\sqrt{\dfrac{A}{\pi}}$

$\qquad = 2\sqrt{\dfrac{36.8}{3.1416}}$

$\qquad = 2\sqrt{11.7138}$

$\qquad = 2 \times 3.42 = 6.84$ in.

■ **Example 2**

A circular race track has an area of 138,544 sq ft. Find the diameter.

Solution $d = 2\sqrt{\dfrac{A}{\pi}}$

$\qquad = 2\sqrt{\dfrac{138,544}{3.1416}}$

$\qquad = 2\sqrt{44,100}$

$\qquad = 2 \times 210 = 420$ ft

Practice check: Do exercise 32 on the right.

---

32. Find the diameter of a circle whose area is 52.4 sq in.

The answer is in the margin on page 287.

■ **Problems 8–21**

Use the space provided below to find the diameter of the following circles whose areas are given.

1. Find the diameter of a circle whose area is 100 m².

2. Find the diameter of a circle whose area is 5,000 sq ft.

3. Find the diameter of a circle whose area is 5.78 sq in.

4. Find the diameter of a circle whose area is 1.19 sq in.

5. Find the diameter of a circle whose area is 30.5 sq in.

6. Find the diameter of a circle whose area is 15.625 cm².

7. Find the diameter of a circle whose area is $30\frac{3}{4}$ sq ft.

8. What is the diameter of a circular disc whose area is $5\frac{1}{4}$ sq in.?

9. The area of a circular drumhead is 82.6 sq in. Find the diameter of the drum.

10. The area of a circular ice skating rink is approximately 15,000 sq ft. What is its diameter?

11. A round steel rod can safely bear a pull of 16,000 lb per square inch. What must be the diameter of a rod to carry 125,000 lb?

12. What must be the diameter of a rivet to have a cross-sectional area of 0.6 sq in.?

13. A copper wire is required to have a cross-sectional area of $\frac{1}{2}$ sq in. What must be the diameter?

14. What must be the diameter of a pipe to have an area of 4.78 sq in.?

Answer to exercise 32

$$d = 2\sqrt{\frac{A}{\pi}}$$
$$= 2\sqrt{\frac{52.4}{3.1416}}$$
$$= 2\sqrt{16.679}$$
$$= 2 \times 4.084$$
$$= 8.168 \text{ in.}$$

15. What must be the diameter of a cylinder if the base area must be 125 sq in.?

8–22 FINDING THE DIAMETER OF A CIRCLE WHEN THE AREA IS GIVEN: METHOD 2

To find the diameter of a circle when the area A is known, use the formula

$$d = \sqrt{\frac{A}{0.7854}}$$

This formula indicates the following operations.

1. Divide the area by 0.7854.

2. Find the square root of the quotient.

■ **Example 1**

Find the diameter of a circle whose area is 25.5 sq in.

Solution $d = \sqrt{\dfrac{A}{0.7854}} = \sqrt{\dfrac{25.5}{0.7854}}$

$= \sqrt{32.47} = 5.70$ in.

■ **Example 2**

A platform of a merry-go-round has an area of 177 sq ft. What is the diameter?

Solution $d = \sqrt{\dfrac{A}{0.7854}}$

$= \sqrt{\dfrac{177}{0.7854}}$

$= \sqrt{225} = 15$ ft

Practice check: Do exerise 33 on the right.

33. Find the diameter of a circle whose area is 406.25 cm².

The answer is in the margin on page 290.

■ **Problems 8–22**

Find the diameter of the following circles using Method 2. Use the space provided to work out the answers.

1. Find the diameter of a circle whose area is 80 sq in.

2. Find the diameter of a circle whose area is 1,000 sq ft.

3. Find the diameter of a circle whose area is 14.62 sq in.

4. Find the diameter of a circle whose area is 3.35 sq in.

5. Find the diameter of a circle whose area is 41.8 sq in.

6. Find the diameter of a circle whose area is 7.5 sq in.

7. Find the diameter of a circle whose area is $33\frac{1}{8}$ cm^2.

8. Find the diameter of a circle whose area is $538\frac{1}{16}$ sq in.

9. What must be the diameter of a steel wire to have a cross-sectional area of $\frac{1}{8}$ sq in.?

10. A circular wooden post is required to have a cross-sectional area of 20 sq in. What must the diameter be?

11. What must be the diameter of a steel rod whose cross-sectional area is $2\frac{1}{2}$ sq in.?

12. The base area of a cylindrical tank is 250 sq ft. What is its diameter?

13. The area of a circular table top is 9.6 sq in. A contractor needs to cut a piece of Plexiglass to cover the table. What would the diameter be for the cut?

14. The cross-sectional area of a drain pipe is 28 sq ft. What is the diameter?

15. The area of the opening in a round flue lining is 113 sq in. What is the diameter?

8–23   FINDING THE DIAMETER OF A CIRCLE EQUAL IN AREA TO THE COMBINED AREAS OF TWO OR MORE CIRCLES

This involves the following processes.

1. Find the area of each circle.
2. Find the sum of the areas.
3. Find the diameter of the circle whose area is equal to the sum of the areas of the individual circles.

■ **Example 1**

Find the diameter of a circle whose area is equal to the combined areas of a 3-in. circle and a 4-in. circle.

Solution
$$\text{Area of 3-in. circle} = 3.1416 \times 1.5 \times 1.5 = 7.07 \text{ sq in.}$$
$$\text{Area of 4-in. circle} = 3.1416 \times 2 \times 2 = \underline{12.56 \text{ sq in.}}$$

$$\text{Combined areas} = 19.63 \text{ sq in.}$$

$$d = 2\sqrt{\frac{A}{\pi}} = 2\sqrt{\frac{19.63}{3.1416}} = 2\sqrt{6.2484} = 2 \times 2.50 = 5$$

The diameter of the required circle is 5 in.

Practice check: Do exercise 34 on the right.

■ **Example 2**

Find the diameter of a circle whose area is equal to the combined areas of three 5-in. circles and two 4-in. circles.

34. Find the diameter of a circle whose area is equal to the combined areas of a 20-in. circle and a 24-in. circle.

The answer is in the margin on page 291.

Answer to exercise 33

$d = \sqrt{\dfrac{A}{0.7854}}$

$\quad = \sqrt{\dfrac{406.25}{0.7854}}$

$\quad = \sqrt{517.252}$

$\quad = 22.74 \text{ cm}$

35. Find the diameter of a circle whose area is equal to the combined areas of two 10-in. circles and three 2-in. circles.

The answer is in the margin on page 294.

36. Find the diameter of a circle whose area is equal to the difference between the area of a 3-in. circle and the area of a 6-in. circle.

The answer is in the margin on page 294.

Solution

$$\text{Area of 5-in. circle} = 3.1416 \times 2.5 \times 2.5$$
$$= 19.64 \text{ sq in.}$$

$$\text{Area of three 5-in. circles} = \quad 3 \times 19.64 = 58.90 \text{ sq in.}$$

$$\text{Area of 4-in. circle} = 3.1416 \times 2 \times 2$$
$$= 12.5664 \text{ sq in.}$$

$$\text{Area of two 4-in. circles} = 2 \times 12.5664 = \underline{25.1328 \text{ sq in.}}$$

$$\text{Combined areas} = 84.03 \text{ sq in.}$$

$$d = 2\sqrt{\frac{A}{\pi}} = 2\sqrt{\frac{84.03}{3.1416}} = 2\sqrt{26.7475} = 2 \times 5.17$$

$$= 10.34$$

The diameter of the required circle is 10.34 in.

Practice check: Do exercise 35 on the left.

■ **Example 3**

Find the diameter of a circle whose area is equal to the difference between the area of an 8-in. circle and the area of a 4-in. circle.

Solution $\text{Area of 8-in. circle} = 3.1416 \times 4 \times 4$
$$= 50.26 \text{ sq in.}$$

$$\text{Area of 4-in. circle} = 3.1416 \times 2 \times 2$$
$$= 12.57 \text{ sq in.}$$

$$\text{Difference between the two areas} = 50.26 - 12.57$$
$$= 37.69 \text{ sq in.}$$

$$d = 2\sqrt{\frac{A}{\pi}} = 2\sqrt{\frac{37.69}{\pi}} = 2\sqrt{12} = 6.93 \text{ in.}$$

The diameter of the required circle is 6.93 in.

Practice check: Do exercise 36 on the left.

■ **Problems 8–23**

Use your knowledge of finding diameters to solve the following problems. Work out the answers in the space provided below.

1. Find the diameter of a circle equal in area to a 6″ circle and a 4″ circle.

2. Find the diameter of a circle equal in area to the combined areas of a 2″ circle, a 3″ circle, and a 5″ circle.

3. Find the diameter of a circle whose area is equal to the area of six circles, each $1\frac{1}{2}''$ in diameter.

4. What must be the diameter of a circle whose area is equal to the sum of the areas of three $1''$ circles and three $\frac{1}{2}''$ circles?

Answer to exercise 34

Area of 20-in circle
$$= 3.1416 \times 10 \times 10$$
$$= 314.16 \text{ sq in.}$$

Area of 24-in circle
$$= 3.1416 \times 12 \times 12$$
$$= 452.39 \text{ sq in.}$$

Combined areas $= 766.55$ sq in.

$$d = 2\sqrt{\frac{A}{\pi}} = 2\sqrt{\frac{766.55}{3.1416}}$$
$$= 2\sqrt{244}$$
$$= 2 \times 15.62$$
$$= 31.24 \text{ in.}$$

---

5. What must be the diameter of a circle whose area is equal to the sum of the areas of ten $2\frac{1}{2}''$ circles?

6. Two $3''$ pipes run together into one pipe. What should be the diameter of the large pipe to carry off the flow from the other two pipes?

7. A $2''$ pipe and another pipe are used to carry off the flow from a $3''$ pipe. Find the size of the other pipe.

8. How many $1''$ pipes are required to carry as much water as a $2''$ pipe?

9. How many air ducts $10''$ in diameter are required to have the same cross-sectional area as one air duct of rectangular cross section $24'' \times 40''$?

10. How many copper wires of No. 20 B. & S. gage ($0.032''$-diameter) are required to have the same cross-sectional area as a No. 0000 B. & S. gage wire whose diameter is $0.46''$?

8-24 SHORT METHOD OF COMPARING AREAS OF CIRCLES

It is convenient to use a shorter method of comparing the areas of circles. The area of a 5-in. square is equal to the combined areas of a 4-in. square and a 3-in. square (see Fig. 8-54).

$$3^3 + 4^2 = 5^2$$

$$9 + 16 = 25$$

To find the size of a square equal in area to the combined areas of two or more squares, we add those areas and find the square root of the sum.

$$\sqrt{3^2 + 4^2} = \sqrt{25} = 5 \text{ in.} \qquad \text{the size of square required}$$

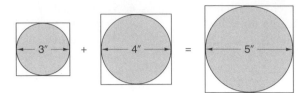

Figure 8–54

This same rule holds for circles; that is, *the areas of circles bear the same ratio to each other as the squares of their diameters.* The area of a 5-in. circle is equal to the combined areas of a 4-in. circle and a 3-in. circle. Instead of actually computing the areas of the circles, square their diameters and get the square root of the sum of the squares.

■ Example

Find by the short method the diameter *d* of a circle whose area is equal to the combined areas of a 6-in. circle, an 8-in. circle, and a 9-in. circle.

37. Find by the short method the diameter of a circle whose area is equal to the combined areas of a 10-in. circle, a 4-in. circle, and a 12-in. circle.

The answer is in the margin on page 294.

Solution
$$
\begin{aligned}
d &= \sqrt{6^2 + 8^2 + 9^2} \\
&= \sqrt{36 + 64 + 81} \\
&= \sqrt{181} \\
&= 13.45 \text{ in.}
\end{aligned}
$$

A circle 13.45 in. in diameter has an area equal to the sum of the areas of a 6-in. circle, an 8-in. circle, and a 9-in circle.

Practice check: Do exercise 37 on the left.

■ Problems 8–24

Find the diameter of a circle equal in area to the sum of the areas of the circles in each problem of Problem Set 8–23 by using the short method of comparing areas of circles. Compare the results with those obtained by the long method.

8–25   AREAS OF RING SECTIONS

Rule _____

To find the area of the ring shown in Fig. 8–55, subtract the area of the inside circle from the area of the outside circle.

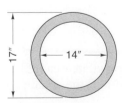

Figure 8–55

■ **Example 1**

Find the area of the cross section of a pipe whose inside diameter is 14 in. and outside diameter 17 in.

Solution   Area of 17-in. circle $= 3.1416 \times 8\frac{1}{2} \times 8\frac{1}{2} = 226.98$ sq in.
             Area of 14-in. circle $= 3.1416 \times 7 \times 7 \quad = \underline{153.94 \text{ sq in.}}$
                                           Net area $= \quad 73.04$ sq in.

The area of the ring section is 73.04 sq in.

■ **Example 2**

A circular flower bed 14 ft in diameter is surrounded by a gravel footpath that is 5 ft wide. Find the area of the path.

Solution   Area of 24-ft bed and path $= 3.1416 \times 12^2 = 452.39$ sq ft
             Area of 14-ft bed $= 3.1416 \times 7^2 = \underline{153.94 \text{ sq ft}}$
                                     Net area $= 298.45$ sq ft

Explanation   The diameter from the center of the garden to the outside border of the path is 14 ft + 5 ft + 5 ft = 24 ft.

Practice check: Do exercise 38 on the right.

38. Find the area of the cross section of a pipe whose inside diameter is 24 in. and outside diameter 20 in.

The answer is in the margin on page 295.

■ **Problems 8–25**

Using the space provided, find the areas of the following ring sections.

1. Find the area of a flat ring whose outside diameter is 11″ and inside diameter 7″.

2. Find the area of cross section of a brass pipe $2\frac{1}{2}″$ inside diameter if the metal is $\frac{1}{4}″$ thick.

3. How many square feet of pavement are there in a path around a fountain 32′ in diameter if the path is 6′6″ wide?

4. What is the area of cross section of a hollow cast-iron column if the outside diameter is 14″ and the metal is $1\frac{1}{4}″$ thick?

5. At $24.95 per square yard, find the cost of paving a circular walk whose inside diameter is 125′. The walk is 8′6″ wide.

6. Find the area of a washer whose inside diameter is $\frac{1}{2}″$ and whose outside diameter is $\frac{5}{8}″$.

Answer to exercise 35

Area of a 10-in. circle
$$= 3.1416 \times 5 \times 5$$
$$= 78.54 \text{ sq in.}$$

Area of 2-in. circle
$$= 3.1416 \times 1 \times 1$$
$$= 3.1416$$

Area of two 10-in. circles
$$= 2 \times 78.54$$
$$= 157.08 \text{ sq in.}$$

Area of three 2-in. circles
$$= 3 \times 3.1416$$
$$= 9.42 \text{ sq in.}$$

Combined areas = 166.5 sq in.

$$d = 2 \sqrt{\frac{166.5}{3.1416}} = 2\sqrt{53}$$
$$= 2 \times 7.28$$
$$= 14.56 \text{ in.}$$

Answer to exercise 36

Area of 3-in. circle
$$= 3.1416 \times 1.5 \times 1.5$$
$$= 7.07 \text{ sq in.}$$

Area of 6-in. circle
$$= 3.1416 \times 3 \times 3$$
$$= 28.27 \text{ sq in.}$$

Difference = 28.27 − 7.07
$$= 21.2 \text{ sq in.}$$

$$d = 2 \sqrt{\frac{21.2}{3.1416}} = 2 \times \sqrt{6.75}$$
$$= 5.2 \text{ in.}$$

Answer to exercise 37

$$d = \sqrt{10^2 + 4^2 + 12^2}$$
$$= \sqrt{100 + 16 + 144}$$
$$= \sqrt{260}$$
$$= 16.12 \text{ in.}$$

7. How many square feet of sod are required for a circular lawn 150′ in diameter if the center of the lawn is occupied by a fountain 35′ in diameter?

8. What is the cross-sectional area of a pipe whose inside and outside diameters are, respectively, $3\frac{1}{2}''$ and $3\frac{3}{4}''$?

9. A flat steel ring $12\frac{1}{2}''$ in diameter has a $2\frac{1}{2}''$ hole in the center. What is its area?

10. What is the cross-sectional area of a steel collar $\frac{3}{16}''$ thick whose inside diameter is $3\frac{3}{8}''$?

8–26   SHORT METHOD OF FINDING THE AREAS OF RING SECTIONS

There is a short method for finding the area of a ring section.

Rule

Multiply the average circumference of the ring by the thickness of the ring.

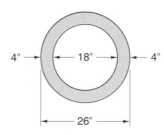

Figure 8–56

In Fig. 8–56 the inside diameter is 18 in. and the outside diameter is 26 in. The thickness of the ring is therefore $\frac{26 - 18}{2} = \frac{8}{2} = 4$ in. The average diameter is half the sum of the inside diameter and the outside diameter: $\frac{26 + 18}{2} = \frac{44}{2} = 22$ in. The circumference of a 22-in. circle is $3.1416 \times 22 = 69.12$ in.

$$\text{Area of ring section} = \text{Average circumference} \times \text{Thickness}$$
$$= 69.12 \times 4$$
$$= 276.46 \text{ sq in.}$$

■ **Example**

Find the area of a flat ring with a 26-in. inside diameter and a 33-in. outside diameter.

Solution

$$\text{Area} = \text{Average circumference} \times \text{Thickness}$$

$$\text{Average diameter} = \frac{26 + 33}{2} = \frac{59}{2} = 29\tfrac{1}{2} \text{ in.}$$

$$\text{Thickness} = \frac{33 - 26}{2} = \frac{7}{2} = 3\tfrac{1}{2} \text{ in.}$$

$$\text{Area} = 3.1416 \times 29\tfrac{1}{2} \times 3\tfrac{1}{2}$$
$$= 324.4 \text{ sq in.}$$

Practice check: Do exercise 39 on the right.

■ **Problems 8–26**

Find the area of each ring section in Problem Set 8–25 by using the short method. Compare the results with those obtained by the longer method.

Answer to exercise 38

Area of 24-in. circle
     = 3.1416 × 12 × 12
     = 452.39 sq in.

Area of 20-in. circle
     = 3.1416 × 10 × 10
     = 314.16 sq in.

Net area = 138.23 sq in.

39. Find the area of a flat ring 17 cm inside diameter and 25 cm outside diameter.

The answer is in the margin on page 297.

8–27   ARCS AND SECTORS OF CIRCLES

A part of a circle such as AOB in Fig. 8–57 is called a **sector** of the circle. A part of the circumference such as AB is called an **arc.** The sector is bounded by the two radii OA and OB and the arc AB. The angle AOB at the center is called the **central angle** of the sector. A central angle is measured by the same number of degrees as its intercepted arc. A 40° sector means a sector whose central angle is 40°. Since 1 degree of arc means $\frac{1}{360}$ of the circumference of the circle, the length of a 40° arc is $\frac{40}{360}$ of the circumference of the circle.

$$\text{Length of arc} = \pi d \left(\frac{x}{360°} \right)$$

where x is the number of degrees of the arc.

$$\text{Area of sector} = \pi r^2 \left(\frac{x}{360°} \right)$$

where x is the number of degrees of the sector.

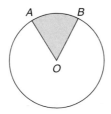

Figure 8–57

■ **Example 1**

Find the length of a 35° arc of a 16-in. circle.

40. Find the length of a 15° arc of a 22-in. circle.

The answer is in the margin on page 298.

Solution   Circumference of 16-in. circle = 3.1416 × 16 = 50.26 in.

Length of 35° arc = $\frac{35}{360}$ × 50.26 = 4.89 in.

Practice check: Do exercise 40 on the left.

■ **Example 2**

Find the area of a 40° sector of a 16-in. circle.

Solution   Area of 16-in. circle = 3.1416 × 8 × 8 = 201.06 sq in.

Area of 40° sector = $\frac{40}{360}$ × 201.06 = 22.34 sq in.

41. Find the area of an 80° sector of a 22-cm circle.

The answer is in the margin on page 298.

Explanation   Since a 1° sector contains $\frac{1}{360}$ of the area of the circle, a 40° sector contains $\frac{40}{360}$ of the area of the circle.

Practice check: Do exercise 41 on the left.

■ **Problems 8–27**

Use the space provided below to find **(a)** the length of arc and **(b)** the area of sector for each of the following.

1. 60° sector of a 9″ circle

2. 27° sector of a $3\frac{1}{2}″$ circle

3. 94° sector of a 45′ circle

4. 118°30′ sector of a $4\frac{1}{2}″$ circle

5. 36°30′ sector of a circle $2′8\frac{1}{2}″$ in diameter

6. 90° sector of a 20″ circle

7. 45° sector of a 24″ circle

8. 30° sector of a 3′ circle

9. 15° sector of a circle 18″ in diameter

10. 40°45′ sector of a circle 2′4″ in diameter

11. 170°10′ sector of a 30-cm circle

12. 89° sector of a 15′9″ circle

13. 230°30′ sector of a 22¾″ circle

14. 53°45′ sector of a 45.6-cm circle

15. 352°20′ sector of a 10′4″ circle

8–28   CIRCLES AND REGULAR FIGURES

A circle is inscribed in another figure when it touches each of the sides of that figure. If a circle is inscribed in a figure, then the sides of that figure are **tangent** to the circle. In Fig. 8–58 circles are inscribed in an equilateral triangle, a square, a regular hexagon, and a regular octagon.

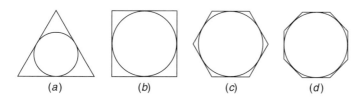

(a)   (b)   (c)   (d)

Figure 8–58

A circle is circumscribed about a figure when all the corners of that figure touch the circle. In Fig. 8–59 the circle is circumscribed about the triangle, square, and so forth. The figures are inscribed in the circle.

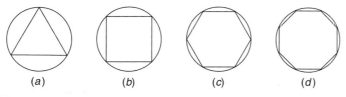

(a)   (b)   (c)   (d)

Figure 8–59

Most of the relations between the circle and the inscribed and circumscribed figures are evident from an inspection of the figure. In Fig. 8–58(*b*) the diameter of the inscribed circle equals the side of the square. In Figs. 8–58(*c*) and 8–58(*d*), the diameters of the inscribed circles are equal to the distance across the flats of the regular hexagon and the regular octagon, respectively. In Fig. 8–59(*b*) the diameter of the circumscribed circle is equal to the diagonal of the square; in Figs. 8–59(*c*) and 8–59(*d*), the diameters of the circles are equal to the distance across the corners of the regular hexagon and the regular octagon, respectively.

Figure 8–60 illustrates the relation between an equilateral triangle and its inscribed and circumscribed circles. The diameter *DB* of the outer circle is divided by the construction into four equal parts: *DC = CO = OA = AB*.

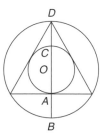

Figure 8–60

The diameter *AC* of the inscribed circle equals $\frac{2}{3}$ of the altitude of the triangle.

Also, *AD*, the altitude of the triangle, equals $\frac{3}{4}$ of the diameter of the circumscribed circle.

The expression 2:3:4 states the relation between these figures.

■ Example 1

Find the area of a square inscribed in a 9-in. circle (see Fig. 8–61).

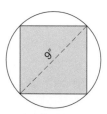

Figure 8–61

Solution   The diagonal of the square is equal to the diameter of the circle, 9 in.

$$\text{Side of square} = 0.707 \times d = 0.707 \times 9 = 6.363 \text{ in.}$$

$$\text{Area of square} = 6.363 \times 6.363 = 40.49 \text{ sq in.}$$

Practice check: Do exercise 42 on the left.

■ **Example 2**

Find the area of a circle inscribed in a regular octagon with $3\frac{1}{2}$-in. sides (see Fig. 8–62).

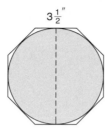

$3\frac{1}{2}''$

Figure 8–62

Solution   The diameter of the circle is equal to the distance across the flats of the octagon.

$$\text{Flats of octagon} = 2.414 \times s = 2.414 \times 3\frac{1}{2} = 8.45 \text{ in.}$$

$$\text{Area of circle} = 3.1416 \times 4.23 \times 4.23 = 56.2 \text{ sq in.}$$

Practice check: Do exercise 43 on the right.

43. Find the area of a circle inscribed in a regular octagon with $9\frac{1}{4}$-cm sides.

The answer is in the margin on page 301.

■ **Example 3**

Find the area of an equilateral triangle inscribed in a $5\frac{1}{4}$-in. circle (see Fig. 8–63).

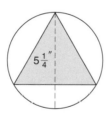

$5\frac{1}{4}''$

Figure 8–63

Solution   The altitude of the triangle is equal to $\frac{3}{4}$ of the diameter of the circle.

$$\text{Altitude of triangle} = \frac{3}{4} \times 5\frac{1}{4} = 3.94 \text{ in.}$$

$$\begin{aligned}\text{Area of triangle} &= 0.577 \times a^2 \\ &= 0.577 \times 3.94 \times 3.94 = 8.96 \text{ sq in.}\end{aligned}$$

Practice check: Do exercise 44 on the right.

44. Find the area of an equilateral triangle inscribed in an $11\frac{3}{4}$-in. circle.

The answer is in the margin on page 301.

■ **Problems 8–28**

Do the following problems. Using the space provided below, draw a neat diagram for each and label all known parts. Organize your work as shown in the example problems. Refer to the Table of Constants on page 244 for the needed ratios.

1. Find the area of a square inscribed in a $6\frac{1}{2}''$ circle.

2. Find the area of an equilateral triangle inscribed in a 35' circle.

3. Find the area of a regular hexagon inscribed in a $1\frac{1}{4}''$ circle.

4. Find the area of a regular octagon inscribed in a 4″ circle.

5. Find the area of a circle inscribed in a $3\frac{1}{2}''$ square.

6. Find the area of a circle inscribed in an equilateral triangle whose sides are 4.72″.

7. Find the area of a circle inscribed in a regular octagon with $1\frac{1}{2}''$ sides.

8. Find the area of a circle inscribed in a regular hexagon with $5\frac{3}{4}''$ sides.

9. Find the flats of a regular hexagon inscribed in a $\frac{3}{4}''$ circle.

10. Find the flats of a regular octagon inscribed in a 12″ circle.

11. A circle is inscribed in a regular hexagon with $2\frac{1}{2}''$ sides. What is the diameter of the circle?

12. Eight holes equally spaced are drilled on the circumference of a $12\frac{1}{2}''$ circle. What is the straight-line distance between the centers of the holes?

13. What is the distance between holes (center to center) if three holes, equally spaced, are drilled on the circumference of a $3\frac{1}{4}''$ circle?

14. Find the depth of cut required to mill a square end on a $1\frac{3}{4}''$ round shaft.

15. Find the depth of cut required to mill a triangular end on a round shaft $\frac{7}{8}''$ in diameter.

8-29   SEGMENTS OF CIRCLES

A portion of the area of a circle such as the shaded part in Fig. 8–64 is called a *segment* of a circle. It is bounded by an arc and a straight line that joins the ends of the arc. A straight line such as AB that joins the ends of an arc is called a *chord*. The angle $B\hat{O}A$ formed at the center of the circle by the radii to the ends of the arc is called a *central angle.* A 65° segment means a segment whose central angle is 65°.

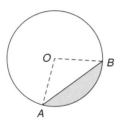

Figure 8–64

By means of the relations between circles and inscribed figures presented in the previous section, the area of certain segments can be found.

■ Example

Find the area of a 90° segment of a 24-in. circle (Fig. 8–65).

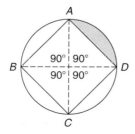

Figure 8–65

Solution   The diagonal of the square is equal to the diameter of the circle, 24 in.

$$\text{Side of square} = 0.707 \times d = 0.707 \times 24 \text{ in.} = 16.97 \text{ in.}$$

$$\text{Area of square} = 16.97 \times 16.97 = 288.0 \text{ sq in.}$$

$$\text{Area of 24-in. circle} = 3.1416 \times 12 \times 12 = 452.4 \text{ sq in.}$$

Subtract the area of the square from the area of the circle and divide by 4.

$$\text{Area of circle} = 452.4 \text{ sq in.}$$
$$\text{Area of square} = \underline{288.0 \text{ sq in.}}$$
$$\text{Area of 4 segments} = 164.4 \text{ sq in.}$$

$$\text{Area of one segment} = \frac{164.4}{4} = 41.10 \text{ sq in.}$$

45. Find the area of a 90° segment of a 45-in. circle.

The answer is in the margin on page 304.

Explanation   There are four 90° angles in a circle. Drawing these four angles and drawing chords between the ends of the arcs, construct the square $ABCD$. Subtracting the area of this square from the area of the circle will leave four segments like the one whose area we are trying to find. Dividing the result by 4 gives the area of one segment.

Practice check: Do exercise 45 on the left.

■ Problems 8–29

Find the areas of the following segments of circles. Use the space provided below to draw net diagrams. Organize your work as shown in the example problem.

1. Find the area of a segment bounded by a chord and a 120° arc of an 8″ circle.

2. Find the area of a segment bounded by a chord and a 90° arc of a 16″ circle.

3. Find the area of a segment bounded by a chord and a 60° arc of a 25″ circle.

4. Find the area of a segment bounded by a chord and a 45° arc of a 36″ circle.

5. Find the area of a segment bounded by a chord and a 120° arc of a 7-cm circle.

6. Find the area of a segment bounded by a chord and a 45° arc of an 18-in. circle.

7. In milling a square on a $1\frac{1}{2}''$ rounded shaft, how many square inches of cross-sectional area are removed?

8. A corner of a square metal ventilating tube 8 in. on a side is placed at the center of a 16-in. circle. What is the area formed by the diagonal of the tube?

8–30   THE ELLIPSE

The main components of the **ellipse** (see Fig. 8–66) are the **major axis,** AB (also called the long diameter), the **minor axis,** CD (also called the short diameter), the circumference or perimeter, and the area.

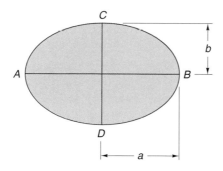

Figure 8–66

The area of the ellipse is found by the formula

$$\text{Area} = \pi ab$$

($a = \frac{1}{2}$ the long diameter and $b = \frac{1}{2}$ the short diameter.)

■ **Example 1**

Find the area of an ellipse 16 in. long and 12 in. wide.

Solution   Area $= \pi ab$

$$a = \tfrac{16}{2} = 8 \qquad b = \tfrac{12}{2} = 6$$

$$\text{Area} = 3.1416 \times 8 \times 6 = 150.8 \text{ sq in.}$$

Practice check: Do exercise 46 on the right.

46. Find the area of an ellipse 22 in. long and 18 in. wide.

The answer is in the margin on page 305.

To find the perimeter of an ellipse, use the following approximate formula.

$$\text{Perimeter} = \pi\sqrt{2(a^2 + b^2)}$$

This formula shows that the perimeter is equal to π times the square root of twice the sum of a^2 and b^2. The several steps in the process are as follows.

1. Substitute in the formula the values of π, a, and b.

2. Square a and b.

3. Add these squares.

4. Multiply the sum of the squares by 2.

5. Find the square root of the product.

6. Multiply the square root of the product by 3.1416.

■ **Example 2**

Find the perimeter of an ellipse with a long diameter of 18 in. and a short diameter of 14 in. (Fig. 8–67). Use 3.1416 for π.

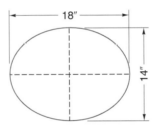

Figure 8–67

Answer to exercise 45

Side of square $= 0.707 \times 45$
$= 31.815$ in.

Area of square $= 31.815^2$
$= 1,012.19$ sq in.

Area of 45-in. circle
$= 3.1416 \times 22.5^2$
$= 1,590.435$ sq in.

Area of 4 segments
$= 578.245$ sq in.

Area of one segment
$= 144.56$ sq in.

Solution   Perimeter $= \pi\sqrt{2(a^2 + b^2)}$
$= 3.1416\sqrt{2(9^2 + 7^2)}$
$= 3.1416\sqrt{2(81 + 49)}$
$= 3.1416\sqrt{2 \times 130}$
$= 3.1416\sqrt{260}$
$= 3.1416 \times 16.1245$
$= 50.66$ in.

47. Find the perimeter of an ellipse with a long diameter of 30 cm and a short diameter of 25 cm.

The answer is in the margin on page 306.

Practice check: Do exercise 47 on the left.

■ **Problems 8–30**

Find the areas and perimeters of the ellipses in the following chart. Use 3.14 for π. Round each answer to the nearest hundredth. Try to estimate the answer before solving the problem.

	Ellipses					
	Long Diameter	**Short Diameter**	*a*	*b*	**Area**	**Perimeter**
1.	8″	5″	_____	_____	_____	_____
2.	36″	10″	_____	_____	_____	_____
3.	6½″	4¼″	_____	_____	_____	_____
4.	7¾″	5½″	_____	_____	_____	_____
5.	4.2″	2.6″	_____	_____	_____	_____
6.	10.8″	5.4″	_____	_____	_____	_____
7.	65′	43′	_____	_____	_____	_____
8.	5′2″	3′4″	_____	_____	_____	_____
9.	16½″	14.8″	_____	_____	_____	_____
10.	5.62″	3.57″	_____	_____	_____	_____

Answer to exercise 46

Area = 3.1416 × 11 × 9
= 311.02 sq in.

Use the space provided below to work out the answers to the following problems.

11. What is the area of an insulated cover for an elliptical-shaped hot tub 7 ft long and 5 ft wide?

12. The base of the Colosseum in Rome is in the form of an ellipse. The ellipse is 189 m long and 155 m wide. Find its area and circumference.

8–31   SUMMARY OF FORMULAS FOR PLANE FIGURES

Right Triangle   (Fig. 8–68)

Altitude (*a*) $a = \sqrt{c^2 - b^2}$

Base (*b*) $b = \sqrt{c^2 - a^2}$

Hypotenuse (*c*) $c = \sqrt{a^2 + b^2}$

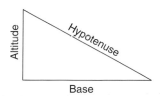

Base

Figure 8–68

Triangle   (Fig. 8–69)

Area (A) $A = \frac{1}{2}ba$

Side (s) $A = \sqrt{s(s - a)(s - b)(s - c)}$

Altitude (a)

Base (b)

Hypotenuse (c)

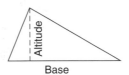

Figure 8–69

Rectangle   (Fig. 8–70)

Area (A) $A = lw$

Length (l)

Width (w)

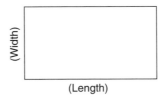

Figure 8–70

Trapezoid   (Fig. 8–71)

Area (A) $A = \left(\dfrac{b_1 + b_2}{2}\right)a$

Altitude (a) Average width $= \dfrac{b_1 + b_2}{2}$

Upper base (b_1)

Lower base (b_2)

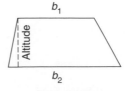

Figure 8–71

Circle (Fig. 8–72)

Circumference (C) $C = \pi d$

Area (A) $A = \pi r^2$

$$A = \frac{\pi d^2}{4}$$

Diameter (d) $d = \dfrac{C}{\pi}$

Radius (r)

pi (π)

Arc length (l_a) $l_a = \left(\dfrac{\text{deg}_a}{360}\right) C$

Arc degrees (deg_a)

Sector area (A_s) $A_s = \left(\dfrac{\text{deg}_s}{360}\right) A$

Sector degrees (deg_s)

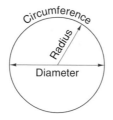

Figure 8–72

Flat Ring (Fig. 8–73)

Area (A) $A = \left(\dfrac{d_1 + d_2}{2}\right) t$

Inside diameter (d_1) Average circumference $= \dfrac{d_1 + d_2}{2}$

Outside diameter (d_2)

Thickness (t)

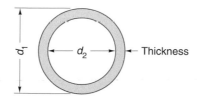

Figure 8–73

Ellipse   (Fig. 8–74)

Area (A)	$A = \pi ab$
$\frac{1}{2}$ long diameter (a)	$a = \dfrac{A}{\pi b}$
$\frac{1}{2}$ short diameter (b)	$b = \dfrac{A}{\pi a}$
Perimeter (P)	$P = \pi\sqrt{2(a^2 + b^2)}$
pi (π)	

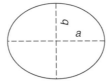

Figure 8–74

■ Self-Test

Use the space provided below to do the following problems. Check your answers with the answers in the back of the book. Use the Table of Constants (page 244) and the Summary of Formulas for Plane Figures (pages 305–308) when taking this test. Use 3.1416 for π and round your answers to the nearest hundredth. Try to estimate the answer before solving the problem.

1. Find the altitude of an equilateral triangle whose side is 6 in.

2. Find the area of a square whose diagonal is 6 in.

3. Find the distance across the flats of a regular hexagon whose diagonal is 4 in.

4. Find the circumference of a 6-in. circle.

5. Find the area of a 10-in. circle

6. Find the diameter of a circle that is equal in area to two 10-in. circles.

7. Find the area of a flat ring whose outside diameter is 1 in. and whose inside diameter is $\frac{1}{2}$ in.

8. Find the area of a 60° sector of a 6-in. circle.

9. Find the area of an ellipse with a 12-in. long diameter and a 10-in. short diameter.

10. Find the perimeter of an ellipse with a 6-in. long diameter and a 4-in. short diameter.

■ Chapter Test

Use the Table of Constants (page 244) and the Summary of Formulas for Plane Figures (pages 305–308) when taking this test. Use 3.1416 as an approximation for π. Use the space provided below to make a neat diagram to accompany each problem. Organize your work.

1. Find the altitude of an equilateral triangle whose side is 9″.

2. Find the area of an equilateral triangle whose altitude is 12″.

3. Find the area of a square whose diagonal is 8″.

4. Find the distance across the flats of a regular hexagon whose side is 10″.

5. Find the diagonal of a regular octagon whose side is 20″.

6. Find the area of a trapezoid whose upper base is 10″, lower base 14″, and altitude 8″.

7. A screwdriver 15″ long is drawn to $\frac{1}{4}$ scale. What will be the length of the line in the drawing that represents the length of this screwdriver?

8. Find the circumference of a $10\frac{1}{2}″$ circle.

9. Find the diameter of a circle whose perimeter is 30″.

10. Find the area of a $5\frac{1}{2}''$ circle.

11. Find the diameter of a circle whose area is 1,000 sq in.

12. Find the diameter of a circle whose area is equal to the sum of the areas of two 5″ circles.

13. Find the area of a flat ring whose outside diameter is 12″ with inside diameter 10″.

14. Find the length of the arc of a 60° sector of a 10″ circle.

15. Find the area of the sector described in Problem 14.

16. Find the area of a square inscribed in a 15″ circle.

17. Find the area of a square circumscribed about a 15″ circle.

18. Find the area of a segment bounded by a chord and a 120° arc of a 12″ circle.

19. Find the area of an ellipse with a 15″ long diameter and a 10″ short diameter.

20. Find the perimeter of the ellipse described in Problem 19.

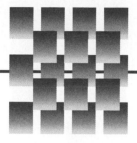

CHAPTER 9

Solids

9–1 DEFINITIONS

A solid has three dimensions—length, width, and height. This chapter discusses the volume and the surface area of some of the common solids that appear frequently in practical work. The vocational and technical student is expected to understand these solids.

9–2 PRISMS AND CYLINDERS

Prisms are solid figures with parallel edges and uniform cross section. If the edges are perpendicular to the bases, the prisms are called *right prisms* [Fig. 9–1(*a–h*)].
    When the base of a right prism is a circle, the figure is a *cylinder.*

9–3 VOLUMES OF PRISMS

The cubic content or capacity of a solid figure is called the *volume.* To find the volume of any prism or cylinder, multiply the area of the base by the height. The formula is

$$\text{Volume} = \text{Area of base} \times \text{Height}$$

■ **Example 1**
Find the volume of a rectangular prism 16 in. high whose base is a $9\frac{1}{2}$-in. square (Fig. 9–2).

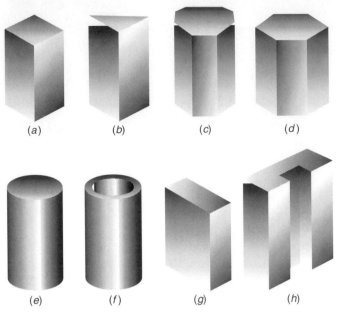

Figure 9–1

1. Find the volume of a rectangular prism 8 ft high whose
base is a $2\frac{1}{4}$-ft square.

The answer is in the margin on
page 314.

Solution

$$\text{Volume} = \text{Area of base} \times \text{Height}$$

$$\text{Area of base} = 9\frac{1}{2} \times 9\frac{1}{2} = 90.25 \text{ sq in.}$$

$$\text{Volume} = 90.25 \times 16$$
$$= 1{,}444 \text{ cu in.}$$

Figure 9–2

Practice check: Do exercise 1 on the left.

■ **Example 2**

Find the volume of a cylinder $4\frac{1}{2}$ in. in diameter and 5 in. long (Fig. 9–3).

Figure 9–3

Solution

$$\text{Volume} = \text{Area of base} \times \text{Height}$$

$$\text{Area of base} = \pi r^2$$
$$= 3.1416 \times 2\frac{1}{4} \times 2\frac{1}{4} = 15.90 \text{ sq in.}$$

$$\text{Altitude} = 5 \text{ in.}$$

$$\text{Volume} = 15.90 \times 5$$
$$= 79.5 \text{ cu in.}$$

2. Find the volume of a cylinder
$3\frac{3}{4}$ in. in diameter and 9 in.
long.

The answer is in the margin on
page 314.

Practice check: Do exercise 2 on the left.

■ Example 3

Find the volume of the solid shell of a hollow cylinder $16\frac{1}{2}$ in. high (Fig. 9–4); the inside diameter is $8\frac{1}{2}$ in. and the outside diameter 10 in.

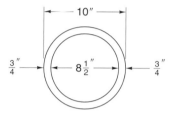

Figure 9–4

Solution

$$\text{Volume} = \text{Area of base} \times \text{Height}$$
$$\text{Area of base (flat ring)} = \text{Average circumference} \times \text{Thickness}$$
$$= 3.1416 \times 9\frac{1}{4} \times \frac{3}{4}$$
$$= 21.8 \text{ sq in.}$$
$$\text{Volume} = 21.8 \times 16\frac{1}{2}$$
$$= 359 \text{ cu in.}$$

Practice check: Do exercise 3 on the right.

3. Find the volume of a solid shell of a hollow cylinder $10\frac{1}{2}$ yd high. The inside diameter is 6 yd and the outside diameter $12\frac{1}{2}$ yd.

The answer is in the margin on page 315.

■ Problems 9–3

Use the space provided to solve the following problems.

1. Find the volume of a rectangular prism whose base is $12'' \times 12''$ and height $24''$.

2. Find the volume of a rectangular prism whose base is $12'' \times 4''$ and height $36''$.

3. Find the volume of a rectangular prism whose base is $3\frac{1}{2}'' \times 5''$ and height $8''$.

4. Find the volume of a rectangular prism whose base is $4\frac{1}{2}'' \times 4\frac{1}{2}''$ and height $6''$.

5. Find the volume of a rectangular prism whose base is $5\frac{1}{4}'' \times 6\frac{3}{4}''$ and height $8\frac{1}{2}''$.

6. Find the volume of a rectangular prism whose base is $1'2'' \times 1'$ and height $1'6''$.

7. Find the volume of a rectangular prism whose base is $1'3\frac{1}{2}'' \times 1'4\frac{1}{2}''$ and height $1'6''$.

8. Find the capacity in gallons of a rectangular tank whose base is $3' \times 3'$ and height $5'$. (1 gallon = 231 cu in.)

9. Find the capacity in gallons of a rectangular tank whose base is $1'6'' \times 2'6''$ and height $3'$. (1 gallon = 231 cu in.)

10. Find the capacity in gallons of a rectangular tank whose base is $2'5\frac{1}{2}'' \times 2'7''$ and height $3'6\frac{1}{2}''$. (1 gallon = 231 cu in.)

Solve the following problems involving the volume of prisms whose bases are not rectangles. (Use the Table of Constants on page 244 to find the area of the base.)

11. Find the volume of a prism 10″ high if the base is a hexagon 5″ across flats.

12. Find the volume of a prism $8\frac{1}{2}''$ high if the base is a hexagon $5\frac{1}{2}''$ across flats.

13. Find the volume of a prism $5\frac{1}{4}''$ high if the base is a hexagon $8\frac{1}{2}''$ across flats.

14. Find the volume of a prism $6\frac{1}{2}''$ high if the base is a hexagon whose diagonal is 6″.

15. Find the volume of a prism whose base is an equilateral triangle with 5″ sides and whose altitude is 15″.

16. Find the volume of a prism whose base is an equilateral triangle with $4\frac{3}{4}''$ sides and whose height is 20″.

17. Find the volume of a prism whose base is an equilateral triangle with $5\frac{3}{4}''$ sides and whose height is $15\frac{1}{2}''$.

18. Find the volume of a prism whose base is an equilateral triangle with $1'2''$ sides and whose height is $3'3''$.

Answer to exercise 3

Area of base $= 3.1416 \times \dfrac{37}{4} \times \dfrac{13}{4}$

$= 94.4$ sq yd

Volume $= 94.4 \times \dfrac{21}{2}$

$= 991.2$ cu yd

19. A prism whose base is an octagon with $15''$ sides is $4\frac{1}{2}''$ high. Find its volume.

20. A prism whose base is an octagon with $4\frac{1}{2}''$ sides is $10\frac{1}{4}''$ high. Find its volume.

Find the capacity of the following prisms. (1 cu ft = 1,728 cu in.; 1 gallon = 231 cu in.)

21. How many cubic feet are there in a cylinder $16''$ in diameter and $42'8''$ long?

22. Find the capacity in gallons of an elliptical tank $31''$ long if the base is an ellipse $21'' \times 17''$.

23. Find the capacity in gallons of a cylindrical tank $14'$ in diameter and $11'6''$ high.

24. Find the capacity in gallons of an elliptical tank $12'6''$ long if the long diameter of the ellipse is $5'$ and the short diameter is $3'4''$.

25. Find the capacity in gallons of a cylindrical tank $8'7''$ in diameter and $10'6''$ high.

Solve the following problems.

26. How many cubic inches of metal are there in a length of pipe $9'6''$ long if the inside diameter is $6''$ and the metal is $\frac{7}{16}''$ thick?

27. Find the weight of a bar of brass $5''$ in diameter and $6'2''$ long. (Brass weighs 512 lb per cu ft.)

28. Find the weight of water in the tank described in Problem 23. (Water weighs 62.5 lb per cu ft.)

29. Find the weight of water in the tank described in Problem 24.

30. Find the weight of water in the tank described in Problem 25.

9–4   FINDING THE HEIGHT OF A PRISM OR A CYLINDER

When the volume of a prism or cylinder and the area of the base are known, the height can be found by using the following formula.

$$\text{Height} = \frac{\text{Volume}}{\text{Area of base}}$$

■ **Example 1**

Find the height of a cylinder 26 in. in diameter to contain 6,500 cu in.

Solution   Area of base $= \pi r^2 = 3.1416 \times 13 \times 13 = 530.93$ sq in.

$$\text{Height} = \frac{\text{Volume}}{\text{Area of base}} = \frac{6,500}{530.93} = 12.24 \text{ in.}$$

■ **Example 2**

A cylindrical iodized-salt container $3\frac{1}{4}$ in. in diameter has a volume of 44.59 cu in. Find the height.

Solution   Area of base $= \pi r^2 = 3.1416 \times 1.625^2$
$= 8.296$ sq in.

$$\text{Height} = \frac{\text{Volume}}{\text{Area of base}} = \frac{44.59}{8.296}$$
$$= 5.375 \text{ in. or } 5\frac{3}{8} \text{ in.}$$

■ **Example 3**

Find the length of a cylindrical rail tank car 8 ft in diameter to have a capacity of 10,630 gallons.

Solution   Volume $= 10,630 \times 231 = 2,455,530$ cu in.
                         $= 1,421$ cu ft

            Area of base $= \pi r^2 = 3.1416 \times 4^2$
                            $= 50.265$ sq in.

            Height $= \dfrac{\text{Volume}}{\text{Area of base}} = \dfrac{1,421}{50.265}$

                      $= 28.27$ ft

Practice check: Do exercise 4 on the right.

4. Find the height of a cylinder 48 in. in diameter to contain 10,400 cu in.

The answer is in the margin on page 319.

■ **Problems 9–4**

Use the space provided below to find the height of the following.

1. How high must a cylindrical tank 8′3″ in diameter be in order to have a capacity of 2,000 gallons?

2. A gasoline tank whose cross section is an ellipse 13″ × 10″ is required to contain 16 gallons. What must be its length?

3. A cylindrical container 7″ in diameter is required to have a capacity of 1 gallon. How high must it be?

4. Find the height of a 5-gallon cylindrical container if the diameter is 12″.

5. Find the depth of a tank 4′10″ square to have a capacity of 1,000 gallons.

6. A milk container $3\frac{1}{2}″$ square is required to hold 1 quart. Find the height.

7. How high must a cylindrical can $4\frac{1}{2}″$ in diameter be in order to contain 2 quarts?

8. What should be the depth of the water in an octagonal tank 10′6″ across flats in order to provide for 10,000 gallons?

9. Find the depth of a rectangular vat 2′6″ × 5′10″ to contain 500 gallons.

10. The volume of a steel shaft 6″ in diameter is 2,560 cu in. Find its length.

9–5   FINDING THE AREA OF THE BASE OF A PRISM OR A CYLINDER

When the volume and the height are known, divide the volume by the height. The formula is

$$\text{Area of base} = \frac{\text{Volume}}{\text{Height}}$$

■ **Example 1**

A cylindrical tank 11 ft high has a volume of 1,050 cu ft. Find the area of the base.

Solution $\text{Area of base} = \dfrac{\text{Volume}}{\text{Height}}$

$$= \frac{1{,}050}{11} = 95.5 \text{ sq ft}$$

If the diameter of the cylinder is required, use the formula

$$d = 2\sqrt{\frac{A}{\pi}}$$

In this example,

$$d = 2\sqrt{\frac{95.5}{3.1416}} = 2\sqrt{30.3985} = 2 \times 5.51 = 11.02 \text{ ft}$$

■ **Example 2**

A rectangular metal can is 6 in. long, 4.5 in. wide, and 8.5 in. high. Does it hold 1 gallon?

Solution   Volume = 6 × 4.5 × 8.5
$\qquad\qquad\qquad\quad$ = 229.5 cu in.

The can does not hold 1 gallon.

Explanation   There are 231 cu in. in 1 gallon. Thus, 229.5 cu in. are not enough.

Practice check: Do exercise 5 on the left.

---

5. A cylindrical tank 14 m high has a volume of 1,540 m³. Find the area of the base and the diameter.

The answer is in the margin on page 320.

■ Problems 9–5

Use the space provided below to find the area of the base of the following prisms.

1. Find the area of the base of a 25-gallon tank 22″ high.

2. If the tank in Problem 1 is a cylinder, find the diameter of the base.

3. If the tank in Problem 1 has a square base, find the length of the sides.

4. A cylindrical tank 6′8″ high is required to contain 200 gallons. What should be its diameter?

5. What should be the diameter of a cylindrical container 10″ high to have a capacity of 1 gallon?

6. Find the diameter of a 1-quart cylindrical container 8″ high.

7. A square water tank 12′8″ high has a capacity of 5,000 gallons. Find the length of the sides.

8. Find the diameter of an oil drum $31\frac{1}{2}$″ high to contain 63 gallons.

9. A cylindrical 10-gallon container is 20″ high. Find the diameter.

10. A circular vat 18″ deep is required to have a capacity of 350 gallons. Find its diameter.

Answer to exercise 4

Area of base = 3.1416×24^2
= 1,809.56 sq in.

Height = $\dfrac{10,400}{1,809.56}$
= 5.75 in.

9–6 LATERAL SURFACES OF PRISMS AND CYLINDERS

Lateral surface is the area of the *sides* of the prism. A sheet of paper that can be wrapped around the prism or cylinder to cover the outside will be a rectangle as high as the prism or cylinder and as long as its perimeter (see Fig. 9–5). Thus, the formula for the lateral surface of a prism is

$$\text{Lateral surface} = \text{Perimeter of base} \times \text{Height}$$

Answer to exercise 5

Area of base $= \dfrac{1{,}540}{14}$

$ = 110 \text{ m}^2$

$d = 2\sqrt{\dfrac{110}{3.1416}}$

$ = 2\sqrt{35}$

$ = 2 \times 5.92$

$ = 11.84 \text{ m}$

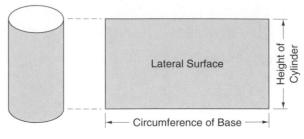

Figure 9–5

■ **Example 1**

Find the lateral surface of a cylinder 6 in. in diameter and 12 in. high.

Solution   Lateral surface = Perimeter of base $\times$ Height

Perimeter of base $= 3.1416 \times 6 = 18.84$ in.

Lateral surface $= 18.84 \times 12$

$ = 226.08$ sq in.

Practice check: Do exercise 6 on the left.

6. Find the lateral surface of a cylinder 8 in. in diameter and 22 in. high.

The answer is in the margin on page 322.

■ **Example 2**

Find the lateral surface of a prism whose height is $5\frac{1}{2}$ in. and whose base is a hexagon $11\frac{1}{2}$ in. across flats.

Solution   Lateral surface = Perimeter of base $\times$ Height

Side of hexagon $= 0.577 \times f = 0.577 \times 11\frac{1}{2} = 6.64$ in.

Perimeter of base $= 6 \times 6.64 = 39.84$ in.

Lateral surface $= 39.84 \times 5\frac{1}{2}$

$ = 219.12$ sq in.

Practice check: Do exercise 7 on the left.

7. Find the lateral surface of a prism whose height is $7\frac{1}{4}$ m and whose base is an octagon $25\frac{1}{2}$ m across flats.

The answer is in the margin on page 322.

■ **Problems 9–6**

Use the space provided below to solve the following problems.

1. Find the lateral surface of a cylinder 18″ in diameter and 12″ high.

2. Find the lateral surface of a cylinder $8\frac{1}{2}$″ in diameter and 1′4″ high.

3. Find the lateral surface of a prism with a 16″ square base and 8′4″ height.

4. Find the lateral surface of a prism with a $1'4\frac{1}{2}$″ square base and $3'6\frac{1}{2}$″ height.

5. Find the lateral surface of a triangular prism with 14″ sides and 32″ height.

6. Find the lateral surface of a triangular prism with $8\frac{1}{2}$″ sides and $12\frac{1}{2}$″ height.

7. Find the lateral surface of a prism 7″ high. The base is a hexagon with $1\frac{1}{2}$″ sides.

8. Find the lateral surface of a prism 6′10″ high. The base is a hexagon with 2′6″ sides.

9. Find the lateral surface of an octagonal prism 2′ high. The base is an octagon 1′8″ across flats.

10. Find the lateral surface of an octagonal prism 2′4″ high. The base is an octagon 2′4″ across flats.

11. At $1.55 per square yard find the cost of painting the sides of a cylindrical gas tank 62′ high and 85′ in diameter.

12. A sheet metal duct is 28″ in diameter and 65′6″ long. How many square feet of metal are required for this duct, making no allowances for joints?

13. The sides and bottom of a vat 4′2″ square and 6′8″ high are lined with sheet lead. How many square feet of lead are required?

14. A rectangular tank 4′10″ by 6′8″ and 5′6″ high is lined (sides and bottom) with sheet copper. How many square feet of copper are required?

15. A sheet metal duct of rectangular cross section 22″ by 40″ is 38′6″ long. Find the number of square feet of metal required.

9–7   PYRAMIDS AND CONES

Figures like those shown in Figs. 9–6(*a–c*) are called *pyramids.* When the base is a circle, the figure is called a *cone* (Fig. 9–7).

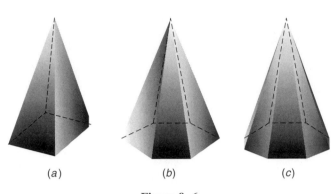

(*a*)   (*b*)   (*c*)

Figure 9–6   **Figure 9–7**

9–8   VOLUMES OF PYRAMIDS AND CONES

Figure 9–8 shows a comparison between a cylinder and a cone of the same base and altitude. *The volume of the cone is one-third the volume of the cylinder.* The volume of the cylinder is equal to the product of the area of the base and the altitude; the volume of the cone equals $\frac{1}{3}$ times the area of the base times the height. The same rule applies to pyramids. Thus,

$$\text{Volume of pyramid or cone} = \tfrac{1}{3} \times \text{Area of base} \times \text{Height}$$

Figure 9–8

■ **Example 1**

Find the volume of a cone 8 in. high; the base is 6 in. in diameter.

Solution   Volume = $\frac{1}{3}$ × Area of base × Height

Area of base = 3.1416 × 3 × 3 = 28.26 sq in.

Volume = $\frac{1}{3}$ × 28.26 × 8
       = 75.4 cu in.

Practice check: Do exercise 8 on the left.

8. Find the volume of a cone 12 ft high; the base is 7 ft in diameter.

The answer is in the margin on page 324.

■ **Example 2**

Find the volume of a pyramid $27\frac{1}{2}$ in. high; the base is a rectangle 19 in. × 16 in.

Solution

Volume = $\frac{1}{3}$ × Area of base × Height

Area of base = $19 \times 16 = 304$ sq in.

Volume = $\frac{1}{3} \times 304 \times 27\frac{1}{2}$
= 2,786.7 cu in.

Practice check: Do exercise 9 on the right.

9. Find the volume of a pyramid $42\frac{1}{4}$ m high; the base is a rectangle 23 m × 18 m.

The answer is in the margin on page 325.

■ **Problems 9–8**

Use the space provided below to solve the following problems.

1. Find the volume of a cone 36″ high if the base is 15″ in diameter.

2. Find the volume of a cone 10″ high if the base is $14\frac{1}{2}$″ in diameter.

3. Find the volume of a pyramid whose base is a 28″ square and whose height is 12″.

4. Find the volume of a pyramid whose base is a $15\frac{1}{2}$″ square and whose height is $18\frac{1}{2}$″.

5. Find the volume of a hexagonal pyramid 10″ high; the base is a hexagon with 5″ sides.

6. Find the volume of a hexagonal pyramid $9\frac{1}{2}$″ high; the base is a hexagon with 5″ sides.

7. Find the volume of a pyramid whose base is an equilateral triangle with 30″ sides and whose height is 56″.

8. Find the volume of a pyramid whose base is an equilateral triangle with 1′2″ sides and whose height is 2′5″.

9. Find the volume of a pyramid whose base is an octagon 6″ across flats and whose height is $8\frac{1}{2}$″.

10. Find the volume of a pyramid whose base is a rectangle $16\frac{1}{2}$″ × $12\frac{1}{2}$″ and whose height is 5′2″.

Answer to exercise 8

Area of base = 3.1416 × 3.5²
        = 38.48 sq ft

Volume = $\frac{1}{3}$ × 38.48 × 12
        = 153.92 cu ft

11. Find the volume of an elliptical cone $18\frac{1}{2}''$ high; the base is an ellipse $39'' \times 56''$.

12. A quantity of sand dumped upon the ground assumed a conical shape with a base 32′ in diameter and a height of 11′. How many cubic yards of sand are there in the pile?

13. A quantity of sand dumped upon the ground assumed a conical shape with a base 40′6″ in diameter and a height of 15′6″. How many cubic yards of sand are there in the pile?

14. An oil container in the shape of an inverted cone is $9\frac{1}{2}''$ high and has a base whose diameter is 11″. Find the capacity in gallons.

15. If cast iron weighs 0.26 lb per cubic inch, what is the weight of a casting in the shape of a square pyramid with a $6\frac{1}{2}''$ square base and a $3\frac{1}{4}''$ height?

9–9   LATERAL SURFACES OF PYRAMIDS AND CONES

In the square pyramid shown in Fig. 9–9, the lateral surface is made up of four equal isosceles triangles like *ADE*. The base of each triangle is a side of the base of the pyramid. We can obtain the area of one of these triangles and multiply it by 4 to get the lateral surface of the pyramid.

Area of triangle = $\frac{1}{2}$ × Base × Altitude

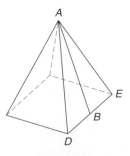

Figure 9–9

If the pyramid in Fig. 9–9 is 12 in. high and the base is a 10-in. square, then the base of triangle *ADE* is 10 in., but the height of the triangle is the line *AB*, called the ***slant height*** of the pyramid.

Figure 9–10 shows the pyramid with one-quarter of it removed to display the actual height *AC* and the slant height *AB*. The triangle *ABC* (Fig. 9–11) is a right triangle with the slant height *AB* as the hypotenuse. The height *AC* of this right triangle is 12 in., the height of the pyramid. The base *BC* of the triangle is half the distance across the square; $\frac{1}{2}$ of 10 in. = 5 in. = *BC*. The hypotenuse $AB = \sqrt{AC^2 + BC^2} = \sqrt{12^2 + 5^2} = \sqrt{144 + 25} = \sqrt{169} = 13$ in., the slant height.

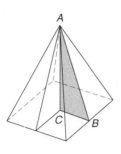

Figure 9–10

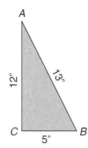

Figure 9–11

This slant height *AB* is the altitude of the triangle *ADE*, which is one of the sides of the pyramid.

$$\text{Area of triangle } ADE = \tfrac{1}{2}ba = \tfrac{1}{2} \times 10 \times 13 = 65 \text{ sq in.}$$

The lateral surface of the pyramid is 4 times the area of one of the sides.

$$4 \times 65 = 4 \times \tfrac{1}{2} \times 10 \times 13 = 260 \text{ sq in.}$$

But 4 × 10 = 40 in., the perimeter of the base. The lateral surface, therefore, equals $\frac{1}{2}$ × the perimeter of the base × the slant height = $\frac{1}{2}$ × 40 × 13 = 260 sq in. The formula for finding the lateral surface of a pyramid is

$$\text{Lateral surface} = \tfrac{1}{2} \times \text{Perimeter of base} \times \text{Slant height}$$

In each instance the slant height of the pyramid is the hypotenuse of a right triangle whose altitude is the height of the pyramid. The base of this right triangle is shown in Fig. 9–10. In the cone, the base of this triangle is one-half the diameter of the base of the cone. In the hexagonal pyramid and the octagonal pyramid, the base of this triangle is one-half the distance across the flats of the base of the pyramid. In the triangular pyramid the base of this triangle is one-third the height of the triangular base.

■ **Example 1**

Find the lateral surface of a pyramid whose base is a hexagon 9$\frac{1}{2}$ in. across flats and whose height is 16 in.

Solution   Lateral surface = $\frac{1}{2}$ × Perimeter of base × Slant height

Side of hexagon = $0.577 \times f = 0.577 \times 9\frac{1}{2} = 5.48$ in.

Perimeter = $6 \times 5.48 = 32.89$ in.

Slant height = $\sqrt{4.75^2 + 16^2} = \sqrt{22.56 + 256}$

$= \sqrt{278.56} = 16.69$ in.

Lateral surface = $\frac{1}{2} \times 32.89 \times 16.69 = 274.47$ sq in.

Practice check: Do exercise 10 on the left.

■ Example 2

Find the lateral surface of a cone 1 ft 9 in. high; the base is 6 ft 8 in. in diameter.

Solution   Lateral surface = $\frac{1}{2}$ × Perimeter of base × Slant height

Perimeter of base = $3.1416 \times 80 = 251.2$ in.

Slant height = $\sqrt{40^2 + 21^2} = \sqrt{1600 + 441}$

$= \sqrt{2041} = 45.2$ in.

Lateral surface = $\frac{1}{2} \times 251.2 \times 45.2$
$= 5{,}677$ sq in.

Practice check: Do exercise 11 on the left.

■ Problems 9–9

Use the space provided below to solve the following problems.

1. Find the lateral surface of a square pyramid with a 4″ square base and a height of 10″.

2. Find the lateral surface of a square pyramid with a $4\frac{1}{2}$″ square base and a height of $10\frac{1}{2}$″.

3. Find the lateral surface of a pyramid whose base is an octagon with 4″ sides and whose height is 6″.

4. Find the lateral surface of a cone whose base is $8\frac{1}{2}$″ in diameter and whose height is 18″.

5. Find the lateral surface of a pyramid whose base is an equilateral triangle with 9″ sides and whose height is 12″.

6. Find the lateral surface of a pyramid whose base is a rectangle 14″ × 20″ and whose height is 12″.

---

10. Find the lateral surface of a pyramid whose base is a hexagon $6\frac{1}{4}$ m across flats and whose height is 22 m.

The answer is in the margin on page 326.

---

11. Find the lateral surface of a cone 4 ft 3 in. high; the base is 9 ft 7 in. in diameter.

The answer is in the margin on page 326.

7. What is the height of a cone whose base is 36′ in diameter and whose slant height is 24′?

8. Find the lateral surface of a pyramid whose base is a hexagon with 10″ sides and whose height is 30″.

9. The top of a copper boiler is conical in shape with a base 12′3″ in diameter and a height of 3′6″. How many square feet of copper were required to make it, allowing 3% for seams?

10. How many square feet of tin are required to cover a roof in the shape of a square pyramid 9′4″ high, with the base a 28′ square?

9–10   FRUSTUMS OF PYRAMIDS AND CONES

When a pyramid or a cone is cut at any point below the apex by a plane parallel to the base, the portion below the cutting plane is called a ***frustum*** of the pyramid or cone. Figure 9–12 shows various frustums.

The volume of the frustum is obtained by the following formula.

$$V = \tfrac{1}{3}h(B + b + \sqrt{B \times b})$$

where V is the volume, h is the height, B is the area of the large base, and b is the area of the small base. The formula says that the volume equals $\tfrac{1}{3}$ times the height

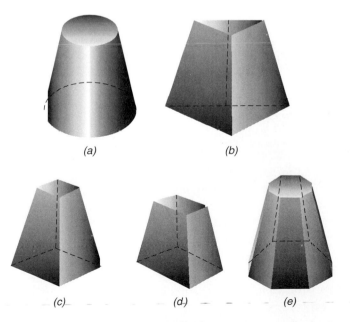

(a)   *(b)*

(c)   *(d)*   *(e)*

Figure 9–12

times the sum of three things: the area of the large base, the area of the small base, and the square root of their product. The process consists of the following steps.

1. Compute the area of B, the large base.
2. Compute the area of b, the small base.
3. Multiply B by b and find the square root of the product.
4. Add the three quantities in the parentheses.
5. Multiply $\frac{1}{3}$ of the height by the sum of the three quantities in the parentheses.

■ Example 1

Find the volume of the frustum of a cone 15 in. high if the upper base is 16 in. in diameter and the lower base 9 in. in diameter.

Solution $V = \frac{1}{3} \times h \times (B + b + \sqrt{B \times b})$

B = area of 16-in. circle $= 3.1416 \times 8 \times 8 = 201.0$ sq in.

b = area of 9-in. circle $= 3.1416 \times 4\frac{1}{2} \times 4\frac{1}{2} = 63.6$ sq in.

$V = \frac{1}{3} \times 15 \times (201.0 + 63.6 + \sqrt{201.0 \times 63.6})$

$\quad = \frac{1}{3} \times 15 \times (201.0 + 63.6 + \sqrt{12783.6})$
$\quad = \frac{1}{3} \times 15 \times (201.0 + 63.6 + 113.1)$
$\quad = \frac{1}{3} \times 15 \times 377.7$
$\quad = 1{,}888.5$ cu in. $= 1.09$ cu ft

This formula is the general formula and applies to all frustums of cones and pyramids regardless of the shape of the base. For the frustum of the cone, however, the formula can be simplified and much time saved in the computation. The simplified formula for finding the volume of the frustum of a cone follows.

$$V = 0.262h(D^2 + d^2 + Dd)$$

12. Find the volume of the frustum of a cone 21 ft high if the upper base is 24 ft in diameter and the lower base 11 ft in diameter.

The answer is in the margin on page 330.

where V = volume
$\quad D$ = diameter of large base
$\quad d$ = diameter of small base
$\quad h$ = height

Practice check: Do exercise 12 on the left.

■ Example 2

Using the simplified formula, find the volume of the frustum of a cone with bases 6 in. and 10 in. in diameter and 12 in. high.

13. Using the simplified formula, find the volume of the frustum of a cone with bases 18 cm and 22 cm in diameter and 24 cm high.

The answer is in the margin on page 330.

Solution $V = 0.262h(D^2 + d^2 + Dd)$
$\qquad = 0.262 \times 12(10^2 + 6^2 + 10 \times 6)$
$\qquad = 0.262 \times 12(100 + 36 + 60)$
$\qquad = 0.262 \times 12 \times 196$
$\qquad = 616.2$ cu in.

Practice check: Do exercise 13 on the left.

■ **Problems 9–10**

Use the space provided below to solve the following problems.

1. Find the volume of the frustum of a square pyramid 15″ high; the upper base is a 42″ square and the lower base a 26″ square.

2. Find the volume of the frustum of a hexagonal pyramid 7″ high; the upper base is a hexagon with 3″ sides and the lower base a hexagon with 8″ sides.

3. Find the volume of the frustum of a cone 10″ high; the lower base is 3″ in diameter and the upper base 6″ in diameter.

4. Find the volume of the frustum of an elliptical pyramid 8″ high; the upper base is an ellipse 7″ × 9″ and the lower base an ellipse 14″ × 18″.

5. Find the capacity in gallons of a pail 11″ deep with a top diameter of 10″ and a bottom diameter of 8″.

6. Find the volume of a frustum of a rectangular pyramid $4\frac{1}{2}$″ high; the upper base is a rectangle 12″ × 16″ and the lower base a rectangle 9″ × 12″.

7. Find the capacity in cubic feet of a square coal hopper in the shape of a frustum of a square pyramid 13′0″ high; the upper base is a 22′ square and the lower base a $5\frac{1}{2}$′ square.

8. An oil measure in the shape of a frustum of a cone is $11\frac{1}{2}$″ deep, with a top diameter of $5\frac{1}{4}$″ and a bottom diameter of $7\frac{1}{2}$″. How many quarts of oil can it contain?

9. A concrete base for a heavy machine has the form of a frustum of a rectangular pyramid. The top is 16′4″ × 4′8″ and the bottom is 32′8″ × 9′4″. The depth is 3′8″. How many cubic yards of concrete were used in building this foundation?

10. A basin in the form of a frustum of an elliptical pyramid 8″ deep is 18″ × 15″ at the top and 12″ × 10″ at the bottom. What is its capacity in gallons?

Answer to exercise 12

B = area of 24-ft circle
   = 3.1416×12^2
   = 452.39 sq ft

b = area of 11-ft circle
   = 3.1416×5.5^2
   = 95.03 sq ft

$V = \frac{1}{3} \times 21 \times (452.39 + 95.03$
    $+ \sqrt{452.39 \times 95.03})$
  $= \frac{1}{3} \times 21 \times (452.39 + 95.03$
    $+ \sqrt{42992.16})$
  $= \frac{1}{3} \times 21 \times (452.39 + 95.03$
    $+ 207.35)$
  $= \frac{1}{3} \times 21 \times 754.77$
  $= 5,283.39$ cu ft

Answer to exercise 13

$V = 0.262 \times 24(22^2 + 18^2$
    $+ 22 \times 18)$
  $= 0.262 \times 24(484 + 324$
    $+ 396)$
  $= 0.262 \times 24(1,204)$
  $= 0.262 \times 28,896$
  $= 7,570.75$ cm^3

9–11   FINDING THE HEIGHT OF THE FRUSTUM OF A PYRAMID OR A CONE

If, in the designing of a vessel, it is required to find the height for a given capacity, the volume formula is modified to read as follows.

$$h = \frac{3V}{B + b + \sqrt{B \times b}}$$

■ Example 1

Find the height of a frustum of a cone to contain 2,000 cu in. The upper base is 16 in. in diameter and the lower base 12 in. in diameter.

Solution $h = \dfrac{3V}{B + b + \sqrt{B \times b}}$

B = area of 16-in. circle = $3.1416 \times 8 \times 8 = 201.0$ sq in.

b = area of 12-in. circle = $3.1416 \times 6 \times 6 = 113.0$ sq in.

$$\sqrt{Bb} = \sqrt{201.0 \times 113.0} = 150.7$$

$$h = \frac{3 \times 2,000}{201.0 + 113.0 + 150.7}$$

$$= \frac{6,000}{464.7}$$

$$= 12.9 \text{ in.}$$

The simplified formula for finding the height of the frustum of a cone is

$$h = \frac{V}{0.262(D^2 + d^2 + Dd)}$$

■ Example 2

Find the answer to Example 1 by means of the simplified formula.

Solution $h = \dfrac{V}{0.262(D^2 + d^2 + Dd)}$

$$= \frac{2,000}{0.262(16^2 + 12^2 + 16 \times 12)}$$

$$= \frac{2,000}{0.262 \times 592}$$

$$= \frac{2,000}{155.1} = 12.9 \text{ in.}$$

14. Find the height of a frustum of a cone to contain 3,500 m^3. The upper base is 32 m in diameter and the lower base 24 m in diameter.

The answer is in the margin on page 332.

Practice check: Do exercise 14 on the left.

■ Problems 9–11

Use the space provided below to find the height of the following frustums.

1. Find the height of the frustum of a square pyramid to contain 400 cu in., the upper base to be an 8″ square, and the lower base a 10″ square.

2. Find the height of the frustum of a cone to contain 2,772 cu in., the top to be 18″ in diameter, and the bottom $13\frac{1}{2}$″ in diameter.

3. Find the height of a frustum of a hexagonal pyramid that is to contain 200 cu in.; the upper base is a hexagon with 2″ sides and the lower base a hexagon with 5″ sides.

4. Find the height of the frustum of a rectangular pyramid that has a volume of 6,000 cu ft, an upper base 30′ × 15′, and a lower base 40′ × 20′.

5. What must be the height of a gallon measure with a top diameter of $4\frac{1}{4}$″ and a bottom diameter of $7\frac{1}{4}$″?

6. A bin in the form of the frustum of a square pyramid is required to have a capacity of 8,000 bushels. What must be its height if the top is a 50′ square and the bottom a 35′ square? (1 bushel = 2,150.4 cu in.)

7. Find the depth of a 10-gallon pail with a top diameter of 14″ and a bottom diameter of 10″.

8. A hopper whose top is 10′ square and whose bottom is 2′ square is required to have a capacity of 500 cu ft. Find the height.

9. A cast-iron driver in the form of a square pyramid is used in a pile-driving machine. It is 35 in. square at the bottom and 15.5 in. square at the top, and it contains 10,000 cu in. Find the height of the driver.

10. A freeway overpass is supported by a concrete column having the shape of a frustum of a cone. The bottom of the column has a 15-ft diameter, the top has a 6-ft diameter, and the volume is 1,400 cu ft. Find the height of the column.

9–12 FINDING THE LATERAL SURFACE OF THE FRUSTUM OF A CONE OR A PYRAMID

Figure 9–13 shows the lateral surface of the frustum of a square pyramid to be made up of four trapezoids like *EFHG*. If the upper base is a 9-in. square, the lower base a 14-in. square, and the height 5 in., then each side is a trapezoid with bases of 9 in. and 14 in. and a height *AB* that is the slant height of the frustum. Figure 9–14 shows a cross section through the figure as shown by the dotted lines in Fig. 9–13. The slant height *AB* is the hypotenuse of the right triangle *ABC*, in which the height is the height of the frustum, 5 in., and the base, is half the difference between the upper and lower bases of the trapezoid.

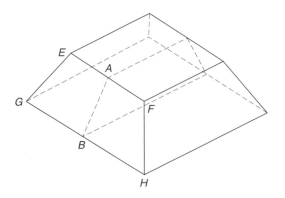

Figure 9–13   **Figure 9–14**

$$\text{Base } BC = \frac{14 - 9}{2} = \frac{5}{2} = 2\tfrac{1}{2} \text{ in.}$$

$$\text{Slant height } AB = \sqrt{5^2 + (2\tfrac{1}{2})^2} = \sqrt{25 + 6.25} = \sqrt{31.25} = 5.59 \text{ in.}$$

$$\text{Average width of } EFHG = \frac{9 + 14}{2} = 11.5 \text{ in.}$$

Area of trapezoid *EFGH* (one side of frustum)
$$= \text{Average width} \times \text{Slant height}$$
$$= 11.5 \times 5.59 = 64.285 \text{ sq in.}$$

Lateral surface = 4 × Area of one side
$$= 4 \times 64.285 = 257.14 \text{ sq in.}$$

Another method of finding the lateral surface is by means of the following formula.

Lateral surface = Average perimeter of bases × Slant height

Perimeter of upper base = 4×9 = 36 in.
Perimeter of lower base = $4 \times 14 = \underline{56 \text{ in.}}$
Sum of perimeters = 92 in.

The average perimeter is $\frac{92}{2}$ or 46 in., which is the same as 4 times the average width of one side.

Lateral surface = 46 × 5.59 = 257.14 sq in.

■ Example

Find the lateral surface of the frustum of a cone 6 in. high with upper and lower bases 11 in. and 7 in. in diameter, respectively (Fig. 9–15).

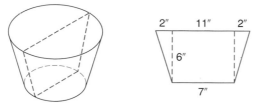

Figure 9–15

Solution   Lateral surface = Average perimeter of bases × Slant height

Perimeter of upper base = 3.1416 × 11 = 34.56 in.
Perimeter of lower base = 3.1416 × 7 = 21.99 in.
                                              56.55 in.

$$\text{Average perimeter} = \frac{56.55}{2} = 28.28 \text{ in.}$$

Altitude of right triangle = 6 in.

$$\text{Base of right triangle} = \frac{11 - 7}{2} = \frac{4}{2} = 2 \text{ in.}$$

$$\text{Slant height} = \sqrt{6^2 + 2^2} = \sqrt{36 + 4} = \sqrt{40} = 6.32 \text{ in.}$$

Lateral surface = 28.28 × 6.32
                      = 178.73 sq in.

Practice check: Do exercise 15 on the right.

15. Find the lateral surface of the frustum of a cone 8 ft high with upper and lower bases 15 ft and 9 ft in diameter, respectively.

The answer is in the margin on page 335.

■ Problems 9–12

Use the space provided below to find the lateral surface areas of the following frustums.

1. Find the lateral surface of the frustum of a square pyramid 16′ high if the upper base is a 28′ square and the lower base an 18′ square.

2. Find the lateral surface of the frustum of a hexagonal pyramid $9\frac{1}{2}''$ high, the upper base of which is a hexagon with $3\frac{1}{2}''$ sides and the lower base a hexagon with 9″ sides.

3. Find the lateral surface of the frustum of an octagonal pyramid 20″ high if the upper base is an octagon with 14″ sides and the lower base an octagon with 22″ sides.

4. Find the lateral surface of the frustum of a cone 6′10″ high if the upper base is 7′2″ in diameter and the lower base 9′4″ in diameter.

5. Find the lateral surface of the frustum of a triangular pyramid 12″ high whose upper base is an equilateral triangle with 10″ sides and whose lower base is an equilateral triangle with 20″ sides.

6. In the following chart, each set of figures, **(a)** to **(i)**, represents a frustum of a cone with dimensions as given. Find the lateral surface for each. Give all answers in square feet, to one decimal place.

Frustums of Cones			
Large Diameter	**Small Diameter**	**Height**	**Lateral Surface (sq ft)**
(a) 1′10″	1′4″	1′6″	_____
(b) 32″	18″	6″	_____
(c) 14″	6″	12″	_____
(d) 8′4″	2′8″	3′6″	_____
(e) 12″	9″	11″	_____
(f) 3′5″	3′2″	3′3″	_____
(g) 12″	8″	4″	_____
(h) 24″	18″	21″	_____
(i) 14′8″	10′6″	8′4″	_____

Figure 9–16

9–13   SPHERES

The volume of a sphere or ball may be found by the formula

$$V = \tfrac{4}{3}\pi r^3$$

in which r is the radius (see Fig. 9–16).

■ Example 1

Find the volume of a 6-in. sphere.

Solution $V = \frac{4}{3}\pi r^3$

                $= \frac{4}{3} \times 3.1416 \times 3^3$

                $= \frac{4}{3} \times 3.1416 \times 27 = 113.1$ cu in.

■ Example 2

Find the volume of a 32.5-in. sphere.

Solution $V = \frac{4}{3}\pi r^3$

                $= \frac{4}{3} \times 3.1416 \times 16.25^3$

                $= \frac{4}{3} \times 3.1416 \times 4{,}291$

                $= 17{,}974.16$ cu in.

■ Example 3

How many gallons of water can be poured into a globe 20 in. in diameter?

Solution $V = \frac{4}{3}\pi r^3$

                $= \frac{4}{3} \times 3.1416 \times 10^3$

                $= \frac{4}{3} \times 3.1416 \times 1{,}000$

                $= 4{,}188.8$ cu in.

$$\text{Gallons in the globe} = \frac{4{,}188.8}{231} = 18.13 \text{ gallons}$$

Practice check: Do exercise 16 on the right.

■ Problems 9–13

Use the space provided below to find the volumes of the following spheres and hemispheres. (A hemisphere equals one-half of a sphere.) Use 3.1416 for π.

1. A 12″ sphere   **2.** A 1″ sphere

3. A 2″ sphere   **4.** A 3″ sphere

5. A 15″ sphere   **6.** A 30′ sphere

Answer to exercise 15

Total base perimeters = 75.39 ft

Average perimeter = 37.7 ft

Altitude of right triangle = 8 ft

Base of right triangle = 3 ft

Slant height = 8.5 ft

Lateral surface = 320.5 sq ft

16. Using a calculator, find the volume of a 22-yd sphere.

The answer is in the margin on page 337.

7. A sphere $\frac{3}{4}''$ in diameter

8. An 18' sphere

9. A 44'' sphere

10. A sphere $1\frac{1}{2}''$ in diameter

11. A hemisphere 11'' in diameter

12. A hemisphere 3'4'' in diameter

13. A hemisphere 16' in diameter

14. A 1'6'' sphere

15. A $2\frac{1}{2}''$ sphere

16. A $3\frac{1}{4}''$ sphere

17. A $1'2\frac{1}{2}''$ sphere

18. A 16.5'' sphere

19. A 25' sphere

20. A 34.25'' sphere

21. A sphere $\frac{7}{8}''$ in diameter

22. A sphere $2\frac{1}{8}''$ in diameter

23. A hemisphere $10\frac{1}{2}''$ in diameter   **24.** A hemisphere $2'6\frac{1}{2}''$ in diameter

25. A hemisphere $3.75'$ in diameter

Answer to exercise 16

$$V = \tfrac{4}{3} \times 3.1416 \times 11^3$$
$$= 5{,}575.29 \text{ cu yd}$$

 9–14 USING THE CALCULATOR TO FIND VOLUMES OF SPHERES

Examine your calculator and see if you have a y^x function key. If you do, then you can find volumes of spheres with the following sequence of key-ins.

■ **Example**

Find the volume of a sphere 12 in. in diameter (Fig. 9–17). Round off your answer to one decimal place.

6 in.

Volume $= \frac{4}{3}\pi r^3$

Figure 9–17

Solution   Volume $= \frac{4}{3}\pi r^3$

The Display Shows

Turn on the calculator	$0.$
Enter 4	$4.$
Press ÷	$4.$
Enter 3	$3.$
Press ×	1.333333333
Press π	3.141592654
Press ×	4.188790205
Enter 6	$6.$
Press y^x	$6.$
Enter 3	$3.$ $y^x = 6^3$
Press =	904.7786842 (answer)

Round off this answer to 904.8 cu in.

Practice check: Do exercise 17 on the right.

---

17. Find the volume of a sphere 34 m in diameter. Round off your answer to one decimal place.

The answer is in the margin on page 339.

■ **Problems 9–14**

Using a calculator, do all the problems from Problem Set 9–13 again. Get some experience using the y^x key. Compare your answers to the answers you wrote on pages 335–337.

9–15   FINDING THE SURFACE AREA OF A SPHERE

The surface area of a sphere is equal to 4 times the area of a circle with the same diameter.

$$\text{Surface area of sphere} = 4\pi r^2$$

■ **Example 1**

Find the surface area of a 6-in. sphere.

Solution $\begin{aligned} \text{Surface area} &= 4 \times \pi \times r^2 \\ &= 4 \times 3.1416 \times 3^2 \\ &= 4 \times 3.1416 \times 9 = 113.09 \text{ sq in.} \end{aligned}$

■ **Example 2**

Find the surface area of a 32.5-in. sphere.

Solution $\begin{aligned} \text{Surface area} &= 4 \times \pi r^2 \\ &= 4 \times 3.1416 \times 16.25^2 \\ &= 4 \times 3.1416 \times 265.06 \\ &= 3{,}318.3 \text{ sq in.} \end{aligned}$

■ **Example 3**

How many square feet of roofing material would be needed to cover a hemisphere dome with a 600-ft diameter?

Solution $\begin{aligned} \text{Surface area} &= 4 \times \pi r^2 \\ &= 4 \times 3.1416 \times 300^2 \\ &= 4 \times 3.1416 \times 90{,}000 \\ &= 1{,}130{,}973.3 \text{ sq ft} \end{aligned}$

$$\begin{aligned} \text{Dome surface area} &= \frac{1{,}130{,}973.3}{2} \\ &= 565{,}486.65 \text{ sq ft} \end{aligned}$$

18. Find the surface area of a 14-cm sphere.

The answer is in the margin on page 340.

Practice check: Do exercise 18 on the left.

■ **Problems 9–15**

Use the space provided below to find the surface area of the following spheres.

1. A sphere 1″ in diameter

2. A sphere 2″ in diameter

3. A $1\frac{1}{2}''$ sphere

4. A $3\frac{1}{4}''$ sphere

Answer to exercise 17

20,579.6 m³

5. A sphere 100′ in diameter

6. A sphere 50′ in diameter

7. A sphere 1.4″ in diameter

8. A sphere 2.6″ in diameter

9. A 22″ sphere

10. A 44″ sphere

11. A 2′6″ sphere

12. A 4′4″ sphere

13. A 3′2″ sphere

14. An 8′8″ sphere

15. A 1.375″ sphere

16. A 3.125″ sphere

17. A $1'\frac{1}{2}''$ sphere

18. A $6'\frac{1}{4}''$ sphere

19. A $4'6\frac{1}{2}''$ sphere

20. A $24'8\frac{3}{4}''$ sphere

9–16   VOLUME OF A RING

To find the volume of a ring of circular cross section, as shown in Fig. 9–18, multiply the circumference of the center line circle by the area of cross section.

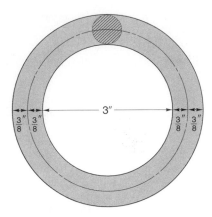

Figure 9–18

■ Example

The ring in Fig. 9–18 has an inside diameter of 3 in., and the cross section of the ring is a $\frac{3}{4}$-in. circle. Find its volume.

Solution   Volume = Circumference of center line circle × Area of cross section

Diameter of center line circle = $3 + \frac{3}{8} + \frac{3}{8} = 3\frac{3}{4}$ in.

Circumference of center line circle = $3.1416 \times 3\frac{3}{4} = 11.78$ in.

Area of cross section = $3.1416 \times \frac{3}{8} \times \frac{3}{8} = 0.44$ sq in.

Volume = 11.78×0.44
       = 5.18 cu in.

19. A ring has an inside diameter of 5 in., and the cross section of the ring is a $\frac{5}{8}$-in. circle. Find its volume.

The answer is in the margin on page 342.

Practice check: Do exercise 19 on the left.

■ Problems 9–16

Using the space provided, find the volume of the following rings. Use 3.1416 for π.

1. The rim of a handwheel of circular cross section if the outside diameter is 10″ and the inside diameter 9″

2. The rim of a handwheel of circular cross section if the outside diameter is 30 ft and the inside diameter 25 ft

3. A ring of circular cross section if the inside diameter is 5″ and the ring is $\frac{7}{8}$″ thick

4. A ring of circular cross section if the inside diameter is 15 cm and the ring is 0.125 cm thick

5. A ring of circular cross section if the outside diameter is $6\frac{1}{2}$″ and the ring is $\frac{3}{4}$″ thick

6. A ring of circular cross section if the outside diameter is $8\frac{1}{2}$″ and the ring is $\frac{5}{8}$″ thick

7. A ring having a circular cross section with an inside diameter of 7″ and a thickness of $\frac{1}{2}$″

8. A ring having a circular cross section with an inside diamter of 13″ and a thickness of $\frac{1}{2}$″

9. The rim of a handwheel with an outside diameter of 8″ and $\frac{3}{4}$″ thick

10. The rim of a handwheel with an outside diameter of 14″ and $\frac{3}{8}$″ thick

---

9–17   VOLUMES OF COMPOSITE SOLID FIGURES

Many objects are made up of parts, the volumes of which can be found by the rules studied in the preceding sections. In cases of this sort, the volume of each part is found and added to find the total volume of the object.

■ **Example 1**

Find the volume of the steeple-head rivet in Fig. 9–19.

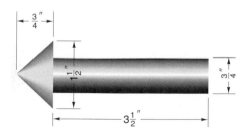

Figure 9–19

Solution

Volume of head (cone) $= \frac{1}{3} \times \frac{3}{4} \times 3.1416 \times \frac{9}{4} \times \frac{9}{4} = 0.442$ cu in.
Volume of body (cylinder) $= \frac{7}{2} \times 3.1416 \times \frac{3}{8} \times \frac{3}{8}$ $\quad = \underline{1.547}$ cu in.
Total $= 1.989$ cu in.

Practice check: Do exercise 20 on the right.

20. Find the volume of the following figure.

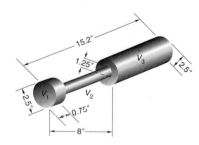

The answer is in the margin on page 343.

■ **Example 2**

Find the volume of the taper bushing shown in Fig. 9–20.

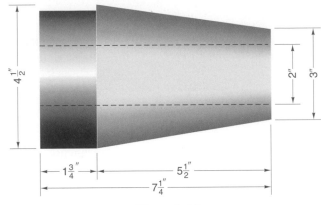

Figure 9–20

Solution

Volume of cylindrical part
$$= 3.1416 \times 2\tfrac{1}{4} \times 2\tfrac{1}{4} \times 1\tfrac{3}{4} \qquad\qquad = 27.8 \text{ cu in.}$$

Volume of tapered part
$$= 0.262h(D^2 + d^2 + Dd)$$
$$= 0.262 \times 5\tfrac{1}{2}[(4\tfrac{1}{2})^2 + 3^2 + 4\tfrac{1}{2} \times 3]$$
$$= 0.262 \times 5\tfrac{1}{2}(20.25 + 9 + 13.5)$$
$$= 0.262 \times 5.5 \times 42.75 \qquad\qquad = \underline{61.6 \text{ cu in.}}$$

Gross volume = 89.4 cu in.

Volume of cylindrical hole through the bushing
$$= 3.1416 \times 1 \times 1 \times 7\tfrac{1}{4} \qquad\qquad = \underline{22.8 \text{ cu in.}}$$

Net volume or volume of the object = 66.6 cu in.

Explanation   First consider the entire object as a solid; then subtract the volume of the cylindrical hole.

Practice check: Do exercise 21 on the left.

21. Find the volume of the casting shown in the following figure.

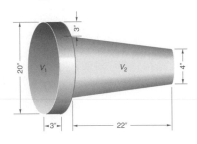

The answer is in the margin on page 344.

■ **Problems 9–17**

Use the space provided below to find the volume of the following composite figures.

1. Find the volume of the conehead rivet shown in Fig. 9–21.

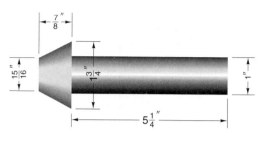

Figure 9–21

Answer to exercise 20

$V_1 = 3.1416 \times 1.25^2 \times 0.75$
$\quad = 3.68$ cu in.

$V_2 = 3.1416 \times 0.625^2 \times 7.25$
$\quad = 8.9$ cu in.

$V_3 = 3.1416 \times 1.25^2 \times 7.2$
$\quad = 35.34$ cu in.

$V_1 + V_2 + V_3 = 47.92$ cu in.

2. Find the volume of the countersunk rivet shown in Fig. 9–22.

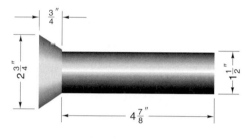

Figure 9–22

3. Find the volume of the hollow cast-iron column shown in Fig. 9–23.

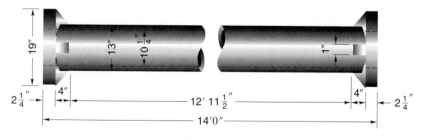

Figure 9–23

Answer to exercise 21

$V_1 = 3.1416 \times 10^2 \times 3$
$\quad = 942.48$ cu in.

V_2 is a volume of a frustum of cone. Thus,

$B = 3.1416 \times 7^2$
$\quad = 153.9$ sq in.

$b = 3.1416 \times 2^2$
$\quad = 12.57$ sq in.

$V_2 = \frac{1}{3} \times 22(153.9 + 12.57$
$\qquad + \sqrt{153.9 \times 12.57})$
$\quad = 1{,}543.3$ cu in. or 10.7 cu ft

4. Find the capacity of the oil can shown in Fig. 9–24.

Figure 9–24

5. Find the capacity of a cylindrical pail with a hemispherical bottom if the diameter is 10″ and the overall depth is 14″.

9–18   WEIGHTS OF MATERIALS

Table 7 at the back of the book gives the weight per cubic inch and per cubic foot of various materials. The weight of any object may be found by computing its volume in cubic inches or cubic feet and multiplying this volume by the unit weight, that is, the weight per cubic inch or cubic foot of that material.

■ **Example 1**

Find the weight of an iron casting whose volume is 7.35 cu ft.

22. Find the weight of an iron casting whose volume is 21.17 cu in.

The answer is in the margin on page 346.

Solution $7.35 \times 450 = 3{,}307.5$ lb

Explanation   Table 7 in the back of this book shows that the weight of cast iron is 450 lb per cu ft. Therefore, the weight of the casting is 7.35×450 or 3,307.5 lb.

Practice check: Do exercise 22 on the left.

■ **Example 2**

Find the weight of a sheet of steel 6 ft 8 in. long, 2 ft 4 in. wide, and $\frac{1}{4}$ in. thick (see Fig. 9–25).

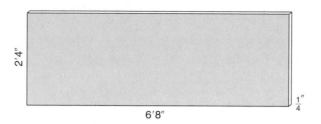

Figure 9–25

Solution $80 \times 28 \times \frac{1}{4} \times 0.283 = 158.5$ lb

Explanation   The product $80 \times 28 \times \frac{1}{4}$ gives the volume of the sheet in cubic inches. Multiplying by 0.283, the weight per cubic inch of steel, gives the weight of the sheet of steel.

Practice check: Do exercise 23 on the right.

23. Find the weight of a sheet of steel 5 ft 9 in. long, 3 ft 3 in. wide, and $\frac{1}{4}$ in. thick.

The answer is in the margin on page 347.

■ **Example 3**

Find the weight of the rim of a cast-iron flywheel of the following dimensions: inside diameter 48 in., outside diameter $56\frac{1}{2}$ in., face $6\frac{1}{2}$ in. wide (see Fig. 9–26).

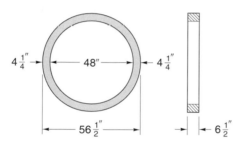

Figure 9–26

Solution $3.1416 \times 52\frac{1}{4} \times 4\frac{1}{4} \times 6\frac{1}{2} \times \frac{450}{1728} = 1,181$ lb

Explanation   Volume of a ring section
$$= \text{Average circumference} \times \text{Area of cross section}$$

$$\text{Diameter of center line circle} = \frac{48 + 56\frac{1}{2}}{2} = \frac{104\frac{1}{2}}{2} = 52\frac{1}{4} \text{ in.}$$

$$\text{Average circumference} = 3.1416 \times 52\frac{1}{4}$$

$$\text{Area of cross section} = 4\frac{1}{4} \times 6\frac{1}{2}$$

The product of $3.1416 \times 52\frac{1}{4} \times 4\frac{1}{4} \times 6\frac{1}{2}$ gives the volume in cubic inches. Dividing by 1,728 changes this to cubic feet. Since each cubic foot of cast iron weighs 450 lb, we multiply by 450, giving 1,181 lb.

Practice check: Do exercise 24 on the right.

24. Find the weight of a brass ring with an inside diameter of 5 in. and a cross section of $\frac{1}{2}$ in.

The answer is in the margin on page 347.

■ Problems 9–18

Use the space provided below to find the weight of the following. Use Table 7 in the back of this text.

1. Find the weight of a rectangular steel plate $3'3'' \times 2'1''$ and $\frac{3}{4}''$ thick.

2. Find the weight of a 20′ length of wrought-iron pipe whose inside diameter is 3.548″ and outside diameter 4″.

3. Find the weight of 1,000 round U.S. Standard wrought-iron washers whose outside diameter is $3\frac{3}{4}''$, with a $1\frac{3}{4}''$ hole and a thickness of $\frac{11}{64}''$.

4. Find the weight of 500 wrought-iron washers 5″ square and $\frac{3}{8}''$ thick. The hole is $1\frac{1}{2}''$ in diameter.

5. Find the weight of a cast-iron manhole cover with semicircular ends if the overall length is $4'2''$, the width $2'1''$, and the thickness $1\frac{1}{2}''$.

6. Find the weight of a triangular steel plate $8\frac{1}{2}''$ on a side and 1″ thick.

7. Find the weight of a steel ball 3″ in diameter.

8. Find the weight of 500 steel bearings 1″ in diameter.

9. Find the weight of 1,000 lead discs $\frac{1}{2}''$ in diameter and $\frac{1}{16}''$ thick.

10. Find the weight of a 4′ length of cast-iron pipe whose inside diameter is $3\frac{1}{2}''$ and whose outside diameter is $4\frac{1}{4}''$.

9–19   WEIGHTS OF CASTINGS FROM PATTERNS

It is possible to estimate very closely the weight of a casting from the weight of the pattern.

■ **Example 1**

Find the weight of a cast-iron casting made from a white pine pattern weighing $9\frac{1}{4}$ lb.

Solution   Volume of pattern $= \dfrac{9\frac{1}{4}}{25} = 0.37$ cu ft

Weight of casting $= 0.37 \times 450 = 166.5$ lb

Explanation   Since 1 cubic foot of white pine weighs approximately 25 lb, the volume of the pattern will be $\dfrac{9\frac{1}{4}}{25}$ or 0.37 of a cubic foot. One cubic foot of cast iron weighs 450 lb. Hence, 0.37 of a cubic foot will weigh 0.37×450 or 166.5 lb.

■ **Example 2**

A pattern made of ash weighs 6 lb 8 oz. What will an aluminum casting made from it weigh?

Solution   Volume of pattern $= \dfrac{6.5}{38} = 0.171$ cu ft

Weight of casting $= 0.171 \times 160 = 27.37$ lb

Practice check: Do exercise 25 on the right.

■ **Problems 9–19**

Using Table 7 and the space provided below, find the weight of the following castings from the weight of the patterns.

1. Find the weight of a brass casting made from a cherry pattern that weighs $7\frac{3}{4}$ lb.

2. A pattern made of white oak weighs 5 lb $7\frac{1}{2}$ oz. What will a copper casting made from it weigh?

3. The ratio of the weight of white pine to cast iron is 1 to 16. If the pattern weighs 4 lb 14 oz, what will an iron casting weigh?

4. Bell metal weighs 14.2 times as much as mahogany. If a pattern of mahogany weighs 8 lb 6 oz, what will the bell metal casting weigh?

Answer to exercise 23

$69 \times 39 \times \frac{1}{4} \times 0.283 = 190.39$ lb

Answer to exercise 24

Diameter of center line circle
$= 5\frac{1}{2}$ in.

Circumference of
center line circle $= 17.28$ in.

Area of cross section
$= 0.196$ sq in.

Volume $= 3.39$ cu in.

Weight $= 1$ lb

---

25. Find the weight of an aluminum casting made from an ash pattern weighing $5\frac{3}{4}$ lb.

The answer is in the margin on page 349.

5. Compute the following weight ratios: **(a)** pine to brass, **(b)** maple to cast iron, **(c)** mahogany to copper, **(d)** pine to bell metal, and **(e)** oak to aluminum.

6. Find the weight of a cast-iron casting made from a maple pattern that weighs $6\frac{1}{2}$ lb.

7. Find the weight of a brass casting made from a white pine pattern that weighs 4 lb 6 oz.

8. Find the ratio of the weight of silver to gold.

9. Find the ratio of the weight of gold to silver.

10. Find the weight of a maple pattern if the brass casting weighs 85 lb.

9–20   BOARD MEASURE

Lumber is measured in terms of board feet. A ***board foot*** of lumber is a piece of wood having an area of 1 sq ft and a thickness of 1 in. Feet board measure may be abbreviated ***fbm.***

From the definition of a board foot, we obtain the following rule.

Rule _____

To find the number of board feet in a piece of lumber, multiply the length in feet by the width in feet by the thickness in inches, counting a thickness less than 1 in. as 1 in.

_____

Lumber is either rough or dressed. ***Rough stock*** is lumber that is not dressed or planed. ***Dressed stock*** is lumber that is planed on one or more sides.

In measuring lumber, we compute the full size, that is, the rough stock required to make the desired piece.

Allowance for Dressing

When lumber is dressed or planed, $\frac{1}{16}''$ is taken off each side if the lumber is less than $1\frac{1}{2}''$ thick. If the lumber is $1\frac{1}{2}''$ or more in thickness, $\frac{1}{8}''$ is taken off each side. Sometimes a little more is planed off in dressing lumber. Lumber for framing houses measures $\frac{1}{2}$ in. less than the name of the piece. Thus, standard 2 in. × 4 in. studs actually measure about $1\frac{1}{2}$ in. × $3\frac{1}{2}$ in.

Standard Lengths and Widths

Building lumber is cut in even lengths, such as 8 ft, 10 ft, 12 ft, 14 ft, and 16 ft, up to 22 ft. Lumber longer than 22 ft is more expensive and more difficult to find than the shorter lengths. Building lumber is also cut in even widths, such as 2 in., 4 in., 6 in., and 8 in., up to 12 in. Lumber wider than 12 in. is also more expensive and more difficult to find. The thickness of building lumber is usually 1 in., 2 in., 3 in., or 4 in.

Hardwoods, such as ash, maple, Honduras mahogany, and walnut, are cut in the widths and lengths that get the most lumber from the tree from which it is cut. This lumber is usually dressed on only two sides. The widths and lengths are not standard. The thickness of this lumber is counted in $\frac{1}{4}$-in. graduations, such as $\frac{1}{2}$ in., $\frac{3}{4}$ in., $\frac{4}{4}$ in., $\frac{5}{4}$ in., and $\frac{6}{4}$ in., up to $\frac{8}{4}$ in. or 2 in. In measuring the width of lumber, a fraction less than $\frac{1}{2}$ in. is disregarded, whereas $\frac{1}{2}$ in. or more is regarded as 1 in. Thus, a board $5\frac{1}{4}$ in. wide is considered 5 in. wide, whereas a board $7\frac{1}{2}$ in. in width is figured as 8 in.

■ Example 1

Find the number of board feet in a piece of lumber 2 in. × 6 in. × 20 ft.

Solution $20 \times \frac{6}{12} \times 2 = 20$ board feet

Explanation Convert the width into feet by dividing 6 in. by 12; that is, 6 in. $= \frac{6}{12}$ of a foot. Multiply the length 20 ft by the width $\frac{6}{12}$ ft by the thickness 2 in., and obtain 20 board feet.

Practice check: Do exercise 26 on the right.

■ Example 2

Find the number of board feet in a piece of lumber $\frac{7}{8}$ in. × 14 in. × 22 ft.

Solution $22 \times \frac{14}{12} \times 1 = 25\frac{2}{3}$ board feet

Explanation Dividing 14 by 12 changes the width to feet. Since the thickness is less than 1 in., it is figured as 1 in. Multiplying the length in feet by the width in feet by the thickness 1 in., we get the board feet in the piece.

Practice check: Do exercise 27 on the right.

■ Example 3

Find the number of board feet in a board $1\frac{1}{2}$ in. × $9\frac{3}{4}$ in. × 16 ft.

Solution $16 \times \frac{10}{12} \times \frac{3}{2} = 20$ board feet

Explanation Since the width is $9\frac{3}{4}$ in., we regard it as 10 in. Dividing 10 in. by 12 reduces it to feet. We then multiply the length in feet by the width in feet by the thickness in inches and obtain the number of board feet in the piece.

Practice check: Do exercise 28 on the right.

Answer to exercise 25

Volume of pattern = 0.151 cu ft

Weight of casting = 24.2 lb

---

26. Find the number of board feet in a piece of lumber 2 in. × 8 in. × 16 ft.

The answer is in the margin on page 351.

---

27. Find the number of board feet in a piece of lumber $\frac{3}{4}$ in. × 4 in. × 12 ft.

The answer is in the margin on page 351.

---

28. Find the number of board feet in a board $1\frac{1}{4}$ in. × $3\frac{3}{4}$ in. × 22 ft.

The answer is in the margin on page 351.

■ **Problems 9–20**

Use the space provided below to find the number of board feet in the following pieces.

1. $2'' \times 4'' \times 12'$ **2.** $2'' \times 6'' \times 10'$ **3.** $1'' \times 8'' \times 14'$ **4.** $2'' \times 4'' \times 8'$

5. $1'' \times 10'' \times 12'$ **6.** $2'' \times 10'' \times 16'$ **7.** $2'' \times 3'' \times 12'$ **8.** $2'' \times 12'' \times 16'$

9. $2'' \times 8'' \times 20'$ **10.** $1'' \times 6'' \times 12'$ **11.** $\frac{3}{4}'' \times 10'' \times 6'$ **12.** $\frac{7}{8}'' \times 10'' \times 10'$

13. $\frac{1}{2}'' \times 8'' \times 16'$ **14.** $1'' \times 8'' \times 8'$ **15.** $2'' \times 3'' \times 16'$ **16.** $2'' \times 2'' \times 12'$

17. $4'' \times 4'' \times 10'$ **18.** $4'' \times 6'' \times 12'$ **19.** $3'' \times 4'' \times 10'$ **20.** $4'' \times 4'' \times 18'$

Find the number of board feet of lumber in the following practical problems.

21. A truckload of lumber is 16′ long, 5′4″ wide, and 4′8″ high. How many board feet does it contain?

22. How many board feet are there in 72 boards $2\frac{1}{4}'' \times 16'' \times 28'$?

23. A building lot 120′ long and 30′ wide is to be enclosed by a fence 10′ high. The boards are 6″ wide and 1″ thick. Bracing pieces 3″ × 4″ × 12′ are spaced 10′ apart. Three longitudinal nailing pieces 2″ × 4″ are used. How many board feet of lumber are required?

24. How many board feet of lumber are required for a boardwalk 200′ long and 10′ wide if 2″-thick planks are used? The planks are to be nailed to three pieces 3″ × 4″ running the full length of the walk.

Answer to exercise 26

$16 \times \frac{8}{12} \times 2 = 21.3$ board ft

Answer to exercise 27

$12 \times \frac{4}{12} \times 1 = 4$ board ft

Answer to exercise 28

$22 \times \frac{4}{12} \times \frac{5}{4} = 9.17$ board ft

25. Find the number of board feet required to build a boardwalk 1 mile long and 12′ wide. The boards are 6″ wide and $1\frac{3}{4}$″ thick, dressed on one side, and spaced $\frac{1}{2}$″ apart. Four planks 3″ × 4″ run the full length of the walk.

26. How many board feet of lumber are required to floor a loft with $1\frac{1}{2}$″ lumber dressed on one side if the floor is 60′6″ long and 22′3″ wide, allowing 2% for squaring?

27. Find the number of board feet required for flooring a circular bandstand 24′ in diameter if $1\frac{3}{4}$″ stock dressed on one side is used. Allow 3% for waste.

28. A porch 50′6″ long and 8′3″ wide is to be floored with 1″ stock planed on one side. If the allowance for squaring is 3%, how many board feet are required?

29. A loading platform is laid with 2″ rough stock. The platform is 42′9″ long and 8′6″ wide. How many board feet are required if 2% is allowed for waste?

30. A factory floor is laid with $1\frac{5}{8}$″ stock dressed on one side. If the floor is 36′6″ × 61′4″, how much lumber is required, allowing 3% for waste?

9–21 FLOORING

Some hardwood flooring is tongued and grooved. In figuring the amount of lumber required for covering surfaces such as the floor or ceiling of a room, one must make allowance for matching as well as for waste. The allowance for matching varies from 24% to 50% of the surface to be covered, depending on the face width of the lumber used. For waste, the usual allowance is 3% to 5%.

The following table shows the percentages customarily allowed for matching.

Face Width of Boards	Percentage Allowed for Matching
$1\frac{1}{2}$ in.	50%
2 in.	$37\frac{1}{2}$%
$2\frac{1}{4}$ in.	$33\frac{1}{3}$%
$3\frac{1}{4}$ in.	24%

The standard thicknesses of hardwood flooring are $\frac{3}{8}$ in., $\frac{1}{2}$ in., $\frac{5}{8}$ in., and $\frac{25}{32}$ in.

■ Example

Find the number of board feet of flooring required for laying a floor $12'6'' \times 21'3''$ with $2\frac{1}{4}''$ matched flooring $\frac{5}{8}''$ thick. Allow 5% for waste.

Solution

$$12.5 \times 21.25 \times 1 = 265.62$$
$$\text{(Allowance for matching) } 33\tfrac{1}{3}\% \text{ of } 265.62 = \ \ 88.54$$
$$\text{(Allowance for waste) } 5\% \text{ of } 265.62 = \ \underline{\ \ 13.28}$$
$$367.44$$

The total flooring required is 368 board feet.

29. Find the number of board feet of flooring required for laying a floor $15'5'' \times 24'1''$ with $2''$ matched flooring $\frac{1}{2}''$ thick. Allow 5% for waste.

The answer is in the margin on page 354.

Explanation   Multiplying the width of the floor by the length gives the area in square feet. Since the thickness of the flooring is less than 1 in., it is figured as 1 in. Hence, multiplying the area of the floor by 1 in. gives 265.62, the number of board feet that would be required if no allowances were made for matching or waste. Since $2\frac{1}{4}$-in. matched flooring is used, the allowance for matching, according to the table, is $33\frac{1}{3}$% of 265.62 or 88.54 fbm. Five percent for waste is 0.05×265.62 or 13.28 fbm. Adding these three quantities gives the total of 367.44 or 368 board feet.

Practice check: Do exercise 29 on the left.

■ Problems 9–21

Use the space provided below to find the amount of flooring required in each of the following problems. Use the foregoing table for tongue-and-groove allowance.

1. Find the number of board feet of flooring required for a room $14'8'' \times 24'0''$ if $2''$ matched flooring $\frac{1}{2}''$ thick is used. Allow 3% for waste.

2. How many board feet of flooring are needed for a room $13'6'' \times 22'9''$ if $2\frac{1}{4}''$ matched boards $\frac{1}{2}''$ thick are used? Allow 5% for waste.

3. An apartment consists of three rooms having the following dimensions: The living room is $11'6'' \times 20'6''$, the bedroom is $10'9'' \times 12'3''$, the dinette is $7'10'' \times 11'8''$, and the foyer is $6'3'' \times 15'6''$. Find the number of board feet required for the floors of the apartment, using $2''$ boards $\frac{5}{8}''$ thick. Allow 3% for waste.

4. A room $14'3'' \times 24'0''$ is to be floored with $2\frac{1}{4}''$ matched lumber $\frac{5}{8}''$ thick. Allowing 3% for waste, how many board feet are required?

5. A room $24'6'' \times 32'6''$ is to be floored with $3\frac{1}{4}''$ matched boards $\frac{25}{32}''$ thick. How many board feet are required if 5% is allowed for waste?

6. At \$1,850.00 per M board feet, that is, 1,000 board feet, find the cost of covering the floor of a room $18'3'' \times 28'6''$. Matched lumber $2\frac{1}{4}''$ wide and $\frac{5}{8}''$ thick is used. Allow 3% for waste.

7. The walls of a dining room $21'3'' \times 12'6''$ are to be wainscoted for a height of $4'$ above the floor. There are two doors $3'0''$ wide. How many board feet of lumber are required if $2''$ boards $\frac{5}{8}''$ thick are used, allowing 2% for waste?

8. A semicircular platform $48'$ in diameter is to be floored with $2\frac{1}{4}''$ stock $\frac{25}{32}''$ thick. Allowing 8% for waste, how many board feet are needed?

9. Compute the amount of lumber required to cover the floor of a dance hall $70' \times 42'$ if $3\frac{1}{4}''$ boards $\frac{5}{8}''$ thick are used. Allow 2% for waste.

10. How many board feet of hardwood flooring are required for a classroom $32'6'' \times 28'9''$ if $2\frac{1}{4}''$ boards $\frac{5}{8}''$ thick are used? Allow 3% for waste.

Answer to exercise 29

$15.417 \times 24.833 \times 1 = 382.85$

Allowance for match
$= 37\frac{1}{2}\%$ of 382.85
$= 143.57$

Allowance for waste
$= 5\%$ of 382.85
$= 19.14$

Total $= 545$ board feet

9–22   SUMMARY OF FORMULAS FOR SOLIDS

Prism and Cylinder

$$\text{Volume} = \text{Area of base} \times \text{Height}$$

$$\text{Area of base} = \frac{\text{Volume}}{\text{Height}}$$

$$\text{Height} = \frac{\text{Volume}}{\text{Area of base}}$$

$$\text{Lateral surface} = \text{Perimeter of base} \times \text{Height}$$

Pyramid and Cone

$$\text{Volume} = \tfrac{1}{3} \times \text{Area of base} \times \text{Height}$$

$$\text{Area of base} = \frac{3 \times \text{Volume}}{\text{Height}}$$

$$\text{Height} = \frac{3 \times \text{Volume}}{\text{Area of base}}$$

$$\text{Lateral surface} = \tfrac{1}{2} \times \text{Perimeter of base} \times \text{Slant height}$$

Frustums of Pyramids and Cones

$$\text{Volume} = \tfrac{1}{3}h(B + b + \sqrt{B \times b})$$

$$h = \frac{3 \times \text{Volume}}{B + b + \sqrt{B \times b}}$$

$$\text{Lateral surface} = \text{Average perimeter of bases} \times \text{Slant height}$$

Sphere

$$\text{Volume} = \tfrac{4}{3}\pi r^3$$

$$\text{Diameter} = \sqrt[3]{\frac{6 \times \text{Volume}}{\pi}}$$

$$\text{Surface area} = 4\pi r^2$$

Solid Ring

$$\text{Volume} = \text{Circumference of the center line circle} \times \text{Area of cross section}$$

Equilateral Triangle, Square, Hexagon, and Octagon

For constants in reference to these figures, see the Table of Constants on page 244.

■ Self-Test

Use the space provided to solve the following problems. Use the tables in the back of the book, the Table of Constants on page 244, and the Summary of Formulas for Solids on pages 305–308. Check your answers with the answers in the back of the book. Try to estimate the answer to the problem before solving it.

1. Find the volume of a rectangular prism whose base is a 4-in. square and whose height is 10 in.

2. Find the volume of a cylinder 9 in. high and 3 in. in diameter.

3. Find the volume of a cone whose base is 4 in. in diameter and whose height is 8 in.

4. Find the lateral surface of a pyramid whose base is a 3-in. square and whose height is 4 in.

5. Find the volume of a 6-in. sphere.

6. Find the volume of a ring whose outside diameter is 4 in. and whose cross section is a 1-in. circle.

7. Find the weight of a steel bar 4 in. square and 20 ft long. Steel weighs 490 lb per cu ft.

8. Find the weight of a cast-iron casting made from a white pine pattern weighing 10 lb. Cast iron weighs 450 lb per cu ft, and white pine weighs 25 lb per cu ft.

9. Find the number of board feet in 48 pieces of 2×6 ceiling joists 14 ft long.

10. Find the number of board feet of flooring required for laying a floor 10 ft by 12 ft with matched flooring $2\frac{1}{4}$ in. wide and $\frac{5}{8}$ in. thick. Allow 33% for matching and 5% for waste.

■ Chapter Test

Use the space provided to solve the following problems. Use the tables in the back of this book, the Table of Constants on page 244, and the Summary of Formulas for Solids on pages 305–308. Organize your work carefully and avoid making arithmetical errors. Estimate the answer before solving each problem.

1. Find the volume of a rectangular prism whose base is $10'' \times 15\frac{1}{2}''$ and height $22''$. (Answer in cu in.)

2. Find the volume of a prism whose base is an equilateral triangle with $10\frac{1}{2}''$ sides and whose height is $16''$. (Answer in cu in.)

3. Find the volume of a cylinder $14''$ in diameter and $10'6''$ long. (Answer in cu ft; 1 cu ft = 1,728 cu in.)

4. Find the weight of a brass bar $6''$ in diameter and $5'2''$ long. (Brass weighs 512 lb per cubic foot.)

5. What should the height of a rectangular prism be if the base is $10'' \times 10''$ and the volume is 1,500 cu in.?

6. A cylindrical 5-gallon container is $12''$ high. Find the diameter. (1 gallon = 231 cu in.)

7. Find the lateral surface of a cylinder $15''$ in diameter and $20''$ high.

8. Find the volume of a cone $18''$ high if the base is $15''$ in diameter.

9. Find the volume of a pyramid whose base is a square $12''$ on a side and whose height is $16''$.

10. Find the lateral surface of the cone in Problem 8.

11. Find the lateral surface of the pyramid in Problem 9.

12. Find the volume of a sphere $8\frac{1}{2}''$ in diameter.

13. Find the volume of a sphere whose radius is $10''$.

14. Find the surface area of a sphere whose diameter is $12''$.

15. Find the volume of a ring of circular cross section if the outside diameter is 6″ and the ring is 1″ thick.

16. The inside dimensions of a rectangular tank are 12″ × 12″ and 20″ long, and the tank is filled with water. What would be the weight of the water? (Water weighs 62.5 lb per cu ft.)

17. How many gallons of water would the tank in Problem 16 hold? (231 cu in. = 1 gallon.)

18. Find the weight of an iron casting whose volume is 7.82 cu ft. (Iron weighs 450 lb per cubic foot.)

19. Find the weight of 500 brass washers 2″ in diameter and $\frac{1}{16}$″ thick. The hole is 1″ in diameter.

20. Find the weight of 500 brass discs 1″ in diameter and $\frac{1}{16}$″ thick.

21. Find the number of board feet in a floor joist 2″ × 10″ × 16′.

22. A truckload of lumber is 16′ long, 7′6″ wide, and 4′6″ high. How many board feet does it contain?

23. How many board feet are in a 4″ × 4″ × 12′-long fir post?

24. Find the number of board feet required for a room 14′6″ × 18′0″ if $2\frac{1}{4}$″ matched flooring $\frac{5}{8}$″ thick is used. Allow $33\frac{1}{3}$% for matching and 3% for waste.

25. At $1,850.00 per thousand board feet, what will be the cost of the flooring in Problem 24?

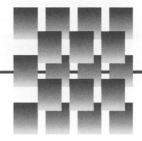

CHAPTER 10

Metric Measure

10–1   INTRODUCTION

Today almost all the industrialized world uses the **metric system** of measurement. The metric system is easier to use (once it is learned) than the **English system** of measure that the United States currently uses. The metric system has 10 as a base, and you can change from one unit to another by shifting the decimal point.

In the United States, Congress has decreed that the changeover to the metric system is voluntary. It is a matter of economics, because goods sold to other countries should be based on their system of measurement. The automobile is a good example. Suppose you were living in France and you purchased an American-manufactured car. All the nuts and bolts would be based on sixteenths of an inch, while all your wrenches and sockets would be based on millimeters. Hardly any of your tools would fit the parts of your new car. You would need two sets of tools. You would also not find it easy to get replacement parts, because the thread sizes in France are based on millimeters, and metric bolts are not compatible with American thread standards. You can see that it would be a lot of trouble for you to maintain the car, and you would probably be reluctant to purchase American-made goods unless their parts were compatible with the parts and tools available to you. The same analogy carries over to other American-made trade goods. Other countries are not going to convert to the measurement system based on inches, feet, yards, miles, and so on, because they find that the metric system is easier to use and is more logical. The metric system is their official system of measurement and incidentally is also our *official* system of measurement. The French developed the metric system and in 1799 adopted it as their official system of measurement. Its use became mandatory in France in 1837.

The unit of length in the metric system is the meter. Today the meter is defined as the distance light travels in a vacuum in 1/299,792,458 of a second. Scientists are able to use this as a measure, and therefore the length of a meter can be duplicated. For most applications the length of the meter is taken as 39.37 inches (see Fig. 10–1). That is,

1 meter = 39.37 inches = 3.28084 feet = 1.09361 yards

359

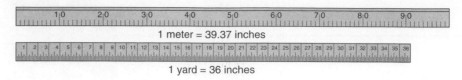

1 meter = 39.37 inches

1 yard = 36 inches

Figure 10–1

The metric unit of area is the square meter. A square meter is the area of a square whose sides are each 1 meter long.

The unit of volume is the cubic meter. A cubic meter is the volume of a cube whose length, width, and height are each 1 meter long. The liter is 0.001 part of a cubic meter, that is, a cube whose length, width, and height are each 0.1 meter long. Although the liter is not an official unit of the International System of Units, it is commonly used to measure the volume of fluids.

The standard unit of weight (technically mass) is the kilogram, a cylinder of platinum–iridium alloy kept at the International Bureau of Weights and Measures in France. There is a duplicate of this kilogram mass standard at the National Bureau of Standards in Washington, D.C.

Measurement of small items is usually done in millimeters (mm). For example, the width of film in 35 mm cameras is 35 millimeters—that is the reason for the name of this type of camera. The measure of nuts and bolts in foreign-made cars is also in millimeters. You can see from these two examples that smaller items are usually measured in millimeters. The standard for measurements on shop drawings is millimeters. All measurements are given in millimeters on shop drawings, no matter how large—this is an industry standard.

If the number of millimeters becomes larger, then it becomes more common to use centimeters (cm). A centimeter is about the width of the fingernail on your small finger. The width and length of this page could be given in centimeters.

Larger items, such as the room size in houses and the sizes of rugs, and so on, use meters (m) as a measure. The length of a competitive swimming pool is 25 meters. Track events use 50 meters, 100 meters, 400 meters, and so forth, for the sprints and distance events. Scientists use 300,000,000 meters per second as the velocity of light.

Distances between two cities use kilometers (km). The speedometer on some cars is calibrated in kilometers per hour, and the odometer registers the number of kilometers the car has been driven.

Time measure is still in our present system, although there is a suggestion to base time units on tens rather than use sixty as a base. Time will probably not be changed to a metric base in the near future.

■ **Example 1**

Find the total length of the part shown in Fig. 10–2. (All measurements are given in millimeters.)

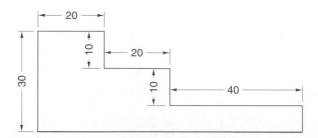

Figure 10–2

Solution   20 mm + 20 mm + 40 mm = 80 mm

Explanation   The total length is the sum of all the dimensions in the horizontal direction for this piece. There are three dimensions to be added: 20 mm + 20 mm + 40 mm = 80 mm.

Practice check: Do exercise 1 on the right.

■ **Example 2**

The distance from Philadelphia to New York is about 150 kilometers. If a salesman has to commute daily from Philadelphia to New York and back, how many kilometers does he travel in 20 days?

Solution   2 × 150 km/day × 20 days = 6,000 km

Explanation   Since he must travel to New York and back to Philadelphia each day, he travels 300 kilometers each day. In 20 days he travels 20 × 300, or 6,000 km.

Practice check: Do exercise 2 on the right.

■ **Problems 10–1**

Use your knowledge of the metric system to answer the following questions.

1. The thickness of a dime is about **(a)** 1 mm, **(b)** 1 cm, **(c)** 1 m, or **(d)** 1 km.

2. The length of a baseball bat is about **(a)** 1 mm, **(b)** 1 cm, **(c)** 1 m, or **(d)** 1 km.

3. If a high school senior boy was 150 cm tall, would the basketball coach think he was tall?

4. The height of a classroom is about **(a)** 3 mm, **(b)** 3 cm, **(c)** 3 m, or **(d)** 3 km.

5. The distance from home plate to first base on an official baseball field is about **(a)** 27 mm, **(b)** 27 cm, **(c)** 27 m, or **(d)** 27 km.

6. The distance from Philadelphia to Pittsburgh is about **(a)** 500 mm, **(b)** 500 cm, **(c)** 500 m, or **(d)** 500 km.

---

1. Find the distance around the piece shown in the following figure.

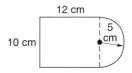

The answer is in the margin on page 363.

---

2. The width of a door is about **(a)** 9.0 m, **(b)** 0.009 km, **(c)** 90 cm, or **(d)** 0.9 m.

The answer is in the margin on page 363.

7. Our inch is a little greater than **(a)** 25 mm, **(b)** 25 cm, **(c)** 25 m, or **(d)** 25 km.

8. Find the total length of the piece shown in Fig. 10–3. (All dimensions are in millimeters.)

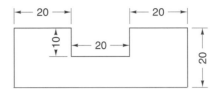

Figure 10–3

9. The distance from the Earth to the sun is about 150,000,000 km. Light travels about 300,000,000 meters per second. How many minutes does it take the light from the sun to reach the Earth using these measurements? (1,000 meters = 1 kilometer.)

10. Film is manufactured in wide rolls and then slit to make the narrower widths to fit modern cameras. If the film is manufactured in 7-meter-width rolls, how many slits would be needed to make 35-millimeter roll film from the entire 7-meter width? (This question is a little bit tricky.)

The ruler in Fig. 10–4 is divided into millimeters. It is a metric ruler. Each division is a millimeter, and each number represents a centimeter (10 millimeters = 1 centimeter).

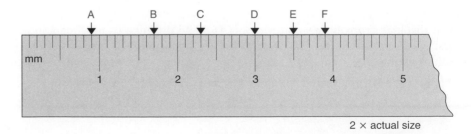

Figure 10–4

11. What is the measure at *A* in milli-meters?

12. Give the measure at *B*.

13. What is the measure at *C*?

14. Give the measure at *D*.

15. How many millimeters does the mark at *E* represent?

16. What is the measure at *F*?

17. What is the measure of *A* added to the measure at *C*?

18. Is 10 times the measure at *C* greater than or less than a meter?

19. How would you express 100 times the measure at *E*?

20. Subtract the measure at *B* from the measure at *F*.

10–2   UNITS OF LENGTH

The metric unit of length is the meter. The following table shows how the meter is subdivided. The units are related to one another by factors of 10.

Metric Linear Measure
1,000 meters = 1 kilometer (km)
100 meters = 1 hectometer (hm)
10 meters = 1 dekameter (dam)
1 meter = 1 meter (m)
0.1 meter = 1 decimeter (dm)
0.01 meter = 1 centimeter (cm)
0.001 meter = 1 millimeter (mm)

To change a unit of length to the next larger unit of length, divide by 10; that is, move the decimal point one place to the left. To change a unit of length to the next smaller unit of length, multiply by 10; that is, move the decimal point one place to the right.

The commonly used units of length are the millimeter (mm), the centimeter (cm), the meter (m), and the kilometer (km).

■ Example 1

Change 80 millimeters (mm) to centimeters.

Solution   80 mm = 8 cm

3. Change 50 millimeters (mm) to decimeters (dm).

The answer is in the margin on page 366.

Explanation   Since centimeters is the next larger unit to millimeters, divide by 10 by moving the decimal point one place to the left, thus converting millimeters to centimeters.

Practice check: Do exercise 3 on the left.

■ Example 2

Change 15.4 centimeters (cm) to millimeters (mm).

Solution   15.4 cm = 154 mm

4. Change 75.6 decimeters (dm) to millimeters (mm).

The answer is in the margin on page 366.

Explanation   Since the millimeter is the next smaller unit to centimeters, multiply by 10 by moving the decimal point one place to the right, thus converting the centimeters to millimeters.

Practice check: Do exercise 4 on the left.

■ Example 3

Change 8,465 meters (m) to kilometers (km).

Solution   8,465 m = 846.5 dam = 84.65 hm = 8.465 km

5. Change 11,786 meters (m) to kilometers (km).

The answer is in the margin on page 366.

Explanation   For each step up, divide by 10 by moving the decimal point one place to the left. Since kilometers are three steps greater than meters, the decimal point is moved three places to the left, effectively dividing by 1,000.

Practice check: Do exercise 5 on the left.

■ Problems 10–2

Use the space provided below to write the following lengths in meters.

1. 5 millimeters (mm)   **2.** 4.1 kilometers (km)

3. 6.3 hectometers (hm)

4. 2 decimeters (dm)

5. 276 millimeters (mm)

6. 25.4 centimeters (cm)

7. 47,635 millimeters (mm)

8. 15.3 millimeters (mm)

9. 0.05 kilometer (km)

10. 0.8 decimeter (dm)

Do the following problems. Be sure to include the units of measure in your answers.

11. What is the total length of eight pieces of steel, each 21.6 cm long?

12. What is the total length of 25 pieces of drill rod, each 12.95 cm long?

13. To find the circumference of a circle, multiply the diameter by 3.14. Find the circumference of a circle whose diameter is 25 cm.

14. A planer takes a 2-mm cut on a piece of steel 100 mm thick. What is the remaining thickness?

15. A steel casting 22.2 mm thick is finished by taking a 0.5-mm cut. What is the final thickness?

16. From a bar of brass 42 cm long the following three pieces are cut: 31.75 mm, 88.9 mm, and 85.5 mm. What is the final length of the bar, allowing 1.55 mm for each cut?

17. The diagonal of a square is found by dividing a side by 0.707. What is the diagonal of a square whose sides are each 15 cm?

18. A space 12 m long is to be divided into 16 equal parts. How long is each part?

19. The diameter of a circle can be found by dividing the circumference by 3.14. What is the diameter of a circle whose circumference is 79.75 mm?

20. How many holes spaced 36.5 mm center to center can be drilled in an angle iron 57.4 cm long, allowing 31.5-mm end distances?

10–3   UNITS OF AREA

The unit of area is the square meter (Fig. 10–5).

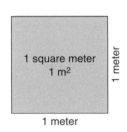

Figure 10–5

The following table shows how the units of area are subdivided. Since area is the product of the length and the width, the units are related to one another by a factor of 100. There are two common measures of area called ***are*** and ***hectare.***

Metric Square Measure
1,000,000 square meters = 1 square kilometer (km²)
10,000 square meters = 1 square hectometer (hm²)
100 square meters = 1 square dekameter (dam²)
1 square meter = 1 square meter (m²)
0.01 square meter = 1 square decimeter (dm²)
0.0001 square meter = 1 square centimeter (cm²)
0.000001 square meter = 1 square millimeter (mm²)

These measures are names of a measure of area so that the term *square* is not used with them. The *hectare* (ha) is equal to 1 square hectometer (hm^2), and the *are* is equal to 1 square dekameter (dam^2).

To change a unit of square measure to the next larger unit of square measure, divide by 100; that is, move the decimal point two places to the left. To change a unit of square measure to the next smaller unit of square measure, multiply by 100; that is, move the decimal point two places to the right.

The commonly used units are square millimeter, square centimeter, square meter, and square kilometer. The hectare is used in land measure and is equivalent to 10,000 square meters.

■ **Example 1**

Change 50,000 square centimeters (cm^2) to square meters (m^2).

Solution $50,000 \ cm^2 = 5 \ m^2$

Explanation   Since square meters is two units larger than square centimeters, divide by 100 twice. That is, divide by 10,000 by moving the decimal point four places to the left, thus converting square centimeters to square meters.

Practice check: Do exercise 6 on the right.

■ **Example 2**

Change 7.75 hectares (ha) to square meters.

Solution $7.75 \ ha = 77,500 \ m^2$

Explanation   Since square meters (m^2) is two units smaller than square hectometers (hm^2), which in land measure is called hectares (ha), multiply by 100 twice. That is, multiply by 100,000 by moving the decimal point four places to the right and adding zeros that are necessary, thus converting square hectometers to square meters.

Practice check: Do exercise 7 on the right.

6. Change 240,000 square millimeters (mm^2) to square meters (m^2).

The answer is in the margin on page 369.

7. Change 52.25 square dekameters (dam^2) to square decimeters (dm^2).

The answer is in the margin on page 369.

■ **Problems 10–3**

Use the space provided below to change the following to the units indicated.

1. $176.5 \ cm^2 =$ _____ dm^2 **2.** $47 \ cm^2 =$ _____ dm^2

3. $86.2 \ mm^2 =$ _____ cm^2 **4.** $4.67 \ dm^2 =$ _____ cm^2

5. 0.752 m^2 = _____ cm^2

6. 0.75 km^2 = _____ m^2

7. 0.005 m^2 = _____ mm^2

8. 976.3 mm^2 = _____ m^2

9. 0.56 km^2 = _____ ha

10. 46,255 m^2 = _____ ha

Do the following problems. Be sure to include the units of measure in your answers.

11. Find the floor area of a room 5 m long and 3.25 m wide.

12. Find the area of the walls of a room 2.5 m high if the room is 6 m long and 3.75 m wide. Make no allowances for door or window openings.

13. What is the total surface area of a cubical box if the box is 5.5 cm on an edge?

14. Find the cost of covering a floor 4.2 m by 6.5 m with carpet costing $12.25 per square meter.

15. A building lot has an area of 500 m^2. If its width is 15 m, how long is the lot?

16. Find the area of a right triangle if the altitude is 5 cm and the base is 4 cm.

17. Find the area of an isosceles triangle if one of the equal sides is 7.8 cm and the base is 6 cm.

18. The area of a square is 156.25 mm². What is the length of its side?

19. Find the area of a right triangle if one side is 20 mm and the hypotenuse is 25 mm.

20. The area of a circle is equal to 3.1416 times the radius squared. Find the area of a circle whose radius is 7.5 mm.

Find the area of the following composite figures. Using the space provided, make a sketch of the figure and show how you have divided it into smaller figures whose area can be found. Label all parts of your computation.

21.

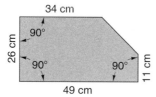

22.

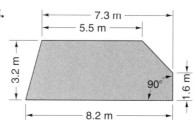

23.

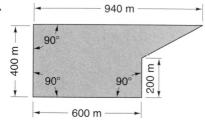

24.

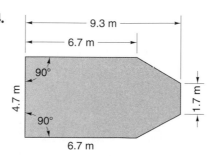

25.

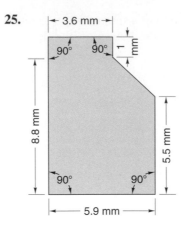

26.

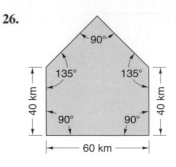

27.

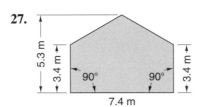

28. Find the area of the ring.

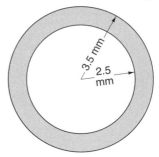

29. Find the area of the shaded portion.

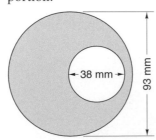

30. Find the area of the 60° sector of the circle whose radius is 37 mm as shown.

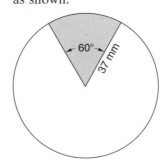

10–4   UNITS OF VOLUME

The unit of volume measure is the cubic meter (Fig. 10–6).

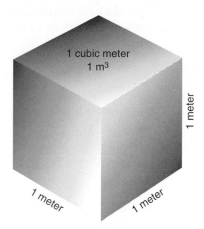

1 cubic meter
1 m³

1 meter

1 meter

1 meter

Figure 10–6

The following table shows how the units of volume are subdivided. Since volume is the product of the length, the width, and the height, the units are related to one another by a factor of 1,000. The liter is a common term for a cubic decimeter and is frequently used when measuring volumes of liquids. The cubic meter is the largest volume measure used in this book.

Metric Volume Measure
1 cubic meter = 1 cubic meter (m³)
0.001 cubic meter = 1 cubic decimeter (dm³)
0.000001 cubic meter = 1 cubic centimeter (cm³)
0.000000001 cubic meter = 1 cubic millimeter (mm³)

To change a unit of volume measure to the next larger unit of volume measure, divide by 1,000 by moving the decimal point three places to the left. To change a unit of volume measure to the next smaller unit of volume measure, multiply by 1,000 by moving the decimal point three places to the right.

The commonly used units of measure are the cubic millimeter (mm³); the cubic centimeter (cm³); the liter (*l*), which is a cubic decimeter (dm³); and the cubic meter (m³).

■ **Example 1**

Change 587.3 cubic millimeters (mm³) to cubic centimeters (cm³).

Solution   587.3 m³ = 0.5873 cm³

Explanation   Since cubic centimeters is the next larger unit to cubic millimeters, divide by 1,000 by moving the decimal place three places to the left and renaming cubic millimeters to cubic centimeters.

Practice check: Do exercise 8 on the right.

8. Change 84,671.4 cubic centimeters (cm³) to cubic meters (m³).

The answer is in the margin on page 373.

9. Change 24.06 cubic meters (m³) to cubic centimeters (cm³).

The answer is in the margin on page 374.

■ Example 2

Change 1.5 cubic meters (m³) to liters (*l*), that is, cubic decimeters (dm³).

Solution $1.5 \text{ m}^3 = 1,500 \ l = 1,500 \text{ dm}^3$

Explanation   Since cubic decimeters is the next smaller unit to cubic meters, multiply by 1,000 by moving the decimal point three places to the right and renaming the cubic meters to liters.

Practice check: Do exercise 9 on the left.

■ Example 3

Find the volume of a rectangular prism whose base is a rectangle 5.5 cm by 5.25 cm and whose height is 6.5 cm.

Solution $\begin{aligned} V &= l \times w \times h \\ &= 5.5 \times 5.25 \times 6.5 \\ &= 187.6875 \text{ cm}^3 \approx 187.7 \text{ cm}^3 \end{aligned}$

10. Find the volume of a cylinder 2.04 m in diameter and 10.6 m long.

The answer is in the margin on page 374.

Explanation   Volume is the product of length, width, and height. Since the measures are all in the same units, multiply the three together and round off the answer to 187.7 cm³.

Practice check: Do exercise 10 on the left.

■ Problems 10–4

Use the space provided below to change the following to the units indicated.

1. 3.5 dm³ = _____ m³   **2.** 4.8 dm³ = _____ cm³

3. 487 mm³ = _____ cm³   **4.** 0.6 cm³ = _____ mm³

5. 0.56 m³ = _____ *l*   **6.** 763.2 *l* = _____ m³

7. 547,345 mm³ = _____ m³   **8.** 0.005 *l* = _____ mm³

9. 0.0067 cm³ = _____ mm³   **10.** 1.35 m³ = _____ mm³

Find the volumes of the following solids. (Use 3.1416 for π.)

11. Find the volume of a rectangular prism 42 cm high whose base is a square 24 cm on each edge.

12. Find the volume of a cylinder 12 cm in diameter and 18 cm long.

Answer to exercise 8

84,671.4 cm³ = 0.0846714 m³

Two steps up means to divide by 1,000,000 or move the decimal point six places to the left.

13. Find the volume of the solid shell of the hollow cylinder in Fig. 10–7.

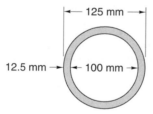

Figure 10–7

14. Find the volume of a rectangular prism whose base is 30.5 cm × 37.5 cm and whose height is 45.7 cm.

15. Find the volume of a cylindrical tank in cubic meters if the diameter of the base is 2.45 m and the height is 7.6 m.

16. Find the volume of a prism 30.2 cm high whose base is a regular hexagon 2.5 cm across the flats. (The area of a hexagon is $0.866f^2$.)

17. Find the volume of a tank 7.5 m high if the base is a regular octagon whose side is 2.75 m long. (The area of a regular octagon is $4.828s^2$.)

18. Find the volume of a prism whose base is an equilateral triangle with 127 mm sides and whose height is 35 cm.

19. Find the volume (in liters) of a tank whose height is 122 cm and whose base is a rectangle 12.7 cm $\times$ 75 cm (1 l = 1 dm^3).

20. Find the volume of a cylindrical tank (in liters) if the height is 2.5 m and the diameter of the base is 1.25 m.

Use the space provided below to do the following problems.

21. Find the capacity in liters of an elliptical tank 4 m long if the base is an ellipse 55 cm $\times$ 50 cm.

22. Find the weight of a brass bar 75 mm in diameter and 250 mm long (brass weighs 8.2 g per cm^3).

23. A rectangular prism with a square base 70 mm on edge is to hole 1 liter of fluid. How tall must the container be?

24. Find the depth of a rectangular vat 2.8 m $\times$ 4.7 m to contain 100 m^3 of fluid.

25. The volume of a steel shaft 75 mm in diameter is 93.75 cm^3. What is its length?

26. Find the lateral surface of a cylinder 36 cm in diameter and 24 cm high.

27. The base of a cylindrical tank is a circle whose area is 188.4 dm^2; the tank has a capacity of 2.826 m^3. What is the lateral surface area?

28. Find the volume of a right circular cone 1 m high whose base is 50 cm in diameter.

29. A quantity of sand dumped upon the ground assumed a conical shape with a circumference about the base of 31.4 m and a height of 3.3 m. How many cubic meters of sand are in the pile?

30. If the sand in Problem 29 weighs 1,600 kg per m^3, what is the weight of the pile?

10–5   UNITS OF WEIGHT

The standard unit of weight is the kilogram, the cylindrical platinum–iridium bar at the International Bureau of Weights and Measures in Sevres, France. Since 1883 this bar has been defined as the international standard of the kilogram. It is the only metric unit defined by an artifact. Scientists are working on a new standard for the kilogram that can be replicated anywhere and is not dependent upon an artifact. Studies are currently underway using carbon and silicon atoms.

The following table shows how the units of weight are subdivided. The units are related to one another by a factor of 10.

Metric Weight (Mass) Measure
1,000 kilograms = 1 metric ton (T)
1,000 grams = 1 kilogram (kg)
100 grams = 1 hectogram (hg)
10 grams = 1 dekagram (dag)
1 gram = 1 gram (g)
0.1 gram = 1 decigram (dg)
0.01 gram = 1 centigram (cg)
0.001 gram = 1 milligram (mg)

To change a unit of weight to the next larger unit of weight, divide by 10; that is, move the decimal point one place to the left. To change a unit of weight to the next smaller unit of weight, multiply by 10; that is, move the decimal point one place to the right. The metric ton measure in our table is three steps up from the kilogram.

The commonly used units of weight are the milligram (mg), the centigram (cg), the gram (g), the kilogram (kg), and the metric ton.

■ **Example 1**

Change 5,648 milligrams to grams.

Solution   5,648 mg = 5.648 g

11. Change 70,146 centigrams (cg) to grams (g).

The answer is in the margin on page 378.

Explanation   Since grams are three units larger than milligrams, divide by 10 three times. That is, divide by 1,000 by moving the decimal point three places to the left, thus converting milligrams to grams.

Practice check: Do exercise 11 on the left.

■ **Example 2**
Change 3.22 centigrams (cg) to milligrams (mg).

Solution   3.22 cg = 32.2 mg

12. Change 25.06 hectograms (hg) to centigrams (cg).

The answer is in the margin on page 378.

Explanation   Since milligrams is the next smaller unit to centigrams, multiply by 10 by moving the decimal point one place to the right, and rename the centigrams to milligrams.

Practice check: Do exercise 12 on the left.

■ **Problems 10–5**
Use the space provided to write the following weights as grams.

1. 4.2 dekagrams (dag)   **2.** 2.3 hectograms (hg)

3. 45.8 decigrams (dg)   **4.** 39.2 milligrams (mg)

5. 0.78 decigram (dg)   **6.** 5.4 decigrams (dg)

7. 8.7 centigrams (cg)   **8.** 1.3 kilograms (kg)

9. 0.09 kilogram (kg)   **10.** 1 metric ton (T)

Do the following problems. Be sure to include the units of measure in your answers.

11. What is the sum of the following weights: 3.4 g, 4.9 g, 18.7 g, 3.05 g?

12. What is the sum of the following weights: 7.6 g, 5.9 dg, 4.7 dg, 3.5 g?

13. Distribute 56 g into eight equal weights.

14. What is the weight of 47 bolts if each bolt weighs 97.3 mg?

15. A piece of steel plate weighs 1 kg. If 25 holes are drilled through the steel plate where the amount of material removed for each hole is 2.67 g, what is the final weight of the steel plate?

16. What is the weight of 687 castings if each casting weighs 3.45 kg?

17. What is the total weight of 45 pieces of steel where each piece weighs 6.85 kg and 35 pieces of brass where each piece weighs 10.5 kg?

18. What is the sum of the following weights: 2.4 kg, 4.8 hg, 8.7 dag, 9.2 hg, and 8.8 hg?

19. If a brick measured 5.7 cm × 9.5 cm × 20.3 cm, what would it weigh if bricks weighed 2.4 grams per cm³?

20. How many bricks in Problem 19 would it take to weigh 1 metric ton?

10–6   CONVERTING ENGLISH LENGTH TO METRIC LENGTH

One meter is equal to 39.37 in. One meter is equal to 100 cm. Hence, 1 cm is equal to 0.3937 in. Using ratio and proportion, we can find the number of centimeters in 1 in. (Fig. 10–8).

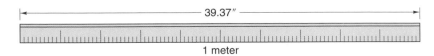

1 meter

Figure 10–8

$$0.3937 \text{ in.} : 1 \text{ cm} = 1 \text{ in.} : x \text{ cm}$$

To find the extreme (x), multiply the means (1 cm and 1 in.), and divide by the other extreme (0.3937).

$$\frac{1 \times 1}{0.3937} = x$$
$$2.54 = x$$

Therefore, 2.54 cm is equal to 1 in.

■ **Example 1**

How many millimeters are in 1 in.?

Solution   2.54 cm = 1 in.
                25.4 mm = 1 in.

Explanation   Since millimeters is the next smaller unit to centimeters, multiply by 10 by moving the decimal point one place to the right, and rename centimeters to millimeters.

Practice check: Do exercise 13 on the left.

■ **Example 2**

A basketball player is 6′7″ tall. How tall is he in meters and parts of meters?

Solution   6′7″ = (6 × 12)″ + 7″ = 79″
                2.54 × 79″ = 200.66 cm
                200.66 cm = 2.0066 m ≈ 2.01 m

Explanation   Change the 6′7″ to inches. Then multiply the 79″ by the number of centimeters in each inch; that is, 2.54 × 79″ = 200.66 cm. Change the centimeters to meters and round off the result to the nearest centimeter.

Practice check: Do exercise 14 on the left.

13. How many decimeters are in 1 in.?

The answer is in the margin on page 380.

14. A building lot has a frontage of 50′6″. How long is the frontage in meters?

The answer is in the margin on page 380.

■ **Example 3**

How many centimeters are in 1 ft?

Solution
$$1'' = 2.54 \text{ cm}$$
$$1' = 12''$$
$$2.54 \times 12 = 30.48 \text{ cm}$$

Explanation   In 1 in. there are 2.54 cm; hence, there are $12 \times 2.54 = 30.48$ cm in 1 ft.

Practice check: Do exercise 15 on the right.

15. How many centimeters are there in 1 yard?

The answer is in the margin on page 381.

The following table summarizes English-to-metric length conversions.

Summary of Converting English Measure to Metric Measure	
1 inch = 2.54 cm = 25.4 mm 1 foot = 12 inches = 30.48 cm	1 yard = 36 inches = 91.44 cm = 0.9144 m 1 mile = 5,280 feet = 1,609.344 m

■ **Problems 10–6**

Change the English measure in the following table to millimeters, centimeters, and meters as indicated.

	English Measure to Metric Measure			
	English Measure	**Millimeters (mm)**	**Centimeters (cm)**	**Meters (m)**
1.	4″	_____	_____	_____
2.	1′6″	_____	_____	_____
3.	8″	_____	_____	_____
4.	6′8″	_____	_____	_____
5.	$4\frac{1}{2}''$	_____	_____	_____
6.	$2\frac{3}{4}''$	_____	_____	_____
7.	$\frac{1}{32}''$	_____	_____	_____
8.	$1'7\frac{1}{2}''$	_____	_____	_____
9.	$2'3\frac{1}{4}''$	_____	_____	_____
10.	Your height	_____	_____	_____

Answer to exercise 13

2.54 cm = 1 in.
0.254 dm = 1 in.

Answer to exercise 14

$(50 \times 12)'' + 6'' = 606''$
$2.54 \times 606'' = 1{,}539.24$ cm
$1{,}539.24$ cm $= 15.3924$ m
≈ 15.39 m

Use the space provided below to work out the answers to the following problems.

11. What is the length in centimeters of the piece of stock required for eight taper pins, each $6\frac{1}{2}''$ long, allowing $\frac{1}{8}''$ waste for each cut?

12. In Fig. 10–9, what is the total length of the bolt in centimeters?

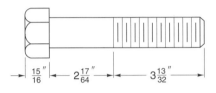

Figure 10–9

13. Find the difference in millimeters between the diameters at the ends of the tapered piece shown in Fig. 10–10.

Figure 10–10

14. A planer takes a $\frac{3}{32}''$ cut on a piece of steel $\frac{3}{8}''$ thick. What is the remaining thickness in millimeters?

15. The diameter of a circle is $1''$. What is its circumference in centimeters? (Use 3.14 for π.)

16. The diameter of a circle is $6\frac{1}{2}''$. What is its circumference in centimeters?

17. The circumference of a circle is $26.69''$. What is its radius in centimeters?

18. A common brick measures $2\frac{1}{4}'' \times 3\frac{3}{4}'' \times 8''$. What is its measure in metric units?

19. The length of a compact car is 164''. What is its length in meters?

20. An engine has a stroke of 3.5''. What is its stroke in metric measure?

10-7 CONVERTING ENGLISH AREA TO METRIC AREA

Area measure is the product of length and width. If the length and the width are in inches, then the area is in square inches. If the length and the width are in centimeters, then the area is in square centimeters. To change from the English area measure to the metric area measure, change the length and width measures to the same metric units and find their product. Metric conversion tables are given in the Appendix of this book.

■ **Example 1**

Find the area of a rectangle in square centimeters if its length is $4\frac{1}{2}$ in. and its width is $3\frac{1}{2}$ in.

Solution
$2.54 \times 4\frac{1}{2} = 2.54 \times 4.5 = 11.430$ cm
$2.54 \times 3\frac{1}{2} = 2.54 \times 3.5 = 8.890$ cm
$8.890 \times 11.430 = 101.6127$ cm^2
$= 101.6$ cm^2

Explanation   Change both of the inch measures to centimeters by multiplying them by 2.54 (the number of centimeters in 1 in.). Then find the product of the two; that is, $8.890 \times 11.430 = 101.6127$. The answer is in square centimeters and is rounded off to one decimal place.

Practice check: Do exercise 16 on the right.

16. Find the area of a rectangle in square centimeters if its length is 20 in. and its width is $12\frac{3}{4}$ in.

The answer is in the margin on page 383.

■ **Example 2**

Find the area of a circle in square centimeters if the radius of the circle is 1.875''.

Solution
$A = \pi r^2$
$= 3.1416 \times (2.54 \times 1.875)^2$
$= 71.25$ cm^2

Explanation   Use 3.1416 as the approximation for π and change the 1.875'' to centimeters by multiplying it by 2.54. This product is the radius in centimeters. Square the radius and multiply by 3.1416. The answer is rounded off to two decimal places.

Practice check: Do exercise 17 on the right.

17. Find the area of a circle in centimeters if its radius is $6\frac{1}{4}$ in.

The answer is in the margin on page 383.

■ **Problems 10–7**

Find the area (in square centimeters) of the rectangles in the following chart. Change the length and the width to centimeters; then find the area to one decimal place accuracy.

			Rectangles		
	Length	**Width**	**Length (cm)**	**Width (cm)**	**Area**
1.	3″	2″	_____	_____	_____
2.	6″	4″	_____	_____	_____
3.	3.5″	3″	_____	_____	_____
4.	6.7″	4.5″	_____	_____	_____
5.	$3\frac{1}{4}''$	2″	_____	_____	_____
6.	$2\frac{3}{4}''$	$1\frac{1}{2}''$	_____	_____	_____
7.	1′2″	9″	_____	_____	_____
8.	$1'4\frac{1}{2}''$	$8\frac{1}{2}''$	_____	_____	_____
9.	$1'4\frac{1}{4}''$	1′3″	_____	_____	_____
10.	$2'6\frac{5}{8}''$	$1'3\frac{1}{2}''$	_____	_____	_____

Use the space provided below to find the following areas. Use metric units in your answer. Select the units that would be reasonable for the particular problem.

11. Find the area of a page of a book $7\frac{1}{2}''$ long and 5″ wide.

12. Find the area of a room 16′4″ long and 10′8″ wide.

13. How many square meters are there in a sidewalk 198′ long and 4′ wide?

14. Find the area of the walls of a room if the walls are 8′ high and the room is 18′ long and 12′ wide. Make no allowance for doors or windows.

15. Allowing 2.5 m² per child, how many children can safely play in a playground 200′ long and 140′ wide?

16. How many square centimeters are there in the six sides of a box $8\frac{1}{2}'' \times 6'' \times 3\frac{1}{2}''$?

17. How many square meters of sheet-rock are needed for all the walls of a garage 15′6″ × 8′9″ and 8′3″ high? (Subtract 6 m² as door and window allowance.)

18. Find the cost of covering a floor 13′9″ × 21′6″ with carpet costing $15.25 per square meter.

Answer to exercise 16

$$2.54 \times 20 = 50.8 \text{ cm}$$
$$2.54 \times 12.75 = 32.385 \text{ cm}$$
$$A = 50.8 \times 32.385 = 1{,}645.158 \text{ cm}^2$$
$$\approx 1{,}645.2 \text{ cm}^2$$

Answer to exercise 17

$$2.54 \times 6.25 = 15.875 \text{ cm}$$
$$A = 3.1416 \times 15.875^2$$
$$= 791.7 \text{ cm}^2$$

19. Find the area of a circle whose radius is 12′.

20. Find the area of a 60° sector of a circle whose radius is 8½″.

10–8 CONVERTING ENGLISH AREA TO METRIC AREA USING CONSTANTS

If we have an area in square yards and we wish to change the measure to square meters, we can use the following procedure. A square meter in English units is a square whose sides are each 39.37″ long. A square meter is equal to 1,549.9969 sq in. A square yard is a square whose sides are each 36″. A square yard is equal to 1,296 sq in. A square yard is smaller than a square meter. The ratio of the square yard to the square meter is

$$\frac{1{,}296}{1{,}549.9969} = 0.8361307$$

To change square yards to square meters, multiply the square yards by 0.83613 (rounded off to the nearest hundred thousandth).

To find the ratio of the square foot to the square meter, compare the number of square inches in a square foot with the number of square inches in a square meter. The ratio is

$$\frac{144}{1{,}549.9969} = 0.0929034$$

To change square feet to square meters, multiply the square feet by 0.09290 (round off the ratio to the nearest hundred thousandth).

To change square inches to square centimeters, compare the square inch to the number of square inches in a square centimeter; that is, 1 cm² = 0.3937″ × 0.3937″. The ratio is

$$\frac{1}{0.15499969} = 6.4516$$

To change square inches to square centimeters, multiply by 6.4516.

The following table summarizes using constants to convert English area to metric area.

Summary of Converting English Area to Metric Area
1 sq in. = 6.4516 cm^2
1 sq ft = 0.09290 m^2
1 sq yd = 0.83613 m^2
1 acre = 0.4047 hectare (ha)

■ Example 1

Change 950 sq yd to square meters.

Solution $950 \times 0.83613 = 794.3235$
$= 794.3$ m^2

18. Change 1,567 sq yd to square meters.

The answer is in the margin on page 387.

Explanation   Multiply the square yards by the constant 0.83613 to change square yards to square meters, and round the answer to one decimal place.

Practice check: Do exercise 18 on the left.

■ Example 2

Change 2.5 sq ft to square meters.

Solution $2.5 \times 0.09290 = 0.23225$
$= 0.23$ m^2

19. Change 20.75 sq ft to square meters.

The answer is in the margin on page 387.

Explanation   Multiply the square feet, 2.5, by the constant 0.09290 to change square feet to square meters. Round the answer to two decimal places.

Practice check: Do exercise 19 on the left.

■ Example 3

Change 4.75 sq in. to square centimeters.

Solution $4.75 \times 6.4516 = 30.6451$
$= 30.65$ cm^2

20. Change $93\frac{1}{4}$ sq in. to square centimeters.

The answer is in the margin on page 387.

Explanation   Multiply the square inches by 6.4516 to change square inches to square centimeters. Round the answer to two decimal places.

Practice check: Do exercise 20 on the left.

■ Problems 10–8

Use the constants in this section to change the following square measures. Use the space provided to work out the answers.

1. Change 1,985 sq yd to square meters.

2. Change 27.4 sq yd to square meters.

3. Change 456.87 sq yd to square meters.

4. Change 45.8 sq ft to square meters.

5. Change 8.75 sq ft to square meters.

6. Change 109.25 sq ft to square meters.

7. Change 40 sq in. to square centimeters.

8. Change 1.5 sq in. to square centimeters.

9. Change 29.75 sq in. to square centimeters.

10. Change 34,450 sq yd to hectares. (1 hectare = 10,000 sq m.)

11. Change 9,280 sq yd to hectares.

12. Change 4,840 sq yd (the area of an acre) to ares. (1 are = 100 sq m.)

13. Change 10,502 sq yd to ares.

14. Change 20 acres to ares.

15. Change 5,000 acres to hectares.

10-9   CONVERTING ENGLISH VOLUME TO METRIC VOLUME

Volume measure is the product of the length, the width, and the height. If these measures are in inches, then the volume is in cubic inches; if these measures are in centimeters, then the volume is in cubic centimeters. To change from the English volume measure to the metric volume measure, change the length, width, and height measures to the same metric units and find their product.

■ Example 1

Find the volume in cubic centimeters of a rectangular prism whose height is 3.5″ and whose base is a rectangle 1.6″ by 2.2″.

Solution
$$V = l \times w \times h$$
$$= (2.2 \times 2.54) \times (1.6 \times 2.54) \times (3.5 \times 2.54)$$
$$= 5.588 \times 4.064 \times 8.890$$
$$= 201.88862 \text{ cm}^3 \approx 201.9 \text{ cm}^3$$

21. Find the volume in cubic centimeters of a rectangular prism whose height is 22.3″ and whose base is a rectangle 10.4″ by 15.5″.

The answer is in the margin on page 388.

Explanation   Change the length, width, and height inch measures to centimeters by multiplying each by 2.54, the number of centimeters in 1 inch. The length, width, and height are now in centimeters. Multiply the length, width, and height. The product is 201.88862 cm³. Round this product to the nearest tenth, that is, 201.9 cm³.

Practice check: Do exercise 21 on the left.

If the volume of a solid is known in English units, the metric equivalent of the volume can be found by using constants that change English volume units to metric volume units, as summarized in the following table.

Summary of Converting English Volume to Metric Volume
1 cu in. = 16.387 cm³
1 cu ft = 28.317 dm³
= 0.0283 m³
1 cu yd = 0.7646 m³

■ Example 2

Change 500 cu in. to cubic centimeters.

Solution $500 \times 16.387 = 8{,}193.5 \text{ cm}^3$

22. Change 825 cu ft to cubic meters.

The answer is in the margin on page 388.

Explanation   To change cubic inches to cubic centimeters, multiply by the constant 16.387, since the constant table shows that there are 16.387 cm³ in every cubic inch.

Practice check: Do exercise 22 on the left.

■ Problems 10–9

Find the volume in cubic centimeters of the rectangular prisms in the following chart. Use 3.1416 as an approximation for π when finding the volume of the cylinders.

Rectangular Prisms

	English Measures			Metric Measures			
	Length	**Width**	**Height**	**Length**	**Width**	**Height**	**Volume**
1.	4″	3″	5″	_____	_____	_____	_____
2.	3.7″	2″	4″	_____	_____	_____	_____
3.	$2\frac{1}{2}″$	$2\frac{1}{2}″$	3″	_____	_____	_____	_____
4.	7.5″	4.2″	4.2″	_____	_____	_____	_____
5.	1′	8″	9″	_____	_____	_____	_____
6.	1′3″	1′1″	1′6″	_____	_____	_____	_____

Answer to exercise 18

$1,567 \times 0.83613 = 1,310.2157$
$\approx 1,310.2 \, \text{m}^2$

Answer to exercise 19

$20.75 \times 0.09290 = 1.927675$
$\approx 1.9 \, \text{m}^2$

Answer to exercise 20

$93.25 \times 6.4516 = 601.6117$
$\approx 601.6 \, \text{cm}^2$

Find the volume in cubic centimeters of the cylinders in the following chart. Use 3.14 as an approximation for π.

Cylinders

	English Measures		Metric Measures		
	Diameter	**Height**	**Diameter**	**Height**	**Volume**
7.	4″	3″	_____	_____	_____
8.	2.8″	4″	_____	_____	_____
9.	$3\frac{1}{2}″$	$2\frac{1}{2}″$	_____	_____	_____
10.	7.6″	3.5″	_____	_____	_____
11.	1′	8″	_____	_____	_____
12.	1′6″	1′2″	_____	_____	_____

Use the constants shown in the table summarizing conversion of English volume to metric volume to find the following volumes in metric measure.

13. The volume of a rectangular prism is 1,444 cu in. What is its volume in cubic centimeters?

14. The volume of a right circular cylinder is 80 cu in. What is its volume in cubic centimeters?

15. The volume of a tank is 5 cu ft. What is its volume in liters? (1 cubic decimeter = 1 liter.)

16. What is the capacity of a gas tank in liters if it holds 20 gallons? (1 gallon = 231 cu in.)

Answer to exercise 21

$V = (22.3 \times 2.54) \times (10.4 \times 2.54)$
$\qquad \times (15.5 \times 2.54\,)$
$\quad = 58,907.561$
$\quad \approx 58,907.6 \text{ cm}^2$

Answer to exercise 22

$825 \times 0.0283 = 23.3475$
$\qquad\qquad\quad \approx 23.3 \text{ m}^3$

17. A concrete truck holds 13 cu yd of concrete. How many cubic meters is this?

18. A circular vat has a capacity of 575 gallons. What is its capacity in cubic meters?

19. A concrete walkway is 4 ft wide, 80 ft long, and 6 in. deep. How many cubic meters of concrete were needed for this walk?

20. A swimming pool is 40 ft wide and 75 ft long. The bottom slants from 4 ft deep at the shallow end to 12 ft deep at the deep end so that a cross section along the 75-ft length is in the shape of a trapezoid and a cross section across the 40-ft width is a rectangle. How many cubic meters of water would it hold if it were filled?

10–10   CONVERTING ENGLISH WEIGHT TO METRIC WEIGHT

One kilogram equals 2.20462 lb, rounded off to 2.2 lb for calculations that do not demand great accuracy. One pound equals 0.453592 kg, rounded off to 0.454 kg for calculations that do not demand great accuracy.

■ **Example 1**

Change 3.5 lb to kilograms.

Solution $3.5 \times 0.454 = 1.5890 \text{ kg}$
$\qquad\qquad\qquad\quad = 1.6 \text{ kg}$

Explanation   To change pounds to kilograms, multiply by the number of kilograms in 1 lb, that is, 0.454. Round off the product to 1.6 kg.

■ **Example 2**

The weight of a certain amount of water is 12 lb. Find its weight in kilograms.

Solution $12 \times 0.454 = 5.448$
$\qquad\qquad\qquad \approx 5.4 \text{ kg}$

■ **Example 3**

A sack of tomatoes weighs 16.16 lb. How much does a half sack of tomatoes weigh in kilograms?

23. Change 23.6 lb to kilograms.

The answer is in the margin on page 390.

Solution $\frac{1}{2} \times 16.16 \times 0.454 = 3.66832$
$\qquad\qquad\qquad\qquad\quad \approx 3.67 \text{ kg}$

Practice check: Do exercise 23 on the left.

■ **Problems 10-10**

Use the space provided below to find the following weights in kilograms.

1. Change 8.75 lb to kilograms.

2. Change $5\frac{1}{2}$ lb to kilograms.

3. Find the sum of the following weights in kilograms: $2\frac{1}{2}$ lb, 4.375 lb, 1.875 lb, 3.25 lb.

4. Distribute 64 lb into 10 equal weights in metric measure.

5. What is the weight in grams of 20 bolts if each bolt weighs 1.1 oz?

6. What is the weight in kilograms of 875 castings if each casting weighs 21.5 lb?

7. Twelve hundred and fifty (1,250) castings weigh 5,625 lb. What is the average weight of each casting in kilograms?

8. One cubic foot of water weighs 62.5 lb. What is its weight in kilograms?

9. One metric ton is 1,000 kg. How many metric tons are there in 275 short tons? (1 short ton = 2,000 lb.)

10. One long ton weighs 2,240 lb. How many metric tons are there in 8,753 long tons?

11. What is your weight in kilograms?

12. A baby weighs between 6 and 9 lb at birth. What is this range in kilograms?

10–11   CONVERTING ENGLISH TEMPERATURE AND METRIC TEMPERATURE

Water boils at 212°F and freezes at 32°F on the Fahrenheit temperature scale. The °F is called "degrees Fahrenheit." In the metric system, water boils at 100°C and freezes at 0°C. This is called the Celsius temperature scale, and the °C is called "degrees Celsius." At 4°C, water is denser than water at 0°C (ice), and therefore ice floats on the denser water and fish and other aquatic life are able to live through the cold weather.

In comparing the two scales, we see that there are 108 divisions in the Fahrenheit scale that cover the same range as 100 divisions in the Celsius scale (Fig. 10–11). If 32 is subtracted from the degrees in the Fahrenheit scale, then both scales start at 0°, the temperature at which water freezes. Using these two ideas, we can make the conversions °F to °C and °C to °F. To change °F to °C, subtract 32 from the °F and multiply the difference by the ratio $\frac{100}{180}$. This ratio is equal to $\frac{5}{9}$. To change °C to °F, multiply the °C by $\frac{9}{5}$ and then add 32.

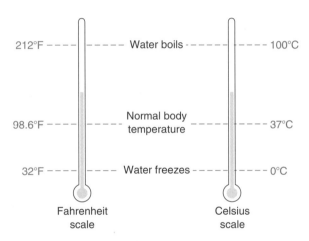

Figure 10–11

■ **Example 1**

Change 68°F to °C.

Solution $(68 - 32) \times \frac{5}{9} = 36 \times \frac{5}{9} = 20°C$

Explanation   Subtract 32 from the °F and then multiply the difference by $\frac{5}{9}$.

Practice check: Do exercise 24 on the left.

■ **Example 2**

Change 50°C to °F.

Solution $(50 \times \frac{9}{5}) + 32 = 122°F$

Explanation   Multiply the °C by $\frac{9}{5}$ and add 32 to that product.

Practice check: Do exercise 25 on the left.

■ **Example 3**

Change 14°F to °C.

Solution $(14 - 32) \times \frac{5}{9} = -18 \times \frac{5}{9} = -10°C$

Explanation   The answer is negative and indicates that the temperature is below the freezing point of water (0°C). The procedure is to subtract 32 from the °F and then multiply the difference by $\frac{5}{9}$. The difference is always negative in temperatures below 32°F.

Practice check: Do exercise 26 on the right.

26. Change 5°F to °C.

The answer is in the margin on page 393.

■ **Problems 10–11**

Change to the temperature scale as indicated.

1. 115°F to °C

2. 35°F to °C

3. 210°F to °C

4. 20°F to °C

5. 0°F to °C

6. 20°C to °F

7. 45°C to °F

8. 10°C to °F

9. 90°C to °F

10. 30°C to °F

11. A dangerously high fever is 105°F. How many degrees Celsius is this?

12. The temperature in Death Valley on July 10, 1913, was 134°F. How many degrees Celsius is this?

13. The temperature is 3°C, and the weather is clear. What would you wear to go for a walk?

14. Would you go ice skating or swimming if the outside temperature was 29°C?

Answer to exercise 24

$(50 - 32) \times \frac{5}{9} = 18 \times \frac{5}{9}$
                    $= 10°C$

Answer to exercise 25

$(65 \times \frac{9}{5}) + 32 = 117 + 32$
                        $= 149°F$

15. On one Presidents' Day the temperature fell to $-40°F$ at Point Barrow, Alaska. How many degrees Celsius was the temperature in Point Barrow that day?

■ Self-Test

Use the space provided below to do the following problems. Check your answers with the answers in the back of the book. Estimate the answer to each problem before solving it.

1. Change 4.8 km to meters.

2. Find the area of a right triangle whose altitude is 12 cm and whose base is 5 cm.

3. Find the volume of a rectangular prism whose base is 18.5 by 16.5 cm and whose height is 12 cm.

4. Distribute 81 cg into 90 equal weights.

5. The radius of a circle is 4 in. What is its circumference in centimeters?

6. What is the total surface area in cubic centimeters of a cube 3 in. on an edge?

7. Change 2.5 sq in. to square centimeters.

8. What is the volume of a cube 6 in. on an edge in metric measure?

9. Change 50 lb to kilograms.

10. Change 100°F to °C.

■ Chapter Test

Use the space provided below to do the following problems. Use the constants developed in this chapter. Make sure that you use units that are reasonable for the problem. Try to estimate the answer before solving each problem.

1. What is the total length of 12 pieces of brass, each 13.4 cm long?

2. A shaper takes a 1.2-mm cut on a piece of steel 1.27 cm thick. What is the remaining thickness?

3. The following pieces were cut from a 6-m length of copper tubing: five pieces 30 cm long and four pieces 42.5 cm long. What is the remaining length of the copper tubing?

4. A space 10 m long is to be divided into 14 equal parts. How long is each part?

5. Find the floor area of a house 7.5 m by 15.5 m.

6. Find the cost of covering a floor 5.2 m by 6 m with a rug at $15.75 per square meter.

7. A building lot has an area of 665 m². What is its length if its width is 17.5 m?

8. Find the area of a triangle if its base is 27 cm and its altitude is 13.5 cm.

9. Find the area of the splice plate shown in Fig. 10–12.

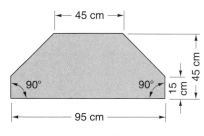

Figure 10–12

10. Find the area of the plate shown in Fig. 10–13.

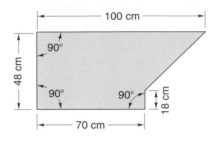

Figure 10–13

11. Find the area of the ring shown in Fig. 10–14.

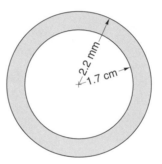

Figure 10–14

12. Find the volume of a rectangular prism whose base is 27 cm by 35.6 cm and whose height is 40 cm.

13. Find the volume of a cylindrical tank if the radius of the base is 1.5 m and the height is 3 m.

14. Find the volume in liters of a tank whose base is a square 60 cm on a side and whose height is 40 cm.

15. Distribute 72 kg into 100 equal weights.

16. What is the weight of 150 bolts if each weighs 35 g?

17. What is the length in centimeters of 12 dowels, each 4.5″ long?

18. The circumference of a circle is 30 in. What is its radius in centimeters?

19. How many square meters are there in a sidewalk 70′ long and 3.5′ wide?

20. Change 15 sq in. to square centimeters.

21. Find the area in square centimeters of a circle whose radius is 4.6 in.

22. Find the volume in cubic centimeters of a right circular cylinder whose height is 5″ and the radius of whose base is 2″.

23. 456 similar castings weigh a total of 2,052 lb. What is the average weight of each casting in kilograms?

24. Change 40°F to °C.

25. Change 70°C to °F.

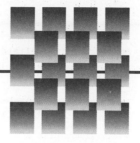

CHAPTER 11

Graphs

11–1   DEFINITIONS

When there are two sets of related facts, the relations between them can often be shown clearly by means of diagrams, or **graphs.** Such graphs are frequently used in all types of business and technical work. Squared paper, sometimes called **cross-section paper,** is generally used for the purpose, and convenient values are assigned to the horizontal and vertical divisions. Two lines at right angles to each other are selected as **axes** or reference lines; the horizontal line is called the **X-axis,** and the vertical line the **Y-axis.** The intersection of these two lines is called the **origin** (see Fig. 11–1). Points are plotted in relation to the two axes. These points are then

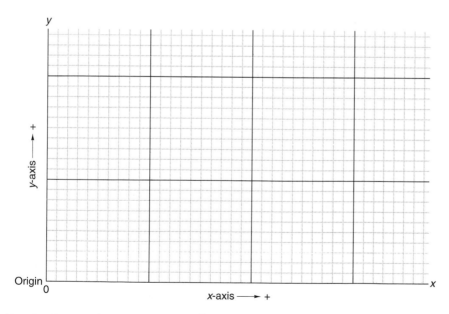

Figure 11–1

connected by straight lines or curves, forming a broken line or a continuous curve, which shows graphically the relations that exist and the changes that take place.

11–2   TYPES OF GRAPHS

There are many kinds of graphs, each adapted to a particular purpose. In general, however, they may be divided into two types: (1) where no causal relation exists and (2) where causal relation does exist.

Where no causal relation exists, that is, where the graph is merely a picture of a changing condition, is the first type of graph. Graphs showing changes of temperature or of population within a given period are examples of this type. The method of construction and the use of such graphs are illustrated by the following example.

■ Example 1

The population of a certain town, taken at intervals of 10 years, varied as shown in the following table. Draw a graph showing the growth of this town.

Year	Population
1910	12,000
1920	16,000
1930	22,400
1940	26,000
1950	27,000
1960	27,200
1970	29,500
1980	29,200
1990	28,900

Construction   Draw the horizontal and vertical reference lines as shown in Fig. 11–2. On the horizontal line, or X-axis, mark off the years 1930, 1920, and so on; on the vertical line, or Y-axis, mark off the populations by thousands. Distances on the vertical lines above the X-axis now represent populations, and distances on the horizontal lines to the right of the Y-axis represent years. At the intersection of the 1920 year line with the 16,000 population line, locate the first point. Locate other points at the intersection of the 1930 year line with the 22,400 population

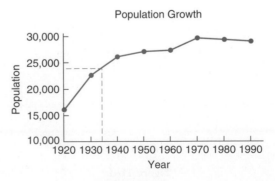

Figure 11–2

line, the 1940 line with the 26,000 line, and so forth. Connecting successive points with straight lines produces a broken line, which is the required graph of this information.

Note that the 10,000 line is used as the base line of this graph. This is done merely to save space, since nothing is gained by starting at zero, which is the real base line.

Explanation In this instance there is no necessary relation between the two sets of figures used in plotting the graph. The values do not depend on any law or rule, and the graph is merely a picture of the increase in the population. Instead of making a study of the tabulated figures, we can see at a glance how the population has varied from 1920 to 1990.

Use of the Graph In connecting successive points by straight lines, we assume that the change from one point to the next has been uniform. To find the probable population for any year other than those given, we locate a point on the graph corresponding to that year. For example, to find the probable population in the year 1926, we note that the 1926 year line intersects the graph at a point whose height represents a population of approximately 19,700. We therefore say that the population of the town in the year 1926 was about 19,700.

Practice check: Do exercise 1 on the right.

Where causal relation does exist, that is, where the figures are related to each other by some definite rule or formula, is the second type of graph. Graphs showing the electrical resistance of a wire in relation to its diameter and graphs showing the variation in the surface speed of an emery wheel due to variation in the number of revolutions per minute are examples of this type. The following examples will illustrate the construction and use of such graphs.

■ **Example 2**

The surface speed in feet per minute of a 6-in. emery wheel, at various speeds, is given in the accompanying table. Draw a graph showing the relation between the surface speed and the revolutions per minute.

Revolutions per Minute	Surface Speed in Feet per Minute
0	0
500	785.4
1,000	1,570.8
1,500	2,356.2
2,000	3,141.6
2,500	3,927.0
3,000	4,712.4

Construction Mark off the rpm on the *X*-axis and the surface speed in feet per minute on the *Y*-axis. The first point is at the origin, zero; the next point is at the intersection of the 500-rpm line with the 785.4-ft-per-minute line; the third point

1. The following table shows the average temperature for 12 months in a small western town. Draw a graph showing the temperature ranges in this town.

Temperature (°F)	Month
34°	Jan.
36°	Feb.
40°	March
52°	April
61°	May
67°	June
72°	July
77°	Aug.
70°	Sept.
56°	Oct.
44°	Nov.
40°	Dec.

The answer is in the margin on page 401. Your graph may vary.

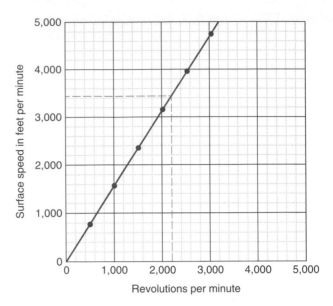

Figure 11–3

is at the intersection of the 1,000-rpm line with the 1,570.8-ft-per-minute line, and so on. Drawing a line through these points gives the required graph (see Fig. 11–3).

Explanation   This graph is a straight line beginning at zero and passing through all the plotted points. This is always true when the two sets of figures used in plotting the graph are related to each other by a simple proportion. The two figures used in plotting any point form the same ratio as those used for any other point; thus, 500 : 785.4 :: 1,000 : 1,570.8 :: 1,500 : 2,356.2, and so forth. When we know that such a proportion exists, we can plot the graph by locating only two points and drawing a straight line through them, continuing the line as far as required. All of the points will lie on this line. It is advisable, however, in cases of this sort, to select the two points as far apart as the limits of the graph will permit. If possible, the divisions on the X-axis should equal the divisions on the Y-axis. The graph then shows the rate of change between the two.

Use of the Graph   This graph may be used for finding the surface speed produced by a given number of revolutions per minute or for finding the revolutions per minute required to produce a given surface speed. To find the surface speed resulting from 2,200 rpm, for example, we note the intersection of the 2,200-rpm line with the graph and read the corresponding value of the surface speed, 3,440 ft per minute, on the Y-axis. To find the number of revolutions per minute required to give a surface speed of 3,000 ft per minute, we note the intersection of the 3,000-ft-per-minute line with the graph and read the corresponding value of the revolutions per minute, 1,910, on the X-axis.

Practice check: Do exercise 2 on the left.

■   **Example 3**

The following table shows the electrical resistance in ohms per thousand feet of copper wires of different diameters. Plot a graph showing how the resistance of the wire varies with the diameter.

2. At a constant rotational speed of 1,000 rpm, the surface speed of certain wheels can be determined according to their diameters. Construct a graph from the following table.

Diameter of Wheels (in.)	Surface Speed (rpm)
5	1,300
8	2,150
11	2,875

The answer is in the margin on page 402. Your graph may vary.

Diameter of Wire in Mils	Resistance in Ohms per 1,000 ft
36	8.04
51	4.01
72	2.00
102	1.00
144	0.50
204	0.25

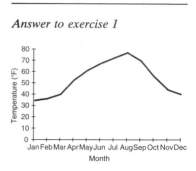

Answer to exercise 1

Construction   Mark off the diameters in mils on the *X*-axis and the resistances in ohms per 1,000 ft on the *Y*-axis. The intersection of the 36-mil line with the 8.04-ohm line is the first point; the second point is the intersection of the 51-mil line with the 4.01-ohm line, and so on. We obtain the graph by drawing a *smooth curve* through all the points (Fig. 11–4).

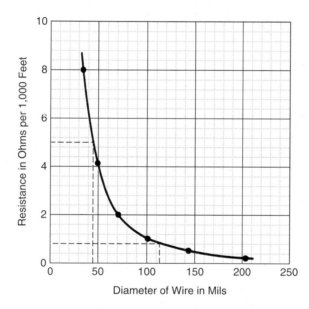

Figure 11–4

Explanation   Since the several resistances are related to their respective diameters by a certain rule or formula (in this instance, the law of inverse squares), there will be no sudden breaks or changes in the direction of the next point, as we can see by plotting a number of intermediate points. It is, however, unnecessary to do this since a smooth curve through the plotted points will also pass through all the intermediate points. The application of this rule enables us to reduce considerably the labor involved in plotting a graph representing a lengthy tabulation. For plotting we can select every third or fifth point and by drawing a smooth curve through them obtain the required graph.

Use of the Graph   Values on this graph are found in the same way as on the graph of Example 2. To find the diameter of a wire whose resistance is 5 ohms per 1,000 ft, we note the intersection of the 5-ohm line with the graph and read the corresponding value of the diameter, 46 mils, on the *X*-axis. To find the resistance

3. Plot a curve showing the relationship between the whole numbers from 0 to 10 and their squares.

The answer is in the margin on page 404. Your graph may vary.

per 1,000 ft of a wire whose diameter is 115 mils, we note the intersection of the 115-mil line with the graph and read the corresponding value of the resistance, 0.8 ohm, on the *Y*-axis.

Practice check: Do exercise 3 on the left.

11–3   USE OF GRAPHS IN EXPERIMENTAL WORK

In making tests of materials and in experimental work of various kinds, we frequently find that we obtain points on the graph that do not seem to lie on the straight line or curve that we know correctly represents the conditions involved. Such points are accidental and may be due to inaccuracies in the measuring instruments or in the measurements, or to small imperfections in the material. When this happens, we disregard the *off* points and draw a smooth curve approximating the position of the majority of the points. Let us take, for example, a tension test of a steel wire where we know that below the elastic limit, the load applied and the elongation are directly proportional to each other. This condition is represented by a straight line. We therefore draw a straight line approximating as nearly as possible the position of the majority of the points, disregarding the off points as inaccurate and not representing the actual conditions.

11–4   TWO OR MORE GRAPHS COMBINED

A combination of two or more graphs is often of great value in showing the relations existing between two or more series of figures. A number of distinct sets of points may be plotted in reference to the same pair of axes, each set of points determining a separate graph, as the following example illustrates.

■ Example
The average daily register and attendance of a certain school are given by the following table. Plot a graph showing these statistics.

Answer to exercise 2

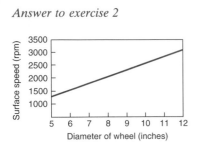

Year	Register	Attendance
1960	711	683
1963	818	783
1966	901	827
1969	915	895
1972	961	933
1975	972	950
1978	1,003	978
1981	1,115	1,052
1984	1,008	982
1987	920	902
1990	864	833

Construction   Taking 600 pupils as a convenient base line, in Fig. 11–5 we plot two points on each vertical year line for which we have data, one point for the register and one for the attendance. Two distinct sets of points are thus obtained and lines

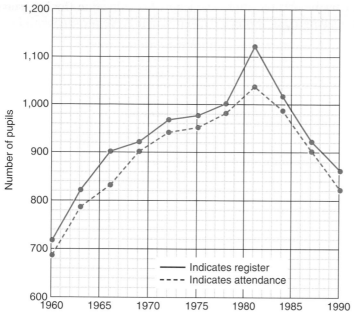

Figure 11–5

drawn through them. To distinguish the two lines, we use a full line for the register and a dotted line for the attendance with an explanatory note, as in Fig. 11–5. Sometimes different colors are used to distinguish one graph from another.

Practice check: Do exercise 4 on the right.

4. The following table lists the numbers of new AIDS cases in the United States from 1985 to 1994. Plot the graph showing these statistics. (Source: National Center for Health Statistics.)

Year	Men	Women
1985	7,539	520
1988	27,106	3,040
1989	29,666	3,380
1990	36,475	4,560
1991	37,722	5,373
1992	39,223	5,980
1993	86,469	16,113
1994	49,887	10,693

The answer is in the margin on page 405. Your graph may vary.

■ **Problems 11–1 to 11–4**

Construct graphs from the given information. Choose your divisions on the *X*-axis and *Y*-axis, and connect the points with a smooth curve. Be neat as well as accurate.

1. The average temperature in New York City at different times of the year is given in the accompanying table. Plot a graph showing the variation of temperature during the year.

Month	Temperature (°F)
January	30
February	31
March	38
April	48
May	59
June	68
July	74
August	72
September	66
October	56
November	44
December	34

Answer to exercise 3

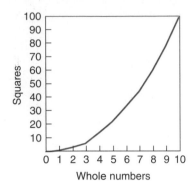

Whole numbers

2. Illustrate by means of a graph the income tax collections for the years indicated in the following data. (Source: U.S. Internal Revenue Service.)

Year	Collections (Dollars)
1890	142,594,697
1900	295,316,108
1910	289,957,220
1920	4,256,456,929
1930	3,040,145,733
1940	11,487,934,290
1950	38,957,131,768
1960	91,774,802,823
1970	195,722,096,497
1980	359,927,162,449
1986	412,162,490,003
1992	1,120,799,558,875

3. The weather report for a certain town shows the temperature, taken at 3-hour intervals, as follows.

Time	Temperature (°F)
12 A.M.	27
3 A.M.	24
6 A.M.	22
9 A.M.	25
12 P.M.	30
3 P.M.	38
6 P.M.	38
9 P.M.	37
12 A.M.	34

Draw a graph showing the variation in temperature during the 24-hour period. Find from the graph the probable temperature at 7 A.M., at 2 P.M., and at 8 P.M.

4. In a tension test of a bar of wrought iron, the elongation of the bar for applied loads was as indicated in the following table. Plot a curve showing the relation between the load applied and the elongation of the bar.

Answer to exercise 4

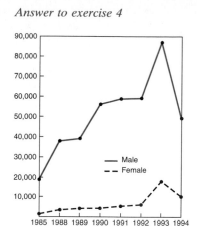

Load (lb)	Elongation (in.)	Load (lb)	Elongation (in.)
200	0.012	2,600	0.230
500	0.035	2,700	0.272
1,000	0.065	2,800	0.357
1,500	0.140	2,900	0.418
2,000	0.172	3,000	0.473
2,500	0.215		

5. Illustrate by means of a graph the U.S. population by Official Census every tenth year starting in 1900 and ending in 1990.

Year	Population
1900	76,212,168
1910	92,228,496
1920	106,021,537
1930	123,202,624
1940	132,164,569
1950	151,325,798
1960	179,323,175
1970	203,302,031
1980	226,542,203
1990	248,709,873

6. From the information in the following table of wages, construct three graphs showing the rise in average union hourly scale from 1960 to 1995 of the building trade; of the printing trade; of the local trucking trade.

	Building		Printing		Local Trucking	
Year	Journeymen	Helpers and Laborers	Book and Job	Newspapers	Drivers	Local Transit Workers
1960	$ 3.86	$ 2.88	$ 3.08	$ 3.48	$ 2.68	$ 2.37
1965	4.64	3.54	3.58	3.94	3.26	2.88
1970	6.54	4.86	4.65	5.13	4.41	4.03
1975	9.32	7.06	6.86	7.57	6.87	6.25
1980	9.94	7.60	7.37	9.14	7.99	7.02
1985	12.31	10.53	9.77	11.02	9.50	8.41
1990	15.20	12.23	11.37	13.12	13.75	12.55
1995	20.60	16.26	17.27	18.60	17.78	15.75

7. The following results were obtained when 10 samples of a certain cement were prepared and tested for tensile strength at different time intervals.

Age (Weeks)	Tensile Strength (lb/sq in.)
1	120
2	150
3	175
4	195
5	210
6	223
7	233
8	242
9	250
10	255

Draw a graph showing the relation between the age of the cement and its tensile strength.

8. The following table shows the number of tons of freight carried in the United States during certain years.

Year	Tons of Freight
1890	631,740,000
1895	686,615,000
1900	1,081,983,000
1905	1,427,732,000
1910	1,849,900,000
1915	1,802,018,000
1920	2,234,548,000

Draw a graph showing the growth of the freight traffic in the United States from 1890 to 1920.

9. The following table gives the normal weight of children from birth until they are 16 years old. Using the same pair of axes, draw a graph showing the weight increase for boys and for girls. From the graph find the normal weight of boys, and the normal weight of girls, 7 years old and $13\frac{1}{2}$ years old.

Age	Normal Weight (kg)	
	Boys	**Girls**
Birth	3.43	3.25
6 months	7.27	7.04
1 year	9.32	9.00
2 years	12.0	11.6
3 years	14.2	13.6
4 years	15.9	15.5
5 years	18.7	18.1
6 years	20.9	19.9
8 years	24.8	24.0
10 years	30.3	29.1
12 years	36.3	37.0
14 years	45.1	45.6
16 years	56.2	51.4

10. The following table shows the minimum gross income needed to borrow money to buy an $80,000 home at various interest rates. The figures are based on a 30-year loan and on monthly payments that are less than 28% of the borrower's gross income. Property taxes and insurance are not included as part of the monthly payment. Make a graph of the change in income needed as the mortgage interest rate changes.

Interest Rate (%)	$80,000 Loan
8	$25,158
8.5	26,386
9	27,588
9.5	28,830
10	30,089
10.5	31,363
11	32,652
11.5	33,953
12	35,267
12.5	36,592
13	37,927

11. A water wheel was tested for efficiency at different speeds, and the following results were obtained.

Speed (rpm)	Efficiency (%)
50	18
100	37
150	51
200	63
250	68
300	$68\frac{1}{2}$
350	65
400	57

Plot the efficiency curve and find the greatest efficiency and the speed at which it occurs.

11–5   CIRCLE GRAPHS

An effective way of showing the percentage distribution of a given quantity is by means of a *circle graph.* The following example illustrates the method.

■ **Example**

The cost of governing a certain city amounted to $24,652,000, apportioned as follows: education, $4,659,228; police and fire departments, $2,317,288; health, hospital, and sanitation, $2,859,632; pensions, $2,243,322; debt service, $4,511,316; others, $8,061,204. Plot a circle graph showing the percentage allotments.

Solution

Education

$$= \frac{4,659,228}{24,652,000} = \ 18.9\% \text{ of total} = 0.189 \times 360° = \ 68.04°$$

Police and fire

$$= \frac{2,317,288}{24,652,000} = \ \ 9.4\% \text{ of total} = 0.094 \times 360° = \ 33.84°$$

Health, etc.

$$= \frac{2,859,362}{24,652,000} = \ 11.6\% \text{ of total} = 0.116 \times 360° = \ 41.76°$$

Pensions

$$= \frac{2,243,322}{24,652,000} = \ \ 9.1\% \text{ of total} = 0.091 \times 360° = \ 32.76°$$

Debt service

$$= \frac{4,511,316}{24,652,000} = \ 18.3\% \text{ of total} = 0.183 \times 360° = \ 65.88°$$

Others

$$= \frac{8,061,204}{24,652,000} = \ \underline{32.7\%} \text{ of total} = 0.327 \times 360° = \underline{117.72°}$$

$$\text{Total} = 100\% \qquad\qquad\qquad\qquad 360°$$

Construction   Compute the percentage of the total that is spent for each item. Since there are 360° in the circumference of a circle, the amount spent on education, which is 18.9% of the total, is represented by a sector whose arc is 18.9% of 360°, or 68.04°. Since 9.4% of the total is allotted to the police and fire departments, that item is represented by a sector whose arc is 9.4% of 360°, or 33.84°, and so forth. With a protractor lay off the respective arcs on the circumference of a circle and draw the radii, forming the sectors as shown in Fig. 11–6.

Practice check: Do exercise 5 on the right.

5. The following table lists the credit-hour requirements for an associate degree in industrial mechanics at a certain technical college. Draw a circle graph representing the requirements.

Courses	Credit Hours
English	15
Mathematics	15
Applied science	10
Technical courses	30
Supporting courses	10
General education	10
Total =	90

The answer is in the margin on page 411. Your graph may vary.

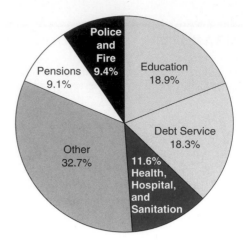

Figure 11–6

■ **Problems 11–5**

Draw a circle graph illustrating the following data. (Use a compass and a protractor for this work.

1. Show by means of a circle graph the percentage of the total subscribed to the International Bank by each of the countries listed in the following table.

Country	Percentage of Total
United States	41.4
United Kingdom	17.0
China	7.8
France	5.9
India	5.2
Canada	4.2
Netherlands	3.6
Other countries	14.9
	Total = 100.0%

2. The world's major producers of primary energy in 1992 were as shown in the following table. Draw a circle graph using this information. (Source: Energy Information Administration.)

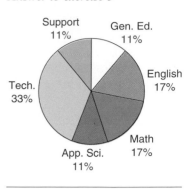

Country	Energy Produced (Quadrillion Btu)
United States	66.68
Russia	45.66
China	30.18
Saudi Arabia	20.63
Canada	14.36
United Kingdom	9.23
Iran	8.53
Mexico	7.76
India	6.94
Norway	6.8
Venezuela	6.8

3. In 1971 the personal expenditures of the population of the United States were $103.5 billion for durable goods such as automobiles, furniture, and the like; $278.1 billion for nondurable goods such as food, clothing, and gas; and $283.3 billion for services such as housing, transportation, and household operation. Draw a circle graph showing the relationships of personal expenditures. (Source: U.S. Department of Commerce.)

4. The following table gives the production of agricultural commodities in 1986 in bushels. Construct a circle graph showing what percentage of the total each item constituted. (Source: U.S. Department of Commerce.)

Commodity	Bushels
Corn	8,252,834,000
Wheat	2,086,780,000
Soybeans	1,764,112,000
Oats	1,185,936,000

5. The outlay for education in a city in 1995 was 29.5% of the entire budget; for police and fire, 13.6%; health, hospital, and sanitation, 9.7%; pensions, 5.2%; debt service, 9%; others, 33%. Illustrate the given percentages by means of a circle graph.

11–6   BAR GRAPHS

Another method of comparing quantities is by means of a ***bar graph,*** illustrated in the following example.

■ **Example**

The United States budget in millions of dollars for the years indicated is as follows. (Source: U.S. Bureau of Budget.)

Year	Budget
1940	6,879
1950	40,940
1960	92,470
1970	198,686
1980	579,011
1990	1,031,308

Draw a bar graph showing the budget increase in these 10-year periods.

Construction   On the horizontal axis, mark off the millions of dollars, and on the vertical axis mark off the years as shown in Fig. 11–7. The length of each bar represents the millions of dollars budgeted in that year. The graph shows at a glance how the U.S. budget has increased.

The graph can also be constructed with the bars drawn vertically as shown in Fig. 11–8. In this instance, the years are marked off on the horizontal axis and the millions of dollars on the vertical axis.

Practice check: Do exercise 6 on the left.

6. Draw a bar graph showing years of expected life at birth as listed in the following table. (Source: U.S. Department of Health and Human Services.)

Year of Birth	Life Expectancy
1920	54.1
1930	59.7
1940	62.9
1950	68.2
1960	69.7
1970	70.8
1980	71.0
1990	75.4

The answer is in the margin on page 414. Your graph may vary.

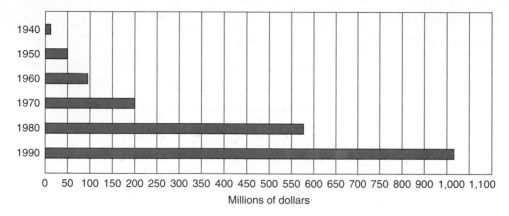

Millions of dollars

Figure 11–7

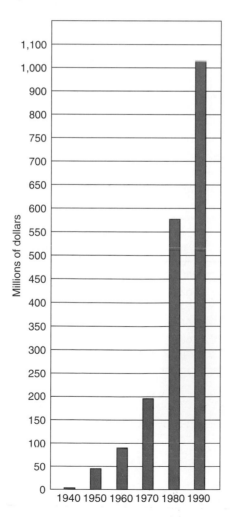

Millions of dollars

Figure 11–8

Answer to exercise 6

■ Problems 11–6

Draw bar graphs showing the following data.

1. Show by means of a bar graph the amount of Pennsylvania anthracite coal produced as given in the following table. (Source: U.S. Bureau of Mines.)

Year	Thousands of Tons
1945	54,934
1950	44,077
1955	26,205
1960	18,817
1965	14,866
1970	9,729
1975	10,144
1980	10,367
1985	7,222
1990	3,441

2. The following table shows the wool produced in millions of pounds for the years 1960 to 1970. Draw a vertical bar graph showing these data. (Source: U.S. Department of Agriculture.)

Year	Wool Produced (Millions of Pounds)
1960	298.9
1962	261.2
1964	237.4
1966	219.2
1968	198.1
1970	176.9

3. Illustrate the following data by a bar graph. (Source: Interstate Commerce Commission.)

Year	Railroad Domestic Freight Traffic (Millions of Ton-Miles)
1965	708,700
1970	771,168
1975	759,000
1980	932,000
1985	895,000
1990	1,071,000

■ Self-Test

Do the following problems. Plan your graph, and decide how the horizontal and vertical axes will be labeled. Choose an appropriate scale.

1. The following table shows the public debt of the United States in billions of dollars from 1940 to 1994. Draw a graph showing these data. (Source: Bureau of Public Debt.)

Year	Debt (Billions of Dollars)
1940	43.0
1950	256.1
1960	284.1
1970	370.1
1980	907.7
1990	3,233.3
1994	4,692.9

2. Johnston Consulting Services projects that its profits will increase $1,000 per year. Construct a graph using the data in the following table.

Year	Profit
1	9,000
2	10,000
3	11,000
4	12,000
5	13,000

3. The following table shows in inches the radii of circles with different areas. Plot a graph showing how the radius varies with the area.

Area (in.²)	Radius (in.)
3.14	1
7.01	1.5
12.57	2
19.63	2.5
28.27	3
38.48	3.5

4. The numbers of new AIDS cases in 1994 in the United States are shown in the following table. Draw a circle graph using this information. (Source: U.S. Department of Health and Human Services.)

Category	Deaths
Male	49,887
Female	10,693
Children	721
Total	61,301

5. The following table shows the average size in acres of U.S. farms for the years 1940 to 1994. Construct a bar graph illustrating this information. (Source: U.S. Department of Agriculture.)

Year	Average Size of Farms (Acres)
1940	174
1950	215
1960	302
1970	390
1980	445
1994	471

■ Chapter Test

Do the following problems. Plan your graph, and decide how the horizontal and vertical axes will be labeled. Choose an appropriate scale.

1. The following table shows U.S. bank failures from 1940 to 1994. Draw a graph showing these data.

Year	Bank Failures
1940	48
1960	2
1970	8
1980	11
1990	168
1994	13

2. A truck driver averages 50 mi/hr. The following table shows how miles traveled varies with hours. Draw a graph using the data in the table.

Time	Miles Traveled
1	50
2	100
3	150
4	200
5	250

3. The following table shows the height of a ball (in feet) when thrown upward at a velocity of 110 ft/sec at various times (in seconds). Construct a graph showing how the height varies with the amount of time.

Time (sec)	Height (ft)
0	0
1	84
2	136
3	156
4	144
5	100

4. The world commercial catch of fish, crustaceans, and mollusks by ocean during 1993 is shown in the following table. Draw a circle graph using this information. (Source: National Marine Fisheries Service.)

Ocean	Amount
Pacific Ocean	53,569
Atlantic Ocean	23,391
Indian Ocean	7,289
	Total = 84,249

5. The following table shows the percentage of U.S. population that is foreign born for the years 1900 to 1994. Construct a bar graph illustrating this information. (Source: U.S. Bureau of the Census.)

Year	Percentage
1900	13.6
1920	13.2
1940	8.8
1950	6.9
1970	4.8
1990	7.9
1994	8.7

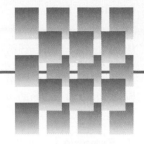

CHAPTER 12

Measuring Instruments

12–1   THE MICROMETER

For making fine measurements, an instrument called the *micrometer* is used. This instrument consists essentially of a 40-pitch screw and a thimble whose circumference is divided into 25 equal parts (see Fig. 12–1).

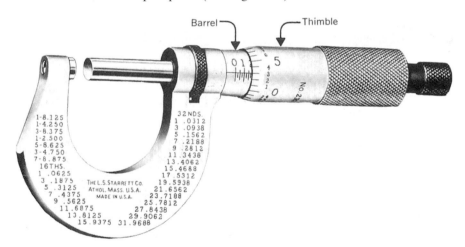

Figure 12–1   (Courtesy of the L. S. Starrett Co., Athol, Mass.)

Since the pitch of the screw is $\frac{1}{40}$ of an inch, each complete turn of the thimble advances the spindle $\frac{1}{40}$ in., or 0.025 in. The barrel or body of the instrument is therefore calibrated by dividing the inch into 40 equal parts, each 0.025 in. For convenience in reading, every fourth division, representing 0.100 in., is numbered, so that number 1 represents 0.100 in.; number 2, 0.200 in.; and so on. The reading is taken at the edge of the thimble. The first mark after the zero represents 0.025 in.; the second mark after the zero, 0.050 in.; the first mark after the 4, 0.425 in.; the third mark after the 8, 0.875 in.; and so forth. To read between these figures, the beveled edge is divided into 25 equal parts, each representing $\frac{1}{25}$ of 0.025 in., which is 0.001 in. When the zero on the thimble coincides with the gage line on the barrel, the reading is simply so many complete turns as shown by the barrel reading. If the zero on the thimble does not coincide with the gage line on the barrel, the line that does coincide must be found and that many thousandths must be added to the barrel reading. For example, if the line marked 15 coincides with the gage line, add 0.015 in.

■ Example 1

What is the reading when the edge of the thimble is between the 0.125 and the 0.150 lines on the barrel and the 18 line on the thimble is the coinciding line?

Solution   Micrometer reading = Barrel + Thimble
$$= 0.125 \text{ in.} + 0.018 \text{ in.} = 0.143 \text{ in.}$$

Practice check: Do exercise 1 on the left.

■ Example 2

Find the reading if the thimble edge is between 0.850 and 0.875 on the barrel and the 3 line coincides.

Solution   Micrometer reading = Barrel + Thimble
$$= 0.850 \text{ in.} + 0.003 \text{ in.} = 0.853 \text{ in.}$$

Practice check: Do exercise 2 on the left.

■ Example 3

Set the micrometer for 0.064 in.

Solution   Micrometer reading = Barrel + Thimble
$$0.064 \text{ in.} = 0.050 \text{ in.} + 0.014 \text{ in.}$$

Bring the thimble edge to 0.050; then turn the thimble to 14.

Practice check: Do exercise 3 on the left.

■ Problems 12–1

Supply the micrometer readings for each of the micrometer settings in the following chart.

	Thimble Is between	Coinciding Line on Thimble	Micrometer Reading
1.	0 and 0.025	8	_____
2.	0.125 and 0.150	18	_____
3.	0.275 and 0.300	17	_____
4.	0.850 and 0.875	16	_____
5.	0.125 and 0.150	5	_____
6.	0.975 and 1.000	1	_____
7.	0.025 and 0.050	22	_____
8.	0.375 and 0.400	4	_____
9.	0.150 and 0.175	2	_____
10.	0.650 and 0.675	16	_____

1. What is the reading when the edge of the thimble is between the 0.275 and 0.300 lines on the barrel, and the 22 line on the thimble is the coinciding line?

The answer is in the margin on page 424.

2. Find the reading if the thimble edge is between 0.550 and 0.575 on the barrel and the 5 line coincides.

The answer is in the margin on page 424.

3. Set the micrometer for 0.336 in.

The answer is in the margin on page 424.

Set the following micrometer readings.

11. 0.109″   **12.** 0.073″   **13.** 0.218″   **14.** 0.813″   **15.** 0.401″

16. $\frac{3}{8}''$ **17.** $\frac{1}{4}''$ **18.** $\frac{53}{64}''$ **19.** $\frac{21}{32}''$ **20.** $\frac{9}{16}''$

12–2   THE TEN-THOUSANDTHS MICROMETER

When more accurate measurements arc required, we employ a micrometer that has an extra scale added to the barrel, enabling us to read to ten-thousandths of an inch. This scale consists of a series of lines on the barrel, parallel to the axis of the barrel (see Fig. 12–2). A part of the circumference of the barrel equal to 9 of the thimble divisions is divided into 10 equal parts. Each part therefore represents $\frac{1}{10}$ of 0.009 in., or 0.0009 in. The difference between one of these divisions and a thimble division is 0.001 in. − 0.0009 in. = 0.0001 in. We find the number of ten-thousandths of an inch to add to the barrel and thimble readings by looking for the line of this auxiliary scale that exactly coincides with a thimble line. A scale of this sort which depends on the *difference between dimensions* for its principle is called a ***vernier scale.***

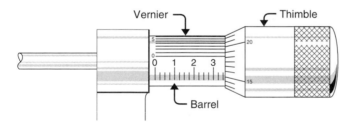

Figure 12–2

The reading on a ten-thousandths micrometer consists of three parts:

1. The barrel, on which each division represents 0.025 in.

2. The thimble, on which each division represents 0.001 in.

3. The vernier, on which each division represents 0.0001 in.

■ **Example**

The thimble is between 0.075 and 0.100, the gage line on the barrel is between 19 and 20 on the thimble, and line 8 on the vernier coincides with a line on the thimble. What is the reading?

Solution   Micrometer reading = Barrel + Thimble + Vernier
                                    = 0.075 in. + 0.019 in. + 0.0008 in.
                                    = 0.0948 in.

Practice check: Do exercise 4 on the right.

4. The thimble is between 0.225 and 0.250, the gage line on the barrel is between 13 and 14 on the thimble, and line 3 on the vernier coincides with a line on the thimble. What is the reading?

The answer is in the margin on page 426.

Answer to exercise 1

0.275 in. + 0.022 in. = 0.297 in.

Answer to exercise 2

0.550 in. + 0.005 in. = 0.555 in.

Answer to exercise 3

Bring the thimble edge to 0.325;
then turn the thimble to 11.

■ **Problems 12–2**

Supply the micrometer reading for each of the micrometer settings in the following chart.

	Thimble Is between	**Gage Line Is between Thimble Lines**	**Coinciding Vernier Line**	**Micrometer Reading**
1.	0.125 and 0.150	16 and 17	3	_____
2.	0.925 and 0.950	1 and 2	1	_____
3.	0 and 0.025	9 and 10	2	_____
4.	0 and 0.025	24 and 25	9	_____
5.	0.025 and 0.050	14 and 15	7	_____
6.	0.800 and 0.825	11 and 12	8	_____
7.	0.625 and 0.650	9 and 10	2	_____
8.	0.975 and 1.000	18 and 19	3	_____
9.	0.850 and 0.875	7 and 8	6	_____
10.	0.225 and 0.250	3 and 4	5	_____

Set the following micrometer readings.

11. 0.9815″   **12.** 0.2713″   **13.** 0.1043″   **14.** 0.0008″

15. 0.2108″   **16.** 0.0315″   **17.** 0.0071″

18. 0.6835″   **19.** $\frac{3}{16}″$ **20.** $\frac{31}{64}″$

12–3   THE VERNIER CALIPER

Another instrument that depends on the vernier principle is the vernier caliper (see Fig. 12–3). This instrument consists essentially of a main scale with a fixed jaw and an auxiliary or vernier scale attached to the sliding jaw. On the main scale each inch is divided into 40 parts, so that each part is $\frac{1}{40}$ in., or 0.025 in. For convenience in reading, every fourth division is numbered 1, 2, 3, and so on, indicating 0.100 in., 0.200 in., 0.300 in., and so forth.

    The zero of the sliding scale is the index or gage line, and its location in reference to the main scale determines part of the reading. The remainder of the reading is determined by the vernier scale, which is constructed as follows. A length equal to 24 main scale divisions, or 0.600 in., is divided into 25 equal parts on the

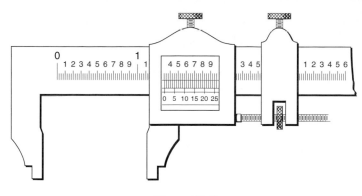

Figure 12–3

vernier scale; every fifth division is numbered to facilitate reading. Each space on the vernier is therefore equal to $\frac{1}{25}$ of 0.600 in., or 0.024 in. The difference between a scale division and a vernier division is 0.025 in. − 0.024 in. = 0.001 in. Figure 12–4 shows an enlarged view of the scale and vernier showing a zero reading, the zero of the vernier coinciding with the zero of the scale. If the sliding jaw is now moved until the first division of the vernier coincides with the first scale division, we have moved 0.001 in. because that is the difference between a vernier division and a scale division. If we move the sliding jaw until the fifth line coincides, we have moved 0.005 in. and set the caliper for 0.005-in. reading. The vernier number that coincides with a scale division shows the number of thousandths of an inch to add to the scale reading.

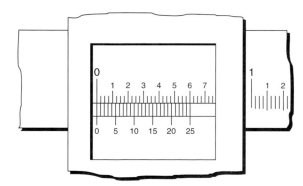

Figure 12–4

■ **Example 1**

The index is between 0.275 in. and 0.300 in. on the scale, and line 23 on the vernier coincides with a scale line. What is the reading?

Solution   Reading = Scale + Vernier
                    = 0.275 in. + 0.023 in. = 0.298 in.

Practice check: Do exercise 5 on the right.

■ **Example 2**

Set the vernier caliper for 0.131 in.

Solution   0.131 = 0.125 + 0.006. Bring the index over 0.125 in. Then move the sliding jaw until line 6 on the vernier coincides with a scale line.

Practice check: Do exercise 6 on the right.

5. The index is between 0.525 in. and 0.550 in. on the scale, and line 18 on the vernier coincides with a scale line. What is the reading?

The answer is in the margin on page 427.

6. Set the vernier caliper for 0.746 in.

The answer is in the margin on page 427.

■ Problems 12–3

Supply the reading for each setting of the vernier caliper in the following chart.

	Index Is between Lines	Coinciding Vernier Line	Reading
1.	0.125 and 0.150	16	_____
2.	0.175 and 0.200	9	_____
3.	0.250 and 0.275	18	_____
4.	0.900 and 0.925	12	_____
5.	0.975 and 1.000	6	_____
6.	0.350 and 0.375	1	_____
7.	0.100 and 0.125	13	_____
8.	0.000 and 0.025	5	_____
9.	0.075 and 0.100	24	_____
10.	0.050 and 0.075	11	_____

Set the vernier caliper for the following measurements.

11. 0.020″	**12.** 0.218″	**13.** 0.173″	**14.** 0.101″
15. 0.908″	**16.** 1.003″	**17.** 0.125″	
18. 0.099″	**19.** $\frac{3}{4}$″	**20.** $\frac{3}{16}$″	

12–4   THE PROTRACTOR

Figure 12–5 shows an ordinary protractor used for measuring and laying out angles. In use, the base line 0° to 180° is laid on one of the lines, with the center of the

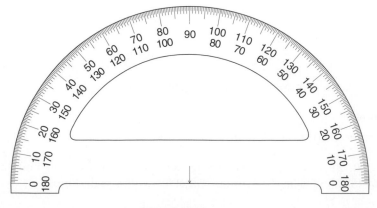

Figure 12–5

protractor at the point that is to be the vertex of the required angle. A mark is made at the point on the circumference where the required number of degrees is indicated. A line is then drawn between these two points. The illustration shows a typical plastic protractor.

12–5 THE VERNIER PROTRACTOR

For more accurate work, a steel protractor is used, carrying a vernier scale in addition to the regular scale (see Fig. 12–6). In this particular instrument, the smallest division on the main scale is $\frac{1}{2}°$, or 30′. By means of the vernier scale, we can divide one of these divisions into 10 parts, enabling us to read to 3′. Figure 12–7 shows an enlarged view of the vernier scale in relation to the main scale. A length of arc equal to 9 of the smallest scale divisions, or $9 \times 30' = 270'$, is divided into 10 equal parts on the vernier scale, making each division equal to $\frac{1}{10}$ of 270, or 27′. The difference between a scale division and a vernier division is $30' - 27' = 3'$.

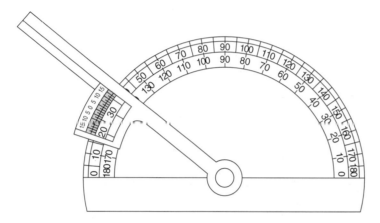

Figure 12–6

When the vernier zero, which is the index, exactly coincides with a scale division, the scale division shows the reading. But when the index does not exactly coincide with any scale division, we take the lower scale division and add the vernier reading as indicated by the coinciding vernier line.

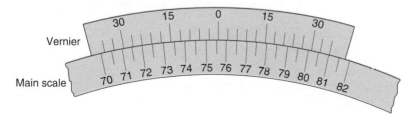

Figure 12–7

■ Example 1

The index is between 27°30′ and 28°, and the 21 on the vernier is the coinciding line. What is the reading?

Solution   Angle = Scale + Vernier
             = 27°30′ + 21′ = 27°51′

Practice check: Do exercise 7 on the right.

Answer to exercise 5

0.525 in. + 0.018 in. = 0.543 in.

Answer to exercise 6

0.746 in. = 0.725 in. + 0.021 in.

Bring the index over to 0.725 in. Then move the sliding jaw until line 21 on the vernier coincides with a scale line.

7. The index is between 87°40′ and 88°, and the 16 on the vernier is the coinciding line. What is the reading?

The answer is in the margin on page 430.

8. Set the vernier protractor for 18°05′.

The answer is in the margin on page 430.

■ **Example 2**

Set the vernier protractor for 72°42′.

Solution   Set the index between 72°30′ and 73°, and move the vernier until the 12 line on the vernier coincides with a scale division.

Practice check: Do exercise 8 on the left.

■ **Problems 12–5**

Supply the reading for each of the vernier protractor settings in the following chart.

	Index Is between	Coinciding Vernier Line	Reading
1.	12°00′ and 12°30′	15	_____
2.	91°30′ and 92°00′	21	_____
3.	11°30′ and 12°00′	27	_____
4.	60°00′ and 60°30′	3	_____
5.	27°00′ and 27°30′	0	_____
6.	41°00′ and 41°30′	15	_____
7.	125°30′ and 126°00′	27	_____
8.	16°30′ and 17°00′	6	_____
9.	3°30′ and 4°00′	21	_____
10.	75°00′ and 75°30′	18	_____

Set the vernier protractor for the following angles.

11. 27°54′   **12.** 2°15′   **13.** 45°45′   **14.** 15°15′   **15.** 52°03′

16. 12°12′   **17.** 9°27′   **18.** 55°57′   **19.** 21°33′   **20.** 91°12′

12–6   THE PLANIMETER

The *planimeter* is an instrument used for measuring irregular areas such as indicator card diagrams. Figure 12–8 shows the Amsler Polar Planimeter, one of the simplest forms of this instrument.

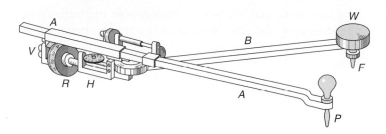

Figure 12–8

The following are the essential parts and their functions.

The long beam *AA* carrying at one end the tracing point *P*.

The short beam *B*, pivoted at one end so as to allow free movement at the joint and carrying at the other end a needle point *F*, which acts as a fixed center around which the entire instrument turns.

A small weight *W* to hold this needle point in place.

The registering apparatus.

The rolling disk *R* of the registering apparatus rotates as the entire instrument is dragged around after the tracing point *P*, which is tracing the outline of the figure whose area is being measured. One complete turn of the rolling disk indicates an area of 10 sq in. The circumference of the disk is therefore divided into 10 parts, each representing 1 sq in. Each space is further divided into 10 parts, each division representing $\frac{1}{10}$ of a square inch. Hundredths of a square inch are read by means of the vernier *V*, which is placed next to the roller. The zero of this vernier is the index or gage line of the instrument.

Complete turns of the roller, each representing 10 sq in., are counted by means of the horizontal disk *H*, which is connected to the roller by means of a worm and gear. Ten complete turns of the roller are required to produce one complete turn of the counter. Its surface is therefore divided into 10 parts, each representing 10 sq in.

12–7   USE OF THE PLANIMETER

The planimeter is used as follows. Figure 12–9 shows a view of the planimeter in reference to the figure whose area is being measured. The fixed point *F* is located conveniently near the figure. The tracing point *P* is placed at a marked point on

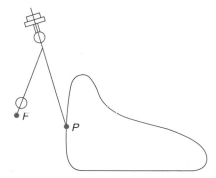

Figure 12–9

Answer to exercise 7

$87°40' + 16' = 87°56'$

Answer to exercise 8

Set the index between 18° and 18°30', and move the vernier until the 5 line on the vernier coincides with the scale division.

the perimeter of the figure and is carefully drawn over the entire perimeter until it returns to the starting point. This motion of the tracing point causes the roller to rotate and register the area. Readings are taken before and after the operation, and the difference between the readings is the area of the figure. To minimize errors it is customary to repeat the operation several times and take the average of the results.

The initial reading is taken with the tracer P at the starting point, and the final reading is taken when the tracer has returned to the starting point.

The following example will illustrate the method of reading the planimeter. If the gage line of the counter is between 6 and 7, the reading is between 60 and 70 sq in.; that is, it is 60 plus, each division on the counter representing 10 sq in. If the vernier index is between 71 and 72 on the roller, the roller reading is between 7.1 and 7.2 sq in., that is, 7.1 plus. If line 6 on the vernier coincides with a roller line, the vernier reading is 0.06 sq in.

$$\text{Total reading of instrument} = \text{Counter} + \text{Roller} + \text{Vernier}$$
$$= 60 + 7.1 + 0.06 = 67.16 \text{ sq in.}$$

If 67.16 sq in. is the initial reading and 73.81 is the final reading, the area of the figure measured is $73.81 - 67.16 = 6.65$ sq in.

■ Problems 12–7

Supply the planimeter reading for each set of conditions in the following chart.

	Counter Is between	Roller Is between	Coinciding Vernier Line	Reading
1.	2 and 3	56 and 57	2	_____
2.	9 and 10	31 and 32	3	_____
3.	5 and 6	00 and 01	1	_____
4.	6 and 7	66 and 67	5	_____
5.	5 and 6	45 and 46	8	_____
6.	0 and 1	98 and 99	9	_____
7.	3 and 4	11 and 12	6	_____
8.	2 and 3	03 and 04	1	_____
9.	7 and 8	33 and 34	4	_____
10.	0 and 1	15 and 16	5	_____

■ Self-Test

Do the following problems. Check your answers with the answers in the back of the book.

Supply the micrometer reading for each micrometer setting in the following chart.

	Thimble Is between	**Coinciding Line on Thimble**	**Micrometer Reading**
1.	0.150 and 0.175	15	_____
2.	0.725 and 0.750	6	_____

Set the following micrometer readings.

3. 0.2823 in.

4. $\frac{7}{16}$ in.

Supply the reading for each setting of the vernier caliper in the following chart.

	Index Is between Lines	**Coinciding Vernier Line**	**Reading**
5.	0.800 and 0.825	4	_____
6.	0.025 and 0.050	14	_____

Set the vernier protractor for the following angles.

7. 51°41′

8. 13°19′

Supply the planimeter reading for each set of conditions in the following chart.

	Counter Is between	**Roller Is between**	**Conciding Vernier Line**	**Reading**
9.	7 and 8	55 and 56	3	_____
10.	4 and 5	13 and 14	6	_____

■ Chapter Test

Set the following micrometer readings.

1. 0.222 in.

2. $\frac{5''}{8}$

Supply the micrometer reading for each micrometer setting in the following chart.

	Thimble Is between	Gage Line Is between Thimble Lines	Coinciding Vernier Line	Micrometer Reading
3.	0.700 and 0.725	6 and 7	2	_____
4.	0.250 and 0.275	8 and 9	5	_____

Set the vernier caliper for the following measurements.

5. 0.315 in.   **6.** $\frac{3}{8}''$

Supply the reading for each of the vernier protractor settings in the following chart.

	Index Is between	Coinciding Vernier Line	Reading
7.	50°00′ and 50°30′	6	_____
8.	18°30′ and 19°00′	14	_____

Supply the planimeter reading for each set of conditions in the following chart.

	Counter Is between	Roller Is between	Conciding Vernier Line	Reading
9.	9 and 10	32 and 33	4	_____
10.	2 and 3	17 and 18	9	_____

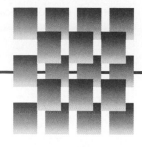

CHAPTER 13

Geometrical Constructions

13–1 APPLICATIONS OF GEOMETRY

Many of the fundamental facts of geometry as related to practical work have already been studied in the chapters on areas and volumes, and the space available in this book will not permit a more complete discussion. There are, however, certain geometrical constructions that are common and of great value in the solution of problems. Some of these will be discussed in the following pages.

The work in this chapter requires the use of the following instruments: a compass, a pair of dividers, a rule or straightedge, and a scale.

13–2 BISECTING A LINE SEGMENT

To **bisect** simply means to cut or divide exactly in half, as shown in the following example.

■ Example
Bisect the line segment *AB* in Fig. 13–1.

Solution   With a radius more than half the length of *AB*, use *A* as a center and draw the arc *CF*; with the same radius, use *B* as center and draw the arc *ED*. Draw a line through the points where these two arcs intersect. This line, crossing *AB* at *H*, cuts *AB* into two equal parts, making *H* the **midpoint.**

Practice check: Do exercise 1 on the right. (NOTE: *Because answers will vary, no marginal answers are provided in this chapter.*)

1. With straightedge and compass, practice bisecting a line.

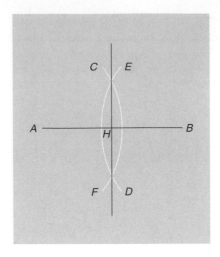

Figure 13–1

13–3　BISECTING AN ANGLE

■ **Example**

Bisect the angle *ABC* in Fig. 13–2.

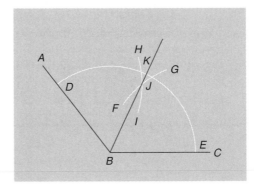

Figure 13–2

2. With straightedge and compass, practice bisecting an angle.

Solution　With any radius, such as *BD*, use *B* as a center and draw the arc *DE*. With the same radius, use *E* as a center and draw the arc *FG*. With the same radius again, use *D* as a center and draw the arc *HI*. Arcs *FG* and *HI* intersect at *J*. The line you draw through *J* and *B* bisects angle *ABC*, which means that angle *ABJ* = angle *CBJ*.

Practice check: Do exercise 2 on the left.

13–4　BISECTING AN ARC

The method used for bisecting an arc is exactly the same as that used for bisecting an angle. In Fig. 13–2, the line *JB* cuts the arc *DE* into two equal arcs, *DK* and *EK*.

13–5 CONSTRUCTING A PERPENDICULAR TO A LINE AT A GIVEN POINT ON THE LINE

A *perpendicular* is a line or a line segment that creates right angles to another line or line segment.

■ Example

From point P construct a perpendicular to the line AB in Fig. 13–3.

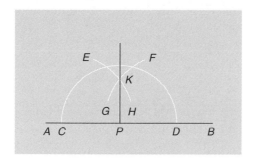

Figure 13–3

Solution   Using P as a center and any convenient radius, draw the semicircle CD. With a radius somewhat larger than PC and with C as a center, draw the arc EH. With the same radius and D as a center, draw the arc GF. These arcs intersect at K. The line through K and P is the required perpendicular. Therefore, angles APK and BPK each equal 90°.

Practice check: Do exercise 3 on the right.

3. With straightedge and compass, practice constructing a perpendicular to a line.

13–6 CONSTRUCTING A PERPENDICULAR TO AN ENDPOINT OF A LINE SEGMENT

■ Example

Construct a perpendicular to the line AB at A in Fig. 13–4.

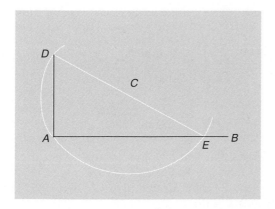

Figure 13–4

Solution 1   Take any point as C, not on the line AB, as a center and CA as a radius, and draw the arc EAD, as shown in Fig. 13–4. From E draw the line EC and prolong it until it cuts the arc at D. Draw a line through A and D. This is the required perpendicular. Therefore, angle $DAE = 90°$.

Solution 2   Lay off on AB (Fig. 13–5) a distance $AC = \frac{4}{4}$ in. $= 1$ in. With a radius of $\frac{5}{4}$ in. $= 1\frac{1}{4}$ in. and using C as a center, draw the arc DE. With a radius of $\frac{3}{4}$ in. and A as a center, draw the arc FG. These two arcs intersect at H. The line through H and A is the perpendicular required. Therefore, angle $HAC = 90°$.

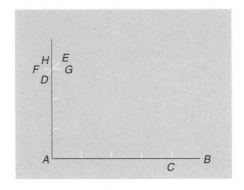

Figure 13–5

4. With straightedge and compass, practice constructing a perpendicular to a given line segment at a given endpoint of the line segment.

Practice check: Do exercise 4 on the left.

13–7   CONSTRUCTING A PERPENDICULAR TO A LINE THROUGH A POINT NOT ON THE LINE

■ **Example**

Construct the perpendicular to the line BC in Fig. 13–6 that passes through point A.

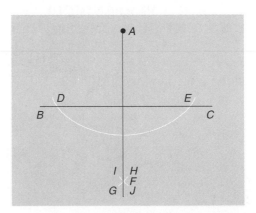

Figure 13–6

5. With straightedge and compass, practice constructing the perpendicular to line BC that passes through point A in Fig. 13–6.

Solution   Using A as a center and any convenient radius, draw an arc cutting line BC at D and E. Using the same radius and D as a center, draw the arc GH. Using E as a center and the same radius, draw the arc IJ. Arcs GH and IJ intersect at F. The line through A and F is the required perpendicular.

Practice check: Do exercise 5 on the left.

13–8 CONSTRUCTING A LINE PARALLEL TO ANOTHER LINE

■ **Example**

Through the point *A* construct a line parallel to *BC* in Fig. 13–7.

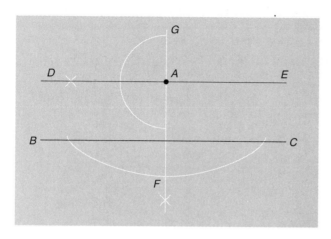

Figure 13–7

Solution   By the method of Section 13–7 draw the line *FG* perpendicular to *BC*. Then by the method of Section 13–5 draw the line *DE* through *A* and perpendicular to *FG*. Then *DE* is parallel to *BC* and is the required line.

Rule

If two lines are perpendicular to the same line and are coplanar, then they are parallel.

Practice check: Do exercise 6 on the right.

6. With straightedge and compass, practice constructing a line parallel to *BC* through point *A* in Fig. 13–7.

13–9 DIVIDING A LINE SEGMENT INTO A NUMBER OF EQUAL PARTS

■ **Example**

Divide line segment *AB* into seven equal parts in Fig. 13–8.

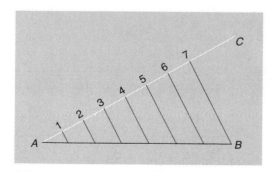

Figure 13–8

7. With straightedge and compass, practice dividing a line segment into 10 equal parts.

Solution   Draw the line AC at an angle to AB, not exceeding 45°. Lay off, with the dividers or scale, seven equal parts on the line AC. Connect the last point, 7, with B. Draw lines through the other points parallel to 7B. The intersections of these lines with line segment AB divide AB into seven equal parts.

Practice check: Do exercise 7 on the left.

13–10   CONSTRUCTING AN ANGLE EQUAL TO A GIVEN ANGLE

■   **Example**

Construct an angle equal to the angle ABC in Figs. 13–9 and 13–10.

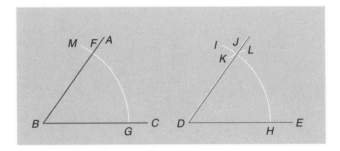

Figure 13–9

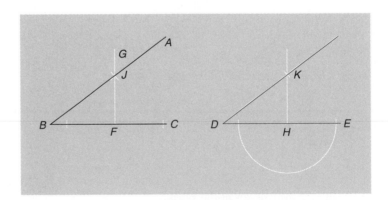

Figure 13–10

Solution 1   For Fig. 13–9, draw the line DE as shown. Then, with B as a center and any radius, draw the arc MG cutting the line AB at F. Using the same radius and D as a center, draw the arc HI. Using FG as a radius and H as a center, draw the arc KL intersecting HI at J. Draw a line through J and D. The angle JDE is equal to ABC and is the required angle.

Solution 2   For Fig. 13–10, at any point F on the line BC, erect a perpendicular FG intersecting the line BA at J. Draw the line DE and lay off DH equal to BF. Erect a perpendicular to DE at H. On this perpendicular, lay off a distance HK equal to FJ. Draw a line through K and D, giving the required angle KDE.

8. With straightedge and compass, practice constructing an angle equal to another angle.

Practice check: Do exercise 8 on the left.

13–11 CONSTRUCTING AN EQUILATERAL TRIANGLE OF GIVEN SIZE

■ **Example**

Construct an equilateral triangle with sides equal to *AB* in Fig. 13–11.

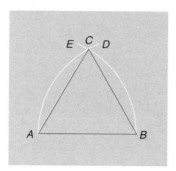

Figure 13–11

Solution   With *A* as a center and *AB* as a radius, draw the arc *BE*. With *B* as a center and the same radius, draw the arc *AD* intersecting arc *EB* at *C*. Join *A* to *C* and *B* to *C*, obtaining the required equilateral triangle *ABC*.

Practice check: Do exercise 9 on the right.

9. With straightedge and compass, practice constructing an equilateral triangle with sides equal to a given line segment.

13–12 CONSTRUCTING A CIRCLE THROUGH THREE GIVEN POINTS

■ **Example**

Draw a circle through the points *A*, *B*, and *C* in Fig. 13–12.

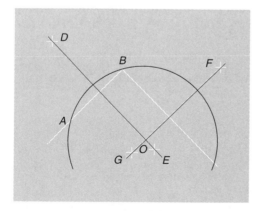

Figure 13–12

Solution   Connect the points *A* and *B*, and *B* and *C*. Draw *DE* bisecting *AB*; draw *FG* bisecting *BC*. The intersection of these two lines at *O* is the center of the circle. With *O* as center and radius *OA*, draw the required circle through the three given points.

Practice check: Do exercise 10 on the right.

10. With straightedge and compass, practice drawing a circle through three given points.

13–13   FINDING THE CENTER OF A CIRCLE OR AN ARC

■ **Example**

Find the center of the arc *ABC* in Fig. 13–13.

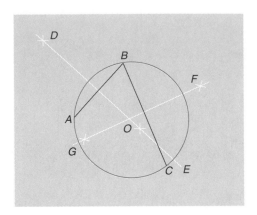

Figure 13–13

Solution   Draw any two chords of the arc as *AB* and *BC* as shown in the figure. Then bisect each chord. The intersection *O* of the bisecting lines *DE* and *FG* is the center of the circle.

NOTE: *This construction is similar to that in Section 13–12.*

11. With straightedge and compass, practice finding the center of an arc.

Practice check: Do exercise 11 on the left.

13–14   INSCRIBING A SQUARE IN A CIRCLE

■ **Example**

Inscribe a square in the circle in Fig. 13–14.

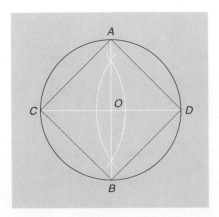

Figure 13–14

Solution   Draw the diameters *AB* and *CD* at right angles to each other. Connecting the four points *A*, *C*, *B*, and *D* gives the required square *ACBD*.

Practice check: Do exercise 12 on the right.

12. With straightedge and compass, practice inscribing a square in a circle.

13–15   CONSTRUCTING A SQUARE OF A GIVEN SIZE

■ **Example**

Construct a square with sides equal to length *AB* in Fig. 13–15.

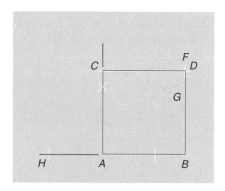

Figure 13–15

Solution   Extend line *DA* to *H*. Erect a perpendicular to line *AB* at *A*. With *A* as center and *AB* as radius, draw an arc cutting this perpendicular at *C*. With the same radius and *C* as center, draw arc *FG*. With the same radius and *B* as center, draw an arc cutting *FG* at *D*. Connect points *ACDB* to produce the required square.

Practice check: Do exercise 13 on the right.

13. With straightedge and compass, practice constructing a square equal to a given length.

13–16   CONSTRUCTING A SQUARE EQUAL IN AREA TO THE SUM OR DIFFERENCE OF TWO GIVEN SQUARES

■ **Example 1**

Construct a square equal in area to the *sum* of the two squares *ABCD* and *EFGH* in Fig. 13–16.

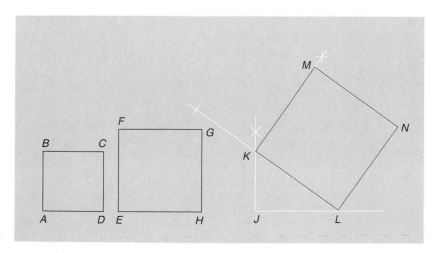

Figure 13–16

Solution   Construct a right triangle whose base *JL* is equal to *EH* and whose altitude *JK* is equal to *AB*. On the hypotenuse *KL* construct a square *KMNL*, which is the required square.

■ Example 2

Construct a square equal in area to the *difference* between the two squares *ABCD* and *EFGH* from Fig. 13–16.

Solution   Construct the right triangle *RST* as shown in Fig. 13–17 with *ST* equal to *EH*, the side of the larger square, as hypotenuse, and *RT* equal to *AD*, the side of the smaller square, as base. Lay off length *RT* equal to *AD* on straight line *XY*. Draw a perpendicular at *R*. With radius *EH* and *T* as center, draw an arc cutting the perpendicular at *S*. On the side *RS* construct the required square *QPSR*.

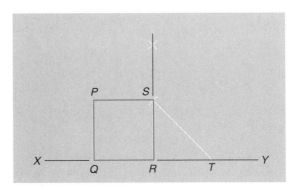

Figure 13–17

14. With straightedge and compass, practice constructing a square equal in area to the sum of two given squares or the difference between two given squares.

Practice check: Do exercise 14 on the left.

13–17   CONSTRUCTING A CIRCLE EQUAL IN AREA TO THE SUM OR DIFFERENCE OF THE AREAS OF TWO GIVEN CIRCLES

■ Example

Construct a circle equal in area to the combined areas of a $\frac{1}{2}$-in. circle and a $\frac{3}{4}$-in. circle as shown in Fig. 13–18.

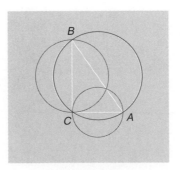

Figure 13–18

Solution Construct the right triangle *ACB*, making *CA* = $\frac{1}{2}$ in. and *CB* = $\frac{3}{4}$ in. The circle drawn with the hypotenuse *AB* as diameter is equal in area to the sum of the areas of the two circles drawn with the sides *CA* and *CB* as diameters. Also, either of the side circles is equal to the difference between the hypotenuse circle and the other side circle.

Practice check: Do exercise 15 on the right.

15. With straightedge and compass, practice constructing a circle equal in area to the combined areas of a $\frac{3}{4}$-in. circle and a 1-in. circle.

13–18   INSCRIBING A HEXAGON IN A CIRCLE

■ **Example**

Inscribe a hexagon in the circle in Fig. 13–19.

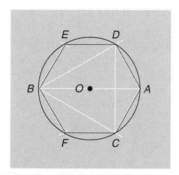

Figure 13–19

Solution *Draw the diameter AB.* With a radius *OA* equal to the radius of the circle and with *A* as a center, draw arcs cutting the circle at *C* and *D*; with the same radius and *B* as a center, draw arcs cutting the circle at *E* and *F*. Connect the successive points, producing the hexagon *BEDACF*.

Practice check: Do exercise 16 on the right.

16. With straightedge and compass, practice inscribing a hexagon in a circle.

13–19   INSCRIBING AN EQUILATERAL
TRIANGLE IN A CIRCLE

■ **Example**

Inscribe an equilateral triangle in the circle in Fig. 13–19.

Solution Using the same construction as in Section 13–18, draw lines connecting the alternate points *B*, *D*, and *C*, producing the equilateral triangle *BDC*.

Practice check: Do exercise 17 on the right.

17. With straightedge and compass, practice inscribing an equilateral triangle in a circle.

13–20   CONSTRUCTING A HEXAGON WHOSE
SIDES WILL BE A GIVEN LENGTH

■ **Example**

Construct a hexagon with sides equal to the given length *AB* in Fig. 13–20.

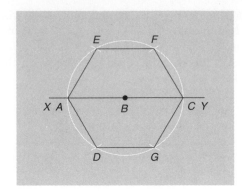

Figure 13-20

18. With straightedge and compass, practice constructing a hexagon with sides equal to a given length.

Solution   On line *XY* in Fig. 13–20, lay off the required length *AB*. With *B* as center and *AB* as radius, draw a circle through *A* cutting *XY* at *C*. With the same radius and *A* as center, draw arcs cutting the circle at *D* and *E*. With the same radius and *C* as center, draw arcs cutting the circle at *F* and *G*. Connecting the successive points gives the required hexagon *AEFCGD*.

Practice check: Do exercise 18 on the left.

13–21   CONSTRUCTING A HEXAGON WITH ONE OF THE SIDES ON A GIVEN LINE

■ **Example**

Construct a hexagon on a given line with sides equal to the given length *AB* in Fig. 13–21.

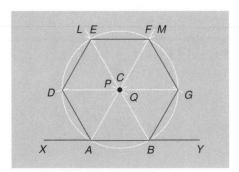

Figure 13-21

Solution   On the line *XY* in Fig. 13–21, lay off the length *AB*. With *A* as center and *AB* as radius, draw the arc *PQ*. With the same radius and *B* as center, draw an arc cutting the first arc at *C*. With the same radius and *C* as center, draw a circle through *A* and *B*. Draw the line *AM* through *A* and *C*, cutting the circle at *F*; draw the line *BL* through *B* and *C*, cutting the circle at *E*. With the same radius and *A* as center, draw an arc cutting the circle at *D*; with the same radius and *B* as center, draw an arc cutting the circle at *G.* Connecting the successive points gives the required hexagon *ADEFGB.*

Practice check: Do exercise 19 on the right.

19. With straightedge and compass, practice constructing a hexagon on a given line with sides equal to a given length.

13–22 INSCRIBING AN OCTAGON IN A CIRCLE

■ **Example**

Inscribe an octagon in the circle in Fig. 13–22.

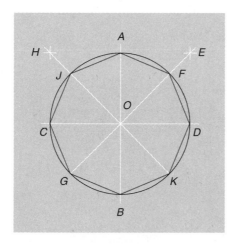

Figure 13–22

Solution   Draw the diameters *AB* and *CD* at right angles to each other. Bisect angle *AOC* with line *HK,* cutting the circle at *J* and *K.* Bisect angle *AOD* with line *EG,* cutting the circle at *F* and *G.* Connecting the successive points gives the octagon *AFDKBGCJ.*

Practice check: Do exercise 20 on the right.

20. With straightedge and compass, practice inscribing an octagon in a circle.

13–23 INSCRIBING AN OCTAGON IN A SQUARE

■ **Example**

Inscribe an octagon in the square in Fig. 13–23.

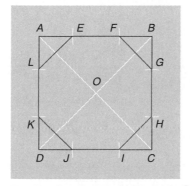

Figure 13–23

21. With straightedge and compass, practice inscribing an octagon in a square.

Solution   Draw the diagonals *AC* and *DB*, intersecting at *O*. With radius *OA* equal to half the diagonal distance and *A* as center, draw arcs cutting the sides of the square at *F* and *K*. Using the same radius, repeat this operation at the other corners of the square, *B*, *C*, and *D*. Connecting successive points produces the octagon *EFGHIJKL*.

Practice check: Do exercise 21 on the left.

13–24   CONSTRUCTING AN OCTAGON OF A GIVEN SIZE

■ **Example**

Construct an octagon with sides equal to *AB* in Fig. 13–24.

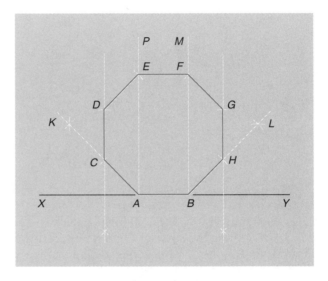

Figure 13–24

22. With straightedge and compass, practice constructing an octagon with sides equal to a given line segment.

Solution   Erect perpendiculars to line *XY* at *A* and *B*. Bisect the angles *PAX* and *MBY*. With *AB* as radius and *A* and *B* as centers, draw arcs cutting the bisectors at *C* and *H*. Draw perpendiculars to *XY* through *C* and *H*. With the same radius and *C* and *H* as centers, draw arcs cutting these perpendiculars at *D* and *G*. With the same radius and *D* and *G* as centers, draw arcs cutting *AP* and *BM* at *E* and *F*. Connecting the successive points gives the required octagon *ACDEFGHB*.

Practice check: Do exercise 22 on the left.

13–25   CONSTRUCTING A PENTAGON

■ **Example**

Inscribe a pentagon in the circle in Fig. 13–25.

Solution   Draw the diameters *AB* and *CD* perpendicular to each other. Bisect *OD* at *E*. With *E* as a center and a radius equal to *EA*, describe the arc *AF*. The line

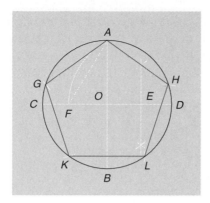

Figure 13–25

AF is the side of the inscribed pentagon. Starting at *A* with a radius equal to the line *AF*, strike arcs around the circumference, giving the points *A*, *G*, *K*, *L*, and *H*. Connecting these points produces the pentagon *AGKLH*.

Practice check: Do exercise 23 on the right.

23. With straightedge and compass, practice inscribing a pentagon in a circle.

13–26   CONSTRUCTING A TANGENT TO A CIRCLE

A *tangent* is defined as a line that intersects with a circle at one, and only one, point.

■ **Example**

Draw a tangent to the circle at the point *A* in Fig. 13–26.

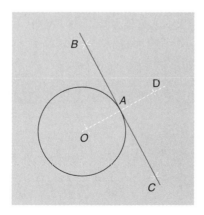

Figure 13–26

Solution   Draw the line *OD* through the center of the circle and point *A*. Draw *BC* perpendicular to *OD* at *A* by the method previously studied. Then *BC* is the required tangent.

Practice check: Do exercise 24 on the right.

24. With straightedge and compass, practice drawing a tangent to a circle at a given point.

13-27   CONSTRUCTING A TANGENT TO A CIRCLE THROUGH A POINT OUTSIDE THE CIRCLE

■ **Example**

Construct a tangent to the circle from the point *A* in Fig. 13–27.

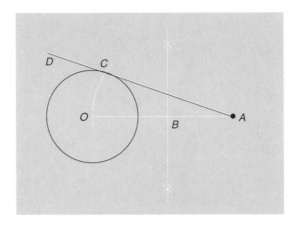

Figure 13–27

25. With straightedge and compass, practice constructing a tangent to a circle from a given point.

Solution   Draw the line *OA* connecting the center of the circle with the point *A*. Bisect the line *OA*, obtaining the midpoint *B*. Using *B* as a center and *BO* as a radius, draw the arc *OC* cutting the given circle at *C*. The line *AD* drawn through *A* and *C* is the required tangent.

Practice check: Do exercise 25 on the left.

13-28   CONSTRUCTING A TANGENT TO TWO CIRCLES OF EQUAL SIZE

■ **Example**

Construct a tangent to the two equal circles shown in Fig. 13–28.

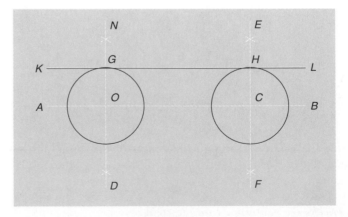

Figure 13–28

Solution   Draw the line AB through the centers O and C of the circles. At O and at C draw the lines ND and EF perpendicular to AB. Draw the line KL through the points G and H, where these perpendiculars intersect the circles. Then KL is the required tangent.

Practice check: Do exercise 26 on the right

26. With straightedge and compass, practice constructing a tangent to two given circles.

13–29   CONSTRUCTING AN INTERNAL TANGENT TO TWO EQUAL CIRCLES

■ **Example**

Construct an internal tangent to the two equal circles shown in Fig. 13–29.

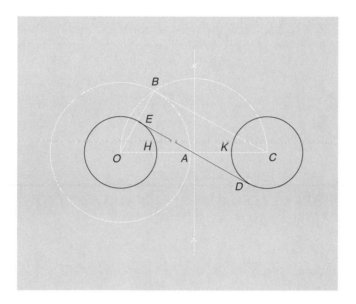

Figure 13–29

Solution   Draw the line OC connecting the centers of the two circles. On this line lay off $HA = OH$. With O as a center and OA as a radius, draw the large circle shown in the figure. Using the method of Section 13–27, draw the tangent CB to the large circle through C. Draw OB intersecting the small circle at E. With E as center and BC as radius, draw an arc cutting circle C at D. Draw ED, the required tangent.

Practice check: Do exercise 27 on the right.

27. With straightedge and compass, practice constructing an internal tangent to two given circles.

13–30   CONSTRUCTING AN EXTERNAL TANGENT TO TWO CIRCLES OF UNEQUAL SIZE

■ **Example**

Construct an external tangent to the two circles in Fig. 13–30.

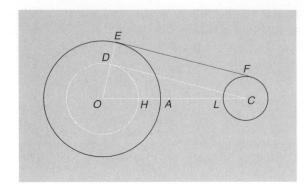

Figure 13–30

Solution Connect the centers of the two circles by the line *OC*. On *OA*, the radius of the large circle, lay off *AH* = *LC*, the radius of the small circle. Using *O* as a center and *OH* as a radius, draw the construction circle shown. By the method of Section 13–27, draw the tangent *CD* to the construction circle and through *C*. Prolong the radius *OD* until it cuts the outer circle at *E*. With *E* as center and *DC* as radius, draw an arc cutting circle *C* at *F*. Connecting *E* and *F* gives the required tangent.

Practice check: Do exercise 28 on the left.

28. With straightedge and compass, practice constructing an external tangent to two given circles.

13–31   CONSTRUCTING AN INTERNAL TANGENT TO TWO UNEQUAL CIRCLES

■ **Example**

Construct an internal tangent to the two circles shown in Fig. 13–31.

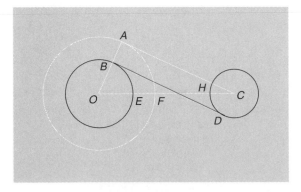

Figure 13–31

Solution Connect the centers by the line *OC*. Lay off on the line of centers the distance *EF* equal to the small radius, making *OF* equal to the sum of the two radii *OE* and *CH*. With *O* as center and radius *OF*, draw the large circle shown in the figure. Draw tangent *CA* to the large circle through point *C*. Connect *O* and *A*, cutting the given circle at *B*. With *B* as center and *AC* as radius, draw an arc cutting circle *C* at *D*. Connecting points *B* and *D* gives the required tangent.

Practice check: Do exercise 29 on the right.

29. With straightedge and compass, practice constructing an internal tangent to two given circles.

13–32   CONSTRUCTING AN ELLIPSE

Example

Construct an ellipse with a long axis of 2 in. and a short axis of $1\frac{1}{4}$ in.

Solution 1   As shown in Fig. 13–32, draw two concentric circles with diameters equal respectively to the long and short axes of the ellipse. Draw two axes *XY* and *LM* at right angles to each other. Draw any radius as *OA*. Through point *A*, where this radius meets the large circle, draw a line *AK* parallel to *LM*. Through point *E*, where radius *OA* crosses the small circle, draw a line *EH* parallel to axis *XY*. The intersection of these two lines *AK* and *EH* at *F* is a point on the perimeter of the ellipse. Other points on the perimeter may be found in a similar manner, and, if a sufficiently large number of points are obtained, the ellipse may be drawn through these points.

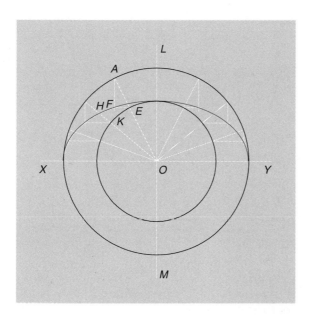

Figure 13–32

Solution 2   As shown in Fig. 13–33, draw the 2-in. and the $1\frac{1}{4}$-in. axes at right angles to each other and bisecting each other. Using *B* as center and *AC*, half the major axis, as radius, draw an arc cutting the major axis at two points *F* and *G*. Insert thumbtacks into the paper at the points *F*, *G*, and *B*; stretch a piece of thread so that it is fairly taut around these three points. Now remove the tack from point *B*, and keeping the string uniformly taut with the point of a pencil, draw the ellipse.

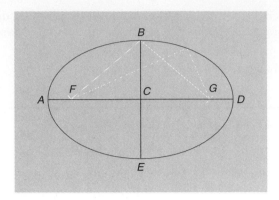

Figure 13–33

30. With straightedge and compass, practice constructing an ellipse with a long axis of 1½ in. and a short axis of 1 in.

Practice check: Do exercise 30 on the left.

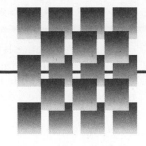

CHAPTER 14

Essentials of Trigonometry

14–1   THE RIGHT TRIANGLE

Every triangle has three sides and three angles. If any three parts are known, at least one of which is a side, then the triangle can be constructed. In Fig. 14–1, triangle ABC is a right triangle in which angle C is the right angle. The sum of the three angles of any triangle is 180°. Since angle C is a right angle, that is, 90°, then the sum of angle A and angle B is 90°. Therefore, angles A and B are *acute* angles. An acute angle is an angle less than 90°. (Review Section 7–14, page 232.)

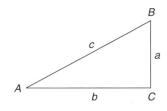

Figure 14–1

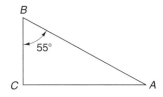

Figure 14–2

■ **Example 1**

In the right triangle of Fig. 14–2, find the value of angle A.

Solution
$$A + B = 90°$$
$$\underline{ B = 55°} \quad \text{Subtract.}$$
Angle $A = 35°$

Explanation   Since angles A and B together are 90°, subtract 55° from 90°, leaving 35° for angle A.

Practice check: Do exercise 1 on the right.

1. In the following right triangle, find the value of angle B if angle $A = 28°50'$.

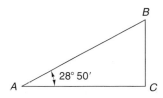

The answer is in the margin on page 456.

2. Using the following right triangle, fill in the blanks in
(a)–(c).

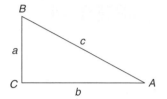

(a) Side a is opposite _____ and
adjacent to _____.

(b) Side b is _____ and _____.

(c) Side c is _____ and called
the _____.

The answers are in the margin
on page 456.

For convenience let a represent the side opposite the acute angle A; b the side opposite the angle B; and c the side opposite the right angle C. The side c, opposite the right angle, is called the *hypotenuse;* the side a, opposite the angle A, is *adjacent to the angle B*; the side b, opposite the angle B, is *adjacent to the angle A*.

Practice check: Do exercise 2 on the left.

In Fig. 14–3, suppose that the measure of side a (the side opposite angle A) is $\frac{3}{4}$ in. and the measure of side b (the side opposite angle B) is $1\frac{1}{2}$ in. The ratio of side a to side b, that is, $\frac{3}{4} : 1\frac{1}{2}$, is 0.5. Thus,

$$\frac{a}{b} = \frac{\frac{3}{4}}{1\frac{1}{2}} = 0.5$$

$$\frac{DE}{AE} = \frac{1\frac{1}{4}}{2\frac{1}{2}} = 0.5$$

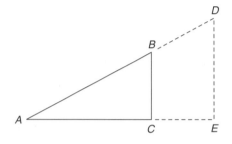

Figure 14–3

In Fig. 14–3, if side b were extended to point E so that the measure of AE were $2\frac{1}{2}$ in., and if side c (the side opposite angle C) were extended to D, a point that meets the perpendicular errected at point E, another right triangle would be formed, triangle ADE. The measure of side DE would be $1\frac{1}{4}$ in. Notice that this ratio of side DE to side AE, that is, $1\frac{1}{4} : 2\frac{1}{2}$, is again 0.5. Notice also that angle A is common to both of these triangles and that the ratios used the side opposite angle A and the side adjacent to angle A. The slant side is called the hypotenuse of the right triangle and is not considered an adjacent side. The value of the ratio of the side opposite the angle A to the side adjacent to the angle A will be found to be the same as long as the angle A remains unchanged. If the angle A were to be increased, the ratio of the opposite side to the adjacent side would no longer be 0.5 but some larger number. If the angle A were to be decreased, the ratio of the two sides would be less than 0.5. Since the size of the angle determines the value of the ratio of the two sides, the ratio may be taken as a measure of the angle.

The ratio obtained by dividing the side opposite the acute angle A by the side adjacent to the angle A is known as the **tangent** to the angle A and is written tan A. Thus,

$$\tan A = \frac{\text{opposite side}}{\text{adjacent side}} = \frac{a}{b}$$

$$\tan B = \frac{\text{opposite side}}{\text{adjacent side}} = \frac{b}{a}$$

For each tangent ratio there is one acute angle degree measurement. For instance, if tan $A = 0.5$, then angle $A = 26°34'$. In Section 14–5, you will learn how to find the angle measurement using a table, given a ratio of an angle.

■ **Example 2**

Find tan A in Fig. 14–4.

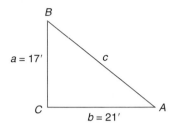

Figure 14–4

Solution $\tan A = \dfrac{a}{b} = \dfrac{17}{21} = 0.8095$

Practice check: Do exercise 3 on the right.

3. Find tan B in the following figure.

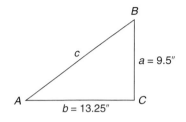

The answer is in the margin on page 457.

Summary

1. Given a and b, find A.

$$\frac{a}{b} = \tan A \quad \therefore \ A = \text{angle whose tangent is } \frac{a}{b}$$

2. Given A and b, find a.

$$\frac{a}{b} = \tan A \quad \therefore \ a = b \times \tan A$$

3. Given A and a, find b.

$$\frac{a}{b} = \tan A \quad \therefore \ b = \frac{a}{\tan A}$$

14–2   TRIGONOMETRIC FUNCTIONS

Instead of determining the value of the angle from the ratio of side a to side b, divide any side by either of the remaining two, and use the value of the ratio thus obtained as a measure of angle A. There are six possible ratios or ***trigonometric functions*** of each angle (see Fig. 14–5).

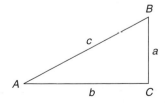

Figure 14–5

$$\text{\textit{sine of angle} } A = \frac{\text{opposite side}}{\text{hypotenuse}} = \frac{a}{c} \quad \text{written } \sin A = \frac{a}{c}$$

$$\text{\textit{cosine of angle} } A = \frac{\text{adjacent side}}{\text{hypotenuse}} = \frac{b}{c} \quad \text{written } \cos A = \frac{b}{c}$$

$$\text{\textit{tangent of angle} } A = \frac{\text{opposite side}}{\text{adjacent side}} = \frac{a}{b} \quad \text{written } \tan A = \frac{a}{b}$$

$$\text{\textit{cotangent of angle} } A = \frac{\text{adjacent side}}{\text{opposite side}} = \frac{b}{a} \quad \text{written } \cot A = \frac{b}{a}$$

$$\text{\textit{secant of angle} } A = \frac{\text{hypotenuse}}{\text{adjacent side}} = \frac{c}{b} \quad \text{written } \sec A = \frac{c}{b}$$

$$\text{\textit{cosecant of angle} } A = \frac{\text{hypotenuse}}{\text{opposite side}} = \frac{c}{a} \quad \text{written } \csc A = \frac{c}{a}$$

For angle B the six functions are as follows.

$$\sin B = \frac{b}{c} \qquad \tan B = \frac{b}{a} \qquad \sec B = \frac{c}{a}$$

$$\cos B = \frac{a}{c} \qquad \cot B = \frac{a}{b} \qquad \csc B = \frac{c}{b}$$

Practice check: Do exercise 4 on the left.

4. Using the following right triangle, state the trigonometric functions in (a)–(d).

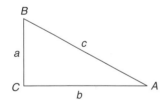

(a) $\dfrac{b}{c} =$ **(b)** $\dfrac{c}{a} =$

(c) $\dfrac{b}{a} =$ **(d)** $\dfrac{c}{b} =$

The answers are in the margin on page 459.

14–3   USE OF TABLES

This section refers to Appendix Table 2, Natural Trigonometric Functions. This table gives the value of the natural trigonometric functions for every angle from $0°00'$ to $90°$ in $10'$ increments. It is found on pages 685 through 689.

To find the functions of an angle less than $45°$, look for the degrees and minutes in the left-hand column of the table, and look for the function at the top of the page. Go down the function column until you reach the row that contains the desired degrees.

■ **Example 1**

Find the sine of $39°50'$.

Procedure   On page 689 find $39°$ in **boldface** type on the left-hand side of the page. Go down further to the $50'$ row. This is the row for $39°50'$. Read at the top of the page the word *sin* (the usual abbreviation for *sine*). Go down the "sin" column until you reach the row for $39°50'$. The 0.6406 found there is the sine of $39°50'$.

Therefore, $\sin 39°50' = 0.6406$.

■ **Example 2**

Find the tangent of $84°10'$.

Procedure   On page 685 find $84°$ in the right-hand column of the table. (Note that all degrees greater than $45°$ and less than $90°$ are in the right-hand column of the table.) Now go *up* the column until you reach $84°10'$. This is the row for $84°10'$. Now read at the bottom of the page the word *tan* (the usual abbreviation for *tangent*). Go up the "tan" column until you reach the row for 84°10'. The 9.788 found there is the tangent of 84°10'.

Therefore, tan 84°10' = 9.788.

Practice check: Do exercise 5 on the right.

Answer to exercise 3

tan B = 1.3947

■ **Example 3**

Find the cosine of 23°15'.

Procedure   23°15' is not in the table, since the table is in degrees and 10' increments. Find the value for 23°15' by the method known as interpolation. Since 23°15' is halfway between 23°10' and 23°20', assume that the tabular values for 23°15' are also halfway between the values of 23°10' and 23°20'. We find cosine 23°10' in the table as 0.9194 and the cosine of 23°20' in the table as 0.9182. (Note that the cosine ratio is a decreasing ratio.) Halfway between 0.9194 and 0.9182 is 0.9189.

Therefore, cos 23°15' is 0.9189.

Practice check: Do exercise 6 on the right.

5. Find the secant of 66°40'.

The answer is in the margin on page 459.

6. Find the sine of 41°45'.

The answer is in the margin on page 459.

■ **Problems 14–3**

Find the value of each of the following from the tables in the back of this book.

1. sin 20°

2. sin 30°

3. sin 25°

4. cos 30°

5. cos 60°

6. cos 45°

7. tan 45°

8. tan 30°

9. tan 60°

10. sin 45°

11. cos 40°10'

12. tan 30°20'

13. sin 45°30'

14. tan 25°10'

15. cos 56°20'

16. cos 70°30'

17. tan 20°40'

18. sin 47°50'

19. cos 37°20'

20. cos 21°20'

Find the value of each of the following by interpolation.

21. sin 30°15′

22. sin 40°25′

23. sin 60°35′

24. cos 28°25′

25. cos 35°35′

26. cos 62°45′

27. tan 20°45′

28. tan 40°55′

29. tan 65°25′

30. sin 80°22′

31. cos 80°42′

32. tan 37°36′

33. sin 75°34′

34. tan 62°27′

35. cos 47°53′

36. tan 27°55′

37. sin 68°31′

38. cos 43°43′

39. tan 35°12′

40. cos 26°5′

14–4   SIN, COS, AND TAN ON THE CALCULATOR

Examine your calculator and see if it has trigonometry function keys. You had to use degrees and decimal parts of a degree to do the previous problems. You can get your displayed answers converted to degrees, minutes, and seconds with the →D.MS key, but you cannot key in degrees, minutes, and seconds. Try some of the problems in this manner.

■   **Example**

What is the sin 47°35′?

Solution

	The Display Shows
Turn on the calculator	0.
Enter 35	35.
Press ÷	35.
Enter 60	60.
Press +	0.583333333
Enter 47	47.
Press =	47.58333333
Press sin	0.738259162 (answer)

This is the trig ratio of the sin 47°35′. Round off to 0.7383.

■ **Problems 14–4**

Use your calculator to do the trig ratios in Problem Set 14–3. Check the answers with those that you got from looking in the tables. Practice will increase your skills.

Answer to exercise 4

(a) cos A and sin B

(b) cosec A and sec B

(c) cot A and tan B

(d) sec A and cosec B

Answer to exercise 5

2.525

Answer to exercise 6

0.6659

14–5   FINDING AN ANGLE CORRESPONDING TO A GIVEN FUNCTION

■ **Example 1**

Find the angle whose sine is 0.5373.

Procedure   Turn to the trigonometry table and search the "sin" column until you find 0.5373. Read across to the left column headed "Degrees" and read 32°30'.
    Therefore, the angle whose sine is 0.5373 is 32°30'.

Practice check: Do exercise 7 on the right.

■ **Example 2**

Find the angle whose tangent is 0.9340.

Procedure   Turn to the trigonometry table and search the tangent column until you find 0.9340. You will note that 0.9340 is not in our table. You will have to interpolate to find the angle whose tangent is 0.9340.

7. Find the angle whose cotangent is 0.1733.

The answer is in the margin on page 462.

By proportion,

$$0.0055 \begin{cases} 0.0015 \begin{cases} 0.9325 = \tan 43°0' \\ 0.9340 = \tan 43°x' \end{cases} x' \\ 0.9380 = \tan 43°10' \end{cases} 10'$$

$$\frac{0.0015}{0.0055} = \frac{x'}{10'}$$

$$\frac{15}{55} = \frac{x'}{10'}$$

$$x \approx 3'$$

Therefore, 0.9340 is $\frac{15}{55}$ of the way from 43°0' to 43°10' in our tangent table, or

$$\frac{15}{55} \times 10 = \frac{30}{21} = 2\frac{8}{11} \quad \text{or almost 3}$$

Therefore, the angle whose tangent is 0.9340 is 43°3'.

Practice check: Do exercise 8 on the right.

8. Find the angle whose cosine is 0.9235.

The answer is in the margin on page 462.

■ **Problems 14–5**

Find the value of angle A in each of the following.

1. sin A = 0.5225

2. sin A = 0.0987

3. cos A = 0.7826

4. cos A = 0.9377

5. tan A = 0.5206

6. tan A = 0.8952

7. $\sin A = 0.7071$ **8.** $\cos A = 0.9689$ **9.** $\tan A = 1.117$

10. $\sin A = 0.9100$ **11.** $\sin A = 0.2840$ **12.** $\cos A = 0.4147$

13. $\tan A = 1.540$ **14.** $\cos A = 0.2164$ **15.** $\tan A = 14.30$

16. $\sin A = 0.3393$ **17.** $\cos A = 0.6428$ **18.** $\tan A = 0.4557$

19. $\cos A = 0.7509$ **20.** $\sin A = 0.4410$

Find the value of angle B in each of the following. Interpolation will be necessary. Round off to the nearest minute.

21. $\sin B = 0.4710$ **22.** $\sin B = 0.2907$ **23.** $\cos B = 0.9894$

24. $\cos B = 0.1968$ **25.** $\tan B = 0.5867$ **26.** $\tan B = 0.2978$

27. $\sin B = 0.6788$ **28.** $\cos B = 0.9166$ **29.** $\tan B = 1.649$

30. $\cos B = 0.3902$ **31.** $\sin B = 0.7067$ **32.** $\cos B = 0.8531$

33. tan B = 1.108   **34.** cos B = 0.4904   **35.** sin B = 0.7161

36. sin B = 0.7731   **37.** cos B = 0.7707   **38.** tan B = 1.047

39. sin B = 0.9801   **40.** cos B = 0.6873

 14–6   USING THE CALCULATOR TO FIND AN ANGLE, GIVEN ITS FUNCTION

If your calculator has the sin key, then it also has the arc sin key. The arc sin means "find an angle whose sine is x." On the calculator the key is usually the second function (2nd F) of the sin key and is denoted sin^{-1}. The cos and tan keys likewise have arc cos and arc tan keys. The cotangent, secant, and cosecant of an angle are not on most calculators. These three functions and their inverses (arc cotangent, and so on) are reciprocals of the functions on the calculator and can easily be evaluated by using the reciprocal key (1/x).

■ **Example**

Find the angle whose sine is 0.5373.

Solution

	The Display Shows
Turn on the calculator	0.
Enter 0.5373	0.5373
Press 2nd F	0.5373
Press sin^{-1}	32.50002661
Press 2nd F	32.50002661 (decimal answer)
Press →D.MS	32.300009 (answer)

This answer is in degrees, minutes, and seconds in the calculator display. The answer is read 32 degrees, 30 minutes, and 00.09 second. Round off this answer to 32°30'. The 00.09 second is the result of the calculator's using a nine-place decimal (0.537300000) when we keyed in a four-place decimal.

Be careful with the results you get on the calculator display. Your answers may be in degrees and decimal parts of a degree (32.50002661 in our example) or in degrees, minutes, and seconds, which looks like the decimal mode but is read *degrees* for all the digits to the left of the decimal point, *minutes* for the next two places to the right of the decimal point, and *seconds* and *hundredths of a second* for the last four places in the display.

■ Problems 14–6

Using a calculator, do Problem Set 14–5 again. Compare your answers with the answers you wrote on pages 459–461.

14–7   SOLUTION OF RIGHT TRIANGLES

The following examples will illustrate the method to be followed in the solution of right triangles.

■ Example 1

Given $A = 36°30'$ and $b = 14$ in., find B, a, and c in Fig. 14–6.

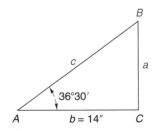

Figure 14–6

Solution   To find B:

$$B = 90 - A = 89°60' - 36°30'$$
$$= 53°30'$$

To find a:

$$a/b = \tan A \quad \therefore \; a = b \times \tan A$$
$$= 14 \times \tan 36°30'$$
$$= 14 \times 0.7400$$
$$= 10.3600 \text{ in.}$$

To find c:

$$b/c = \cos A \quad \therefore \; c = \frac{b}{\cos A}$$
$$= \frac{14}{\cos 36°30'}$$
$$= \frac{14}{0.8039} = 17.4 \text{ in.}$$

Practice check: Do exercise 9 on the left.

Answer to exercise 7

80°10′

Answer to exercise 8

22°34′

$$\frac{0.0004}{0.0011} = \frac{x'}{10'}$$
$$x \approx 4'$$

9. For the following figure, given $B = 25°20'$ and $c = 5.43$ ft, find A, a, and b.

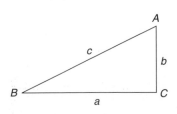

The answers are in the margin on page 464.

Summary

The method of solving a right triangle when an acute angle and one of the sides are given is as follows.

1. To find the unknown acute angle, subtract the given acute angle from 90°.
2. To find an unknown side, write a function of the given angle, employing for that purpose the given side and the required side. From that function obtain an expression for the required side in terms of the given side and the function of the angle. For example:

I. *Given A and a.*

$$\text{To find } b: \quad \frac{b}{a} = \cot A \quad \therefore \ b = a \times \cot A$$

$$\text{To find } c: \quad \frac{a}{c} = \sin A \quad \therefore \ c = \frac{a}{\sin A}$$

II. *Given A and c.*

$$\text{To find } a: \quad \frac{a}{c} = \sin A \quad \therefore \ a = c \times \sin A$$

$$\text{To find } b: \quad \frac{b}{c} = \cos A \quad \therefore \ b = c \times \cos A$$

III. *Given A and b.*

$$\text{To find } a: \quad \frac{a}{b} = \tan A \quad \therefore \ a = b \times \tan A$$

$$\text{To find } c: \quad \frac{b}{c} = \cos A \quad \therefore \ c = \frac{b}{\cos A}$$

If angle B is given, derive angle A by subtracting B from 90° and use Formulas I, II, and III to find the missing sides.

■ **Example 2**

Given $a = 8$ in. and $b = 15$ in., find A, B, and c in Fig. 14-7.

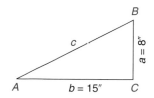

Figure 14-7

Solution   To find A:

$$\tan A = \frac{a}{b} = \frac{8}{15} = 0.5333 \quad \therefore \ A = 28°04'$$

10. For the following figure, given $a = 135$ m and $b = 212$ m, find A, B, and c.

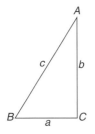

The answers are in the margin on page 466.

To find B:

$$B = 90° - A = 90° - 28°04' = 61°56'$$

To find c:

$$c = \sqrt{a^2 + b^2} = \sqrt{8^2 + 15^2} = 17.00 \text{ in.}$$

Explanation   Since side a (the side opposite angle A) and side b (the side opposite angle B) are known, then the tangent of angle A (the opposite side divided by the adjacent side) can be found, namely, $\frac{8}{15}$. Reducing this to a decimal gives tan A = 0.5333. Referring to a table of natural tangents, find by interpolation that the angle whose tangent is 0.5333 is 28°04'.

Having found angle A, find angle B by subtracting angle A from 90°. Side c was found by using the Pythagorean theorem; that is, the square of the hypotenuse is equal to the sum of the squares of the other two sides. Side c can also be found by writing the sine and cosine functions of angle A, thus:

$$\frac{a}{c} = \sin A \quad \therefore \ c = \frac{a}{\sin A}$$

or

$$\frac{b}{c} = \cos A \quad \therefore \ c = \frac{b}{\cos A}$$

Practice check: Do exercise 10 on the left.

Summary

The method of solving a right triangle when both sides opposite the acute angles are given is as follows.

1. To find angle A, first find tangent A. Then refer to a table of natural tangents, and find the angle whose tangent is tangent A.

2. To find angle B, subtract angle A from 90°.

3. The find c, use the sine or cosine function of angle A, or use the Pythagorean theorem.

$$\frac{a}{c} = \sin A \quad \therefore \ c = \frac{a}{\sin A} \quad \text{or}$$

$$\frac{b}{c} = \cos A \quad \therefore \ c = \frac{b}{\cos A} \quad \text{or}$$

$$c = \sqrt{a^2 + b^2}$$

■   **Example 3**

Given $a = 4.5$ in. and $c = 7.25$ in., find A, B, and b in Fig. 14–8.

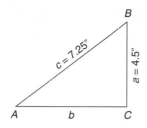

Figure 14–8

Solution   To find A:

$$\sin A = \frac{a}{c} = \frac{4.5}{7.25} = 0.6207 \quad \therefore A = 38°22'$$

To find B:

$$B = 90° - A = 90° - 38°22' = 51°38'$$

To find b:

$$b = \sqrt{c^2 - a^2} = \sqrt{7.25^2 - 4.5^2} = 5.68 \text{ in.}$$

Practice check. Do exercise 11 on the right.

11. For the following figure, given $b = 2.467$ yd and $c = 7.813$ yd, find A, B, and a.

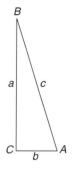

The answers are in the margin on page 467.

Summary

The method of solving a right triangle when one of the sides opposite an acute angle and the hypotenuse are given is as follows.

I. *Given a and c.*

To find angle A, first find $\sin A$ by using $\sin A = \dfrac{a}{c}$; then find angle A whose sine is $\sin A$.

To find angle B, subtract angle A from 90°.

To find b, use one of the following formulas.

$$\frac{b}{c} = \cos A \quad \therefore b = c \times \cos A \quad \text{or}$$

$$\frac{a}{b} = \tan A \quad \therefore b = \frac{a}{\tan A} \quad \text{or}$$

$$b = \sqrt{c^2 + a^2}$$

II. *Given b and c.*

To find angle B, first find $\sin B$ by using $\sin B = \dfrac{b}{c}$; then find angle B whose sine is $\sin B$.

To find angle A, subtract angle B from 90°.

To find a, use one of the following formulas.

$$\frac{a}{c} = \cos B \quad \therefore \ a = c \times \cos B \quad \text{or}$$

$$\frac{b}{a} = \tan B \quad \therefore \ a = \frac{b}{\tan B} \quad \text{or}$$

$$a = \sqrt{c^2 + b^2}$$

■ Problems 14–7

Find the missing parts of the following right triangles.

1. Given $A = 24°$ and $a = 7''$, find B, b, and c.

2. Given $B = 30°40'$ and $a = 15''$, find A, b, and c.

3. Given $A = 70°40'$ and $a = 11.4''$, find B, b, and c.

4. Given $A = 48°10'$ and $b = 54.8''$, find B, a, and c.

5. Given $A = 8°$ and $b = 45''$, find B, a, and c.

6. Given $B = 16°10'$ and $b = 90$ m, find A, a, and c.

7. Given $B = 66°20'$ and $c = 60.55$ cm, find A, a, and b.

8. Given $A = 53°28'$ and $c = 16.5$ mm, find B, a, and b.

9. Given $A = 31°56'$ and $c = 21.7$ m, find B, a, and b.

10. Given $B = 45°45'$ and $a = 32.5$ cm, find A, b, and c.

Find the missing parts of the following right triangles.

11. Given $a = 7.5''$ and $b = 8.75''$, find A, B, and c.

12. Given $a = 9.4''$ and $b = 12.8''$, find A, B, and c.

13. Given $a = 12.8''$ and $b = 16''$, find A, B, and c.

14. Given $a = 6.6''$ and $b = 14.48''$, find A, B, and c.

Answer to exercise 11

$A = 71°35'$

$B = 18°25'$

$a = 7.413$ yd

15. Given $a = 40'$ and $b = 28'$, find A, B, and c.

16. Given $a = 38.3$ cm and $b = 45.6$ cm, find A, B, and c.

17. Given $a = 20$ mm and $b = 28$ mm, find A, B, and c.

18. Given $a = 102$ m and $b = 126$ m, find A, B, and c.

19. Given $a = 16.75$ cm and $b = 24.5$ cm, find A, B, and c.

20. Given $a = 34$ cm and $b = 40$ cm, find A, B, and c.

Find the missing parts of the following right triangles.

21. Given $a = 3.8''$ and $c = 5.2''$, find A, B, and b.

22. Given $b = 18''$ and $c = 22''$, find A, B, and a.

23. Given $b = 92'$ and $c = 124''$, find A, B, and a.

24. Given $a = 15''$ and $c = 25''$, find A, B, and b.

25. Given $b = 7.1''$ and $c = 15.84''$, find A, B, and a.

26. Given $a = 21$ cm and $c = 38$ cm, find A, B, and b.

27. Given $a = 14.3$ cm and $c = 23.4$ cm, find A, B, and b.

28. Given $b = 10.6$ m and $c = 16.2$ m, find A, B, and a.

29. Given $a = 45$ m and $c = 70$ m, find A, B, and b.

30. Given $b = 25.5$ cm and $c = 40$ cm, find A, B, and a.

14–8   ISOSCELES TRIANGLES

A perpendicular dropped from the vertex of an isosceles triangle to the base divides the isosceles triangle into two equal right triangles that can be solved by the methods shown in the preceding section (see Fig. 14–9).

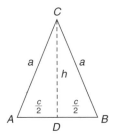

Figure 14–9

■ **Example 1**

Given $a = 16$ in. and $c = 12$ in., find A, B, C, and h in Fig. 14–10.

Solution   See Fig. 14–11.

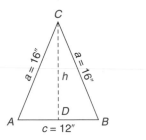

Figure 14–10

Figure 14–11

To find A:

$$\cos A = \frac{c/2}{a} = \frac{6}{16} = 0.3750 \quad \therefore \ A = 67°59'$$

$$B = A = 67°59'$$

To find C, first find $C/2$:

$$C/2 = 90° - A = 90° - 67°59' = 22°01'$$
$$C = 2 \times 22°01' = 44°02'$$

or

$$C = 180° - (A + B)$$
$$= 180° - (67°59' + 67°59')$$
$$= 180° - 135°58' = 44°02'$$

To find h:

$$h = \sqrt{a^2 - (c/2)^2} = \sqrt{16^2 - 6^2} = 14.83 \text{ in.}$$

Practice check: Do exercise 12 on the right.

■ **Example 2**

Given $C = 54°42'$ and $a = 25$ in.. find A, B, c, and h in Fig. 14–12.

Solution   See Fig. 14–13.

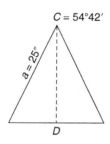

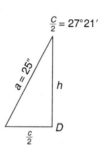

Figure 14–12   **Figure 14–13**

To find A:

$$A = 90° - C/2 = 90° - 27°21' = 62°39'$$
$$B = A = 62°39'$$

To find c, first find $c/2$:

$$\frac{c/2}{a} = \sin 27°21'$$

$$c/2 = a \times \sin 27°21'$$
$$= 25 \times 0.4594 = 11.48 \text{ in.}$$
$$\therefore \ c = 2 \times 11.48 \text{ in.} = 22.96 \text{ in.}$$

To find h:

$$h/a = \cos 27°21'$$
$$\therefore \ h = a \times \cos 27°21'$$
$$= 25 \times 0.8882 = 22.20 \text{ in.}$$

Practice check: Do exercise 13 on the right.

12. For the following figure, given $a = 10$ cm and $c = 6$ cm, find A, B, C, and h.

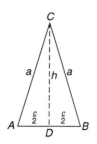

The answers are in the margin on page 471.

13. For the following figure, given $A = 50°10'$ and $a = 16''$, find C, c, and h.

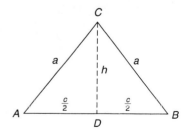

The answers are in the margin on page 472.

■ **Problems 14–8**

Compute the missing dimensions of the isosceles triangles in the following chart.

Isosceles Triangles

	Base	Sides	Altitude	Base Angles	Vertex Angle
1.	12″	10″	_____	_____	_____
2.	10″	13″	_____	_____	_____
3.	8″	_____	_____	60°	_____
4.	_____	15″	_____	30°	_____
5.	18 m	_____	_____	_____	90°
6.	_____	_____	10 cm	45°	_____
7.	_____	_____	8 cm	_____	120°
8.	_____	16 mm	_____	_____	60°
9.	_____	10.8″	_____	64°15′	_____
10.	_____	14.6″	_____	_____	85°30′
11.	18.2″	_____	_____	_____	71°40′
12.	16″	_____	_____	69°45′	_____
13.	24.5 cm	_____	15.6 cm	_____	_____
14.	_____	_____	28 cm	82°30′	_____
15.	_____	32.8 cm	25.4 cm	_____	_____
16.	_____	_____	19.3 mm	_____	100°

■ **Miscellaneous Problems Using Trigonometry**

The following problems can be solved by using the mathematics you have learned so far. The key to solving these problems is organizing your work. Make a neat sketch and label the known parts. Study your sketch and determine what you need to know to find the solution to the problem. Avoid arithmetical errors. Estimate the answer before solving the problem.

1. The equal sides of an isosceles triangle are 25″ long, and the altitude is 14″. Compute the base and the angles.

2. An isosceles triangle has sides of 13″, 13″, and 5″. Compute the angles and the altitude.

3. An isosceles triangle has a base of 4.34 cm and an altitude of 10 cm. Compute the angles and the length of the equal sides.

4. The base angles of an isosceles triangle are each 72°46′, and the altitude is 7.5″. Compute the vertex angle, the base, and the length of the equal sides.

5. The vertex angle of an isosceles triangle is 38°22′, and the equal sides are each 2.5″ long. Find the base angles, the base, and the altitude.

6. Find the perpendicular distance from the center to the side of a hexagon with 5″ sides. (*Suggestion:* Draw *OA* as shown in Fig. 14–14. Angle *OAB* = 60°. Solve for *OB*.)

Figure 14–14

7. Find the included taper angle in a piece of work having a taper of 6.25 cm per meter.

8. Find the included taper angle in a piece of work having a taper of 0.6″ per foot.

9. Figure 14–15 shows a 29° Acme thread. Find the bottom flat of a 1″ pitch thread if the depth is $\frac{1}{2}$ pitch + 0.01″.

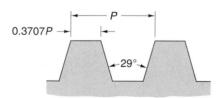

Figure 14–15

10. In the Whitworth thread in Fig. 14–16, the sides make an angle of 55° with each other. The depth of the thread is two-thirds of the depth of the triangle formed by prolonging the sides. Find the depth of a thread of 1″ pitch.

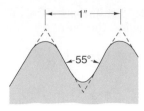

Figure 14–16

11. A road has a rise of 4 m in 100 m. What is the angle of slope with the horizontal?

12. A road has a slope of 1°50′ with the horizontal. What is the rise in 100 m?

13. What is the largest square that can be milled from a circular disk 6″ in diameter?

14. Find the radius of a circle circumscribed about an octagon with 10″ sides.

15. Find the angle of slope of the rafter in Fig. 14–17.

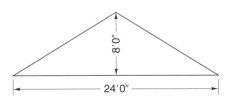

Figure 14–17

16. A chord of a $2\frac{1}{2}''$ circle is 1.028″ from the center. Find its length.

17. The chord of a 19.5-cm circle is 11.25 cm long. Find the central angle and the distance from the center.

18. Lay out an angle of 12°40′ by means of the value of its natural tangent.

19. Compute the missing dimension in the sketch in Fig. 14–18.

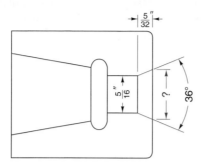

Figure 14–18

20. Compute the missing dimension in the flat-headed screw shown in Fig. 14–19.

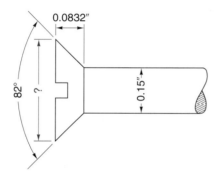

Figure 14–19

21. Figure 14–20 shows a taper gib with a taper of $\frac{3}{16}''$ per foot. Compute the following dimensions:

(a) Thickness at the large end.
(b) Dimension A at the small end.
(c) Width W at the large end.
(d) Width w at the small end.

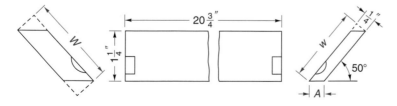

Figure 14–20

22. Twelve bolt holes are to be drilled in a cylinder head on the circumference of a 36-cm circle. What is the straight line distance between centers of holes?

23. If 15 holes are drilled on the circumference of a 27″ circle, find the distance center to center of holes.

24. Nine holes are to be drilled on the circumference of a circle 16.5 cm in diameter. What should be the distance center to center of holes?

25. Find the length of the chord joining two points 119° apart on the circumference of a $3\frac{1}{2}$″ circle.

26. Find the distance between two points 37°30′ apart on the circumference of a circle 2.37 cm in diameter.

27. What should be the check measurement over a pair of $\frac{1}{2}$″ wires on the dovetail slide shown in Fig. 14–21, both inside and outside? The sides form an angle of $62\frac{1}{2}$° with the horizontal. (*Suggestion:* The dashed lines indicate the solution.)

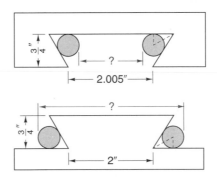

Figure 14–21

28. Figure 14–22(*a*) shows the arrangement for gaging screw threads by the three-wire system. Compute the measurement over the wires for the following sharp V-threads:

(a) $\frac{1}{4}$″ diameter, 20 threads per inch, wires 0.035″ in diameter.
(b) $\frac{9}{16}$″ diameter, 12 threads per inch, wires 0.050″ in diameter.

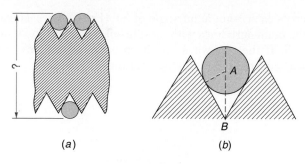

(a) (b)

Figure 14–22

(*Suggestion:* The solution is indicated in Fig. 14–22(*b*). Add twice the distance *AB* to the root diameter, and add half a wire diameter on each side.)

29. Compute the check measurements for the same two screws as in Problem 28 but for American National threads. Use the same wires.

30. Compute the height of a $7\frac{1}{2}''$ sine bar for an angle of 27°55′.

31. A sine bar 10″ long has a difference in elevation of 4.873″ between the two ends. What angle is indicated?

32. A cylindrical gasoline tank 14″ in diameter and 3′1″ long is lying in a horizontal position (Fig. 14–23). A gage inserted through the top indicates a depth of gasoline of $11\frac{1}{4}''$. How many gallons of gasoline are there in the tank?

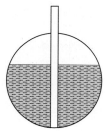

Figure 14–23

33. Two roads cross each other at an angle of 70° (Fig. 14–24). The arc *AB* is drawn tangent to the main curb lines with a radius of 3 m. Find the length of the piece of curbing *AB*. If the roadways are 20 m wide, how many square meters of pavement are required for their intersection?

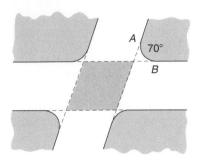

Figure 14–24

34. In a Brown and Sharpe 29° worm thread, find the diameter of the wire laid in the groove that will lie exactly flush with the top of the thread: **(a)** 4 threads per inch, **(b)** 5 threads per inch, **(c)** 3 threads per inch. See Fig. 14–25.

35. In the 29° Acme tap thread, the space *S* equals $0.6293P + 0.0052''$. Find the diameter of a wire that will lie in the groove flush with the top of the thread if the pitch is **(a)** $\frac{1}{8}''$, **(b)** $\frac{1}{6}''$, **(c)** $\frac{1}{10}''$. See Fig. 14–26.

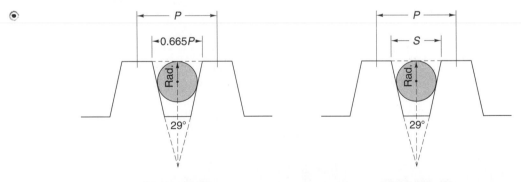

Figure 14–25   **Figure 14–26**

■ Self-Test

Solve the following problems. Use Appendix Table 2, Natural Trigonometric Functions, in the back of this book. Check your answers with the answers given in the back of the book.

1. What is the sin 30°?

Use the right triangle of Fig. 14–27 to do Problems 2–5.

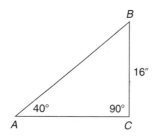

Figure 14–27

2. How large is angle *B*?

3. What is the sine of angle *A*?

4. How long is side *AB*?

5. How long is side *AC*?

Use the isosceles triangle of Fig. 14–28 to do Problems 6–10.

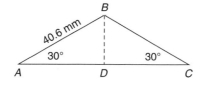

Figure 14–28

6. How long is side *BC*?

7. How long is the altitude of triangle *ABC*?

8. What is the size of the vertex angle *B*?

9. How long is the base *AC*?

10. What is the area of triangle *ABC*?

■ **Chapter Test**

Solve the following problems involving triangles. Make a sketch of each triangle and label all the known parts. Organize your work and avoid arithmetical errors. Use the tables in the back of the book.

1. Find the tangent of 35°10′.

2. Find the sine of 25°25′.

3. Find the angle whose cotangent is 3.867.

4. Find the angle whose cosine is 0.2565.

5. In right triangle *ABC*, angle *A* = 25°, angle *C* = 90°, and side *a* = 15″. Find side *c*.

6. In right triangle *ABC*, angle *B* = 42°20′, angle *C* = 90°, and side *b* = 18″. Find side *a*.

7. In right triangle *ABC*, angle *C* = 90°, side *a* = 24″, and side *b* = 20″. Find angle *A*.

8. Find side *c* in Problem 7.

9. In right triangle ABC, side $a = 19''$, side $c = 28''$, and angle $C = 90°$. Find side b.

10. Find angle A in Problem 9.

11. The two legs of a right triangle are 7 mm and 24 mm. How long is the hypotenuse?

12. In isosceles triangle ABC, angle $A =$ angle B, angle $C = 100°$, and side $c = 16''$. Find side a.

13. In isosceles triangle ABC, angle $A =$ angle B, side $c = 26$ cm, and side $b =$ side $a = 16$ cm. Find angle B.

14. Find angle A in Problem 12.

15. Find the area of a triangle whose sides are 20 cm, 20 cm, and 24 cm. (Use Hero's formula given in Section 7–12.)

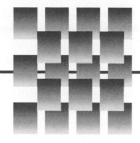

CHAPTER 15

Strength of Materials

15–1 STRESS AND STRAIN

When a load of 100 lb is suspended by a wire, the load tends to stretch or elongate the wire. This tendency of the load to *distort* or *change* the shape of the wire is called **strain.** The wire resists the tendency of the load to pull it apart and in so doing exerts a force opposite to the load. The internal force by which the wire resists the tendency of the outside load to change its shape is called **stress.**

Similarly, a load on a column tends to compress or crush the column. The column reacts against the tendency of the load to crush it and exerts a force opposite to the load. The tendency of the outside load to change the shape of the column is called *strain,* and the internal force by which the column resists the tendency of the outside load to change its shape is called *stress.*

15–2 KINDS OF STRESSES

Three kinds of direct stress may be developed in a body by the application of an outside force or load: *tension, compression,* and *shear.*

Tension

A weight suspended from a rod tends to pull the rod apart. The stress developed in the rod to resist being *pulled* apart is called **tensile stress.** The rod is said to be in **tension.**

Compression

A load on a column tends to crush it. The stress the column develops to resist being *crushed* is called **compressive stress.** The column is said to be in **compression.**

Shear

A rivet connecting two plates will sometimes give way because the plates push so hard on the rivet in opposite directions as to cause one half of the rivet to slide past the other; that is, the two halves of the rivet are separated as if the rivet were cut by a pair of shears (see Fig. 15–1). The stress the rivet develops to resist being cut apart is called *shearing stress.* The rivet is said to be in *shear.*

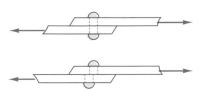

Figure 15–1

Bending stress is a combination of tension and compression. *Twisting* or *torsion* is a form of shearing stress.

15–3   UNIT STRESS

Unit stress is the stress per unit of area and is usually expressed in pounds per square inch. Thus, if a column whose cross-sectional area is 6 sq in. supports a load of 60,000 lb, the unit stress is $\frac{60,000}{6}$ or 10,000 lb per square inch.

15–4   ELASTIC LIMIT

If a suspended rod with a cross-sectional area of 1 sq in. is loaded with a weight, the latter stretches the rod a certain amount, depending upon the length of the rod and the load applied. If the load that is used is not too great, the rod will spring back to its original length when the load is removed. When a sufficiently heavy load is hung from the rod, the rod will not spring back to its original length when the load is removed. There remains a permanent set. So the *elastic limit* is the unit stress beyond which a body will not return to its original shape when the load is removed.

For example, a steel rod of 1-sq-in. cross-sectional area from which a load of 35,000 lb is suspended springs back to its original length when the load is removed but acquires a permanent set when a load of 36,000 lb is applied. Thus, the elastic limit of that rod is 35,000 lb.

15–5   ULTIMATE STRENGTH

The *ultimate* or *breaking strength* of a body is the unit stress that causes that body to break. Thus, it takes almost 50,000 lb to pull apart a wrought-iron bar with a cross-sectional area of 1 sq in. We therefore say that the ultimate tensile strength of wrought iron is 50,000 lb per square inch.

Engineers must know the strength of the materials that enter into the machine or structure under design. To determine the ultimate strength of the materials in common use, specimens are subjected to tensile, compressive, or shearing stresses in testing machines, and the breaking point in each instance is noted. In this way the average values of the elastic limits and ultimate strengths in pounds per square inch were obtained for the materials listed in the following table.

Elastic Limit and Ultimate Strength (lb/sq in.)				
			Ultimate Strength	
Material	Elastic Limit	Tension	Compression	Shear
Timber	3,000	8,000	8,000	500 (with grain) 3,000 (across grain)
Stone			6,000	1,500
Brick		250–300	3,000	1,000
Cast iron	6,000 (tension) 20,000 (comp.)	25,000	90,000	18,000
Wrought iron	28,000	50,000	55,000	40,000
Medium steel	36,000	60,000	60,000	50,000
Concrete, Portland 1:2:4		100–150	2,400	300–500

15–6   SAFETY FACTOR

Although the ultimate compressive strength of steel is 60,000 lb per square inch, in designing a machine or structure the engineer will design each member on the assumption that the material can develop a compressive stress of only 15,000 lb per square inch. In other words, the engineer will make each member of the machine or structure four times as strong as is required in order to keep it from breaking down when loaded. The ratio of the ultimate strength to the actual stress is called the *safety factor.* In the example just cited, the safety factor is $\frac{60,000}{15,000} = 4$. The safety factor ranges from 4 for steel, designed for steady stress, to 35 for brick, designed to withstand shock. A larger safety factor is assumed in the design of a railroad bridge than in the design of an office building because the moving load on a railroad bridge causes the structure to vibrate, whereas for the building, the load is steady.

15–7   WORKING UNIT STRESSES

As a result of investigation and experience, the values in the following table have been adopted as safe working stresses for the materials most commonly employed. Notice that in each instance the working unit stress is well within the elastic limit.

Safe Working Stresses (lb/sq in.)			
Material	Tension	Compression	Shear
Timber	750	750	500
Cast iron	4,000	15,000	3,000
Wrought iron	12,000	12,000	10,000
Medium steel	15,000	15,000	10,000

■ **Example 1**

How many pounds can be safely carried in tension by a steel rod whose cross section is a rectangle 2 in. × $\frac{3}{4}$ in.? (From the preceding table we note that steel has a safe working stress of 15,000 lb per sq inch when in tension.)

Solution   Area of cross section = $2 \times \frac{3}{4} = 1\frac{1}{2}$ sq in.
Carrying capacity in tension = $1\frac{1}{2} \times 15,000 = 22,500$ lb

1. Allowing a tensile stress of 12,000 lb per square inch, how many pounds can be safely carried in tension by a wrought-iron column whose cross section is a rectangle 3 in. × $1\frac{1}{4}$ in.?

The answer is in the margin on page 486.

■ **Example 2**

Allowing a tensile stress of 15,000 lb per square inch, how many pounds can be safely carried by a steel rod $\frac{7}{8}$ in. in diameter?

Solution   Area of cross section = $0.7854 \times 0.875^2 = 0.6013$ sq in.
Safe carrying capacity = $15,000 \times 0.6013 = 9,020$ lb

Practice check: Do exercise 1 on the left.

■ **Example 3**

Allowing a tensile stress of 15,000 lb per square inch, find the diameter of a steel rod to carry 100,000 lb.

Solution   Required area of cross section = $\dfrac{100,000}{15,000} = 6.6667$ sq in.

$$\text{Diameter} = 2\sqrt{\frac{\text{area}}{\pi}} \quad \text{(See Chapter 8 for areas of circles.)}$$

2. Allowing a tensile stress of 4,000 lb per square inch, find the diameter of a cast-iron rod to carry 15,000 lb.

The answer is in the margin on page 486.

$$= 2\sqrt{\frac{6.6667}{3.1416}} = 2\sqrt{2.1221}$$
$$= 2 \times 1.46 = 2.92 \text{ in.}$$

Practice check: Do exercise 2 on the left.

■ **Problems 15–7**

Answer the following questions involving working stresses. Organize and label all work in the space provided.

1. A steel rod has a rectangular cross section $3\frac{1}{2}'' \times 1\frac{3}{4}''$. Allowing a tensile stress of 15,000 lb per square inch, how many pounds can it carry?

2. How many pounds can be safely carried in tension by a steel rod $1\frac{7}{8}''$ in diameter?

3. What must be the diameter of a steel bar to carry safely a tensile force of 65,000 lb?

4. What should be the diameter of a steel bar to carry a load of 60,000 lb, allowing a tensile stress of 16,000 lb per square inch?

5. What should be the diameter of a steel bar to carry safely a load of 72,000 lb in tension, allowing 15,000 lb per square inch?

6. A balcony is supported from the ceiling by means of six wrought-iron rods. Allowing a tensile stress of 12,000 lb per square inch, what size rods must be used if the total weight is 80,000 lb?

7. Allowing a stress of 10,000 lb per square inch, what is the shearing strength of an iron bar $2\frac{1}{2}'' \times \frac{7}{16}''$ in cross section?

8. The ends of a floor beam carrying a total load of 6,400 lb uniformly distributed rest on cast-iron plates placed on the brick. If the plates are $8'' \times 10''$, what is the pressure per square inch on the brick?

9. A loading platform is supported by six square sticks of timber. What must be the size of the supports if the platform is to carry 120,000 lb and the allowable compressive stress is 500 lb per square inch?

10. What is the ultimate compressive strength in pounds per square inch of a brick $8'' \times 4'' \times 2''$ if it gave way in the testing machine under a load of 56,000 lb when lying flat?

15–8   PRESSURE IN PIPES

The tendency of steam or water pressure in a pipe is to burst the pipe longitudinally. The material of the pipe is therefore under tension when the pipe is under pressure. The total pressure on a diametral plane is equal to

$$P = p \times L \times D$$

where P = total pressure

p = pressure per square inch

L = length of pipe in inches

D = diameter of pipe in inches

The tendency of the steam or water pressure to burst the pipe is resisted by the tension in the walls of the pipe (see Fig. 15–2). The total resisting stress in the pipe is equal to

$$S = 2 \times s \times L \times t$$

where S = total resisting stress

s = unit tensile stress

L = length of pipe in inches

t = thickness of pipe

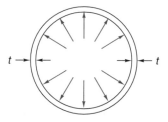

Figure 15–2

For the pipe to withstand bursting, the resisting stress must equal the total pressure. Therefore,

$$2 \times s \times L \times t = p \times L \times D$$

$$\therefore \ t = \frac{p \times L \times D}{2 \times s \times L} = \frac{p \times D}{2 \times s}$$

This formula enables us to find the required thickness of pipe to withstand a given pressure.

To find the safe internal unit pressure in a pipe of given thickness, solve the preceding formula for p.

$$p = \frac{2 \times s \times t}{D}$$

■ Example

Allowing a tensile stress of 5,000 lb per square inch, how thick should a 20-in. wrought-iron steam pipe be to withstand a pressure of 300 lb per square inch?

Solution $t = \dfrac{p \times D}{2 \times s} = \dfrac{300 \times 20}{2 \times 5{,}000} = 0.6$ in.

This is the minimum thickness of pipe required. The next higher standard thickness would be used.

Practice check: Do exercise 3 on the right.

■ **Problems 15–8**

Use the formulas in Section 15–8 to solve the following problems.

1. How thick must a 15″ steam pipe be to withstand a steam pressure of 250 lb per square inch with a unit tensile stress of 5,000 lb per square inch?

2. What may be the maximum pressure per square inch in a steam pipe $\frac{3}{4}''$ thick if its diameter is 16″? Allow a unit tensile stress of 5,500 lb per square inch.

3. A cast-iron water pipe 30″ in diameter is to withstand a pressure of 75 lb per square inch. Allowing a tensile stress of 2,000 lb per square inch, how thick must the pipe be?

4. Allowing a unit tensile stress of 2,000 lb per square inch, what is the maximum allowable pressure per square inch in a cast-iron pipe 15″ in diameter and $\frac{5}{8}''$ thick?

5. Find the maximum allowable pressure per square inch in a steam pipe $\frac{7}{8}''$ thick and 20″ in diameter, allowing a tensile stress of 2,500 lb per square inch.

6. What must be the thickness of a steam pipe 18″ in diameter to withstand a pressure of 100 lb per square inch if the unit tensile stress is 3,500 lb per square inch?

3. Allowing a tensile stress of 2,000 lb per square inch, how thick should a 10-in. wrought-iron steam pipe be to withstand a pressure of 200 lb per square inch?

The answer is in the margin on page 490.

15–9   RIVETED JOINTS

A riveted joint may fail in one of two ways: Either the rivets may be sheared off, or the plates may be crushed by the pressure of the rivets. In the former, the area to be sheared is the cross-sectional area of the rivet. The safe load on the rivet for this type of stress is called the ***shearing value*** of the rivet.

   In order to fail by tearing through the plate, the rivet must crush an area that is a rectangle formed by the thickness of the plate and the diameter of the rivet. The safe load per rivet for this type of stress is called the ***bearing value*** of the rivet.

The common unit stresses allowed for steel plates and rivets are 10,000 lb per square inch for shearing and 15,000 lb per square inch for bearing.

Figure 15–3 shows three types of riveted joints. In (a) and (c) the rivets are said to be in **single shear** because they may be destroyed by shearing at one place only. In (b) the rivets are in **double shear** because they must be sheared off in two places at the same time in order to fail.

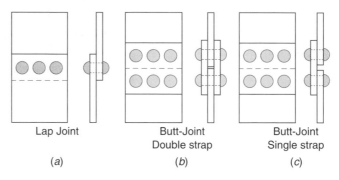

Lap Joint   Butt-Joint   Butt-Joint
                   Double strap   Single strap

(a)   (b)   (c)

Figure 15–3

■ **Example 1**

Figure 15–4(a) shows two plates, each $\frac{1}{2}$ in. thick, connected by one rivet $\frac{7}{8}$ in. in diameter. Find the strength of the joint.

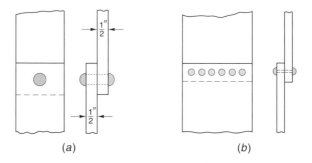

(a)   (b)

Figure 15–4

Solution   Cross-sectional area of rivet $= 3.1416 \times \frac{7}{16} \times \frac{7}{16} = 0.601$ sq in.
                Shearing value of rivet $= 0.601 \times 10,000 = 6,010$ lb
                Bearing area of rivet $= \frac{7}{8} \times \frac{1}{2} = \frac{7}{16} = 0.438$ sq in.
                Bearing value of rivet $= 0.438 \times 15,000 = 6,570$ lb

The smallest value determines the strength of the joint, which is 6,010 lb.

■ **Example 2**

If the plates in Fig. 15–4(b) are connected by six rivets, each $\frac{7}{8}$ in. in diameter, what is the strength of the joint?

Solution $6 \times 6,010 = 36,060$

Explanation   Since one rivet is good for 6,010 lb, six rivets will carry $6 \times 6,010$ or 36,060 lb.

■ **Example 3**

How many $\frac{7}{8}$-in. rivets are required to carry a stress of 100,000 lb through the two plates in Fig. 15–4(*b*)?

Solution $\dfrac{100,000}{6,010} = 16.64$ or 17 rivets

Explanation   We find the number of rivets required by dividing the total stress to be carried by the lowest value of one rivet. You must always round the answer *up* to a whole number of rivets.

Practice check: Do exercise 4 on the right.

■ **Example 4**

In Fig. 15–5 the inside plate is $\frac{3}{4}$ in. thick and the outside plates are each $\frac{1}{2}$ in. thick. The rivet is $\frac{7}{8}$ in. in diameter. What is the strength of the joint?

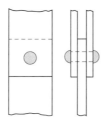

Figure 15–5

Solution   Cross-sectional area of rivet $= 3.1416 \times \frac{7}{16} \times \frac{7}{16} = 0.601$ sq in.
Shearing value of rivet (double shear) $= 2 \times 0.601 \times 10,000$
$= 12,020$ lb
Bearing area of rivet $= \frac{7}{8} \times \frac{3}{4} = \frac{21}{32} = 0.656$ sq in.
Bearing value of rivet $= 0.656 \times 15,000 = 9,840$ lb

The strength of the joint, as determined by the lowest value of the rivet, is 9,840 lb.

Practice check: Do exercise 5 on the right.

■ **Problems 15–9**

Find the solutions to the following problems involving strength of rivets. Organize your work. Use the example problems and the accompanying table as a guide.

Cross-Sectional Area of Rivets			
Diameter (in.)	**Area (sq in.)**	**Diameter (in.)**	**Area (sq in.)**
$\frac{3}{8}$	0.1104	$\frac{3}{4}$	0.4418
$\frac{1}{2}$	0.1963	$\frac{7}{8}$	0.6013
$\frac{5}{8}$	0.3068	1	0.7854

4. Two plates, each $\frac{3}{4}$ in. thick, are connected by four rivets, each $\frac{5}{8}$ in. in diameter.
(a) Find the strength of the joint. **(b)** How many $\frac{5}{8}$-in. rivets are required to carry a stress of 80,000 lb?

The answer is in the margin on page 491.

5. Suppose that in Fig. 15–5 the inside plate is $\frac{7}{8}$ in. thick and the outside plates are each $\frac{1}{2}$ in. thick. The rivet is $\frac{5}{8}$ in. in diameter. What is the strength of the joint?

The answer is in the margin on page 491.

Answer to exercise 3

0.5 in.

1. Two $\frac{5}{8}''$ plates are connected by a lap joint with $\frac{7}{8}''$ rivets. What is the strength of the joint if 12 rivets are used?

2. Find the strength of the joint shown in Fig. 15–6 if eight $\frac{3}{4}''$ rivets are used.

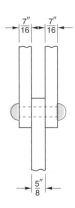

Figure 15–6

3. How many $\frac{5}{8}''$ rivets will be required on each side of the joint shown in Fig. 15–7 if all plates are $\frac{9}{16}''$ thick and the stress to be developed is 60,000 lb?

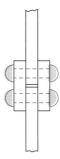

Figure 15–7

4. Find the strength of the joint shown in Fig. 15–8 if ten $\frac{1}{2}''$ rivets are used on each side of the butt joint. All plates are $\frac{1}{2}''$ thick.

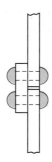

Figure 15–8

5. Find the strength of the joint shown in Fig. 15–9 if the plates are $\frac{1}{2}''$ thick and the straps $\frac{9}{16}''$ thick. Eight $\frac{3}{4}''$ rivets are used on each side of the butt joint.

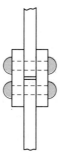

Figure 15–9

6. Compute the strength of the joint in Problem 1 if **(a)** eighteen $\frac{1}{2}''$ rivets are used, **(b)** sixteen $\frac{5}{8}''$ rivets, **(c)** fourteen $\frac{3}{4}''$ rivets, **(d)** ten $1''$ rivets.

7. (a) How many $\frac{7}{8}''$ rivets are required to transmit a stress of 15,000 lb through a lap joint of $\frac{3}{4}''$ plates? **(b)** How many $\frac{1}{2}''$ rivets? **(c)** How many $\frac{5}{8}''$ rivets? **(d)** How many $\frac{3}{8}''$ rivets? **(e)** How many $1''$ rivets?

8. A single-strap butt joint of two $\frac{1}{2}''$ plates is required to transmit 75,000 lb. **(a)** How many $\frac{1}{2}''$ rivets are required if the strap is $\frac{1}{2}''$ thick? **(b)** How many $\frac{5}{8}''$ rivets if the strap is $\frac{9}{16}''$ thick? **(c)** How many $\frac{3}{4}''$ rivets with a $\frac{1}{2}''$ strap? **(d)** How many $\frac{7}{8}''$ rivets with a $\frac{7}{16}''$ strap? **(e)** How many $1''$ rivets with a $\frac{1}{2}''$ strap?

9. Two $\frac{11}{16}''$ plates are connected by a double-strap butt joint with $\frac{1}{2}''$ straps. **(a)** What is the strength of such a joint with eight $\frac{7}{8}''$ rivets on each side of the butt joint? Compute the strength of this joint with **(b)** twenty-four $\frac{1}{2}''$ rivets on each side of the butt joint, **(c)** twenty $\frac{5}{8}''$ rivets on each side, **(d)** eight $1''$ rivets on each side.

10. **(a)** How many $\frac{3}{4}''$ rivets are required on each side of the butt joint in Problem 9 if a stress of 90,000 lb is to be transmitted? **(b)** How many $\frac{1}{2}''$ rivets? **(c)** How many $\frac{5}{8}''$ rivets? **(d)** How many $1''$ rivets?

■ **Self-Test**

Use the space provided to do the following problems. Check your answers with the answers in the back of the book.

1. Allowing a tensile stress of 4,000 lb per square inch, how many pounds can a cast-iron rod carry with a rectangular cross section $4\frac{1}{4}$ in. $\times$ $2\frac{1}{2}$ in.?

2. What should be the diameter of a steel bar to carry a load of 150,000 lb, allowing a tensile stress of 15,000 lb per square inch?

3. Allowing a unit tensile stress of 6,500 lb per square inch, what may be the maximum pressure per square inch in a steam pipe $\frac{1}{2}$ in. thick if its diameter is 20 in.?

4. How many $\frac{5}{8}$-in. rivets are required to transmit a stress of 16,500 lb through a single stress joint of $\frac{3}{4}$-in. plates?

5. Find the strength of a double shear joint if the plates are $\frac{5}{8}$ in. thick and the straps are $\frac{1}{4}$ in. thick. Six $\frac{3}{4}$-in. rivets are used on each side of the butt joint.

■ **Chapter Test**

1. What must be the diameter of a wrought-iron bar to carry safely a tensile force of 80,000 lb?

2. Allowing a stress of 10,000 lb per square inch, what is the shearing strength of a steel bar $3\frac{3}{4}$ in. $\times$ $1\frac{5}{8}$ in. in cross section?

3. How thick must a 20-in. steam pipe be to withstand a steam pressure of 325 lb per square inch with a unit tensile stress of 6,000 lb per square inch?

4. Find the maximum allowable pressure per square inch in a steam pipe $\frac{3}{4}$ in. thick and 15 in. in diameter. Allow a tensile stress of 3,000 lb per square in.

5. How many $\frac{3}{8}$-in. rivets are required to transmit a stress of 15,500 lb through a single strap butt joint of $\frac{3}{4}$-in. plates?

6. Two $\frac{3}{4}$-in. plates are connected by a double strap butt joint with $\frac{1}{2}$-in. straps. Compute the strength of this joint with 16 rivets on each side of the butt joint.

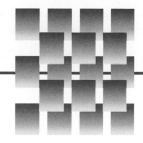

CHAPTER 16

Work and Power

16–1   WORK AND POWER

When we lift a weight of 1 lb a height of 1 ft, we do 1 foot-pound of work. If we exert a pull of 50 lb in dragging a load a distance of 10 ft, we do 10 × 50 or 500 foot-pounds of work.

A *foot-pound* (abbreviated as **ft-lb**) is the unit of work. It is the work done in raising 1 lb a height of 1 ft, or it is the pressure of 1 lb exerted over a distance of 1 ft in any direction.

A freight elevator lifting 2,000 lb a height of 40 ft does 80,000 ft-lb of work whether it takes 2 minutes or 4 minutes to do it. But an elevator that would lift the load in 2 minutes would have twice the *power* of one that would take 4 minutes to do the same amount of work.

To find the quantity of work done by a force, multiply the force by the distance that it has moved the body. That is,

$$\text{Work} = \text{Force} \times \text{Distance}$$

Practice check: Do exercise 1 on the right.

1. A man pushes a handcart forward a distance of 100 ft by exerting a force of 20 lb upon it. How much work does he do?

The answer is in the margin on page 497.

Power is the rate at which work is done. It is measured in foot-pounds per minute. Thus, if an elevator did 80,000 ft-lb of work in 2 minutes, its power would be $\frac{80,000}{2}$ or 40,000 ft-lb per minute. If it took 4 minutes to do it, its power would be $\frac{80,000}{4}$ or 20,000 ft-lb per minute.

$$\text{Power} = \frac{\text{Work}}{\text{Time}}$$

Practice check: Do exercise 2 on the right.

2. A man lifts a 50-lb weight 25 ft in 30 seconds. What is his power?

The answer is in the margin on page 497.

The unit of power is the *horsepower* (abbreviated as **hp**), which is equal to 33,000 ft-lb per minute. It is interesting to note that 33,000 ft-lb per minute is considered more than the work a horse can do. The horse used in the original tests was an unusually strong horse, so three-quarters of the figure used for a horsepower more nearly approximates the work a horse can do.

$$\text{Horsepower} = \frac{\text{Work}}{\text{Time (Minutes)} \times 33,000}$$

■ Example

An engine does 120,000 ft-lb of work in 3 minutes. Compute the horsepower.

Solution $\text{Horsepower} = \dfrac{120,000}{3 \times 33,000} = 1.2$

Explanation   Since the engine takes 3 minutes to do 120,000 ft-lb of work, in 1 minute it will do $\frac{1}{3}$ of 120,000 or 40,000 ft-lb of work.

Since $33,000 \dfrac{\text{ft-lb}}{\text{min}} = 1$ hp, divide by 33,000.

Practice check: Do exercise 3 on the left.

---

3. What is the horsepower rating of an electric hoisting machine that can lift a load of 2,750 lb a distance of 50 ft in 20 seconds?

The answer is in the margin on page 498.

■ Problems 16–1

Use the space provided below to find the amount of work done in the following problems. Use foot-pounds as the unit of work. Convert this work to horsepower when asked to do so.

1. A man carries a load of 50 lb up a flight of stairs 10 ft high. Compute the work done in foot-pounds.

2. Compute the horsepower required for an elevator to lift 2,400 lb a height of 120 ft in $1\frac{3}{4}$ minutes.

3. Find the horsepower of an engine that pumps 30 cu ft of water per minute from a depth of 320 ft. Water weighs 62.5 lb per cubic foot.

4. What must be the horsepower of an engine to lift a girder weighing 7 tons a height of 40 ft in 5 minutes?

5. A man pushing a wheelbarrow a distance of 450 ft exerts a force of 45 lb. How much work does he do?

6. An engine does 84,000 ft-lb of work in $2\frac{1}{2}$ minutes. Compute the horsepower.

16–2 HORSEPOWER OF A STEAM ENGINE

The horsepower of a steam engine actually means the *indicated horsepower*. The formula for finding the indicated horsepower is

$$\text{hp} = \frac{P \times L \times A \times N}{33{,}000}$$

where hp = indicated horsepower
  P = mean effective pressure on the piston in pounds per square inch
    (P is found by means of an indicator card.)
  L = length of stroke in feet
  A = area of piston in square inches
  N = number of strokes per minute
    = 2 × rpm for a double-acting engine

■ Example

Find the horsepower of a 26 in. × 40 in. double-acting engine running at 80 rpm under a mean effective pressure of 75 lb per square inch.

  NOTE: *In specifying the dimensions of a cylinder, the first number gives the diameter or bore; the second number, the stroke.*

Procedure

1. Find the area of the piston in square inches (use the formula $A = 0.7854d^2$).

2. Find the length of the stroke in feet.

3. If the engine is a double-acting engine, multiply the rpm by 2.

4. Substitute the values in the formula for horsepower of steam engines and solve.

Solution $\text{hp} = \dfrac{P \times L \times A \times N}{33{,}000}$

  $P = 75$

  $L = 40 \text{ in.} = 3\frac{1}{3} \text{ ft} = \frac{10}{3} \text{ ft}$

  $A = 26^2 \times 0.785 = 530.66 \text{ sq in.}$

  $N = 80 \times 2 = 160$

  $\text{hp} = \dfrac{75 \times 10 \times 530.66 \times 160}{3 \times 33{,}000} = 643.2 \text{ hp}$

Practice check: Do exercise 4 on the right.

■ Problems 16–2

Use the space provided to find the horsepower of the following steam engines.

Answer to exercise 1

Work = 20 lb × 100 ft
  = 2,000 ft-lb

Answer to exercise 2

Work = 50 lb × 25 ft
  = 1,250 ft-lb

Power = $\dfrac{1{,}250 \text{ ft-lb}}{0.5 \text{ min}}$
  = 2,500 ft-lb per min

4. Compute the horsepower of a 24″ × 32″ double-acting steam engine running at 120 rpm with a mean effective pressure of 65 lb.

The answer is in the margin on page 499.

Answer to exercise 3

$$\text{hp} = \frac{2,750 \times 50}{\frac{1}{3} \times 33,000}$$
$$= 12.5 \text{ hp}$$

1. What is the indicated horsepower of an 18″ × 28″ double-acting steam engine making 100 rpm with a mean effective pressure of 48 lb?

2. Compute the horsepower of a 32″ × 50″ double-acting steam engine running at 75 rpm with a mean effective pressure of 55 lb.

3. A 24″ × 40″ double-acting engine is running at 80 rpm under a mean effective pressure of 50 lb. Find the horsepower.

4. Find the horsepower of a 22″ × 26″ double-acting steam engine making 75 rpm under a mean pressure of 110 lb.

5. Compute the horsepower of a double-acting steam engine with two 16″ × 30″ cylinders if the engine is making 140 rpm under a mean effective pressure of 50 lb.

16–3   HORSEPOWER OF GAS ENGINES

There are many formulas for computing the horsepower of a gas engine. The standard formula for the rated horsepower of a 4-cycle gas engine is that of the Society of Automotive Engineers, formerly the American Licensed Automobile Manufacturers. This formula, now known as the N.A.C.C. (National Automobile Chamber of Commerce) formula, is as follows.

$$\text{hp} = \frac{D^2 \times N}{2.5} = 0.4 \times D^2 \times N$$

where D = diameter of cylinders in inches
       N = number of cylinders

This formula is based on a piston speed of 1,000 ft per minute and a mean effective pressure of 90 lb per square inch.

5. Find the rated horsepower of a 12-cylinder engine with a $3\frac{5}{8}″$ bore and a $4\frac{1}{16}″$ stroke.

The answer is in the margin on page 500.

■ **Example**

Find the horsepower of a $4\frac{1}{2}″ × 6″$, 4-cycle, 4-cylinder gas engine.

Solution $\text{hp} = 0.4 \times D^2 \times N = 0.4 \times 4.5^2 \times 4 = 32.4 \text{ hp}$

Practice check: Do exercise 5 on the left.

■ **Problems 16–3**

Use the space provided and the N.A.C.C. formula to find the horsepower of the following gasoline engines.

1. Find the rated horsepower of a 6-cylinder automobile engine if the cylinders have a $3\frac{5}{16}''$ bore and a $3\frac{1}{2}''$ stroke.

2. A gas engine with $3\frac{1}{4}''$ bore and $4\frac{1}{2}''$ stroke has 8 cylinders. What is the rated or taxable horsepower?

3. Find the rated horsepower of a 12-cylinder engine with a bore of $3\frac{1}{8}''$.

4. A certain V-8 engine has cylinders $3\frac{1}{16}'' \times 3\frac{3}{4}''$. What is the rated horsepower?

5. If the bore of the engine in Problem 4 is increased by $\frac{1}{8}''$, what increase in horsepower will result?

6. Find the horsepower of a 16-cylinder $3'' \times 4''$ engine.

7. Find the horsepower of a 4-cylinder $3\frac{1}{8}'' \times 4\frac{3}{8}''$ engine.

8. How much additional horsepower will be developed by the engine in Problem 7 if the bore is increased by $\frac{3}{16}''$?

Answer to exercise 4

$P = 65$ rpm

$L = 32''$
$\quad = 2\frac{2}{3}$ ft   or $\frac{8}{3}$ ft

$A = 24^2 \times 0.785$
$\quad = 452.16$ sq in.

$N = 120 \times 2$
$\quad = 240$

$hp = \dfrac{65 \times 8 \times 452.16 \times 240}{3 \times 33,000}$

$\quad = 570$ hp

16–4 BRAKE HORSEPOWER

The **brake horsepower** (bhp) is the power delivered by an engine at its flywheel, or the power actually available for use. The brake horsepower of an engine may be determined by the use of a device known as the **Prony brake** (Fig. 16–1). This device consists of a steel or leather belt carrying a number of wooden blocks or shoes clamped around the pulley by means of the bolt *B*. The bolt enables us to vary the friction between the shoes and the rim of the pulley. The pull developed by this friction at a distance *L* from the shaft is measured by the platform scale on which the lever rests. Allowance must be made for the weight of the lever itself. The formula for determining brake horsepower is

$$bhp = \frac{2 \times \pi \times L \times N \times W}{33,000}$$

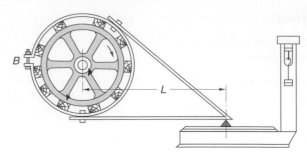

Figure 16–1

where L = length of lever arm in feet
       N = number of revolutions per minute
       W = net force = scale reading minus weight of lever

■ Example

6. Compute the brake horse-power of an engine if the Prony brake lever is 4′6″, a net force on the scale is 26 lb, and the engine speed is 250 rpm.

The answer is in the margin on page 502.

Find the brake horsepower of an engine determined by a Prony brake where L is 4 ft, the engine is running at 400 rpm, and the net force on the scale is 30 lb.

Solution   bhp $= \dfrac{2 \times \pi \times L \times N \times W}{33{,}000}$

$= \dfrac{2 \times 3.1416 \times 4 \times 400 \times 30}{33{,}000} = 9.14$ hp

Practice check: Do exercise 6 on the left.

■ Problems 16–4

Use the space provided to solve the following problems.

1. Calculate the brake horsepower of a gas engine running at 450 rpm if the net scale reading is 25 lb and the lever arm is 4′6″ long.

2. What is the brake horsepower of a steam engine making 120 rpm if the Prony brake lever is 5′0″ long and the net force on the scale is 28 lb?

3. The brake horsepower of an engine making 60 rpm was determined by a Prony brake having a lever arm of 3′4″. Compute the brake horsepower if the net force on the scale was 40 lb.

4. Compute the brake horsepower of a gas engine as determined by a Prony brake having a lever arm of 3′3″, a net force on the scale of 20 lb, and a speed of 400 rpm.

5. An engine making 80 rpm is tested with a Prony brake having a lever arm of 3′6″. What is the brake horsepower if the net force on the scale is 28 lb?

16–5   ELECTRICAL POWER

The unit of electrical power is the *watt* (W). A **watt** is the power used when 1 volt (1 V) causes 1 ampere (1 A) of current to flow. The watts consumed in any circuit are found by multiplying the amperes (current) in the circuit by the volts (pressure) impressed on the circuit. Thus,

$$\text{watts} = \text{amperes} \times \text{volts}$$

$$\text{amperes} = \frac{\text{watts}}{\text{volts}}$$

$$\text{volts} = \frac{\text{watts}}{\text{amperes}}$$

Practice check: Do exercise 7 on the right.

7. At what rate is electric energy being used by an electric appliance that operates on a 110-volt line and draws 5 amperes of current?

The answer is in the margin on page 503.

Kilowatt

Because the watt is too small a unit for convenience in computation, we use the **kilowatt** (kW), which is equal to 1,000 watts. In electrical power machines, 1 horsepower is equal to 746 watts, or in round numbers, 750 watts, which is equal to $\frac{3}{4}$ of a kilowatt.

$$\text{kilowatts} = \frac{\text{amperes} \times \text{volts}}{1,000}$$

■ **Example 1**

A generator delivers a current of 50 amperes at a pressure of 110 volts. What power in kilowatts does it supply?

Solution $\text{kW} = \dfrac{\text{amperes} \times \text{volts}}{1,000} = \dfrac{50 \times 110}{1,000} = 5.5 \text{ kW}$

Practice check: Do exercise 8 on the right.

8. A 550-volt motor draws a current of 20 amperes. What is the power of the motor in kilowatts?

The answer is in the margin on page 503.

Electrical Horsepower

Electric tools such as hand electric drills and saws always have their voltage and amperage ratings listed. The mechanic is interested in the amount of power the tool can deliver, which is easily deduced using the formula that converts watts to horsepower.

The formula for the horsepower of an electric motor is

Answer to exercise 6

$$bhp = \frac{2 \times 3.1416 \times 4.5 \times 250 \times 26}{33,000}$$

$$= \frac{183,783.6}{33,000}$$

$$= 5.57 \, bhp$$

$$hp = \frac{amperes \times volts}{750} = \frac{watts}{750}$$

Thus,

$$watts = hp \times 750$$

■ Example 2

Compute the horsepower input of a motor running on a 110-volt line and taking 20 amperes.

9. A refrigerator motor receives a current of 6 amperes from a 110-volt line. Compute the horsepower.

The answer is in the margin on page 505.

Solution $hp = \dfrac{amperes \times volts}{750} = \dfrac{20 \times 110}{750} = 2.93 \, hp$

Practice check: Do exercise 9 on the left.

■ Problems 16–5

Use the space provided to solve the following problems about electric devices.

1. An electric drill requires 110 volts and uses 2.3 amperes of current. What is its horsepower?

2. An electric saw requires 110 volts and uses 10 amperes of current. What is its horsepower?

3. An electric jigsaw requires 110 volts and uses 2.7 amperes of current. What is its horsepower?

4. An electric router requires 110 volts and uses 3.7 amperes of current. What is its horsepower?

5. If two electric drills have the same voltage requirements and one drill uses twice as much current as the other, how would the horsepower of the two compare?

6. A 100-watt bulb requires 110 volts. How many amperes of current are used by the bulb?

7. A motor taking 5 amperes is running on a 220-volt line. How much power does the motor receive?

8. Find the kilowatt capacity of a generator delivering 12 amperes at 220 volts.

9. How many watts are consumed by an electric iron taking 6.4 amperes on a 110-volt line?

10. What must be the horsepower of a motor for a trolley car if the car requires a current of 50 amperes at 400 volts? What is the horsepower input of the motor?

11. A generator delivers to a motor 200 amperes at 400 volts. What is the horsepower input of the motor?

12. Ten percent of the power delivered to a certain motor is lost by friction. What horsepower will this motor deliver if it receives 25 amperes at 110 volts?

Answer to exercise 7

watts = amperes × volts
= 5 × 110
= 550 watts

Answer to exercise 8

$$kW = \frac{20 \times 550}{1,000}$$
$$= 11\,kW$$

16–6   MECHANICAL EFFICIENCY OF MACHINES

Because of friction losses, and for other reasons, no machine gives out all the power it receives. The power put into a machine is called the ***input.*** The power that the machine delivers is called the ***output.*** The ***efficiency*** of a machine is the ratio of the output and is always expressed in percent.

$$\text{Efficiency} = \frac{\text{Output}}{\text{Input}}$$

$$\text{Output} = \text{Efficiency} \times \text{Input}$$

$$\text{Input} = \frac{\text{Output}}{\text{Efficiency}}$$

Thus, if a motor receives 10 hp and gives out only 8 hp, its efficiency is $\frac{8}{10} = 0.80$, or 80%.

The efficiency of a steam engine is the ratio of brake horsepower to indicated horsepower. If the indicated horsepower is 7.5 and the brake horsepower is only 6, the efficiency is 6/7.5 − 0.80, or 80%.

■ **Example 1**

A motor receiving 2,500 watts has an efficiency of 90%. What horsepower will it deliver?

Solution   Input = 2,500 watts = 2.5 kW
          Output = 90% of input = 0.90 × 2.5 = 2.25 kW
          hp = kW ÷ $\frac{3}{4}$ = 2.25 ÷ $\frac{3}{4}$ = 2.25 × $\frac{4}{3}$ = 3 hp

10. What is the brake horse-power of a steam engine that has an indicated horse-power of 10.4 hp and an efficiency of 80%?

The answer is in the margin on page 506.

Explanation   One kilowatt = 1,000 watts; therefore, 2,500 watts = $\frac{2500}{1000}$ = 2.5 kW. Since the efficiency is 90%, the motor gives out 90% or 0.90 of what it gets. Therefore, the output equals 0.90 × 2.5 or 2.25 kW. One horsepower equals $\frac{3}{4}$ kW; therefore, to convert kilowatt to horsepower, we divide kW by $\frac{3}{4}$ or multiply by $\frac{4}{3}$.

Practice check: Do exercise 10 on the left.

■ **Example 2**

What current does a 5-hp 220-volt motor take if its efficiency is 80%?

Solution   Input = $\frac{\text{Output}}{\text{Efficiency}}$ = $\frac{5}{0.80}$ = 6.25 hp

          Watts = hp × 750 = 6.25 × 750 = 4,687.50

          Current = $\frac{\text{watts}}{\text{volts}}$ = $\frac{4,687.5}{220}$ = 21.31 amperes

11. On what voltage must a 20-hp motor be run if it takes a current of 42.6 amperes and is 90% efficient?

The answer is in the margin on page 506.

Explanation   If the motor is a 5-hp motor with an efficiency of 80%, it means that 5 hp is 80% of the input. Hence, to find the input, we divide the output by the efficiency; that is, 5/0.80 = 6.25 hp.
      To convert the horsepower to watts, we multiply by 750. We find the current by dividing watts by volts.

Practice check: Do exercise 11 on the left.

■ **Example 3**

What is the efficiency of a generator that delivers 60 amperes at 110 volts if the input is 10 hp?

Solution   Efficiency = $\frac{\text{Output}}{\text{Input}}$

          Output = $\frac{60 \times 110}{1,000}$ = 6.6 kW

          Input = 10 × $\frac{3}{4}$ = 7.5 kW

          Efficiency = $\frac{6.6}{7.5}$ = 0.88 = 88%

12. What is the efficiency of a 6.6-hp motor if it takes 44 amperes on a 120-volt line?

The answer is in the margin on page 506.

Explanation   Multiplying 60 amperes by 110 volts gives the output in watts. Dividing watts by 1,000 gives the output in kilowatts. To change the horsepower input to kilowatts, we multiply by $\frac{3}{4}$. We then divide the kilowatts output by the kilowatts input to get the efficiency.

Practice check: Do exercise 12 on the left.

NOTE: *Output and input must be in the same units. That is, to find efficiency, we divide kilowatt output by kilowatt input, or horsepower output by horsepower input.*

■ **Problems 16–6**

Using the space provided below, supply the missing information about the following motors.

1. A motor having an efficiency of 75% receives 6 kW. What horsepower will it deliver?

2. A steam engine whose indicated horsepower is 8.5 hp has an efficiency of 70%. What will be the brake horsepower of the engine?

3. A 110-volt motor taking 20 amperes has an efficiency of 85%. What horsepower will it deliver?

4. An engine supplies 120 hp to a generator delivering 125 amperes at 550 volts. What is the efficiency of the generator?

5. Find the efficiency of a steam engine whose indicated horsepower is 6.2 hp and whose brake horsepower is 4.8 hp.

6. What is the efficiency of a 4.5-hp motor if it takes 32 amperes on a 110-volt line?

7. On what voltage must a 10-hp motor be run if it takes a current of 37.5 amperes and is 86% efficient?

8. What current does a 10-hp, 220-volt motor take if it is 88% efficient?

9. A 12-hp, 110-volt motor has an efficiency of 90%. What current does it require?

10. A generator having an efficiency of 80% receives 60 hp from the driving engine. What current will it deliver at 110 volts?

11. What is the efficiency of a 12-hp motor if it requires 90 amperes at 120 volts?

indicated hp = 10.4 hp

bhp = 80% of indicated hp
  = 0.80 × 10.4
  = 8.32 hp

Answer to exercise 11

$$\text{Input} = \frac{\text{Output}}{\text{Efficiency}}$$
$$= \frac{20}{0.90}$$
$$= 22.2 \text{ hp}$$

watts = hp × 750
  = 22.2 × 750
  = 16,650

$$\text{volts} = \frac{\text{watts}}{\text{amperes}}$$
$$= \frac{16,650}{42.6}$$
$$= 390.8 \text{ volts}$$

Answer to exercise 12

Output = 6.6 hp

$$\text{Input} = \frac{\text{amperes} \times \text{volts}}{750}$$
$$= \frac{44 \times 120}{750}$$
$$= 7.04 \text{ hp}$$

$$\text{Efficiency} = \frac{\text{Output}}{\text{Input}}$$
$$= \frac{6.6}{7.04}$$
$$= 93\tfrac{3}{4}\%$$

■ Self-Test

Use the space provided to solve the following problems involving work and power. Refer to the formulas in the text. Check your answers with the answers in the back of the book.

1. A machine does 64,000 ft-lb of work in 4 minutes. Compute its horsepower.

2. A roofer carries an 80-lb bundle of shingles 12 ft up a ladder. How many foot-pounds of work does he do?

3. A 20″ × 36″ double-acting steam engine is running at 100 rpm under a mean effective pressure of 40 lb. Find the horsepower.

4. Find the brake horsepower of an engine determined by a Prony brake where the lever length is 4 ft, the engine is running at 4,000 rpm, and the net force on the scale is 60 lb.

5. A generator delivers 10 amperes at 120 volts. What power in kilowatts does it supply?

6. A sander draws 3.4 amperes of current at 120 volts. What is its horsepower?

7. How many amperes of current will twelve 100-watt light bulbs draw on a 220-volt line?

8. A 10-hp motor delivers 8.7 hp. What is its efficiency?

9. A motor delivers $\frac{3}{4}$ horsepower and draws 5.9 amperes on a 110-volt line. What is its input horsepower?

10. A shaper requires 110 volts and uses 7.5 amperes. What is its horsepower?

■ Chapter Test

Solve the following problems involving work and power. Refer to the formulas in the text. Organize your work and label all parts of the problem.

1. A man moves 1 ton of coal 12 ft. How much work does he do in foot-pounds?

2. An engine can do 48,000 ft-lb of work in $2\frac{1}{2}$ minutes. Compute its horsepower.

3. What must be the horsepower of a machine to lift a 2-ton girder to a height of 30 ft in 2 minutes?

4. What is the indicated horsepower of a $16'' \times 28''$ double-acting steam engine making 100 rpm with a mean effective pressure of 54 lb?

5. Find the rated horsepower of a 6-cylinder automobile engine if the cylinders have a $3\frac{1}{4}''$ bore and a $3\frac{1}{2}''$ stroke.

6. If the bore in Problem 5 were increased to $3\frac{5}{16}''$, how much additional horsepower would be developed?

7. Calculate the brake horsepower of a gasoline engine running at 1,000 rpm if the net scale reading is 80 lb and the lever arm is 4 ft long.

8. An electric drill requires 110 volts and uses 2.7 amperes of current. What is its horsepower?

9. An electric saw requires 110 volts and is rated at 1 hp. How much current does it draw?

10. An electric jigsaw requires 110 volts and uses 1.8 amperes of current. What is its horsepower?

11. A 150-watt electric bulb requires 110 volts. How many amperes of current are used by the bulb?

12. A motor having an efficiency of 75% receives 5 kW. What horsepower will it deliver?

13. A 110-volt motor taking 5.5 amperes of current has an efficiency of 90%. What horsepower will it deliver?

14. A motor has an input of 5 hp and delivers 4.7 hp. What is its efficiency?

15. What is the efficiency of a 1-hp motor if it takes 9.0 amperes on a 110-volt line?

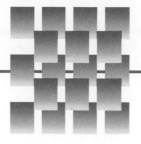

CHAPTER 17

Tapers

17–1 DEFINITIONS

A piece of work that decreases gradually in diameter so that it assumes a conical shape is said to be *tapered.* Figure 17–1 shows a piece of tapered work.

When we speak of the **taper** on a piece of work, we mean the *difference between the diameters* at the ends of the piece. If, as in Fig. 17–1, the large diameter is $1\frac{1}{2}$ in. and the small diameter 1 in., the taper, or difference between the diameters, is $\frac{1}{2}$ in. Therefore, when we say that a piece of work has a taper of $\frac{1}{2}$ in., we mean that the diameter at the large end is $\frac{1}{2}$ in. greater than the diameter at the small end.

Taper is usually designated as a *fraction of an inch per foot of length;* the length is measured along the axis or center line of the piece of work. When we say, for example, that a piece of work has a taper of $\frac{5}{8}$ in. per foot, we mean that there is a difference of $\frac{5}{8}$ in. between the diameters 1 ft apart.

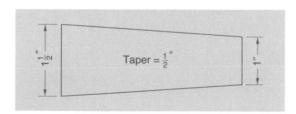

Figure 17–1

Taper may also be specified as a fraction of an inch per inch of length of the work. Thus, we may say that a piece of work has a taper of $\frac{1}{16}$ in. per inch. This means that the diameters 1 in. apart differ by $\frac{1}{16}$ in.

17–2 COMPUTING TAPER AND DIAMETER

In Fig. 17–2, the taper in 2 in. is twice as much as the taper in 1 in.; the taper in 3 in. is three times as much as the taper in 1 in.; and so on. Hence, if we know the taper in a 1-in. length of a piece of work, we can find the taper in any other

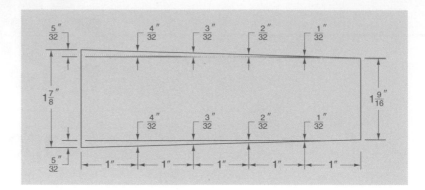

Figure 17–2

length by multiplying the taper per inch by the number of inches in the given length.

1. The taper per inch on a piece of work is 0.25 in. What is the taper in 10 in.?

The answer is in the margin on page 513.

■ **Example 1**

The taper per inch on a piece of work is 0.05 in. What is the taper in 8 in.?

Solution   Taper = 8 × 0.05 = 0.4 in.

Practice check: Do exercise 1 on the left.

If, on the other hand, we are told that the taper in a foot length of a piece of work is $\frac{1}{2}$ in., and we are required to find the taper in a 1-in. length, we divide by 12 because in 1 in. there is $\frac{1}{12}$ as much taper as there is in 12 in.

■ **Example 2**

Find the taper per inch if the taper per foot is $\frac{5}{8}$ in.

2. Find the taper per inch if the taper per foot is $\frac{9}{16}$ in.

The answer is in the margin on page 513.

Solution $\frac{5}{8} \div 12 = \frac{5}{8} \times \frac{1}{12} = \frac{5}{96} = 0.0521$ in.
Taper per inch = 0.0521 in.

Practice check: Do exercise 2 on the left.

■ **Example 3**

Find the taper per foot on the piece of work shown in Fig. 17–3.

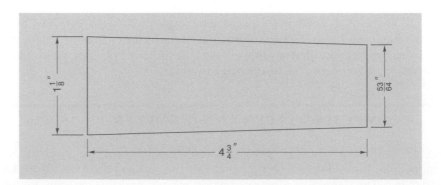

Figure 17–3

Solution   Taper in piece $= 1\frac{1}{8}$ in. $- \frac{53}{64}$ in. $= \frac{19}{64}$ in.

Taper per inch $= \frac{19}{64} \div 4\frac{3}{4}$

$\qquad\qquad\quad = \frac{19}{64} \times \frac{4}{19} = \frac{1}{16}$ in.

Taper per foot $= 12 \times \frac{1}{16} = \frac{12}{16} = \frac{3}{4}$ in.

Explanation   We first find the taper by subtracting the small diameter from the large diameter. We then find the taper per inch by dividing the taper by the length of the piece. Finally, we multiply the taper per inch by 12, giving the taper per foot, $\frac{3}{4}$ in.

The solution may be abbreviated by combining the steps. Thus,

$$\text{Taper} = 1\frac{1}{8} - \frac{53}{64} = \frac{19}{64} \text{ in.}$$

$$\text{Taper per foot} = \frac{19}{64} \div 4\frac{3}{4} \times 12$$

$$= \frac{19}{64} \times \frac{4}{19} \times 12 = \frac{3}{4} \text{ in.}$$

Practice check: Do exercise 3 on the right.

3. Find the taper per foot on the following piece of work.

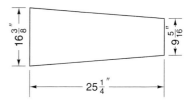

The answer is in the margin on page 513.

■ Example 4

The piece of work shown in Fig. 17–4 has a taper of 0.6 in. per foot. Find the small diameter.

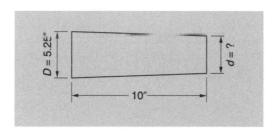

Figure 17–4

Solution   Taper per inch $= \frac{1}{12} \times 0.6 = 0.05$ in.

Taper in 10 in. $= 10 \times 0.05 = 0.50$ in.

Small diameter $=$ Large diameter $-$ Taper

$\qquad\qquad\qquad\quad = 5.25 - 0.50 = 4.75$ in.

Explanation   To find the small diameter, we must subtract the taper in the piece of work from the large diameter. Dividing the taper per foot by 12 gives 0.05 in., the taper per inch. Multiplying this by 10 gives 0.50 in., the taper. Subtracting the taper from the large diameter gives the small diameter, 4.75 in.

Practice check: Do exercise 4 on the right.

4. The following piece of work has a taper of 0.15 in. per foot. Find the small diameter.

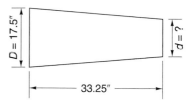

The answer is in the margin on page 514.

■ Example 5

Find the large diameter of a piece of work $7\frac{3}{4}''$ long, having a small diameter of 0.658 in. and a taper of $\frac{1}{16}$ in. per inch, as shown in Fig. 17–5.

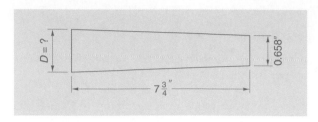

Figure 17–5

5. Find the large diameter of the piece of work shown in the following figure. It is 15″ long and has a small diameter of 3.75″ and a taper of 0.4″ per inch.

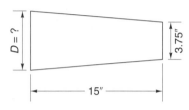

The answer is in the margin on page 515.

Solution

$$\text{Taper} = 7\tfrac{3}{4} \times \tfrac{1}{16} = \tfrac{31}{4} \times \tfrac{1}{16}$$
$$= \tfrac{31}{64} = 0.485 \text{ in.}$$

$$\text{Large diameter} = \text{Small diameter} + \text{Taper}$$
$$= 0.658 + 0.485 = 1.143 \text{ in.}$$

Explanation   The large diameter is equal to the small diameter plus the taper. The taper in the piece that is $7\tfrac{3}{4}$ in. long is $7\tfrac{3}{4}$ times the taper per inch, or $7\tfrac{3}{4} \times \tfrac{1}{16} = 0.485$ in. The large diameter is therefore equal to $0.658 + 0.485 = 1.143$ in.

Practice check: Do exercise 5 on the left.

17–3   AMERICAN STANDARD SELF-HOLDING (SLOW) TAPER SERIES

The taper series most commonly used are the following.

1. Brown and Sharpe
2. Morse
3. $\tfrac{3}{4}$-in.-per-ft series

The Brown and Sharpe taper is $\tfrac{1}{2}$ in. per ft.
The Morse taper is nominally $\tfrac{5}{8}$ in. per ft. The exact taper for each size is as follows.

No. 1: 0.600 in. per ft
No. 2: 0.600 in. per ft
No. 3: 0.602 in. per ft
No. 4: 0.623 in. per ft
No. $4\tfrac{1}{2}$: 0.623 in. per ft
No. 5: 0.630 in. per ft

In the $\tfrac{3}{4}$-in.-per-ft series, the taper in all cases is $\tfrac{3}{4}$ in. per ft.
A less commonly used taper is the *steep taper.* The most common steep taper has a taper of $3\tfrac{1}{2}$ in. per ft.

■ Problems 17–3

Use the space provided below to find the following tapers.

1. What is the taper in the piece of work in Fig. 17–6? What is the taper per inch? What is the taper per foot?

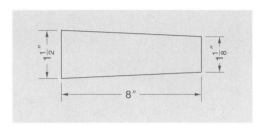

Figure 17–6

2. What is the taper per inch in the Brown and Sharpe? The Morse No. 1? The $\frac{3}{4}$-in.-per-foot series?

3. Compute the small diameter of the lathe center in Fig. 17–7. The taper is Morse No. 2.

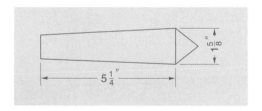

Figure 17–7

4. What is the diameter 4 in. from the small end of the lathe center in Problem 3?

514   Chapter 17 / Tapers

5. Compute the taper per foot on the taper socket in Fig. 17–8.

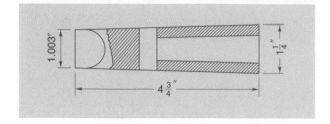

Figure 17–8

6. Compute the taper per foot on the bushing in Fig. 17–9.

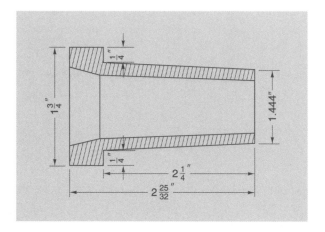

Figure 17–9

7. Compute the taper per foot on the reamer in Fig. 17–10.

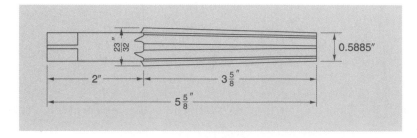

Figure 17–10

8. Compute the taper per foot on the collet in Fig. 17–11.

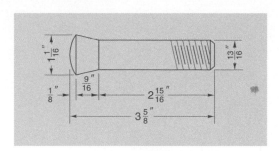

Figure 17–11

9. Compute the taper per foot on the milling machine arbor in Fig. 17–12.

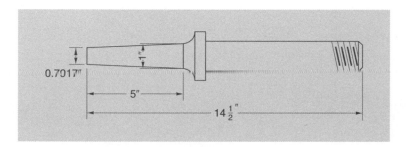

Figure 17–12

10. Compute the taper per foot on the drill socket in Fig. 17–13.

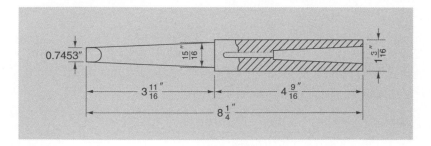

Figure 17–13

11. Compute the taper per foot on the drill shank in Fig. 17–14.

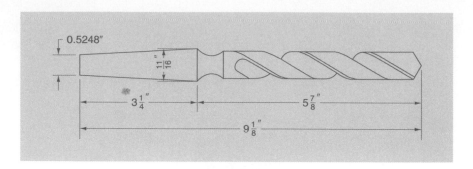

Figure 17–14

12. Compute the taper per foot on the mandrel in Fig. 17–15.

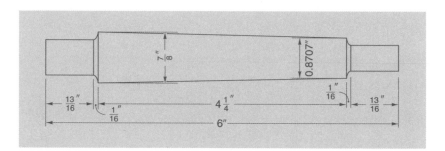

Figure 17–15

13. Compute the taper per foot at both ends of the piston rod in Fig. 17–16.

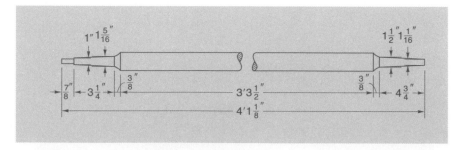

Figure 17–16

14. The inside taper of the taper socket in Fig. 17–17 is Morse No. 5. Compute the small diameter.

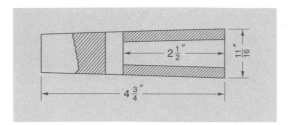

Figure 17–17

15. Compute the large diameter of the inside taper (Brown and Sharpe) of the taper bushing in Fig. 17–18.

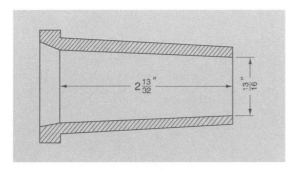

Figure 17–18

16. Compute the small end diameter of the inside taper (Brown and Sharpe) of the drill socket in Fig. 17–19.

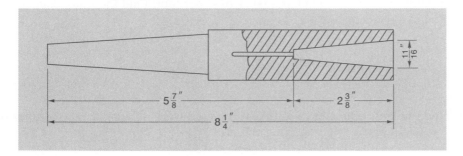

Figure 17–19

17. Find the diameter at the large end of the taper pin reamer in Fig. 17–20. The taper is $\frac{3}{8}''$ per foot.

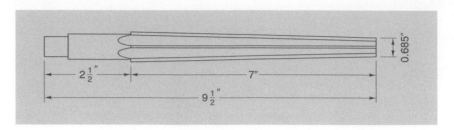

Figure 17–20

18. The crank pin in Fig. 17–21 is to be cut with Morse No. 4 taper. Find the diameter at the small end of the tapered position.

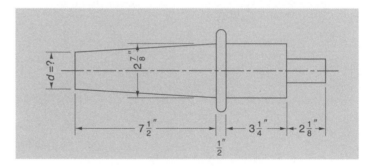

Figure 17–21

19. A piece of work $6\frac{1}{2}''$ long has a taper of $\frac{3}{4}''$ per foot. Find the small diameter if the large diameter is $\frac{1}{2}''$.

20. Find the large diameter of a piece of tapered work $4\frac{1}{8}''$ long, having a taper of $\frac{5}{8}''$ per foot and a small diameter of 0.572''.

21. Find the small diameter of the shaft in Fig. 17–22. The taper is Brown and Sharpe.

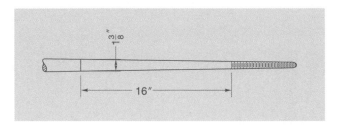

Figure 17–22

17–4   TAPER ANGLE

The taper on a piece of work is sometimes specified by *the angle included between the sides* or the prolongation of the sides of the piece of work. Thus, in Fig. 17–23, the sides *AB* and *CB* meet at *B*, forming the angle *ABC*, which is the *included angle* or the *taper angle.*

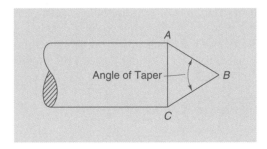

Figure 17–23

In Fig. 17–24, the sides *BA* and *EF* of the piece of tapered work, when prolonged to *H*, form the angle of taper *BHE*.

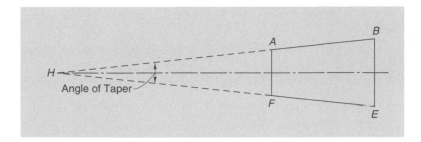

Figure 17–24

■ Example 1

Compute the taper angle in a piece of work 10 in. long, whose small diameter is 2 in. and large diameter $3\frac{1}{2}$ in., as shown in Fig. 17–25.

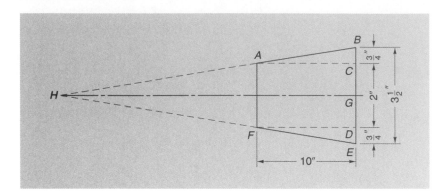

Figure 17–25

6. Compute the taper angle in the piece of work shown in the following figure. It is 15 in. long, and its small diameter is 3 in. and large diameter $4\frac{1}{4}$ in.

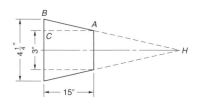

The answer is in the margin on page 522.

Solution   Draw AC parallel to the center line of the piece of work, forming the right triangle ABC. Angle $BAC = BHG$ = half of the taper angle. From trigonometry we know that the tangent of an angle equals $\dfrac{\text{side opposite}}{\text{side adjacent}}$. Therefore,

$$\text{tangent } BAC = \frac{BC}{AC} = \frac{\text{half of taper}}{\text{length}} = \frac{0.75}{10} = 0.0750$$

By referring to a table of natural functions, we find that this tangent corresponds to an angle of 4°17′. Since angle DFE = angle BAC, the included angle at H, which is twice BAC, equals $2 \times 4°17' = 8°34'$.

Practice check: Do exercise 6 on the left.

■ Example 2

Find the angle of taper in a piece of work having a taper of $\frac{5}{8}$ in. per foot, as shown in Fig. 17–26.

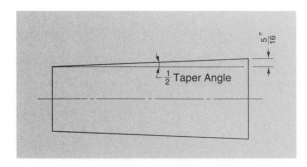

Figure 17–26

Solution   Tangent of $\frac{1}{2}$ taper angle $= \dfrac{\frac{1}{2} \text{ of } \frac{5}{8}}{12}$

$$= \tfrac{1}{2} \times \tfrac{5}{8} \times \tfrac{1}{12} = \tfrac{5}{192} = 0.02604$$

$$\tfrac{1}{2} \text{ taper angle} = 1°29\tfrac{1}{2}'$$

$$\text{Taper angle} = 2 \times 1°29\tfrac{1}{2}' = 2°59'$$

Practice check: Do exercise 7 on the right.

■ **Example 3**

Find the taper per foot if the taper angle is 12°40′, as shown in Fig. 17–27.

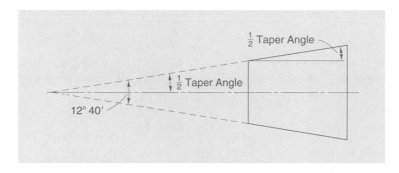

Figure 17–27

Solution $\frac{1}{2}$ of the taper angle $= \dfrac{12°40'}{2} = 6°20'$

Tangent 6°20′ = 0.1110

Taper per inch = 2 × 0.1110 = 0.2220

Taper per foot = 12 × 0.2220 = 2.6640 in.

Explanation   From a table of natural functions, obtain the tangent of half the taper angle. This gives half the taper in 1 in. of length. To get the taper per inch, multiply this result by 2. The taper per foot is 12 times the taper per inch.

Practice check: Do exercise 8 on the right.

■ **Problems 17–4**

Use the space provided to find the following taper angles.

1. Compute the taper angle for the following tapers: Brown and Sharpe, Morse No. 5, $\frac{3}{4}$-in.-per-foot series.

2. What is the included angle of the bevel of the piece shown in Fig. 17–9? The length of the bevel is $\frac{3}{8}''$, the large diameter $1\frac{1}{8}''$, and the small diameter 0.9253″.

7. Find the angle of taper in the following piece of work having a taper of $\frac{3}{8}$ in. per foot.

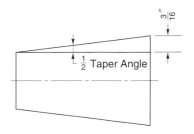

The answer is in the margin on page 523.

8. For the following figure, find the taper per foot if the angle is 8°50′.

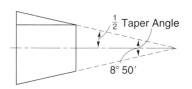

The answer is in the margin on page 523.

3. Compute the taper angle in the tapered portion of the collet in Fig. 17–11.

4. What is the angle of taper of the left-hand end of the piston rod in Fig. 17–16?

5. Find the taper per foot if the taper angle is 5°.

6. Find the taper per inch if the taper angle is $3\frac{1}{2}°$.

7. The taper angle on a piece of tapered work is 7°. What is the taper per foot?

8. What standard taper has a taper angle of 2°59′?

9. The angle with the center line in a piece of tapered work is 2°23′. What is the taper per foot?

10. Compute the taper angle in the bevel gear blank shown in Fig. 17–28.

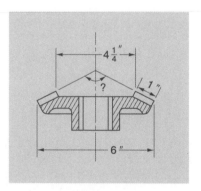

Figure 17–28

11. Find the included or taper angle in the drill chuck in Fig. 17–29.

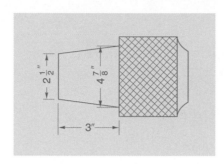

Figure 17–29

12. Compute the included angle in the point of the countersink in Fig. 17–30.

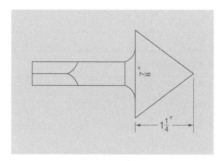

Figure 17–30

17–5 TAPER TURNING BY OFFSETTING THE TAIL STOCK

There are three methods of turning a taper in a lathe. The first method, by offsetting the tail stock, can be used only for outside tapers and only when the taper to be turned is not too great. The effect of offsetting the tail stock will be seen from a study of Fig. 17–31.

For cylindrical turning, the tail stock center T_1 is on the center line of the lathe, that is, directly in line with the head stock center H. When the tail stock is offset for cutting a taper, the center assumes the position T_2, and the piece of work assumes the position $DEBC$. The tool, however, travels in a straight line parallel to the center line of the machine from C to A. The radius of the work is therefore reduced at the tail end by the amount AB, which is equal to the tail stock offset from T_1 to T_2. But, since the work revolves, an equal amount is taken off all around. In other words, the diameter at the tail end is less than the diameter at the head

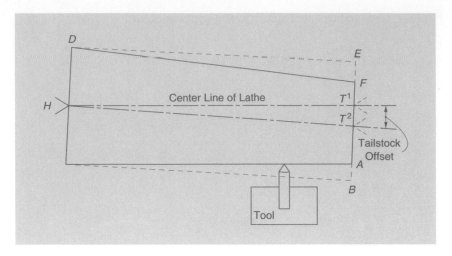

Figure 17–31

end by an amount equal to twice the offset of the tail stock. The tail stock offset is therefore equal to half the total taper.

Rule

The amount of set-over of the tail stock is equal to one-half the total length of the work in inches multiplied by the taper per inch.

■ **Example 1**

Calculate the tail stock offset for turning the taper on the piece in Fig. 17–32.

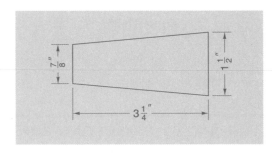

Figure 17–32

Solution   Taper $= 1\frac{1}{2} - \frac{7}{8} = \frac{5}{8}$ in.

Offset $= \frac{1}{2}$ of taper $= \frac{1}{2} \times \frac{5}{8} = \frac{5}{16}$ in.

Practice check: Do exercise 9 on the left.

9. Calculate the tail stock offset turning the taper on the piece in the following figure.

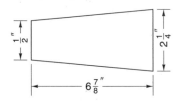

The answer is in the margin on page 526.

■ **Example 2**

Calculate the tail stock offset for turning a taper of $\frac{3}{4}$ in. per foot for a distance of 4 in. on a piece of work 10 in. long, as shown in Fig. 17–33.

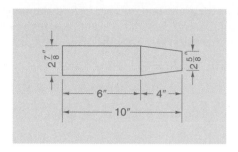

Figure 17–33

Solution
$$\text{Taper per inch} = \tfrac{1}{12} \times \tfrac{3}{4} = \tfrac{1}{16} \text{ in.}$$

$$\text{Taper in whole length of piece of work} = 10 \times \tfrac{1}{16} = \tfrac{5}{8} \text{ in.}$$

$$\text{Tail stock offset} = \tfrac{1}{2} \text{ of } \tfrac{5}{8} = \tfrac{5}{16} \text{ in.}$$

Notice that even though the tapered part is only 4 in. long, the taper for the whole length of the piece is found and the offset is computed as one-half of this overall taper.

Practice check: Do exercise 10 on the right.

■ **Example 3**

Compute the tail stock offset for turning tapers at both ends of the piston rod in Fig. 17–34.

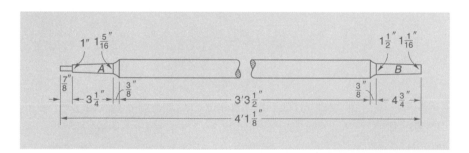

Figure 17–34

Solution   To turn the left end taper at *A*, we must compute the taper for the whole length of the rod as if it were tapered the same amount per inch as the part at *A*. The tail stock will then be offset half of the total taper so computed.

$$\text{Taper in part } A = 1\tfrac{5}{16} - 1 = \tfrac{5}{16} \text{ in.}$$

$$\text{Taper per inch} = \tfrac{5}{16} \div 3\tfrac{1}{4} = \tfrac{5}{16} \times \tfrac{4}{13} = \tfrac{5}{52} \text{ in.}$$

$$\text{Total length} = 4 \text{ ft } 1\tfrac{1}{8} \text{ in.} = 49\tfrac{1}{8} \text{ in.}$$

$$\text{Taper in whole length of rod} = 49\tfrac{1}{8} \times \tfrac{5}{52} = 4.724 \text{ in.}$$

$$\text{Tail stock offset} = \tfrac{1}{2} \text{ of } 4.724 = 2.362 \text{ in. or } 2\tfrac{23}{64} \text{ in.}$$

10. Calculate the tail stock off-set for turning a taper of $\tfrac{3}{8}$ in. per foot for a distance of 5 in. on a piece of work 20 in. long (see the following figure).

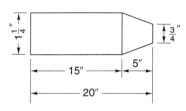

The answer is in the margin on page 527.

To turn the taper at the right end, we again find the taper in the entire length of the rod as if it were tapered the same amount per inch as the right end B. The tail stock offset will be one-half of the total taper.

$$\text{Taper in part } B = 1\frac{1}{2} - 1\frac{1}{16} = \frac{7}{16} \text{ in.}$$

$$\text{Taper per inch} = \frac{7}{16} \div 4\frac{3}{4} = \frac{7}{16} \times \frac{4}{19} = \frac{7}{76} \text{ in.}$$

$$\text{Overall taper} = 49\frac{1}{8} \times \frac{7}{76} = 4.525 \text{ in.}$$

$$\text{Tail stock offset} = \frac{1}{2} \text{ of } 4.525 = 2.263 \text{ in. or } 2\frac{17}{64} \text{ in.}$$

■ Problems 17–5

Use the space provided below to compute the following tail stock offsets.

1. Compute the tail stock offset for turning the taper on the piece of work in Fig. 17–35.

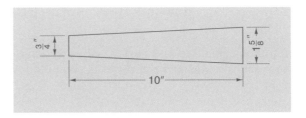

Figure 17–35

2–9. Compute the tail stock offsets for cutting the tapers in Problems 5 through 12 from Problem Set 17–3.

10. Compute the tail stock offsets for turning the tapers at the ends of the piston rod in Fig. 17–36.

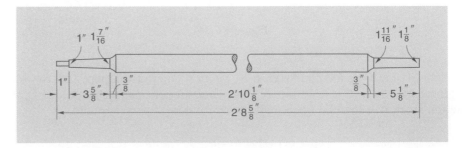

Figure 17–36

17–6   TAPER TURNING BY USING THE COMPOUND REST

This method is used where short tapers or tapers having a considerable angle are to be turned. To set the compound rest for cutting a taper, we must know the included angle, that is, the taper angle (see Fig. 17–37).

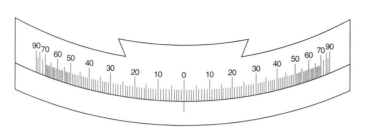

Figure 17–37

■ **Example 1**

Compute the setting of the compound rest for cutting the taper on the piece in Fig. 17–38.

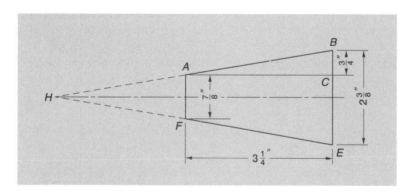

Figure 17–38

Solution   Tangent BAC = tangent of $\frac{1}{2}$ included angle $= \dfrac{BC}{AC} = \dfrac{\frac{3}{4}}{3\frac{1}{4}}$

$$= \tfrac{3}{4} \times \tfrac{4}{13} = \tfrac{3}{13} = 0.23077$$

Angle $BAC = \frac{1}{2}$ taper angle $= 13°0'$

Included angle $= 2 \times 13°0' = 26°0'$

To turn this taper, the compound rest is not set to read the included angle, 26°, but to *90° minus half the included angle;* 90° $- \frac{1}{2}$ of 26° = 90° $- 13° - 77°0'$. See Fig. 17–39.

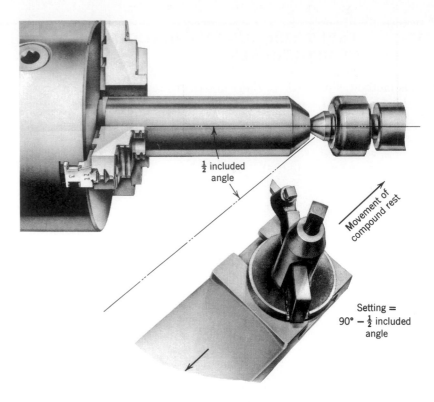

Figure 17–39

11. Compute the setting of the compound rest for cutting the taper on the piece in the following figure.

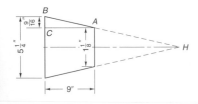

The answer is in the margin on page 530.

The ***complement*** of an angle is equal to 90° minus the given angle. For example, the complement of 30° is 90° − 30° = 60°; the complement of 6°50′ is 90° − 6°50′ = 83°10′.

We can therefore state the rule for setting the compound rest as follows.

Rule

Set the compound rest to the complement of one-half the included angle.

Practice check: Do exercise 11 on the left.

■ **Example 2**

To what angle must the compound rest be set to turn the lathe center, which has an included angle of 60°? See Fig. 17–40.

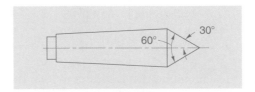

Figure 17–40

Solution   Included angle = 60°

$\frac{1}{2}$ included angle = $\frac{1}{2}$ of 60° = 30°

Setting of compound rest = 90° − 30° = 60°

Practice check: Do exercise 12 on the right.

12. In the following figure, to what angle must the compound rest be set to turn the lathe center, which has an included angle of 50°?

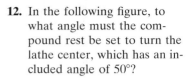

■ **Problems 17–6**

Using the space provided below, answer the following questions involving compound rest settings.

The answer is in the margin on page 531.

1. To what angle must the compound rest be set to cut the tapers and bevels shown in Fig. 17–41?

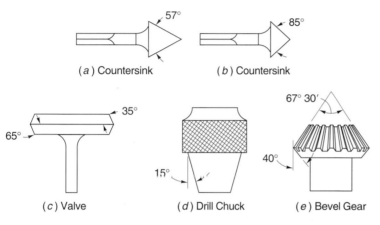

Figure 17–41

2. Compute the angle for cutting the following tapers with the compound rest: Brown and Sharpe, Morse No. 5, $\frac{3}{4}$-in.-per-foot series.

3. Find the compound rest setting for cutting the inside bevel on the taper bushing in Fig. 17–42.

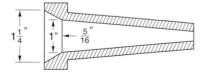

Figure 17–42

4. Figure the compound rest setting for cutting the taper on the collet in Fig. 17–43.

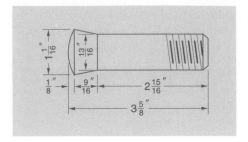

Figure 17–43

5. Compute the compound rest setting for cutting the tapers at both ends of the piston rod in Fig. 17–44.

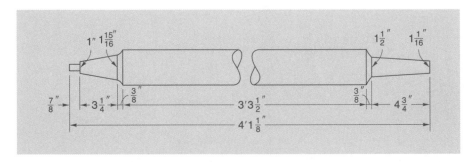

Figure 17–44

6. How would you set the compound rest for cutting the following included angles: 5°; 3½°; 7°30′; 8°20′; 16°50′?

[]

17–7   TAPER TURNING BY USING THE TAPER ATTACHMENT

To turn a taper by means of the taper attachment, we compute the taper per foot, disregarding the length of the tapered portion and the overall length of the work. In one type of taper attachment, the swivel bar that controls the taper is graduated at one end in inches per foot of taper and at the other end in degrees. The attachment can be set for any taper up to 3 in. per foot, which corresponds to a taper angle of 14°15′ (see Fig. 17–45). In another type of attachment, the graduations at one end give the taper in eighths of an inch per foot, and the other end gives it in tenths of an inch per foot.

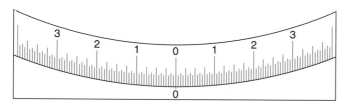

Figure 17–45

■ **Example**

Compute the setting of the taper attachment for turning the taper shown in Fig. 17–46.

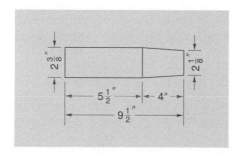

Figure 17–46

Solution   Taper in 4 in. $= 2\frac{3}{8} - 2\frac{1}{8} = \frac{1}{4}$ in.

Taper per inch $= \frac{1}{4} \div 4 = \frac{1}{4} \times \frac{1}{4} = \frac{1}{16}$ in.

Taper per foot $= 12 \times \frac{1}{16} = \frac{3}{4}$ in. or $\frac{6}{8}$ in.

Adjust the taper attachment so that it reads the required taper per foot, namely, $\frac{6}{8}$ in. per foot.

Practice check: Do exercise 13 on the right.

Answer to exercise 12

Included angle $= 50°$

$\frac{1}{2}$ included angle $= 25°$

Setting $= 90° - 25°$
       $= 65°$

13. Compute the setting of the taper attachment for turning the taper shown in the following figure.

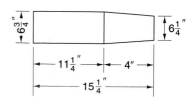

The answer is in the margin on page 533.

■ **Problems 17–7**

Use the space provided below to find the setting of the taper attachment for cutting the following tapers.

1. How would you set the taper attachment for turning a piece of work having a Brown and Sharpe taper? A Morse No. 1? A $\frac{3}{4}$-in.-per foot?

2. How should the taper attachment be set to turn tapers having the following included angles: 4°30′; 3°00′; 5°30′; 7°00′; 9°40′?

3. Compute the setting of the taper attachment for turning the taper on the taper socket in Fig. 17–8.

4. Compute the setting of the taper attachment for turning the outside taper on the bushing in Fig. 17–9.

5. Compute the setting of the taper attachment for turning the taper on the reamer in Fig. 17–10.

6. Compute the setting of the taper attachment for turning the taper on the collet in Fig. 17–11.

7. Compute the setting of the taper attachment for turning the taper on the milling machine arbor in Fig. 17–12.

8. Compute the setting of the taper attachment for turning the outside taper on the drill socket in Fig. 17–13.

9. Compute the setting of the taper attachment for turning the taper on the drill shank in Fig. 17–14.

10. Compute the setting of the taper attachment for turning the taper on the mandrel in Fig. 17–15.

11. Compute the setting of the taper attachment for cutting the tapers at both ends of the piston rod shown in Fig. 17–16.

■ Self-Test

Using the space provided, solve the following problems involving tapers. Check your answers with the answers in the back of the book.

Answer to exercise 13

Taper in 4 in. = $6\frac{3}{4} - 6\frac{1}{4}$
= $\frac{1}{2}$ in.

Taper per inch = $\frac{1}{2} \div 4$
= $\frac{1}{8}$ in.

Taper per foot = $12 \times \frac{1}{8}$
= $\frac{3}{2}$ in. or $1\frac{1}{2}$ in.

1. The taper per inch on a piece of work is 0.02 in. What is its taper in 12 in.?

2. The taper per foot on a piece of work is 0.6 in. What is its taper in 8 in.?

3. What is the taper per inch if the taper per foot is $\frac{3}{4}$ in.?

4. What is the taper per foot if the taper per inch is $\frac{1}{32}$ in.?

5. Find the taper on a piece of work 9 in. long if the diameter of the large end is 2 in. and the diameter of the small end is $1\frac{9}{4}$ in.

6. What is the taper per inch of the piece of work in Problem 5?

7. What is the taper per foot of the piece of work in Problem 5?

8. A tapered piece of work is 14 in. long. What is its taper per inch if the diameter of the large end is 3.562 in. and the diameter of the small end is 3.182 in.?

9. What is the taper angle on a piece of work having a taper of 0.75 in. per foot?

10. What is the taper angle on a piece of work having a taper of 0.0156 in. per inch if the piece is 2 ft long?

■ **Chapter Test**

Use the space provided to solve the following problems involving tapers.

1. The taper per inch on a piece of work is 0.06″. What is the taper in 10″?

2. What is the taper per inch if the taper per foot is $\frac{7}{8}$″?

3. Find the taper per foot on a piece of work if the diameter of the large end is $1\frac{1}{4}$″, the diameter of the small end is $1\frac{7}{64}$″, and the piece is $4\frac{1}{2}$″ long.

4. A piece of work has a taper of 0.55″ per foot. What is its diameter 6″ away from the large end that has a diameter of 2.37″?

5. Find the large diameter of a piece of work $8\frac{3}{4}$″ long, having a small diameter of 0.625″ and a taper of $\frac{1}{16}$″ per inch.

6. The Brown and Sharpe taper is $\frac{1}{2}$″ per foot. What is the taper per inch?

7. Compute the taper angle on a piece of work 20″ long whose small diameter is 2″ and large diameter is $3\frac{1}{4}$″.

8. Find the taper angle in a piece of work having a taper of $\frac{1}{2}$″ per foot.

9. The compound rest is set to the _____ of one-half the included angle.

10. The amount of set-over of the tail stock is equal to _____ (the total length of the work in inches multiplied by the taper per inch).

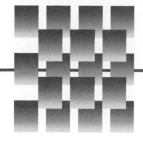

CHAPTER 18

Speed Ratios of Pulleys and Gears

18–1   GEAR TRAINS

In trains of gears, the gears that transmit power are called ***driving gears,*** whereas the others are *followers* or ***driven gears.***

Two meshing gears, such as *A* and *B* in Fig. 18–1, with the same number of teeth, will revolve at the same rate of speed. Assume *A* to be the driving gear and *B* the driven gear. If *A* revolves *clockwise*, that is, in the direction of the hands of a clock, as shown by the arrow, it will cause *B* to revolve in the opposite direction, or counterclockwise.

Let each gear have 20 teeth. When *A* revolves, every tooth on it passes the point *C* and engages a tooth on *B*. When *A* has made one complete turn, that is, when its 20 teeth have passed the point *C*, they will have engaged 20 teeth on gear *B*, thus causing *B* also to make one complete turn.

If, however, *A* has 20 teeth and *B* 40 teeth (Fig. 18–2), in one complete turn of *A* its 20 teeth will pass *C* and will engage 20 teeth on gear *B*. But since gear *B* has 40 teeth, gear *A* will have to make two complete revolutions in order to cause *B* to make one complete turn. In this instance, the gears are said to have a speed ratio of 1 to 2; that is, one revolution of *B* corresponds to two revolutions of *A*. If *A* had 20 teeth and *B* 60 teeth, *A* would make three revolutions while *B* made one. The gears would then have a speed ratio of 1 to 3. In each case the *smaller* gear is the *faster* and the *larger* gear the slower.

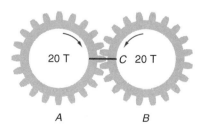

Figure 18–1

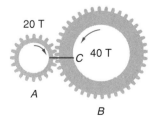

Figure 18–2

The relation existing between the speeds of gears and their numbers of teeth is called an ***inverse ratio;*** that is, the speeds of gears vary not in the same order as the number of their teeth but in the inverse order. In other words, the speed of a gear diminishes rather than increases with an increase in its number of teeth.

The rule governing the speeds of gears can be expressed as follows.

$$\frac{\text{Teeth on driven}}{\text{Teeth on driver}} = \frac{\text{Revolutions of driver}}{\text{Revolutions of driven}}$$

Rule _____

To find the speed of either gear, multiply the size and speed of the other gear and divide by the size of the gear whose speed is sought.

_____

■ **Example 1**

Two meshing gears have 40 and 60 teeth. If the smaller gear makes 120 rpm, how many revolutions per minute will the larger gear make?

Solution　　$\text{rpm of large gear} = \dfrac{\text{size} \times \text{rpm of small gear}}{\text{size of large gear}}$

$$= \frac{40 \times 120}{60} = 80 \text{ rpm}$$

Practice check: Do exercise 1 on the left.

1. Two meshing gears have 50 and 70 teeth. If the smaller gear makes 140 rpm, how many revolutions per minute will the larger gear make?

The answer is in the margin on page 538.

■ **Example 2**

In Fig. 18–3, *A* has 24 teeth and *B* 36 teeth. If *B* makes 100 rpm, how many revolutions per minute will *A* make?

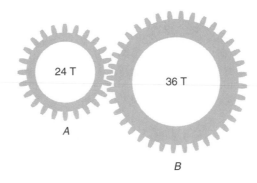

24 T

36 T

A

B

Figure 18–3

2. In Fig. 18–3, suppose *A* has 35 teeth and *B* 60 teeth. If *B* makes 140 rpm, how many revolutions per minute will *A* make?

The answer is in the margin on page 538.

Solution　　$\text{rpm of } A = \dfrac{\text{teeth} \times \text{rpm of } B}{\text{teeth of } A} = \dfrac{36 \times 100}{24} = 150 \text{ rpm}$

Practice check: Do exercise 2 on the left.

Rule _____

To find the number of teeth on either gear, multiply the size and speed of the other gear and divide by the speed of the one whose number of teeth is sought.

_____

■ **Example 3**

The driving gear has 24 teeth and makes 120 rpm. How many teeth must the driven gear have to make 80 rpm?

Solution $\quad$ teeth of driven $= \dfrac{\text{teeth} \times \text{rpm of driver}}{\text{rpm of driven}}$

$$= \frac{24 \times 120}{80} = 36 \text{ teeth}$$

■ **Problems 18–1**

Solve the following problems involving driving and driven gears. Using the space provided, make a sketch and label the known parts.

1. A gear having 120 teeth meshes with one having 24 teeth. If the large gear makes one revolution, how many revolutions will the small gear make?

2. In Fig. 18–4, gear *A* has 36 teeth and *B* has 96 teeth. If gear *A* makes 75 rpm, how many revolutions per minute will gear *B* make?

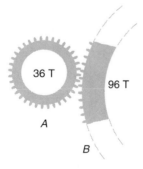

36 T

96 T

A

B

Figure 18–4

3. In Fig. 18–5, gear A has 56 teeth and gear B has 80 teeth. If gear A makes 210 rpm, how many revolutions per minute will gear B make?

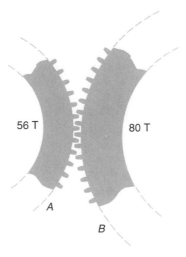

56 T   80 T

A

B

Figure 18–5

4. In Fig. 18–6, gear A has 40 teeth and makes 180 rpm. If gear B is to make 120 rpm, how many teeth must it have?

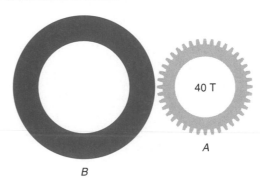

40 T

A

B

Figure 18–6

5. A 48-tooth gear running 320 rpm is required to drive another gear at 240 rpm. How many teeth must the driven gear have?

18–2   IDLERS

If we want to transmit power from shaft A to shaft B in Fig. 18–7, and the distance between the shafts is so great that it would require unusually large gears to bridge the gap, we must resort to idler gears. Gear C in Fig. 18–7 is such an idler.

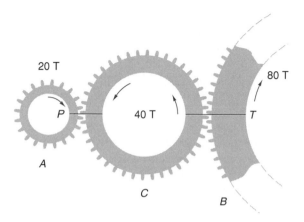

Figure 18–7

Assume that A has 20 teeth, C 40 teeth, and B 80 teeth. When A revolves once, C makes $\frac{1}{2}$ a revolution; and, while C revolves once, B makes $\frac{1}{2}$ a revolution; so that for one revolution of A the gear B makes $\frac{1}{4}$ a revolution. That is, the speed of B is the same as if it were in direct mesh with gear A. Hence, the idler has no effect on the relative speeds of the gears between which it is placed, and it is omitted from all computations.

The only effect of the idler is to reverse the direction of the motion of B. Without the idler, as shown in Figs. 18–3 and 18–4, B turns counterclockwise while A turns clockwise. With the idler interposed between A and B, when A turns clockwise, B also turns clockwise, as can readily be seen by a study of Fig. 18–7.

In the case of two idlers, such as C and D in Fig. 18–8, the direction of rotation of B is the same as when there are no idlers. That is, idler D neutralizes the effect of idler C. In general, the effect of idlers on the direction of motion of a follower or driven gear may be stated thus: *An odd number of idlers between two gears will cause the driven gear to rotate in the same direction as the driver; an even number of idlers will cause the driven gear to rotate in the direction opposite to that of the driver.*

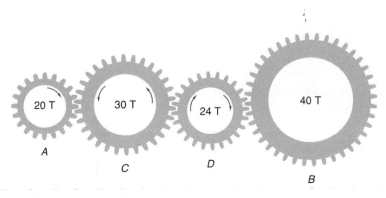

Figure 18–8

18–3   FINDING THE NUMBER OF TEETH FOR A GIVEN SPEED RATIO

The method of computing the number of teeth in gears that will give a desired speed ratio is illustrated by the following example.

■ Example

Find two suitable gears that will give a speed ratio between driver and driven of 2 to 3.

Solution $\dfrac{2}{3} = \dfrac{2 \times 12}{3 \times 12} = \dfrac{24 \text{ teeth on follower}}{36 \text{ teeth on driver}}$

---

3. Find two suitable gears that will give a speed ratio between driver and driven of 3 to 5.

The answer is in the margin on page 542.

Explanation   Express the desired ratio as a fraction and multiply both terms of the fraction by any convenient multiplier that will give an equivalent fraction whose numerator and denominator will represent available gears. In this instance, 12 was chosen as a multiplier, giving the equivalent fraction $\frac{24}{36}$. Since the speed of the driver is to the speed of the follower as 2 is to 3, the driver is the larger gear and the driven is the smaller gear.

Practice check: Do exercise 3 on the left.

■ Problems 18–3

Solve the following problems involving gear trains. Using the space provided, make a sketch of the train and label all the known parts.

1. The speeds of two gears are in the ratio of 1 to 3. If the faster one makes 180 rpm, find the speed of the slower one.

2. The speed ratio of two gears is 1 to 4. The slower one makes 45 rpm. How many revolutions per minute does the faster one make?

3. Two gears are to have a speed ratio of 2.5 to 3. If the larger gear has 72 teeth, how many teeth must the smaller one have?

4. Find two suitable gears with a speed ratio of 3 to 4.

5. Find two suitable gears with a speed ratio of 3 to $5\frac{1}{2}$.

6. In Fig. 18–9, *A* has 24 teeth, *B* has 36 teeth, and *C* has 40 teeth. If gear *A* makes 200 rpm, how many revolutions per minute will gear *C* make?

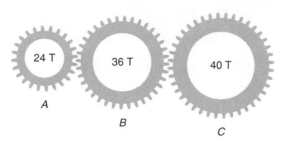

A

B

C

Figure 18–9

7. In Fig. 18–10, *A* has 36 teeth, *B* has 60 teeth, *C* has 24 teeth, and *D* has 72 teeth. How many revolutions per minute will gear *D* make if gear *A* makes 175 rpm?

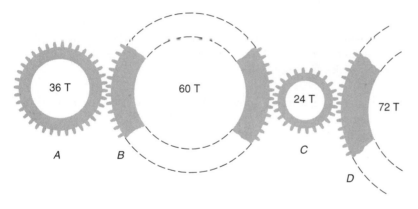

A B C D

Figure 18–10

18–4 COMPOUND GEARING

Figure 18–11 shows a train of compound gears. The driver *A* is in mesh with *B*, making *B* a driven gear. Gear *C*, keyed to the same shaft as *B*, is also a driver and meshes with *D*, so that *D* is a driven gear. Gears *B* and *C*, being keyed to the same shaft, revolve at the same rate of speed.

The following example illustrates the method of finding the speed of a driven gear of a compound train.

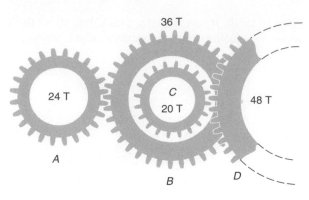

Figure 18–11

■ Example 1

In Fig. 18–11, if gear A makes 72 rpm, how many revolutions per minute will gear D make?

Solution $72 \times \frac{24}{36} \times \frac{20}{48} = 20$ rpm

Explanation   Since B has 36 teeth and A has 24 teeth, B will make $\frac{24}{36}$ as many revolutions as A. That is, $72 \times \frac{24}{36}$ represents the speed of B. But C, being keyed to the same shaft as B, revolves at the same rate of speed. Hence, when A makes 72 revolutions, C will make $72 \times \frac{24}{36}$ revolutions. Since D has 48 teeth, whereas C has 20 teeth, D will make $\frac{20}{48}$ as many revolutions as C, or $72 \times \frac{24}{36} \times \frac{20}{48} = 20$ rpm.

A similar analysis is applicable where the number of teeth of a driven gear is wanted, or the speed or the number of teeth of a driver is desired.

In the solution of the preceding example, notice that all the numbers in the numerator applied to the drivers, whereas those in the denominator applied to the driven gears. Thus, we can state the following rule.

4. Suppose that in Fig. 18–11, $A = 48$ T, $B = 42$ T, $C = 30$ T, and $D = 56$ T. If gear A makes 49 rpm, how many revolutions will gear D make?

The answer is in the margin on page 545.

Rule

To find the revolutions per minute or number of teeth of any driven gear in a train, divide the continued product of the given factors of the driving gears by the continued product of the given factors of the driven gears.

Practice check: Do exercise 4 on the left.

■ Example 2

In Fig. 18–11, if A makes 72 rpm, how many teeth must D have if it is to make 20 rpm?

5. Suppose that in Fig. 18–11, $A = 48$ T, $B = 42$ T, $C = 30$ T, and $D = 56$ T. If A makes 35 rpm, how many teeth must D have if it is to make 40 rpm?

The answer is in the margin on page 545.

Solution $\dfrac{\text{Product of driver data}}{\text{Product of driven data}} = \dfrac{72 \times 24 \times 20}{36 \times 20} = 48$ teeth

To find the number of teeth or revolutions per minute of a driving gear, divide the continued product of all that is known about the driven gears by the continued product of all that is known about the driving gears.

Practice check: Do exercise 5 on the left.

■ Example 3

In Fig. 18–11, how many revolutions per minute must gear A make if gear D is required to make 20 rpm?

Solution $\dfrac{\text{Product of driven data}}{\text{Product of driver data}} = \dfrac{36 \times 48 \times 20}{24 \times 20} = 72 \text{ rpm}$

Practice check: Do exercise 6 on the right.

6. Suppose that in Fig. 18–11, $A = 48$ T, $B = 42$ T, $C = 35$ T, and $D = 56$ T. How many revolutions per minute must gear A make if gear D is required to make 30 rpm?

The answer is in the margin on page 545.

If a certain speed ratio is required between the first driver and the last driven in a compound gear train, we proceed as in the following example.

■ Example 4

Find four suitable gears that will transmit motion from A to D at the ratio of 3 to 5 (see Fig. 18–12).

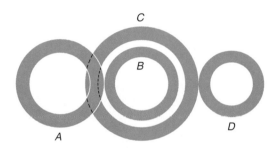

Figure 18–12

Solution $\dfrac{3}{5} = \dfrac{2}{2\frac{1}{2}} \times \dfrac{1\frac{1}{2}}{2} = \dfrac{2 \times 16}{2\frac{1}{2} \times 16} \times \dfrac{1\frac{1}{2} \times 24}{2 \times 24}$

$$= \dfrac{\overset{B \quad D}{32 \times 36} = \text{followers}}{\underset{A \quad C}{40 \times 48} = \text{drivers}}$$

Explanation   Write the desired speed ratio as a fraction, $\frac{3}{5}$. Split each term into two factors, namely, $3 = 2 \times 1\frac{1}{2}$ and $5 = 2\frac{1}{2} \times 2$, giving the two fractions $\dfrac{2}{2\frac{1}{2}}$ and $\dfrac{1\frac{1}{2}}{2}$. Treat each of these fractions separately as in the case of a simple gear train; that is, multiply each term by some convenient multiplier which will give an equivalent fraction whose terms will express available gears. The terms of the fraction $\dfrac{2}{2\frac{1}{2}}$ are multiplied by 16, giving the gears 32 and 40; and the terms of the fraction $\dfrac{1\frac{1}{2}}{2}$ are multiplied by 24, giving the gears 36 and 48. Since speeds of gears are inversely proportional to the number of their teeth, we have the proportion

$$\frac{\text{Speed of driver}}{\text{Speed of driven}} = \frac{\text{Teeth of driven}}{\text{Teeth of driver}}$$

7. Find four suitable gears that will transmit motion from *A* to *D* at the ratio of 2 to 3 in Fig. 18–12.

The answer is in the margin on page 546.

Hence, the two numbers in the numerator of the result represent driven gears in the train, and the two numbers in the denominator represent drivers. In mesh they would appear as in Fig. 18–12, where *A* drives *B* and *C* drives *D*.

Practice check: Do exercise 7 on the left.

■ Problems 18–4

Solve the following problems involving gear trains. Using the space provided, sketch the train described and label the parts with the given information.

1. In Fig. 18–13, gear *A* has 28 teeth, *B* has 48 teeth, *C* has 24 teeth, and *D* has 32 teeth. If gear *A* makes 180 rpm, how many revolutions per minute will gear *D* make?

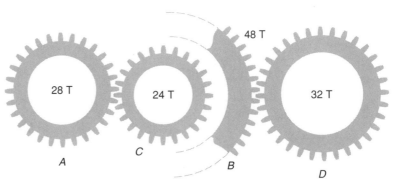

28 T   24 T   48 T   32 T

A   *C*   *B*   *D*

Figure 18–13

2. Figure 18–14 shows a compound gear train. Find the speed ratio between shaft *A* and shaft *D*.

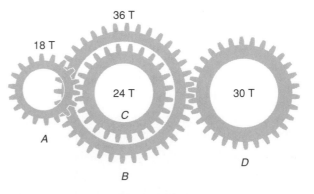

18 T   36 T   30 T
24 T
C
A   *B*   *D*

Figure 18–14

3. If *A* in Fig. 18–15 makes 42 rpm, how many teeth must *D* have in order to make 105 rpm?

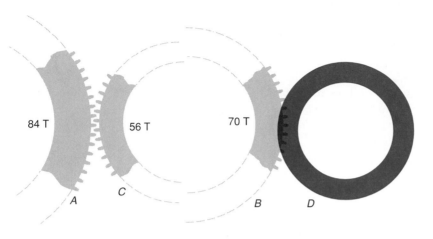

Figure 18–15

Answer to exercise 4

$$49 \times \frac{48}{42} \times \frac{30}{56} = 30 \text{ rpm}$$

Answer to exercise 5

$\dfrac{\text{Product of driver data}}{\text{Product of driven data}}$

$$= \frac{35 \times 48 \times 30}{42 \times 40}$$

$$= 30 \text{ teeth}$$

Answer to exercise 6

$\dfrac{\text{Product of driven data}}{\text{Product of driver data}}$

$$= \frac{42 \times 56 \times 30}{48 \times 35}$$

$$= 42 \text{ rpm}$$

4. In Fig. 18–16, if *A* makes one revolution, how many revolutions will *D* make?

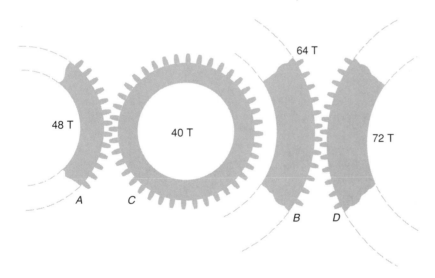

Figure 18–16

Answer to exercise 7

$$\frac{2}{3} = \frac{2}{2} \times \frac{1}{1\frac{1}{2}}$$

$$= \frac{2 \times 18}{2 \times 18} \times \frac{1 \times 24}{1\frac{1}{2} \times 24}$$

$$= \frac{36 \times 24 = \text{followers}}{36 \times 48 = \text{drivers}}$$

5. In Problem 4, what must be the speed of A in order to cause D to make 50 rpm?

6. In Fig. 18–17, how many teeth must gear C have if gear A makes 100 rpm and gear D makes 180 rpm?

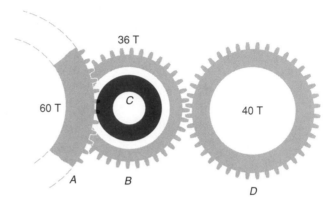

Figure 18–17

7. In Fig. 18–18, gear A has 40 teeth; B, 36 teeth; C, 60 teeth; D, 48 teeth; and E, 72 teeth. How many revolutions per minute will E make if A makes 240 rpm?

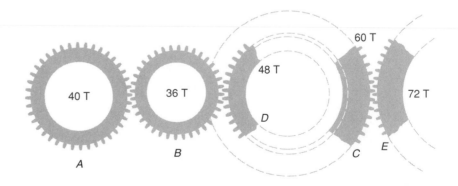

Figure 18–18

8. Gear *E* in Fig. 18–19 is required to make 100 rpm while *A* makes 75 rpm. How many teeth must *E* have if *A* has 24 teeth; *B*, 64; *C*, 32; and *D*, 40 teeth?

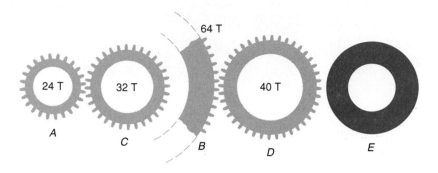

Figure 18–19

9. If a speed ratio of 2 to 5 is required between *A* and *D* in Fig. 18–20, what gears would you use?

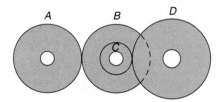

Figure 18–20

18–5   WORM AND GEAR

To find the revolutions per minute of a worm wheel when we know the revolutions per minute of the worm and also whether it is single- or double-threaded, we proceed as in the following example.

■ **Example**

If the worm *A* is triple-threaded and makes 100 rpm, how many revolutions per minute will the worm gear *B* make if it has 50 teeth? (See Fig. 18–21.)

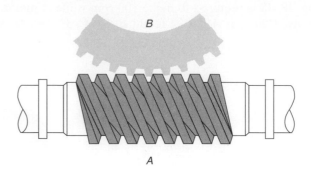

Figure 18–21

Solution $\dfrac{100 \times 3}{50} = 6$ rpm

8. If the worm A in Fig. 18–21 is double-threaded and makes 150 rpm, how many revolutions per minute will the worm gear B make if it has 75 teeth?

The answer is in the margin on page 551.

Explanation   The worm thread is regarded as a gear with one tooth if single-threaded, two teeth if double-threaded, and so on. Hence, to find the revolutions per minute of the worm gear, multiply the revolutions per minute and the number of threads of the worm, and divide by the number of teeth on the wheel.

Practice check: Do exercise 8 on the left.

18–6   TRAINS OF SPUR, BEVEL, AND WORM GEARING

Problems involving speed ratios of bevel gears are solved exactly like problems with spur gears.

■ **Example**

If gear F in Fig. 18–22 makes 240 rpm and has 48 teeth, E has 32 teeth, D has 20 teeth, C has 24 teeth, worm A is double-threaded, and worm gear B has 40 teeth, how many revolutions per minute will gear B make?

Solution   rpm of gear $B = \dfrac{\text{product of driver data}}{\text{product of driven data}}$

$$= \dfrac{240 \times 48 \times 20 \times 2}{32 \times 24 \times 40}$$

$$= 15 \text{ rpm}$$

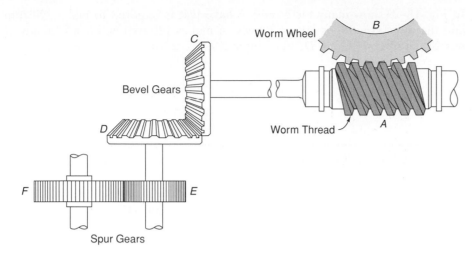

Figure 18–22

■ Problems 18–6

Use the space provided to solve the following problems.

1. In Fig. 18–22, the driving gear *F* has 44 teeth and makes 280 rpm. Gear *E* has 28 teeth; *D*, 36 teeth; and *C*, 24 teeth; and the worm *A* is triple-threaded. If the worm wheel has 32 teeth, how many revolutions per minute will it make?

2. If the single-threaded worm *A* in Fig. 18–22 makes 140 rpm, how many revolutions per minute will the worm wheel *B* make if it has 48 teeth?

3. In Fig. 18–23, gear *A* has 60 teeth and makes 160 rpm. Gear *B* has 28 teeth; *C*, 24 teeth; *D*, 20 teeth; *E*, 36 teeth; and *F*, 32 teeth. How many revolutions per minute will *F* make?

Figure 18–23

4. In Fig. 18–23, how many teeth must F have if it is required to make 320 rpm and the other conditions are as follows: A makes 120 rpm and has 72 teeth; B has 24 teeth; C, 40 teeth; D, 36 teeth; and E, 32 teeth?

18–7   PULLEY TRAINS

The formulas used for gear trains are applicable also to pulley trains, except that for pulleys the diameter is used instead of the number of teeth.

■ **Example 1**

Pulley A in Fig. 18–24 is 8 in. in diameter and makes 120 rpm. How many revolutions per minute does pulley B make if its diameter is 12 in.?

Figure 18–24

9. Suppose that pulley B in Fig. 18–24 is 15 cm in diameter and makes 100 rpm. How many revolutions per minute does pulley A make if its diameter is 10 cm?

The answer is in the margin on page 552.

Solution   rpm of $B = \dfrac{8 \times 120}{12} = 80$

Practice check: Do exercise 9 on the left.

■ **Example 2**

In Fig. 18–25, pulley A is 10 cm in diameter and makes 140 rpm. What must be the diameter of B if it is required to make 100 rpm?

Figure 18–25

10. Suppose that pulley B in Fig. 18–25 is 5 in. in diameter and makes 120 rpm. What must be the diameter of A if it is required to make 150 rpm?

The answer is in the margin on page 552.

Solution   Diameter of $B = \dfrac{10 \times 140}{100} = 14$ cm

Practice check: Do exercise 10 on the left.

■ **Example 3**

In Fig. 18–26, A is 8 in. in diameter, B is 20 in., C is 10 in., and D is 16 in. in diameter. If A makes 200 rpm, how many revolutions per minute will D make?

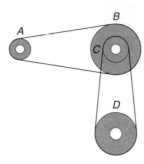

Figure 18–26

Answer to exercise 8

$$\frac{150 \times 2}{75} = 4 \, \text{rpm}$$

Solution   Speed of $D = \dfrac{\text{Product of all driver data}}{\text{Product of all driven data}}$

$$= \frac{200 \times 8 \times 10}{20 \times 16} = 50 \, \text{rpm}$$

Practice check: Do exercise 11 on the right.

■ **Problems 18–7**

Use the space provided to solve the following problems involving pulley trains.

1. The smaller pulley in Fig. 18–27 is 12 cm in diameter and the larger 20 cm. How many revolutions per minute will the larger pulley make if the smaller pulley makes 80 rpm?

Figure 18–27

11. Suppose that in Fig. 18–26, A is 10 in. in diameter, B is 25 in., C is 8 in., and D is 20 in. If A makes 150 rpm, how many revolutions per minute will D make?

The answer is in the margin on page 552.

2. If the large pulley in Fig. 18–28 is 18″ in diameter and makes 60 rpm, what must be the diameter of the small pulley to make 90 rpm?

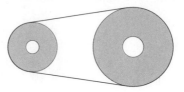

Figure 18–28

3. In Fig. 18–29, pulley A is 14″ in diameter; B, 10″; C, 16″; and D, 12″. If A makes 120 rpm, how many revolutions per minute will D make?

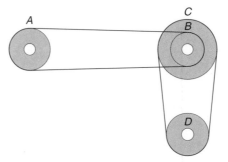

Figure 18–29

4. The pulleys of the train in Fig. 18–30 have the following diameters: A, 12″; B, 20″; C, 8″; D, 16″; E, 10″; F, 12″. If A makes 75 rpm, how many revolutions per minute will F make?

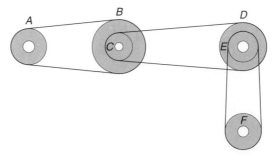

Figure 18–30

5. It is required to transmit power from A to D in Fig. 18–31 at a speed ratio of 3 to 2. Find four suitable pulleys for this train.

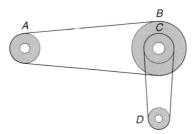

Figure 18–31

6. A cone pulley having steps of 11″, 9″, and 6½″ in diameter, and running at 140 rpm, drives another cone pulley having diameters of 5″, 7″, and 9½″. What speeds can be obtained?

7. A driving pulley 36″ in diameter on a line shaft making 200 rpm is belted to a 16″ pulley on a countershaft. The countershaft has a pulley 14″ in diameter belted to a 6″ pulley on an emery wheel. How many revolutions per minute will the emery wheel make?

8. A countershaft running at 240 rpm has a cone pulley with steps 11″, 8″, and 5″ in diameter driving a cone on a lathe with steps 6″, 9″, and 12″ in diameter. What range of speeds can be obtained?

■ **Self-Test**

Use the space provided to solve the following problems. Check your answers with the answers in the back of the book.

1. Two gears have 30 and 60 teeth. If the small gear makes 180 rpm, what is the rpm of the large gear?

2. If a driving gear has 28 teeth and makes 100 rpm, how many teeth must the driven gear have to make 70 rpm?

3. The speeds of two gears are in the ratio of 1 to 2. If the faster gear makes 180 rpm, what is the speed of the slower gear?

4. The speeds of two gears are in the ratio of 2 to 5. If the slower gear makes 2,800 rpm, what is the speed of the faster gear?

5. If a worm is double-threaded and makes 120 rpm, how many revolutions will the worm gear make if it has 80 teeth?

6. Two pulleys are connected by a vee-belt. The pulleys are 6 cm and 8 cm in diameter. What is the rpm of the larger pulley if the rpm of the smaller is 1,750?

7. A shaper must turn about 10,000 rpm. Its spindle pulley is 1.5 in. in diameter and is to be connected by a vee-belt to a 3,400-rpm motor. What size motor pulley diameter is needed?

8. A step pulley has steps of 2 in., $2\frac{1}{2}$ in., and 3 in. in diameter and, running at 1,725 rpm, drives a 6-in. pulley. What speeds can be obtained at the shaft of the 6-in. pulley?

9. An idler of 20 teeth is placed between a driven gear and a driver gear. Will the rpm of the driven gear be changed by the addition of the idler?

10. If the driver in Problem 9 rotates clockwise, what will be the direction of the driven gear?

■ **Chapter Test**

Solve the following problems involving gears and pulleys. Use the space provided to make a sketch for each problem. Label all known parts. Estimate the answer before solving.

1. Two meshing gears have 20 and 40 teeth. If the smaller gear makes 150 rpm, how many revolutions per minute will the larger gear make?

2. If the gears in Problem 1 have 20 and 35 teeth, and the smaller gear makes 150 rpm, how many rpm will the larger gear now make?

3. If a driving gear has 30 teeth and makes 132 rpm, how many teeth must the driven gear have to make 44 rpm?

4. An idler of 40 teeth is placed between the driven gear and the driver gear in Problem 3. How many rpm will the driven gear now make?

5. The speeds of two gears are in the ratio of 1 to 3. If the faster gear makes 150 rpm, what is the speed of the slower one?

6. In Fig. 18–32, *A* has 30 teeth, *B* has 60 teeth, *C* has 24 teeth, and *D* has 72 teeth. How many rpm will gear *D* make if gear *A* makes 180 rpm?

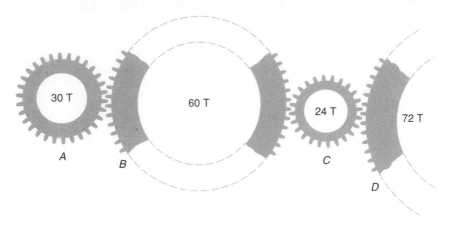

Figure 18–32

7. In Fig. 18–32, under the conditions as stated in Problem 6, how many rpm does gear *C* make?

8. In Fig. 18–32, under the conditions as stated in Problem 6, how many rpm does gear *B* make?

9. In Fig. 18–33, if *A* makes 60 rpm, how many rpm will gear *D* make?

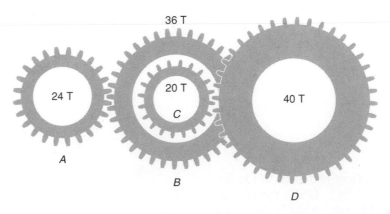

Figure 18–33

10. In Fig. 18–33, what is the speed ratio from *A* to *D*?

11. If a worm is double-threaded and makes 100 rpm, how many rpm will the worm gear make if it has 75 teeth?

12. Two pulleys are connected by a vee-belt. The smaller pulley is 8″ in diameter, the larger pulley is 12″ in diameter, and the larger pulley makes 1,750 rpm. How many rpm does the smaller pulley make?

13. An electric motor makes 1,750 rpm and is to be connected by pulleys and a vee-belt to a saw that must make close to 4,000 rpm. Find two pulley sizes for this job.

14. A step pulley has steps of 3″, $3\frac{1}{2}$″, and 4″ in diameter and, running at 3,400 rpm, drives a 4″ pulley. What speeds can be obtained at the shaft of the 4″ pulley?

15. If instead of the 4″ pulley in Problem 14, another step pulley of 4″, $3\frac{1}{2}$″, and 3″ is the driven pulley, what speeds can be obtained?

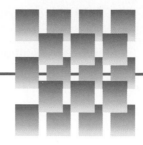

CHAPTER 19

Screw Threads

19–1   INTRODUCTION

Screw threads are used (*a*) to fasten parts together such as with bolts, (*b*) to adjust parts as with levelers on the base of machinery, or (*c*) to transmit power such as is done with a vise. During World War II, the United States, Britain, and Canada became convinced that it would be a great convenience if a standard were adopted for threads. The American Standard Unified Thread was the standard that was adopted.

19–2   PITCH

The ***pitch*** of a screw thread is the distance between corresponding points of two adjacent threads. In Fig. 19–1, the space marked *P* indicates the pitch for the threads shown.

If, for example, the distance between threads is $\frac{1}{8}$ in., we say that the thread is $\frac{1}{8}$-in. pitch. If the pitch is $\frac{1}{8}$ in., there are eight threads to the inch. Hence, a machinist will often speak of an 8-pitch thread instead of a $\frac{1}{8}$-in. pitch thread. In this chapter both methods of designating pitch will be employed, since the meaning in either case is unmistakable.

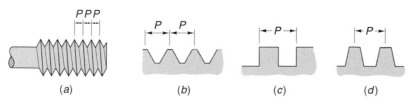

(*a*)   (*b*)   (*c*)   (*d*)

Figure 19–1

From the definition of pitch it follows that

$$\text{Pitch} = \frac{1}{\text{Number of threads per inch}}$$

and

$$\text{Number of threads per inch} = \frac{1}{\text{Pitch}}$$

■ Example 1

Find the pitch of a thread having 12 threads per inch.

Solution $\text{Pitch} = \dfrac{1}{\text{Number of threads per inch}} = \dfrac{1}{12}$ in.

■ Example 2

Find the pitch of a screw having $4\frac{1}{2}$ threads per inch.

Solution $\text{Pitch} = \dfrac{1}{\text{Number of threads per inch}}$

$$= \frac{1}{4\frac{1}{2}} = 1 \div \frac{9}{2} = 1 \times \frac{2}{9} = \frac{2}{9}\text{ in.}$$

Practice check: Do exercise 1 on the left.

1. Find the pitch of a screw having $2\frac{2}{3}$ threads per inch.

The answer is in the margin on page 562.

■ Example 3

A screw has a pitch of $\frac{5}{8}$ in. Find the number of threads per inch.

Solution $\text{Number of threads per inch} = \dfrac{1}{\text{Pitch}} = \dfrac{1}{\frac{5}{8}} = 1 \times \dfrac{8}{5}$

$$= \tfrac{8}{5} = \text{Threads per inch}$$

Practice check: Do exercise 2 on the left.

2. A screw has a pitch of $\frac{9}{16}$ in. Find the number of threads per inch.

The answer is in the margin on page 562.

■ Problems 19–2

Use the space provided to solve the following pitch problems.

1. What is the pitch of a screw having $3\frac{1}{2}$ threads per inch?

2. What is the pitch of a screw having $2\frac{1}{3}$ threads per inch?

3. A screw having five threads per inch has what pitch?

4. A screw having eight threads per inch has what pitch?

5. Find the pitch of a screw having $5\frac{1}{2}$ threads per inch.

6. Find the pitch of a screw having 18 threads per inch.

7. Compute the pitch of a screw with $2\frac{3}{4}$ threads per inch.

8. A $\frac{3}{4}''$-pitch screw has how many threads per inch?

9. A $\frac{2}{5}''$-pitch screw has how many threads per inch?

10. How many threads per inch are on a screw of $\frac{2}{7}''$ pitch?

11. How many threads per inch are on a screw of $\frac{5}{16}''$ pitch?

12. A screw has a pitch of $\frac{3}{8}''$. Find the number of threads per inch.

13. Find the number of threads per inch in a $\frac{7}{16}''$-pitch screw.

14. Find the number of threads per inch in a $\frac{1}{4}''$-pitch screw.

15. The pitch of a thread is $\frac{1}{14}''$. How many threads are there in $1''$?

19–3   LEAD

The *lead* of a screw is the distance the nut advances on the screw in one turn, or the distance the screw would advance if given one complete turn.

In a *single-threaded* screw, the lead is *equal* to the pitch; that is, in one complete turn, the screw advances a distance equal to the pitch. In a *double-threaded* screw, the lead is *twice* the pitch; in a *triple-threaded* screw, the lead is *three times* the pitch; and so forth.

In a screw of $\frac{1}{8}$-in. pitch, the lead has the following values.

Single-threaded: Lead = Pitch = $\frac{1}{8}$ in.

Double-threaded: Lead = 2 × Pitch = $2 \times \frac{1}{8} = \frac{1}{4}$ in.

Triple-threaded: Lead = 3 × Pitch = $3 \times \frac{1}{8} = \frac{3}{8}$ in.

Figure 19–2 illustrates the relation between pitch and lead in a single, double, triple, and quadruple square thread.

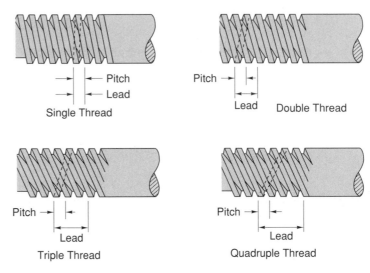

Figure 19–2

■ Problems 19–3

Using the space provided, solve the following lead problems.

1. A single-threaded screw has 12 threads per inch. What is its pitch? What is its lead?

2. A single-threaded screw has $4\frac{1}{2}$ threads per inch. What is its pitch? What is its lead?

3. What is the lead of an 8-pitch double-threaded screw?

4. What is the lead of a 6-pitch double-threaded screw?

5. What is the lead of a triple 8-pitch thread?

6. What is the lead of a triple $\frac{5}{8}$-pitch thread?

7. The lead of a double-threaded screw is $\frac{3}{8}''$. What is its pitch?

8. The lead of a double-threaded screw is $\frac{1}{2}''$. What is its pitch?

9. What is the lead of a $\frac{3}{16}''$-pitch quadruple thread?

10. What is the lead of a $\frac{1}{4}''$-pitch quadruple thread?

11. Find the pitch of a screw with a triple thread whose lead is $\frac{3}{4}''$.

12. Find the pitch of a screw with a triple thread whose lead is $\frac{7}{8}''$.

13. A single thread of $\frac{1}{16}''$-pitch has what lead?

14. A single thread of $\frac{3}{4}''$-pitch has what lead?

15. What is the pitch of a double-threaded screw having a lead of $\frac{3}{32}''$?

19–4   DEFINITIONS APPLYING TO SCREW THREADS

Figure 19–3 illustrates the following terms.

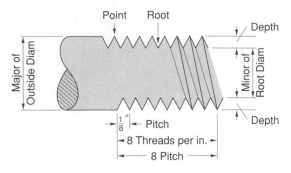

Figure 19–3

The top of the thread is called the ***point*** or ***crest.***

The bottom of the thread is called the ***root.***

The ***depth*** of the thread is the vertical distance between the top and the bottom of the thread. Twice the depth is known as the ***double depth.***

The ***major diameter,*** sometimes referred to as the ***outside diameter,*** is the largest diameter of the thread of the screw or nut.

The ***minor*** or ***root diameter*** is the smallest diameter of the thread of the screw or nut; that is, it is the diameter at the roots or bottoms of the threads. It is evident from the sketch that the root diameter of a thread is equal to the major diameter minus the double depth.

19–5   SHARP V-THREAD

The sharp V-thread is so named because the sides of the thread form the letter V. It is now used for special work where additional thread contact is desired, and it is also used with brass pipe work. The study of the sharp V-thread will make clear the dimensions of the other thread forms discussed in this chapter.

In Fig. 19–4 the horizontal line AB (joining two successive roots) forms, with the sides of the thread, an equilateral triangle ABC. If the pitch of the thread is 1 in., the sides of the triangle will be 1 in. each. The depth of the V-thread H, which

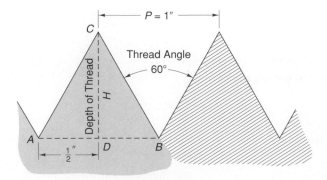

Figure 19–4

is a perpendicular from the crest C to the side AB, divides the triangle into two equal right triangles. In the right triangle ACD,

$$H = \sqrt{\overline{AC}^2 - \overline{AD}^2} = \sqrt{1^2 - (\tfrac{1}{2})^2} = \sqrt{1 - \tfrac{1}{4}} = \sqrt{\tfrac{3}{4}} = 0.866 \text{ in.}$$

That is, the depth of a sharp V-thread of 1-in. pitch is 0.866 in. The depth of a $\frac{1}{2}$-in. pitch V-thread is equal to one-half of 0.866, or 0.433 in.

Rule _____

To find the depth of a sharp V-thread, multiply the pitch by 0.866.

■ **Example**

Find the depth of a V-thread having 16 threads per inch.

Solution   Depth = Pitch $\times$ 0.866 = $\tfrac{1}{16} \times$ 0.866 = 0.0541 in.

Practice check: Do exercise 3 on the right.

3. Find the depth of a V-thread having 6 threads per inch.

The answer is in the margin on page 568.

■ **Problems 19–5**

Using the space provided, find the depth of V-threads having the following numbers of threads per inch.

1. 20   **2.** 18   **3.** 14   **4.** 12   **5.** 11

6. 9   **7.** 8   **8.** 7   **9.** 6   **10.** 5

11. 19   **12.** 24   **13.** 13   **14.** 10   **15.** 15

16. 17   **17.** 22   **18.** 21   **19.** 30   **20.** 32

19–6   DOUBLE DEPTH OF SHARP V-THREAD

Twice the depth of a thread is called the *double depth*. The double depth of a 1-in.-pitch sharp V-thread is $2 \times 0.866 = 1.732$ in.; the double depth of a $\frac{1}{4}$-in.-pitch V-thread is $\frac{1}{4} \times 1.732 = 0.433$ in.

■ Example

Find the double depth of V-threads having 7 threads per inch.

4. Find the double depth of a V-thread having $5\frac{3}{4}$ threads per inch.

The answer is in the margin on page 568.

Solution   Double depth = pitch $\times$ 1.732 = $\frac{1}{7} \times$ 1.732
$$= 0.2474 \text{ in.}$$

Practice check: Do exercise 4 on the left.

19–7   MINOR DIAMETER

Figure 19–3 shows that the minor diameter of a screw thread is equal to the major diameter minus the double depth.

■ Example

Find the minor diameter of a $\frac{3}{8}$-in.–16 V-thread. ($\frac{3}{8}$ in. = major diameter; 16 = number of threads per inch.)

5. Find the minor diameter of a $\frac{7}{16}$-in.–12 V-thread.

The answer is in the margin on page 568.

Solution   Major diameter = $\frac{3}{8}$ = 0.3750 in.
Double depth = $\frac{1}{16} \times$ 1.732 = $\underline{0.1082 \text{ in.}}$
Minor diameter   = 0.2668 in.

Practice check: Do exercise 5 on the left.

■ Problems 19–7

Use the space provided to find the double depth of V-threads with the following numbers of threads per inch.

1. 3   **2.** $3\frac{1}{4}$ **3.** $3\frac{1}{2}$ **4.** 4   **5.** $4\frac{1}{2}$

6. 5   **7.** 8   **8.** 12   **9.** 16   **10.** 20

11. 2   **12.** $2\frac{3}{4}$ **13.** 14   **14.** 6   **15.** $5\frac{1}{4}$

16. 18   **17.** 10   **18.** $4\frac{3}{4}$ **19.** 22   **20.** 30

Find the minor diameters of the following V-thread screws.

21. $\frac{1}{4}''$–20   **22.** $\frac{7}{16}''$–14   **23.** $\frac{9}{16}''$–12   **24.** $\frac{13}{16}''$–10   **25.** $\frac{15}{16}''$–9

26. $1\frac{1}{4}''$–7   **27.** $1\frac{1}{2}''$–6   **28.** $1\frac{3}{4}''$–5   **29.** $2\frac{1}{4}''$–$4\frac{1}{2}$ **30.** $2\frac{7}{8}''$–4

31. $\frac{3}{8}''$–19   **32.** $\frac{3}{4}''$–24   **33.** $\frac{11}{16}''$–15   **34.** $1\frac{5}{8}''$–17   **35.** $3\frac{3}{4}''$–22

36. $1\frac{7}{8}''$–$2\frac{1}{8}$ **37.** $2\frac{1}{4}''$–$5\frac{1}{2}$ **38.** $2\frac{5}{8}''$–21   **39.** $2\frac{3}{4}''$–30   **40.** $1\frac{1}{8}''$–32

19–8   TAP DRILL SIZES

To tap a nut or hole is to cut a thread in it. A tap drill is used for drilling holes before tapping. Although theoretically the diameter of the drill is equal to the minor diameter of the bolt for which the nut is to be used, in practice the tap drills are somewhat larger than the minor diameter of the thread (see Fig. 19–5). The tap drill diameter for a sharp V-thread is equal to the minor diameter plus one-quarter of the pitch.

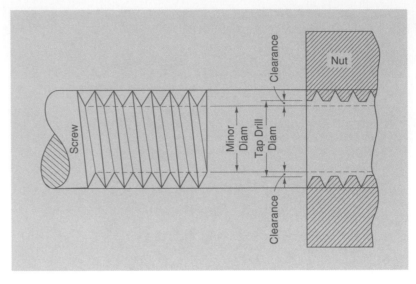

Figure 19–5

■ Example

Find the proper size tap drill for a $\frac{7}{16}$-in.–14 V-thread.

Solution   Major diameter = $\frac{7}{16}$ in.   = 0.4375 in.
Double depth = $\frac{1}{14} \times 1.732 = \underline{0.1237 \text{ in.}}$
Minor diameter   = 0.3138 in.
Clearance = $\frac{1}{4} \times$ pitch
$= \frac{1}{4} \times \frac{1}{14} = \frac{1}{56}$ = $\underline{0.0179 \text{ in.}}$
Drill size   = 0.3317 in.

6. Find the proper size tap drill for a $\frac{1}{4}$-in.–16 V-thread.

The answer is in the margin on page 572.

According to the table of decimal equivalents, the nearest $\frac{1}{64}$ in. to 0.3317 is 0.3281, which is the decimal equivalent of $\frac{21}{64}$ in. Hence, a $\frac{21}{64}$-in. drill is to be used.

Practice check: Do exercise 6 on the left.

19–9   THE UNIFIED THREAD

The Unified thread, the standard thread agreed upon by the United States, Britain, and Canada, follows the shape of the older American Sharp V-thread. The top of the thread is flattened and the bottom of the thread is rounded. The sides of the thread, if extended, would form the Sharp V-thread (see Fig. 19–6). The dimensions of the Unified thread are based on the depth of the sharp V-thread, *H*.

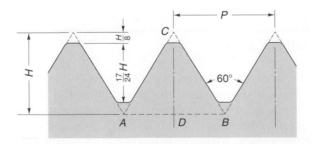

Figure 19–6

The crest of the thread is flattened (it can be rounded) a distance of $H/8$. The root of the thread is $\frac{17}{24}H$ from the crest. The root may be flat or rounded. If the pitch of the Unified thread is 1 in., the distance H will be 0.866 in. as in the sharp V. If the pitch of the Unified thread is $\frac{1}{2}$ in., the distance H will be 0.433, and so on. The depth of the Unified thread is $\frac{17}{24}H$ as shown in Fig. 19–6. For the 1-in. pitch, the depth of the thread is $\frac{17}{24} \times 0.866 = 0.61343$ in. For the $\frac{1}{2}$-in. pitch, the depth of the thread is $\frac{17}{24} \times 0.433 = 0.30671$, and so forth.

Rule

To find the depth of the Unified thread, multiply the pitch by 0.61343.

■ **Example**

Find the depth of a Unified thread having 16 threads per inch.

Solution   Depth = Pitch × 0.61343 = 0.0383

Practice check: Do exercise 7 on the right.

7. Find the depth of a Unified thread having 6 threads per inch.

The answer is in the margin on page 572.

■ **Problems 19–9**

Use the space provided to find the proper tap drill sizes for the following V-threads.

1. $\frac{5}{16}''$–18

2. $\frac{3}{8}''$–16

3. $\frac{1}{2}''$–12

4. $\frac{5}{8}''$–11

5. $\frac{11}{16}''$–11

6. $\frac{3}{4}''$–10

7. $\frac{7}{8}''$–9

8. $1''$–8

9. $1\frac{3}{8}''$–6

10. $2''$–$4\frac{1}{2}$

11. $\frac{3}{8}''$–19

12. $\frac{11}{16}''$–24

13. $1\frac{7}{8}''$–15

14. $\frac{3}{4}''$–17

15. $\frac{1}{4}''$–22

16. $1\frac{3}{4}''$–$2\frac{1}{8}$

17. $\frac{5}{8}''$–$5\frac{1}{2}$

18. $\frac{5}{16}''$–21

19. $\frac{3}{4}''$–7

20. $1\frac{1}{8}''$–5

Find the depth of the Unified threads having the following number of threads per inch.

21. 20

22. 18

23. 16

24. 14

25. 13

26. 12   **27.** 11   **28.** 10   **29.** 9   **30.** 8

31. 5   **32.** 25   **33.** 7   **34.** 15   **35.** 17

36. 19   **37.** 21   **38.** 22   **39.** 24   **40.** 26

19–10   DOUBLE DEPTH OF THE UNIFIED THREAD

The double depth of a 1-in.-pitch Unified thread is $1 \times 2 \times 0.61343 = 1.22686$ in.; the double depth of a $\frac{1}{4}$-in.-pitch Unified thread is $\frac{1}{4} \times 2 \times 0.61343 = 0.30671$ in.

■ **Example**

Find the double depth of Unified threads with 6 threads per inch.

8. Find the double depth of Unified threads with 15 threads per inch.

The answer is in the margin on page 572.

Solution   Double depth = Pitch $\times$ 2 $\times$ 0.61343
                       $= \frac{1}{6} \times 2 \times 0.61343$
                       $= 0.2045$ in.

Practice check: Do exercise 8 on the left.

19–11   MINOR DIAMETER OF UNIFIED THREADS

The minor diameter of the Unified thread is equal to the major diameter of the screw thread minus the double depth.

■ **Example**

Find the minor diameter of a $\frac{1}{4}$–20 Unified thread. ($\frac{1}{4}$ means major diameter $= \frac{1}{4}$ inch; 20 means 20 threads per inch.)

9. Find the minor diameter of a $\frac{1}{2}$–15 Unified thread.

The answer is in the margin on page 572.

Solution   Major diameter $= \frac{1}{4}$ $= 0.25000$ in.
                Double depth $= \frac{1}{20} \times 2 \times 0.61343 = \underline{0.06134 \text{ in.}}$
                Minor diameter $= 0.18866$ in.

Practice check: Do exercise 9 on the left.

■ Problems 19–11

Using the space provided, find the double depth of the Unified threads with the following number of threads per inch.

1. 8   **2.** 9   **3.** 10   **4.** 11   **5.** 12

6. 13   **7.** 14   **8.** 16   **9.** 18   **10.** 20

11. 5   **12.** 25   **13.** 7   **14.** 23   **15.** 17

16. 19   **17.** 21   **18.** 22   **19.** 24   **20.** 26

Find the minor diameter of the following Unified threads.

21. $\frac{5}{16}$–18   **22.** $\frac{3}{8}$–16   **23.** $\frac{7}{16}$–14   **24.** $\frac{1}{2}$–13   **25.** $\frac{9}{16}$–12

26. $\frac{5}{8}$–11   **27.** $\frac{3}{4}$–10   **28.** $\frac{7}{8}$–9   **29.** $\frac{11}{16}$–11   **30.** 1–8

31. $1\frac{3}{8}$–6   **32.** 2–$4\frac{1}{2}$ **33.** $\frac{3}{4}$–17   **34.** $\frac{1}{4}$–22   **35.** $1\frac{1}{8}$–5

19–12   AMERICAN NATIONAL THREAD

There are two series of American National threads: the American National coarse-thread series and the American National fine-thread series.

Answer to exercise 6

Major diameter
$= \frac{1}{4}$ in. = 0.2500 in.
Double depth
$= \frac{1}{16} \times 1.732 = \underline{0.1083 \text{ in.}}$
  Minor diameter   = 0.1417 in.
Clearance
$= \frac{1}{4} \times \frac{1}{16} = \frac{1}{64} = \underline{0.01563 \text{ in.}}$
  Drill size   = 0.1573 in.
               or $\frac{5}{32}$ in.

Answer to exercise 7

0.1022

Answer to exercise 8

0.818 in.

Answer to exercise 9

Major diameter   = 0.50000 in.
Double depth   = $\underline{0.08179 \text{ in.}}$
  Minor diameter = 0.41821 in.

The American National coarse-thread series comprises the former U.S. Standard threads supplemented in the sizes below $\frac{1}{4}$ in. by part of the A.S.M.E. Standards (see Table 8 at the back of the text).

The American National fine-thread series comprises the regular screw thread series of the S.A.E. supplemented in sizes below $\frac{1}{4}$ in. by the A.S.M.E. fine-thread series (see Table 9).

The American National thread is a 60-degree thread with flattened tops and bottoms. If we start with a V-thread and divide the height or depth into eight equal parts, the American National is formed by eliminating one-eighth of the depth at the top and one-eighth of the depth at the bottom. The depth of the resulting thread is therefore equal to three-fourths of the depth of the V-thread of the same pitch.

If, as in Fig. 19–7, the pitch is 1 in., the depth of the V-thread is 0.866 in. and the depth of the corresponding American National thread is $\frac{3}{4}$ of 0.866, or 0.6495 in. The width of the thread at top and bottom is $\frac{1}{8}$ of the pitch.

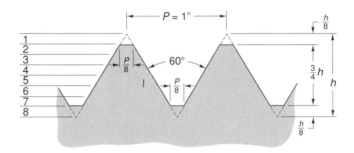

Figure 19–7

To be able to compute the proportions of an American National thread of any pitch, you should memorize the constant 0.6495.

Likewise, you should memorize the constant for the American Standard Unified thread, which is 0.61343.

■ **Example**

Find the depth of an American National thread having 16 threads to the inch.

Solution   Depth = Pitch $\times$ 0.6495 = $\frac{1}{16} \times$ 0.6495
                             = 0.0406 in.

Practice check: Do exercise 10 on the left.

10. Find the depth of an American National thread having 20 threads to the inch.

The answer is in the margin on page 575.

19–13   DOUBLE DEPTH OF AMERICAN NATIONAL THREAD

The double depth of an American National thread of 1-in. pitch is obviously equal to 2 $\times$ 0.6495 in. or 1.299 in. The double depth of an American National thread of any pitch is equal to *the pitch multiplied by* 1.299.

■ **Example**

Find the double depth of an American National thread having 16 threads per inch.

Solution   Double depth = Pitch × 1.299

$$= \frac{1}{16} \times 1.299 = 0.0812 \text{ in.}$$

Practice check: Do exercise 11 on the right.

11. Find the double depth of an American National thread having 20 threads per inch.

The answer is in the margin on page 576.

19–14   MINOR DIAMETER OF AMERICAN NATIONAL THREAD

The minor diameter is equal to the major diameter minus the double depth.

■ Example

Find the minor diameter of a $\frac{5}{8}$-in.–11 American National thread.

Solution   Major diameter = $\frac{5}{8}$in.   = 0.6250 in.
            Double depth = $\frac{1}{11} \times 1.299$ = 0.1181 in.
            Minor diamcter   = 0.5069 in.

Practice check: Do exercise 12 on the right.

12. Find the minor diameter of a $\frac{3}{4}$-in.–15 American National thread.

The answer is in the margin on page 576.

■ Problems 19–14

Using the space provided below, find the single depth of the following American National threads.

1. 7 threads per inch

2. 8 threads per inch

3. $\frac{1}{10}''$ pitch

4. $\frac{1}{12}''$ pitch

5. $5\frac{1}{2}$ threads per inch

6. 13 threads per inch

Find the double depth of the following American National threads.

7. Pitch = $\frac{1}{20}''$

8. $4\frac{1}{2}$ threads per inch

9. Pitch = $\frac{1}{11}''$

10. 9 threads per inch

11. 6 threads per inch

12. 11 threads per inch

Find the minor diameter of the following American National threads.

13. $\frac{1''}{4}$–20   **14.** $\frac{3''}{8}$–16   **15.** $\frac{1''}{2}$–13   **16.** $\frac{3''}{4}$–10   **17.** $1''$–8

18. $1\frac{1}{2}''$–6   **19.** $1\frac{3}{4}''$–5   **20.** $2''$–$4\frac{1}{2}$ **21.** $2\frac{3}{4}''$–4   **22.** $\frac{1''}{8}$–7

19–15   SIZE OF TAP DRILL FOR AMERICAN NATIONAL THREADS

Commercial tap drills are designed to allow an engagement of three-fourths of the full depth of the screw thread. For a nut to engage the screw three-fourths of the depth of the thread, the tap drill diameter must be equal to the minor diameter of the screw plus one-fourth the double depth of the thread.

■ **Example 1**

Find the commercial size tap drill (that is, for $\frac{3}{4}$ engagement) for a $\frac{7}{8}$-in.–9 American National coarse-thread.

Solution   Major diameter = $\frac{7}{8}$ in.   = 0.8750 in.
               Double depth = $\frac{1}{9} \times 1.299$ = 0.1443 in.

               Minor diameter   = 0.7307 in.

               Clearance = $\dfrac{\text{Double depth of thread}}{4}$

                         = $\dfrac{0.1443}{4}$ = 0.0361 in.

               Drill size   = 0.7668 in.

The nearest $\frac{1}{64}$ in. is $\frac{49}{64}$ in. Use a drill $\frac{49}{64}$ in. in diameter.

Practice check: Do exercise 13 on the left.

13. Find the commercial size tap drill for a $1\frac{1}{4}$-in.–6 American National coarse-thread.

The answer is in the margin on page 576.

The *minimum* standard tap drill adopted by the American Standard Association is for a nut that will engage the screw thread five-sixths of the depth of the thread. The tap drill size for such a nut must be equal to the minor diameter of the screw thread plus one-sixth of the double depth of the thread.

■ **Example 2**

Find the tap drill size for a $\frac{7}{8}$-in.–9 American National coarse-thread for $\frac{5}{6}$ engagement.

Solution   Major diameter = $\frac{7}{8}$ in.   = 0.8750 in.
            Double depth = $\frac{1}{9} \times 1.299$ = 0.1443 in.
              Minor diameter   = 0.7307 in.

            Clearance = $\dfrac{\text{Double depth of thread}}{6}$

                      = $\dfrac{0.1443}{6}$ = 0.0241 in.

            Drill size   = 0.7548 in.

The nearest $\frac{1}{64}$ in. is $\frac{3}{4}$ in. Use a drill $\frac{3}{4}$ in. in diameter.

Practice check: Do exercise 14 on the right.

Answer to exercise 10

Depth = $\frac{1}{20} \times 0.6495$
      = 0.0325 in.

14. Find the tap drill size for a $\frac{7}{16}$-in.–12 American National coarse-thread for $\frac{5}{8}$ engagement.

The answer is in the margin on page 578.

19–16   WIDTH OF POINT OF TOOL

The shape of the point of the tool for cutting an American National thread must be the same as the bottom of the thread. The point of the tool therefore is flattened an amount equal to one-eighth of the pitch of the thread to be cut.

■ **Example**

Find the width of tool point for cutting a $\frac{1}{4}$-in.-pitch American National thread.

Solution   Width of tool point = $\frac{1}{8} \times$ pitch = $\frac{1}{8} \times \frac{1}{4} = \frac{1}{32}$ in.

Practice check: Do exercise 15 on the right.

15. Find the width of tool point for cutting a $\frac{1}{2}$-in. pitch American National thread.

The answer is in the margin on page 578.

■ **Problems 19–16**

Using the space provided, find the commercial tap drill size and also the minimum tap drill size for each of the following American National threads.

1. $\frac{5}{16}''$–18   **2.** $\frac{7}{16}''$–14   **3.** $\frac{7}{8}''$–9   **4.** 1″–8   **5.** $1\frac{3}{8}''$–6

6. $1\frac{3}{4}''$–5   **7.** 4″–4   **8.** $1\frac{1}{2}''$–6   **9.** $2\frac{1}{4}''$–$4\frac{1}{4}$ **10.** $2\frac{1}{2}''$–4

Find the width of tool point required to cut the following American National threads.

11. Pitch = $\frac{2}{9}''$ **12.** Pitch = $\frac{1}{4}''$ **13.** Pitch = $\frac{1}{16}''$ **14.** Pitch = $\frac{1}{7}''$

15. Pitch $= \frac{1}{12}''$ **16.** Pitch $= \frac{2}{7}''$

Answer to exercise 11

Double depth $= \frac{1}{20} \times 1.299$
$ = 0.06495$ in.

Answer to exercise 12

Major diameter	$= 0.7500$ in.
Double depth	$= \underline{0.0866 \text{ in.}}$
Minor diameter	$= 0.6634$ in.

Answer to exercise 13

Major diameter	$= 1.2500$ in.
Double depth	$= \underline{0.2165 \text{ in.}}$
Minor diameter	$= 1.0335$ in.
Clearance	$= \underline{0.0541 \text{ in.}}$
Drill size	$= 1.0876$ in.

Use a drill $\frac{3}{32}$ in. in diameter.

19–17   SQUARE THREAD

The square thread (Fig. 19–8) is so called because a cross section of the thread is theoretically a square. The sketch is self-explanatory. The thickness of the thread and the width of the space are each equal to one-half the pitch. In the nut, however, the space is slightly more than half the pitch, to allow for a sliding fit. The depth of the thread, although theoretically equal to half the pitch, is actually cut 0.005 in. deeper than half the pitch.

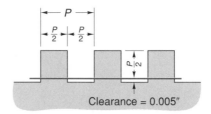

Figure 19–8

The width of the tool point for cutting a square thread is equal to the width of the space, which is half the pitch.

19–18   TAP DRILL FOR SQUARE THREAD

The relation between a square-threaded screw and its nut is shown in Fig. 19–9. Note that there is a clearance of 0.005 in. between the screw threads and the nut threads, both top and bottom. The tap drill therefore must be 2 × 0.005 in., or

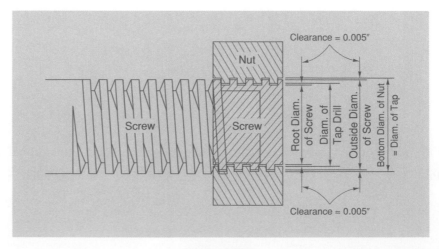

Figure 19–9

0.01 in. larger than the root diameter of the screw, and the tap diameter will be 2×0.005 in., or 0.01 in. larger than the outside diameter of the screw.

■ **Example 1**

Find the double depth of a $\frac{5}{8}$-in.–8 square thread.

Solution   Double depth = Pitch + 2 × Clearance
                              $= \frac{1}{8} + 2 \times 0.005$ in. = 0.125 in. + 0.01 in.
                              = 0.135 in.

Practice check: Do exercise 16 on the right.

■ **Example 2**

Find the root diameter of a $\frac{3}{8}$-in.–16 square-thread screw.

Solution   Screw size = $\frac{3}{8}$ in.   = 0.3750 in.
                Double depth = $\frac{1}{16}$ + 0.01 = 0.0625 + 0.01 = $\underline{0.0725\ \text{in.}}$
                        Root diameter   = 0.3025 in.

Practice check: Do exercise 17 on the right.

■ **Example 3**

Find the size of tap drill for a $\frac{5}{8}$-in.–8 square-thread screw.

Solution   Figure 19–9 shows that the diameter of the tap drill is 2×0.005 in. larger than the root diameter of the thread.

                    Screw size = $\frac{5}{8}$ in.   = 0.6250 in.
                    Pitch = $\frac{1}{8}$ in.   = $\underline{0.1250\ \text{in.}}$
                            Diameter of tap drill = 0.5000 in. = $\frac{1}{2}$ in.

Practice check: Do exercise 18 on the right.

■ **Problems 19–18**

Use the space provided to solve the following problems involving square-threaded screws.

1. Find the double depth of a $\frac{3''}{4}$–6 square thread.

2. Find the double depth of a $\frac{7''}{8}$–$4\frac{1}{2}$ square thread.

3. Find the double depth of a square-thread screw 1″ in diameter with a double thread of $\frac{2''}{5}$ lead.

4. Find the double depth of a square-thread screw 4″ in diameter with three threads per inch.

16. Find the double depth of a $\frac{3}{4}$-in.–6 square thread.

The answer is in the margin on page 579.

17. Find the root diameter of a $\frac{7}{8}$-in.–$6\frac{1}{4}$ square-thread screw.

The answer is in the margin on page 579.

18. Find the size of tap drill for a $\frac{3}{16}$-in.–16 square-thread screw.

The answer is in the margin on page 579.

5. Find the double depth of a $2\frac{1}{2}''$ square-thread screw with four threads per inch.

6. Find the double depth of a square-thread screw $\frac{5}{8}''$ in diameter with a double thread of $\frac{1}{4}''$ lead.

7. Find the double depth of a $1\frac{3}{4}''$ square-thread screw with four threads per inch.

8–14. Find the root diameters of the square-thread screws in Problems 1–7.

15–21. Find the tap drill sizes for the square-thread screws in Problems 1–7.

19–19   THE ACME 29-DEGREE SCREW THREAD

The Acme 29-degree screw thread is best understood by a study of its derivation from the square thread. In Fig. 19–10 we start with a square thread and draw lines making an angle of $14\frac{1}{2}°$ with the vertical sides of the square thread. The sides of the resulting Acme thread will therefore make an angle of 29° with each other. The depth of the Acme thread is 0.01 in. greater than half the pitch of the thread.

The dimensions of the Acme thread for the screw (Fig. 19–11) are as follows.

$$F = \text{Flat on top} = 0.3707 \times \text{Pitch}$$

$$S = \text{Width of space} = 0.6293 \times \text{Pitch}$$

$$T = \text{Bottom flat} = \text{Width of tool point for cutting thread} = 0.3707 \times \text{Pitch} - 0.0052 \text{ in.}$$

$$D = \text{Depth} = P/2 + 0.010 \text{ in.}$$

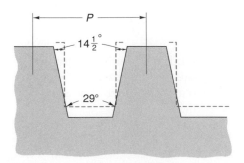

Figure 19–10

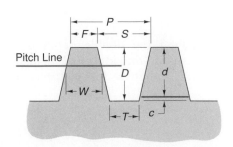

Figure 19–11

■ **Example**

Find the root diameter, top flat, and bottom flat of a $\frac{3}{4}$-in.–8 Acme 29-degree screw thread.

Solution

Screw size $= \frac{3}{4}$ in. $\qquad\qquad\qquad\qquad = 0.7500$ in.
Depth $= P/2 + 0.010$ in. $= \frac{1}{2} \times \frac{1}{8} + 0.010$ in.
$\qquad\qquad = 0.0625 + 0.010 = 0.0725$ in.
Double depth $= 2 \times 0.0725$ in. $\qquad\qquad\underline{= 0.1450\ \text{in.}}$
$\qquad$ Root diameter $\qquad\qquad\qquad\qquad = 0.6050$ in.

$$\text{Top flat} = 0.3707 \times \text{Pitch} = 0.3707 \times \tfrac{1}{8} = 0.0463\ \text{in.}$$

$$\text{Bottom flat} = 0.3707 \times \text{Pitch} - 0.0052$$
$$= 0.3707 \times \tfrac{1}{8} - 0.0052 = 0.0463 - 0.0052$$
$$= 0.0411\ \text{in.}$$

Practice check: Do exercise 19 on the right.

19–20 TAP DRILL AND TAP, ACME THREAD

The relative sizes of the Acme screw and the tap drill and tap are shown in Fig. 19–12. The clearances between the screw threads and the nut threads are all 0.010 in. The tap drill for the nut is therefore 2×0.010, or 0.020 in. larger than the root diameter of the screw, and the diameter of the tap is 0.020 in. larger than the outside diameter of the screw. The dimensions of the threads on the tap are as follows.

$$\text{Top flat} = 0.3707 \times \text{Pitch} - 0.0052\ \text{in.}$$

$$\text{Bottom flat} = 0.3707 \times \text{Pitch} - 0.0052\ \text{in.}$$

$$\text{Depth} = P/2 + 0.020\ \text{in.}$$

Answer to exercise 16

Double depth $= \frac{1}{6} + 0.01$ in.
$\qquad\qquad = 0.1667 + 0.01$ in.
$\qquad\qquad = 0.1767$ in.

Answer to exercise 17

Screw size $\qquad = 0.875$ in.
Double depth $\qquad \underline{= 0.170\ \text{in.}}$
$\quad$ Root diameter $= 0.705$ in.

Answer to exercise 18

Screw size $\quad = 0.1875$ in.
Pitch $\qquad\quad \underline{= 0.0625\ \text{in.}}$
$\quad$ Diameter $= 0.1250$ in.
$\qquad\qquad\quad = \frac{1}{8}$ in.

19. Find the root diameter, top flat, and bottom flat of a $\frac{7}{8}$-in.–4 Acme 29-degree screw thread.

The answer is in the margin on page 581.

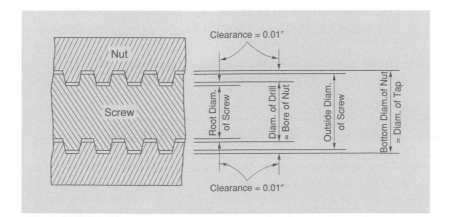

Figure 19–12

■ **Example**

A $\frac{3}{4}$-in. bolt has an Acme 29-degree thread of $\frac{1}{8}$-in. pitch. Find tap drill size, bottom and top flats of tap thread, and outside diameter of tap.

Solution

$$\begin{aligned}
\text{Tap drill size} &= \text{Root diameter} + 2 \times 0.01 \text{ in.}\\
&= 0.605 + 0.02\\
&= 0.625 \text{ in.}
\end{aligned}$$

$$\begin{aligned}
\text{Bottom flat} &= 0.3707 \times \text{Pitch} - 0.0052\\
&= 0.3707 \times \tfrac{1}{8} - 0.0052\\
&= 0.0463 - 0.0052 = 0.0411 \text{ in.}
\end{aligned}$$

$$\text{Top flat} = \text{Same as bottom flat} = 0.0411 \text{ in.}$$

$$\begin{aligned}
\text{Outside diameter of tap} &= \text{Outside diameter of screw} + 2 \times 0.01\\
&= 0.750 + 0.020 = 0.7700 \text{ in.}
\end{aligned}$$

---

20. A $\frac{7}{8}$-in. bolt has an Acme 29-degree thread of $\frac{1}{4}$-in. pitch. Find tap drill size, bottom and top flats of tap thread, and outside diameter of tap.

The answer is in the margin on page 582.

Practice check: Do exercise 20 on the left.

■ **Problems 19–20**

In each of the following Acme 29-degree screw threads, find **(a)** root diameter; **(b)** top flat of screw; **(c)** bottom flat of screw; **(d)** tap drill size; **(e)** bottom flat of tap; **(f)** top flat of tap; **(g)** outside diameter of tap. Use the space provided to work out the answers.

1. $2''$–4   **2.** $1\frac{1}{2}''$–$5\frac{1}{2}$ **3.** $2\frac{1}{8}''$–$3\frac{1}{2}$ **4.** $\frac{1}{2}''$–10   **5.** $1''$–7

6. $\frac{7}{8}''$–9   **7.** $\frac{3}{4}''$–10   **8.** $\frac{5}{8}''$–$\frac{1}{4}$ **9.** $\frac{9}{10}''$–12   **10.** $\frac{1}{2}''$–6

19–21   THE BROWN AND SHARPE 29-DEGREE WORM THREAD

This worm thread has dimensions that are similar to those of the Acme 29-degree screw thread. Section 19–19 gives the dimensions of this 29-degree worm thread.

19–22   METRIC STANDARD SCREW THREADS

The form of the metric thread and the dimensions of the parts are exactly like those of the American National thread; that is, the sides form an angle of 60° and the thread is flattened $h/8$ top and bottom (see Fig. 19–13). Both the diameter and the pitch of the screw are given in millimeters. See Appendix Table 3, Converting Millimeters to Inches and Decimals.

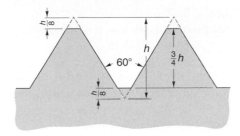

Figure 19–13

■ **Example 1**

Find the double depth of a 6.5-mm-pitch thread.

Solution   6.5 mm = 6.5 × 0.0394 in.   = 0.2559 in.
             Double depth = Pitch × 1.299 = 0.2559 × 1.299
                                          = 0.3324 in.

Practice check: Do exercise 21 on the right.

21. Find the double depth of a 4.25-mm-pitch thread.

The answer is in the margin on page 585.

■ **Example 2**

Find the root diameter of a metric screw having an outside diameter of 10 mm and a pitch of 1.5 mm.

Solution   10 mm = 0.3937 in.; 1.5 mm   = 0.0591 in.
             Outside diameter   = 0.3937 in.
             Double depth = 0.0591 × 1.299 = 0.0767 in.
               Root diameter   = 0.3170 in.

Practice check: Do exercise 22 on the right.

22. Find the root diameter of a metric screw having an outside diameter of 5.2 mm and a pitch of 2.0 mm.

The answer is in the margin on page 585.

■ **Example 3**

What size drill should be used for a metric thread with a 26-mm diameter and a 3-mm pitch?

Solution   Diameter = 26 mm = 1.0236 in.
             Pitch = 3 mm = 0.1181 in.
             Diameter   = 1.0236 in.
             Double depth = 0.1181 × 1.299   = 0.1534 in.
               Root diameter   = 0.8702 in.
             Clearance = $\frac{1}{8}$pitch = $\frac{1}{8}$ × 0.1181 = 0.0148 in.
               Drill size   = 0.8850 in.

The nearest $\frac{1}{64}$ in. above this is $\frac{57}{64}$ in. Hence, a $\frac{57}{64}$-in. drill should be used.

Practice check: Do exercise 23 on the right.

23. What size drill should be used for a metric thread with an 11-mm diameter and a 4-mm pitch?

The answer is in the margin on page 585.

■ **Example 4**

Find the width of tool point for a 7-mm-pitch thread.

24. Find the width of tool point for a 15-mm-pitch thread.

The answer is in the margin on page 586.

Solution   Pitch = 7 mm = 0.2756 in.
          Width of tool point = $\frac{1}{8}$ pitch = $\frac{1}{8} \times 0.2756$
                  = 0.0345 in.

NOTE: *To compute the various parts of a metric thread, first convert the given metric dimensions into inches and then proceed exactly as with American National threads.*

  Practice check: Do exercise 24 on the left.

■ Problems 19–22

Compute the parts of the metric threads in the following chart.

	Diameter (mm)	Pitch (mm)	Double Depth (in.)	Root Diameter (in.)	Drill Size (in.)	Width of Tool Point (in.)
1.	3	0.5	_____	_____	_____	_____
2.	5	0.75	_____	_____	_____	_____
3.	6	1.0	_____	_____	_____	_____
4.	9	1.0	_____	_____	_____	_____
5.	12	1.5	_____	_____	_____	_____
6.	20	2.5	_____	_____	_____	_____
7.	22	2.5	_____	_____	_____	_____
8.	24	3.0	_____	_____	_____	_____
9.	30	3.5	_____	_____	_____	_____
10.	40	4.0	_____	_____	_____	_____

19–23   WHITWORTH STANDARD THREADS

The Whitworth or British standard thread is a 55-degree V-thread with rounded top and bottom. If we start with a 55-degree V-thread, the Whitworth is obtained by eliminating one-sixth of the height of the V both top and bottom and rounding off the thread at the point and at the root as shown in Fig. 19–14. The depth of the resulting thread is therefore two-thirds of the depth of the corresponding V-thread.

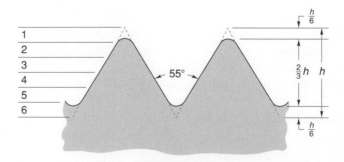

Figure 19–14

The dimensions of the Whitworth standard thread follow.

$$\text{Depth} = \text{Pitch} \times 0.6403$$

$$\text{Radius} = \text{Pitch} \times 0.1373$$

■ **Example 1**

Find the double depth of a Whitworth standard thread having 12 threads per inch.

Solution   Double depth $= 2 \times \frac{1}{12} \times 0.6403 = 0.1067$ in.

Practice check: Do exercise 25 on the right.

■ **Example 2**

Find the root diameter of a $\frac{1}{4}$-in.–20 Whitworth standard thread.

Solution   Outside diameter $= \frac{1}{4}$ in. $= 0.2500$ in.
Double depth $= 2 \times \frac{1}{20} \times 0.6403 = \underline{0.0640 \text{ in.}}$
Root diameter $= 0.1860$ in.

Practice check: Do exercise 26 on the right.

25. Find the double depth of a Whitworth standard thread having 7 threads per inch.

The answer is in the margin on page 587.

26. Find the root diameter of a $\frac{3}{8}$-in.–10 Whitworth standard thread.

The answer is in the margin on page 587.

19–24   TAP DRILL SIZE FOR WHITWORTH THREAD

The diameter of the tap drill used in boring a nut blank for a Whitworth thread is larger by one-eighth of the pitch than the root diameter of the thread.

■ **Example**

Find the proper tap drill size for a $\frac{7}{8}$-in.–9 Whitworth thread.

Solution   Outside diameter $= \frac{7}{8}$ in. $= 0.8750$ in.
Double depth $= 2 \times \frac{1}{9} \times 0.6403 = \underline{0.1423 \text{ in.}}$
Root diameter $= 0.7327$ in.
Clearance $= \frac{1}{8} \times \text{Pitch} = \frac{1}{8} \times \frac{1}{9} = \underline{0.0139 \text{ in.}}$
Drill size $= 0.7466$ in.

The nearest $\frac{1}{64}$ in. above this decimal is $\frac{3}{4}$ in. Hence, a $\frac{3}{4}$-in. drill should be used.

Practice check: Do exercise 27 on the right.

27. Find the proper tap drill size for a $\frac{5}{8}$-in.–12 Whitworth thread.

The answer is in the margin on page 587.

19–25   RADIUS OF TOOL POINT FOR WHITWORTH THREAD

■ **Example**

Find the radius of tool point for cutting a Whitworth standard thread having $4\frac{1}{2}$ threads per inch.

Solution $\text{Pitch} = \dfrac{1}{4\frac{1}{2}} = \dfrac{2}{9}$ in.

$\text{Radius} = \text{Pitch} \times 0.1373 = \frac{2}{9} \times 0.1373$
$= 0.0305$ in.

■ **Problems 19–25**

Use the space provided to find the double depth of the following Whitworth threads.

1. Pitch = $\frac{1}{24}''$ **2.** Pitch = $\frac{1}{10}''$ **3.** 9 threads per inch

4. 6 threads per inch   **5.** $3\frac{1}{2}$ threads per inch

Find the root diameter of the following Whitworth threads.

6. $\frac{3}{8}''$–16   **7.** $\frac{5}{8}''$–11   **8.** $\frac{3}{4}''$–10   **9.** $1\frac{7}{8}''$–$4\frac{1}{2}$ **10.** $3\frac{5}{8}''$–$3\frac{1}{4}$ **11.** $\frac{11}{16}''$–11

Find the proper size drill to be used for each of the following Whitworth threads.

12. $\frac{5}{16}''$–18   **13.** $\frac{1}{2}''$–12   **14.** $\frac{13}{16}''$–10   **15.** $1''$–8   **16.** $1\frac{1}{4}''$–7

17. $1\frac{1}{2}''$–6   **18.** $2''$–$4\frac{1}{2}$ **19.** $2\frac{1}{2}''$–4   **20.** $3''$–$3\frac{1}{2}$ **21.** $5\frac{1}{4}''$–$2\frac{5}{8}$

22–31. Find the radius of tool point for cutting the threads in Problems 12–21.

19–26   AMERICAN STANDARD TAPER PIPE THREADS

Figure 19–15 illustrates the following characteristics of the American Standard taper pipe threads.

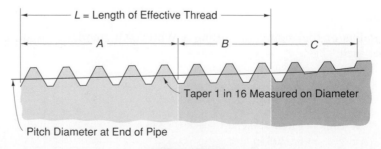

Figure 19–15

There are three varieties of thread on the pipe. At the end of the pipe, in the section marked *A*, the threads are perfect at both top and bottom. The length of section *A* is found by Briggs's formula:

$$A = (0.8D + 4.8) \times P$$

where *D* is the actual external diameter and *P* is the pitch of the thread. The taper of section *A* is $\frac{1}{16}$ in. per inch or $\frac{3}{4}$ in. per foot measured on the diameter. The threads in this section are 60-degree V-threads with top and bottom slightly flattened. The depth of the threads, therefore, instead of being $0.866 \times P$, is only $0.8 \times P$.

In the section marked *B* in Fig. 19–15, there are *two* threads having the same taper at the bottom as the threads in section *A* but with flat tops.

In the section marked *C*, there are three to four imperfect threads with flat tops and bottoms; the imperfections are caused by the chamfer of the threading die.

The value of section *L*, the length of effective thread, which consists of *A* + *B*, is found by the formula

$$L = (0.8D + 6.8)P$$

■ **Example**

Find the depth of a perfect American Standard pipe thread of $\frac{1}{8}$-in. pitch.

Solution   Depth $= 0.8 \times P = 0.8 \times \frac{1}{8} = 0.1$ in.

Practice check: Do exercise 28 on the right.

19–27   LENGTH OF PART HAVING PERFECT THREADS AND LENGTH OF EFFECTIVE THREAD

■ **Example**

A pipe whose actual external diameter is 2.875 in. is threaded with American Standard pipe threads of $\frac{1}{8}$-in. pitch. What is the length of the part with the perfect threads and what is the length of effective thread?

Solution $A = (0.8D + 4.8) \times P$
$= (0.8 \times 2.875 + 4.8) \times \frac{1}{8}$
$= (2.3 + 4.8) \times \frac{1}{8}$
$= 7.1 \times \frac{1}{8} = 0.89$ in.

$L = (0.8D + 6.8)P$
$= (0.8 \times 2.875 + 6.8)P$
$= (2.3 + 6.8)P$
$= 9.1 \times \frac{1}{8} = 1.14$ in.

Practice check: Do exercise 29 on the right.

19–28   THICKNESS OF METAL BETWEEN BOTTOM OF THREAD AND INSIDE OF PIPE IN STRAIGHT PIPE THREAD

The thickness of the metal between the bottom of the thread and the inside of the pipe is equal to half the difference between the root diameter of the thread and the actual inside diameter of the pipe. See Fig. 19–16.

Answer to exercise 21

4.25 mm $= 4.25 \times 0.0394$
$= 0.16745$ in.
Double depth $= 0.16745 \times 1.299$
$= 0.2175$ in.

Answer to exercise 22

2 mm	$= 0.0788$ in.
5.2 mm	$= 0.2049$ in.
Double depth	$= \underline{0.1024\ \text{in.}}$
Root diameter	$= 0.1025$ in.

Answer to exercise 23

Pitch	$= 0.1576$ in.
Diameter	$= 0.4334$ in.
Double depth	$= \underline{0.2047\ \text{in.}}$
Root diameter	$= 0.2287$ in.
Clearance	$= \underline{0.0197\ \text{in.}}$
Drill size	$= 0.2484$ in.

A $\frac{1}{4}$-in. drill should be used.

28. Find the depth of a perfect American Standard pipe thread of $\frac{3}{8}$-in. pitch.

The answer is in the margin on page 588.

29. A pipe whose actual external diameter is 5.325 in. is threaded with American Standard pipe threads of $\frac{3}{8}$-in. pitch. What is the length of the part with the perfect threads and what is the length of the effective thread?

The answer is in the margin on page 588.

Answer to exercise 24

 Pitch − 0.591 in.
 Width = 0.0739 in.

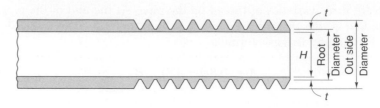

Figure 19–16

30. Find the amount of metal left between the bottom of a $\frac{1}{8}$-in. thread and the inside of the pipe on a pipe having an actual inside diameter of 2.232 in. and an outside diameter of 6 in.

The answer is in the margin on page 588.

■ **Example**

Find the amount of metal left between the bottom of a $\frac{1}{8}$-in. thread and the inside of the pipe on a pipe having an actual inside diameter of 3.548 in. and an outside diameter of 4 in.

Solution

Outside diameter	= 4.000 in.
Double depth = $2 \times \frac{1}{8} \times 0.8$	= 0.200 in.
Root diameter	= 3.800 in.
Inside diameter of pipe	= 3.548 in.
	0.252 in.

Thickness of metal = $\frac{1}{2} \times 0.252 = 0.126$ in.

Practice check: Do exercise 30 on the left.

■ **Problems 19–28**

Use the space provided to find the depth of the following perfect American Standard pipe threads.

1. $\frac{1}{27}''$ pitch   **2.** $\frac{1}{18}''$ pitch

3. $\frac{1}{14}''$ pitch   **4.** $11\frac{1}{2}$ threads per inch

Find the length of the part in which the threads are perfect and the effective length of thread in the following pipes having American Standard threads.

	Actual Outside Diameter	Number of Threads per Inch	Length of Part with Perfect Threads	Effective Length of Thread
5.	0.405″	27	_____	_____
6.	0.675″	18	_____	_____
7.	1.050″	14	_____	_____
8.	1.9″	$11\frac{1}{2}$	_____	_____
9.	4.0″	8	_____	_____

Find the thickness of the metal between the bottom of the thread and the inside of the pipe in each of the following pipes.

	Actual Inside Diameter	Actual Outside Diameter	Number of Threads per Inch	Thickness of Metal
10.	0.364″	0.540″	18	————
11.	0.623″	0.840″	14	————
12.	1.048″	1.315″	$11\frac{1}{2}$	————
13.	3.067″	3.5″	8	————
14.	9″	9.688″	8	————

19–29   LATHE GEARING FOR CUTTING SCREW THREADS

One of the most important uses to which a lathe is put is cutting screw threads. This is accomplished by closing the split nut attached to the carriage so as to engage the lead screw. In cutting a desired number of threads per inch on a screw, it is evident that while the tool travels a distance of 1 in., the piece of work must make a number of revolutions equal to the required number of threads per inch.

The speed of the tool depends on the speed of the lead screw, which in turn depends upon the pitch of the lead screws; the speed of rotation of the work depends upon the speed of the spindle. Hence, if the lathe is so geared that the spindle and the lead screw revolve at the same rate of speed, the number of threads per inch cut on the work will be the same as the number of threads per inch on the lead screw. If, for example, the lead screw has 6 threads per inch, the piece of work will revolve 6 times while the tool travels 1 in., and the resulting screw will have 6 threads per inch.

If we want to cut 12 threads per inch, the piece of work will have to revolve 12 times while the tool travels 1 in., that is, while the lead screw revolves 6 times. But for the work to revolve twice as fast as the lead screw, the spindle must turn twice as fast as the lead screw. This is accomplished by the change gears D and F in Fig. 19–17, which shows the arrangement of the gears in a simple geared lathe.

> A = Spindle gear that turns the work
>
> B and B' = Reverse gears to give direction to the lead screw as right or left hand screw is required.
>
> C = Fixed stud gear
>
> D = Change stud gear
>
> E = Intermediate or idler gear
>
> F = Lead screw gear

Case 1

If the ratio of spindle to stud is 1 to 1, that is, if gear A and gear C have the same number of teeth, then the pitch of the work will be the same as the lead screw pitch if gears D and F are of equal size. If a screw is required having three times as many threads per inch as the lead screw, the spindle will have to revolve three times as fast as the lead screw, and therefore a gear with one-third as many teeth must be placed on the stud as on the lead screw.

Answer to exercise 25

Double depth = 0.1829 in.

Answer to exercise 26

Outside diameter = 0.3750 in.
Double depth   = 0.1281 in.
  Root diameter = 0.2469 in.

Answer to exercise 27

Outside diameter = 0.6250 in.
Double depth   = 0.1067 in.
  Root diameter = 0.5183 in.
Clearance   = 0.0104 in.
  Drill size   = 0.5287 in.

A $\frac{17}{32}$-in. drill should be used.

Answer to exercise 28

Depth = 0.3 in.

Answer to exercise 29

$A = (0.8 \times 5.325 + 4.8) \times \frac{3}{5}$
  = 5.436 in.
$L = (0.8 \times 5.325 + 6.8) \times \frac{3}{5}$
  = 6.636 in.

Answer to exercise 30

Outside diameter   = 6.000 in.
Double depth   = 0.200 in.
  Root diameter   = 5.800 in.
Inside diameter   = 2.232 in.
                    3.568 in.
One-half thickness = 1.784 in.

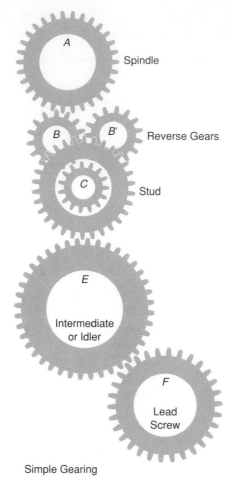

Simple Gearing

Figure 19–17

The preceding may be expressed by the following simple formula.

$$\frac{\text{Lead screw}}{\text{Screw to cut}} = \frac{\text{Stud gear}}{\text{Lead screw gear}}$$

which means that the ratio of the stud gear to the lead screw gear must be the same as the ratio of the lead screw to the screw we wish to cut.

■ **Example 1**

What change gears are required to cut a screw with 10 threads per inch if the lead screw has 6 threads per inch?

Solution $\dfrac{\text{Lead screw}}{\text{Screw to cut}} = \dfrac{6}{10} = \dfrac{6 \times 4}{10 \times 4} = \dfrac{24}{40}$

Explanation   Write a fraction with the number of threads per inch on the lead screw as the numerator and the number of threads per inch on the screw to be cut as the denominator. Since the teeth of the stud gear and the lead screw gear must be in the same ratio as the terms of this fraction, we multiply both terms of the fraction by any multiplier that will give us an equivalent fraction whose terms will represent suitable gears. In this instance, 4 was chosen as the multiplier, giving gears 24 and 40. If the 24-tooth gear is placed on the stud D and the 40-tooth gear on the lead screw F, the ratio of the speed of the stud (and therefore of the spindle) to the speed of the lead screw will be as 40 is to 24, or 10 to 6. Therefore, the piece of work will revolve 10 times to 6 turns of the lead screw, and the result will be a screw having 10 threads per inch.

Any other multiplier might be selected that would give other gears that would produce the same result as 24 and 40. For example,

$$\frac{\text{Lead screw}}{\text{Screw to cut}} = \frac{6}{10} = \frac{6 \times 6}{10 \times 6} = \frac{36}{60}$$

Multiplying both terms of the fraction by 6 gives the gears 36 and 60. Placing the 36 on the stud and the 60 on the lead screw, the speed ratio between the spindle and the lead screw is the same as before, namely, 10 to 6, and we get a screw with 10 threads per inch.

NOTE: *Since idler E does not affect the relative speeds of D and F, it does not enter into the computations.*

Practice check: Do exercise 31 on the right.

31. What change gears are required to cut a screw with 15 threads per inch if the lead screw has 8 threads per inch?

The answer is in the margin on page 591.

Case 2

If spindle gear A and stud gear C do not have the same number of teeth, that is, if the ratio of spindle to stud is not 1 to 1, we proceed as follows in computing the change gears.

1. As in Case 1, write a fraction expressing the ratio of the number of threads on the lead screw to the number of threads to be cut.
2. Multiply this ratio by the ratio of the number of teeth on the stud gear to the number of teeth on the spindle gear.
3. Find suitable gears by multiplying the result of Step 2 by any convenient multiplier.

■ **Example 2**

What change gears are required to cut 16 threads per inch with a lead screw having 6 threads per inch when the spindle gear has 20 teeth and the stud gear has 40 teeth?

Solution

$$\frac{\text{Lead screw}}{\text{Screw to cut}} = \frac{6}{16}$$

$$\frac{\text{Fixed stud gear}}{\text{Spindle gear}} = \frac{40}{20}$$

$$\text{Change gears} = \frac{6}{16} \times \frac{40}{20} = \frac{3}{4}$$

$$\frac{3}{4} \times \frac{12}{12} = \frac{36}{48}$$

$$\text{Stud gear} = 36 \text{ teeth}$$

$$\text{Lead screw gear} = 48 \text{ teeth}$$

Practice check: Do exercise 32 on the right.

32. What change gears are required to cut 20 threads per inch with a lead screw having 12 threads per inch when the spindle gear has 40 teeth and the stud gear has 60 teeth?

The answer is in the margin on page 591.

■ **Problems 19–29**

Using the space provided, find suitable change gears to be used in the cases given. Select gears from the following sct, which is furnished with a certain lathe.

24 28 32 36 40 44 48 52 56 60 64 68 72 76
80 84 88 92 96 100 104 108 112 116 120

1. To cut 8 threads per inch with a 6-pitch lead screw

2. To cut 14 threads per inch with an 8-pitch lead screw

3. To cut 12 threads per inch with an 8-pitch lead screw

4. To cut a thread of $\frac{1}{13}''$ pitch with a lead screw whose pitch is $\frac{1}{6}''$

5. To cut 10 threads per inch with a 4-pitch lead screw

6. Determine the gears for stud and lead screw when 18 threads per inch are to be cut with a lead screw whose pitch is $\frac{1}{8}''$.

7. A lathe having an 8-pitch lead screw is found to have a 32 gear on the stud and an 88 gear on the lead screw. How many threads per inch will it cut on a piece of work?

8. A lathe whose lead screw has 6 threads per inch has a lead screw gear of 36 teeth. What gear must be placed on the spindle to cut a thread of $\frac{1}{4}''$ pitch?

9. It is desired to cut 20 threads per inch on a screw. If the pitch of the lead screw is $\frac{1}{8}''$ and the stud gear has 32 teeth, what gear must be placed on the lead screw?

10. It is found that a lathe having a stud gear of 36 teeth and a lead screw gear of 84 teeth cuts a screw with 7 threads per inch. What is the pitch of the lead screw?

Solve the following problems concerning change gears.

11. If the ratio of spindle to stud is 1 to 3 and the lead screw has 8 threads per inch, what change gears are required to cut 12 threads per inch?

12. Compute the necessary change gears for cutting 20 threads per inch with a lead screw having 8 threads per inch when the spindle gear has 18 teeth and the stud gear has 36 teeth.

Answer to exercise 31

$$\frac{8}{15} = \frac{8 \times 4}{15 \times 4} = \frac{32}{60}$$

Answer to exercise 32

$$\frac{\text{Lead screw}}{\text{Screw to cut}} = \frac{12}{20}$$

$$\frac{\text{Fixed stud gear}}{\text{Spindle gear}} = \frac{60}{40}$$

$$\text{Change gears} = \frac{9}{10}$$

$$= \frac{9}{10} \times \frac{12}{12}$$

$$= \frac{108}{120}$$

Stud gear = 108 teeth
Lead screw gear = 120 teeth

13. Find the change gears for cutting a $\frac{1}{6}''$-pitch screw with a $\frac{1}{8}''$-pitch lead screw when the spindle gear has 16 teeth and the stud gear has 32 teeth.

14. If the spindle gear has 16 teeth and the stud gear 48 teeth, what change gears are required for cutting a screw with 8 threads to the inch if the lead screw has 12 threads to the inch?

15. What gear must be placed on the stud of a lathe if the ratio of spindle to stud is 1 to 2, the lead screw pitch is $\frac{1}{6}''$, an 80-tooth gear is on the lead screw, and a pitch of $\frac{1}{10}''$ is wanted on a screw?

19–30   FRACTIONAL THREADS

To cut a fractional number of threads per inch, follow the same method as for whole threads.

■ **Example**

Find the change gears necessary for cutting $11\frac{1}{2}$ threads per inch with a lead screw having 6 threads per inch.

Solution $$\frac{\text{Lead screw}}{\text{Screw to cut}} = \frac{6}{11\frac{1}{2}} = \frac{6 \times 8}{11\frac{1}{2} \times 8} = \frac{48}{92}$$

$$\text{Stud gear} = 48 \text{ teeth}$$

$$\text{Lead screw gear} = 92 \text{ teeth}$$

Practice check: Do exercise 33 on the right.

33. Find the change gears necessary for cutting $8\frac{3}{4}$ threads per inch with a lead screw having 5 threads per inch.

The answer is in the margin on page 593.

■ **Problems 19–30**

Using the space provided below, find the change gears for cutting the following fractional number of threads per inch.

1. Find the change gears for cutting $4\frac{1}{2}$ threads per inch with an 8-pitch lead screw and a spindle-to-stud ratio of 1 to 1.

2. With a spindle-to-stud ratio of 1 to 2, find the change gears for cutting a screw with $3\frac{1}{2}$ threads per inch, the lead screw having 6 threads per inch.

3. It is desired to cut $5\frac{1}{2}$ threads per inch on a bolt. With a spindle-to-stud ratio of 1 to 1 and a lead screw having 6 threads per inch, what change gears are necessary?

4. What change gears would be required in Problem 3 if the spindle-to-stud ratio were 1 to 2?

5. What change gears would be required in Problem 3 if the spindle-to-stud ratio were 1 to 3?

In Problems 6–10, assume that the spindle-to-stud ratio is 1 to 1 and compute the change gears required.

6. To cut $2\frac{3}{8}$ threads per inch with a 6-pitch lead screw

7. To cut $2\frac{3}{4}$ threads per inch with a 6-pitch lead screw

8. To cut $2\frac{1}{4}$ threads per inch with an 8-pitch lead screw

9. To cut $2\frac{5}{8}$ threads per inch with an 8-pitch lead screw

10. To cut $4\frac{1}{2}$ threads per inch with a 6-pitch lead screw

19–31   USE OF COMPOUND GEARING

Sometimes it is impossible to obtain the required speed ratio between spindle and lead screw for cutting a desired thread by simple gearing with the gears available. In addition, certain cases may require gears with an unusually large number of teeth. To obviate these possibilities, recourse is had to ***compound*** gearing. In this system, the idler E of Fig. 19–17 is replaced by two gears G and H as shown in Fig. 19–18. One of these compound gears meshes with the change gear D, and the other meshes with the lead screw gear F. The following examples will illustrate the method of computing compound gearing.

Answer to exercise 33

$$\frac{\text{Lead screw}}{\text{Screw to cut}} = \frac{5}{8\frac{3}{4}} = \frac{5 \times 8}{8\frac{3}{4} \times 8}$$
$$= \frac{40}{70}$$
Stud gear = 40 teeth
Lead screw gear = 70 teeth

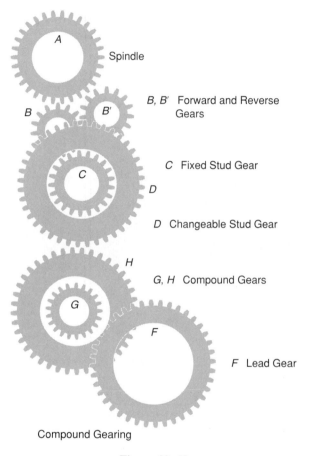

A   Spindle

B, B' Forward and Reverse Gears

C Fixed Stud Gear

D Changeable Stud Gear

G, H Compound Gears

F Lead Gear

Compound Gearing

Figure 19–18

■ **Example 1**

Find the compound gearing required to cut a screw having 35 threads per inch with a lead screw having 6 threads per inch; the ratio of spindle to stud is 1 to 1.

Solution $\dfrac{\text{Lead screw}}{\text{Screw to cut}} = \dfrac{6}{35} = \dfrac{2}{5} \times \dfrac{3}{7} = \dfrac{2 \times 16}{5 \times 16} \times \dfrac{3 \times 8}{7 \times 8} = \dfrac{32}{80} \times \dfrac{24}{56}$

Explanation

1. Express in the form of a fraction the ratio of the number of threads per inch in the lead screw to the number of threads per inch in the screw to be cut. This gives the fraction $\frac{6}{35}$.

2. Separate each term of this fraction into two factors: $6 = 2 \times 3$ and $35 = 5 \times 7$. This gives the two fractions $\frac{2}{5}$ and $\frac{3}{7}$.

3. Treat each of the fractions exactly as in simple gearing; that is, multiply both numerator and denominator by any convenient number that will give an equivalent fraction whose terms will represent suitable gears. In this case the terms of the fraction $\frac{2}{5}$ are multiplied by 16, giving the fraction $\frac{32}{80}$; and the terms of the fraction $\frac{3}{7}$ are multiplied by 8, giving the fraction $\frac{24}{56}$.

4. The numerators 32 and 24 represent driving gears, and the denominators 80 and 56 represent driven gears. Either of the drivers may be placed on the stud, and either of the driven may be placed on the lead screw.

   If any of these four gears are not among those furnished with the lathe, use, instead of 16 and 8, such other multipliers as will give available gears.

   If the ratio of fixed stud gear to spindle is not 1 to 1, multiply the ratio of lead screw to be cut by the ratio of stud to spindle, and then proceed as in Example 1 of compound gearing.

---

34. Find the compound gearing required to cut a screw having 26 threads per inch with a lead screw having 4 threads per inch; the ratio of spindle to stud is 1 to 1.

The answer is in the margin on page 596.

Practice check: Do exercise 34 on the left.

■ **Example 2**

Find the compound gearing required to cut 5 threads per inch on a lathe having a lead screw with 6 threads per inch; the ratio of fixed stud gear to spindle gear is 2 to 1.

Solution

$$\frac{\text{Lead screw}}{\text{Screw to cut}} = \frac{6}{5}$$

$$\frac{\text{Fixed stud gear}}{\text{Spindle gear}} = \frac{2}{1}$$

$$\frac{6}{5} \times \frac{2}{1} = \frac{12}{5}$$

$$\frac{12}{5} = \frac{3}{2} \times \frac{4}{2\frac{1}{2}} = \frac{3 \times 12}{2 \times 12} \times \frac{4 \times 12}{2\frac{1}{2} \times 12} = \frac{36}{24} \times \frac{48}{30}$$

The driving gears are 36 and 48.
The driven gears are 24 and 30.

---

35. Find the compound gearing required to cut 8 threads per inch on a lathe having a lead screw with 10 threads per inch; the ratio of fixed stud gear to spindle gear is 2 to 1.

The answer is in the margin on page 596.

NOTE: *In giving the sizes of the compounding gears, the first mentioned is the driver; the second, the driven gear.*

Practice check: Do exercise 35 on the left.

■ **Problems 19–31**

Using the space provided, find the compound gearing necessary to cut the following threads per inch on the lathe described.

Note: Unless otherwise stated, the ratio of spindle to stud is 1 to 1.

1. Calculate, by compounding, the change gears necessary to cut 20 threads per inch in a lathe with a lead of 6.

2. Find, by compounding, the change gears required for cutting $3\frac{1}{2}$ threads per inch in a lathe with a screw constant of 6.

3. Find, by compounding, the change gears for cutting a $\frac{1}{24}''$-pitch thread on a lathe with a lead of 6.

4. On a lathe having 8 threads per inch on the lead screw, a 72 gear is used on the lead screw and a 36 gear on the stud. If the compound gears are 48 and 24, how many threads per inch will be cut?

5. Find, by compounding, suitable gears to be used in cutting 23 threads per inch with a lead screw having 6 threads per inch.

6. What gears should be used to cut 16 threads per inch on a lathe having 8 threads per inch on the lead screw if the compounding gears are 30 and 60?

7. With a lead screw having 6 threads per inch, a 20-tooth stud gear, and an 80-tooth lead screw gear, what compounding gears could be used to cut 48 threads per inch?

8. It is desired to cut a screw with $5\frac{1}{2}$ threads per inch. If the lead screw has 6 threads per inch and the ratio of spindle to stud is 1 to 2, find the necessary change gears.

9. What change gears are required to cut $3\frac{3}{4}$ threads per inch on a lathe having a lead screw of $\frac{1}{6}''$ pitch if the compound gears are 24 and 48?

10. What change gears are required to cut a double-threaded screw of $\frac{3}{4}''$ lead if the lead screw has 6 threads per inch?

Answer to exercise 34

$$\frac{\text{Lead screw}}{\text{Screw to cut}} = \frac{4}{26} = \frac{2 \times 2}{2 \times 13}$$

$$= \frac{2 \times 8}{2 \times 8} \times \frac{2 \times 4}{13 \times 4}$$

$$= \frac{16}{16} \times \frac{8}{52}$$

Driving gear = 16 and 8
Driven gear = 16 and 52

Answer to exercise 35

$$\frac{\text{Lead screw}}{\text{Screw to cut}} = \frac{10}{8}$$

$$\frac{\text{Fixed stud gear}}{\text{Spindle gear}} = \frac{2}{1}$$

$$\frac{10}{8} \times \frac{2}{1} = \frac{20}{8}$$

$$\frac{20}{8} = \frac{3}{2} \times \frac{6\frac{2}{3}}{4}$$

$$= \frac{3 \times 6}{2 \times 6} \times \frac{6\frac{2}{3} \times 6}{4 \times 6}$$

$$= \frac{18}{12} \times \frac{40}{24}$$

Driving gears = 18 and 40
Driven gears = 12 and 24

---

36. Find the change gears for cutting a thread of 1.6-mm pitch on a lathe whose lead screw pitch is $\frac{1}{5}$ in.

The answer is in the margin on page 598.

---

11. What change gears are required to cut $2\frac{2}{3}$ threads per inch on a lathe with a lead of 6, and with 24 and 48 compound gears?

12. What change gears are required to cut $2\frac{7}{8}$ threads per inch on a lathe with a lead of 6?

13. We are required to cut a triple-threaded screw of $\frac{3}{8}''$ pitch. What change gears are required if the lead screw has 8 threads per inch?

14. We are required to cut a quadruple-threaded screw with a lead of $1\frac{1}{4}''$. Compute the compound gearing required if the lead screw has 6 threads per inch.

19–32   CUTTING METRIC THREADS

To cut a metric thread on a lathe whose lead screw pitch is given in inches, we proceed as follows.

■ **Example**

Find the change gears for cutting a thread of 2.5-mm pitch on a lathe whose lead screw pitch is $\frac{1}{6}$ in.

Solution   First find the number of threads per inch in the screw to be cut (1 in. = 25.4 mm).

$$\text{Number of threads per inch} = \frac{25.4}{2.5} = 10\frac{4}{25}$$

$$\text{Number of threads per inch in the lead screw} = \frac{1}{\frac{1}{6}} = 6$$

$$\frac{\text{Lead screw}}{\text{Screw to cut}} = \frac{6}{10\frac{4}{25}} = \frac{6}{\frac{254}{25}} = 6 \times \frac{25}{254} = \frac{150}{254}$$

$$\frac{150}{254} = \frac{2}{2} \times \frac{75}{127} = \frac{2 \times 24}{2 \times 24} \times \frac{75 \times 1}{127 \times 1} = \frac{48}{48} \times \frac{75}{127}$$

Gears 48 and 75 are drivers.
Gears 48 and 127 are driven.

NOTE: *A special 127-tooth gear is generally used in cutting metric threads on a lathe whose lead screw pitch is given in inches.*

Practice check: Do exercise 36 on the left.

■ Problems 19–32

Find the compound gearing necessary to cut the following metric threads.

1. Compute the gears required to cut a thread of 3-mm pitch with a lead screw of 6.

2. What change gears are necessary by compounding to cut a 3.5-mm-pitch thread with a lead screw of 6?

3. It is required to cut a thread of 4-mm pitch with a lead screw of 8. Find the compound gearing necessary.

4. Calculate, by compounding, the change gears necessary for cutting a thread of 7.5-mm pitch with a lead screw of 6.

5. What change gears are required for cutting a thread of 10-mm pitch with a lead screw of 6?

■ Self-Test

Use the space provided to solve the following problems involving screw threads. Check your answer with the answers in the back of the book.

1. What is the pitch of a screw having 13 threads per inch?

2. A single-threaded screw has 20 threads per inch. How far does it advance if it is turned into a nut for five complete turns?

3. A double-threaded screw has 20 threads per inch. What is its lead?

4. What is the double depth of a Sharp V-thread having 18 threads per inch?

5. What is the double depth of a Unified thread having 18 threads per inch?

6. What is the minor diameter of a $\frac{1''}{4}$–20 American National threaded bolt?

7. What is the tap drill size for the bolt in Problem 6?

8. What is the root diameter of a $\frac{3''}{4}$–6 square thread screw?

9. Find the depth of a perfect American Standard pipe thread of $\frac{1}{14}$ in. pitch.

10. What is the ratio of the change gears required to cut a screw with 9 threads per inch if the lead screw has 6 threads per inch?

■ **Chapter Test**

Solve the following problems involving screw threads. Use the procedures explained in this chapter.

1. What is the pitch of a screw having 20 threads per inch?

2. A single-threaded screw has 13 threads per inch. What is its pitch? What is its lead?

3. Find the depth of a V-thread having 20 threads per inch.

4. Find the double depth of a V-thread having 12 threads per inch.

5. Find the minor diameter of a $\frac{1''}{4}$–20 V-threaded bolt.

6. Find the proper tap drill size for a $\frac{5''}{16}$–18 V-thread.

7. Find the depth of an American Standard thread having 13 threads per inch.

8. Find the minor diameter of a $\frac{1''}{4}$–20 American Standard thread.

9. Find the double depth of a $\frac{3''}{4}$–6 square thread.

10. Find the depth of a perfect American Standard pipe thread of $\frac{1''}{18}$ pitch.

11. What is the ratio of the change gears required to cut a screw with 20 threads per inch if the lead screw has 6 threads per inch?

12. Find the change gear ratio for cutting $4\frac{1}{2}$ threads per inch with an 8-pitch lead screw and a spindle-to-stud ratio of 1 to 1.

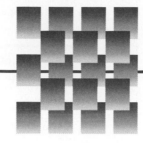

CHAPTER 20

Cutting Speed and Feed

20–1   CUTTING SPEED AND SURFACE, OR RIM, SPEED

In the interest of efficiency and economy, the machinist should run machines at various speeds, depending on the nature of the metal that is being worked, the kind of cut (that is, whether it is a roughing or a finishing cut), the feed and depth of cut, the material and shape of the cutting tool, and the strength and design of the machine. Machinery handbooks contain tables giving the proper speeds at which various metals should be worked on the different machines. Likewise, in the operation of grindstones, emery wheels, and flywheels, it is necessary to know the speed at which they may be run to obtain the best results. The proper speed for grindstones and emery wheels is determined by the strength of the stone and the nature of the grinding to be done.

Problems involving cutting speed or rim speed may be grouped in the following three classes.

1. Calculate the revolutions per minute for a given work diameter.
2. Calculate the proper feed for a roughing cut or a finishing cut.
3. Determine the time that is required to machine a given piece of work.

20–2   CUTTING SPEED ON THE LATHE

The *cutting speed* is the rate at which a tool passes over the work; it is expressed in feet per minute or meters per minute. For the lathe, cutting speed may be defined as the distance on the surface of the work that passes the point of the tool in 1 minute. The cutting speed in feet per minute is found by multiplying the circumference of the work expressed in feet by the number of revolutions per minute. The cutting speed in meters per minute is found by multiplying the circumference of the work expressed in meters by the number of revolutions per minute.

The following chart can be used for determining lathe cutting speeds using a high-speed tool bit. A chart of this type is used as a starting place. Cutting speeds may vary, depending upon the condition of the lathe, the sharpening of the tool

Cutting Speed in Surface Feet and Meters per Minute Using High-Speed Tool Bits				
	Roughing Cut		**Finishing Cut**	
Type of Metal	**ft/min**	**m/min**	**ft/min**	**m/min**
Machine steel	90	27	100	30
Cast iron	70	21	90	27
Brass	160	50	200	60
Aluminum	200	60	300	90

bit, the ridgidity of the work, the kind of coolant used, the depth of cut, and other factors.

If a cutting lubricant is used, then the above speeds can be increased by about 25%. Tungsten carbide tools allow a further increase up to 8 times the speeds shown on the chart. Check with the manufacturer of the tool for the recommended speed.

■ Example 1

Using the chart, find the rpm for machining a roughing cut on a 2-in.-diameter machine steel bar using a high-speed tool bit.

Solution   rpm = cutting speed/circumference
$$= 90/(2\pi/12)$$
$$= 171.9 \quad \text{(Use 180 rpm.)}$$

Explanation   The chart shows that the cutting speed for a roughing cut on machine steel is 90 ft/min. When this cutting speed is divided by the circumference in feet, the rpm is 171.9. Since a lathe probably doesn't have this spindle speed in the gear change box, a nearby setting would have to be used. Quite possibly, 180 rpm would be available. The 2-in. diameter must be changed to feet, and the circumference of the 2-in. bar in feet is $2\pi/12$.

Since the cutting speed is approximate and the choices of the spindle speed are limited, the machinist uses a simplified version of this formula which is easy to use and easy to remember.

$$\text{rpm} = \text{cutting speed} \times 4/\text{diameter in inches}$$

Now our solution is

1. Using the chart, find the rpm for machining a finishing cut on a 4-in.-diameter machine brass bar using a high-speed tool bit.

The answer is in the margin on page 604.

$$\text{rpm} = 90 \times 4/2$$
$$= 180$$

In like manner, if the diameter is in millimeters, then a practical rpm formula is

$$\text{rpm} = \text{cutting speed} \times 320/\text{diameter in mm}$$

Practice check: Do exercise 1 on the left.

■ Example 2

Using the chart, find the rpm for machining a roughing cut on a 50-mm-diameter machine steel bar using a high-speed tool bit.

Solution   rpm = 27 × 320/50
                    = 172.8   (Use 180 rpm.)

Explanation   The chart shows that machine steel is rough cut at 27 meters per minute. When this cutting speed is used in the metric formula, the rpm is 172.8 rpm. That spindle speed is probably not available, but 180 probably is. Therefore, a 50-mm-diameter bar could be safely and speedily cut at 180 rpm.

Practice check: Do exercise 2 on the right.

2. Using the chart, find the rpm for machining a finishing cut on a 75-mm-diameter aluminum bar using a high-speed tool bit.

The answer is in the margin on page 605.

■ Problems 20–2

Use the cutting speed chart and the simplified formulas shown in the example problems to solve the following problems. Use the space provided to work out the solutions, and round up your answers to the nearest multiple of 10.

1. Find the rpm for machining a rough cut on a 1″-diameter machine steel bar using a high-speed tool bit.

2. Find the rpm for machining a finish cut on a 25-mm-diameter machine steel bar using a high-speed tool bit.

3. Find the rpm for machining a finish cut on a $4\frac{1}{2}$″-diameter cast-iron bar using a high-speed tool bit.

4. Find the rpm for machining a rough cut on a 100-mm-diameter cast-iron bar using a high-speed tool bit.

5. Find the rpm for machining a finish cut on a brass bar with a $3\frac{7}{9}$″-diameter using a high-speed tool bit.

6. Find the rpm for machining a rough cut on a brass bar 55 mm in diameter using a high-speed tool bit.

7. Find the rpm for machining a finish cut on an aluminum bar 2.55″ in diameter using a high-speed tool bit.

8. Find the rpm for machining a rough cut on an aluminum bar 63.5 mm in diameter using a high-speed tool bit.

Answer to exercise 1

rpm $= 200 \div \dfrac{4\pi}{12}$

$= 191$ (Use 200 rpm.)

20–3   CUTTING FEED OF A LATHE

The *cutting feed* of a lathe is the distance the tool travels sideways along the work for every revolution. If, for example, the lathe is set for 0.015-in. feed, then the cutting tool will travel along the work 0.015 in. for every revolution of the spindle. In metric, if the lathe is set for 0.40-mm feed, then the cutting tool will travel along the work 0.40 mm for every revolution of the spindle. The feed is usually dependent upon the speed of the lead screw, which is regulated by the gears in the quick-change gear box.

In general, for a given piece of metal, a faster feed is used for roughing cuts, and a slower feed is used for finishing cuts. The roughing cut is to remove excess material as quickly as possible, and the finishing cut is to make the surface of the piece acceptable for the intended purpose of the piece.

The feed is usually less on harder metals and greater on softer materials. The following table will give a starting point for feeds using a high-speed tool bit. If a cutting lubricant is used, then the feed can be increased about 25%. Tungsten carbide tools allow a further increase. The manufacturer can supply a recommended feed for their product.

	Cutting Feed Using a High-Speed Tool Bit			
	Roughing Cut		**Finishing Cut**	
Type of Metal	**in.**	**mm**	**in.**	**mm**
Machine steel	0.010	0.25	0.003	0.08
Cast iron	0.015	0.40	0.005	0.15
Brass	0.020	0.50	0.003	0.08
Aluminum	0.020	0.50	0.005	0.15

The spindle speed and the feed determine how much time is needed to complete an operation. The mechanic has a goal of getting each job done as quickly as possible and thus should adjust the feed and speed to rates that are consistent with getting the job done quickly.

■ **Example**

If the spindle speed of a lathe is 180 rpm and the feed is set at 0.010 in., how long will it take to machine one pass across a 6-in. length of 2-in.-diameter machine steel bar?

Solution

$$\text{Feed} = 0.010 \text{ in.}$$
$$\text{Number of } 0.010 \text{ in. in } 6 \text{ in.} = 600$$
$$\text{Time} = 600/180 = 3\tfrac{1}{3} \text{ minutes}$$

Explanation   The tool is traveling sideways at the rate of 0.010 in. for each turn of the spindle. The tool must travel sideways for 6 in. to complete the pass. Each rotation of the spindle uses up 0.020 in. of the 6-in. length of the bar. There are 600 one hundredths of 1 inch in 6 in. The spindle is rotating at 180 rpm. The time then would be 600 one hundredths divided by 180 rpm.

Practice check: Do exercise 3 on the left.

3. Find the time it would take to machine one pass across a $3\tfrac{3}{4}$-in. length of 3-in.-diameter machine cast-iron bar. The feed is set at 0.015 in., and the spindle speed is 150 rpm.

The answer is in the margin on page 606.

■ Problems 20–3

Use the table on cutting speed and the table on feed to solve the time needed to make one pass in the following problems. In Problems 5–10, use the simplified formula in arriving at the rpm.

$$\text{rpm} = CS \times 4/\text{diameter in inches} \quad \text{or}$$

$$\text{rpm} = CS \times 320/\text{diameter in millimeters}$$

Round up the rpm to the nearest multiple of 10. Use the space provided to work out the answers.

Answer to exercise 2

$$\text{rpm} = 90 \times \frac{320}{75}$$

$$= 384$$

1. Find the time it would take to machine one pass across a $4\frac{1}{2}''$ length of $4''$-diameter machine steel bar. The feed is set at 0.010 in., and the spindle speed is 80 rpm.

2. Find the time it would take to machine one pass across a 100-mm length of 50-mm-diameter machine steel bar. The feed is 0.25 mm, and the spindle speed is 180 rpm.

3. Find the length of time it would take to machine one pass across an aluminum bar that is $5''$ in diameter and $12''$ long. The feed is set at 0.020 in., and the spindle speed is 160 rpm.

4. Find the length of time it would take to machine one pass across a bronze bar that is 75 mm in diameter and 180 mm long. Use a feed of 0.40 mm and a spindle speed of 120 rpm.

5. Find the time necessary to machine a rough cut $6''$ long on a $1''$-diameter machine steel bar using a high-speed tool bit.

6. Find the time necessary to machine a finish cut 140 mm long on a 25-mm-diameter machine steel bar using a high-speed tool bit.

7. Find the time necessary to machine a rough cut 50 mm long on a 100-mm-diameter cast-iron bar using a high-speed tool bit.

8. Find the time necessary to machine a finish cut $8''$ long on a brass bar $3\frac{7}{8}''$ in diameter using a high-speed tool bit.

9. Find the time necessary to machine a rough cut $12''$ long on an aluminum bar $2.55''$ in diameter using a high-speed tool bit.

10. Find the time necessary to machine a finish cut $12''$ long on an aluminum bar $1.55''$ in diameter using a high-speed tool bit.

Answer to exercise 3

Time $= \frac{250}{150} = 1\frac{2}{3}$ minutes

20–4   DRILL PRESS AND MILLING MACHINE

The *cutting speed* of the drill press and milling machine is defined as the rate at which the tool passes over the work. This definition is the same as that given for the lathe, but in this case it is the tool moving and not the work. The same chart can be used. The cutting speed, now a rotational speed, must be adjusted to fit the rpm possible on each machine.

There is no exact rule for determining the speed and feed using cutting tools for a particular material. The best that can be done is to use the charts available to arrive at a starting speed and feed and then adjust the speed and feed according to how the material is cutting. Experience will help to determine a starting speed that will remove material quickly and machine an acceptable finish for the application.

Use the charts in Sections 20–2 and 20–3 for the examples and problems in this chapter.

■ Example 1

Find the rotational speed of a 1-in. drill that is to bore a 1-in. hole in a piece of machine steel.

Solution   The machine steel cutting speed is 90 ft per minute. Thus,

$$\text{rpm} = \text{cutting speed} \times 4/\text{diameter in inches}$$
$$= 90 \times 4/1$$
$$= 360$$

Explanation   The chart shows that machine steel is rough cut at 90 ft/min. Using the simplified rpm formula, the drill speed should be about 360 rpm.

Practice check: Do exercise 4 on the left.

4. Find the rotational speed of a 2-in. drill that is to bore a 2-in. hole in a piece of brass.

The answer is in the margin on page 608.

■ Example 2

Find the rotational speed of a 25-mm milling cutter used to cut a slot in a piece of machine steel.

Solution   The machine steel cutting speed is 27 m/min. Thus,

$$\text{rpm} = \text{cutting speed} \times 320/\text{diameter in mm}$$
$$= 27 \times 320/25$$
$$= 345.6 \quad \text{(Use 360 rpm.)}$$

Explanation   The chart shows that machine steel is rough cut at 27 m/min. Using the simplified rpm formula, the cutter speed should be about 360 rpm. Examples 1 and 2 have about the same cutter speeds because the cutters are about the same size.

Practice check: Do exercise 5 on the left.

5. Find the rotational speed of a 55-mm milling cutter used to cut a slot in a piece of cast iron.

The answer is in the margin on page 608.

■ Problems 20–4

1. Find the rotational speed of a 1″ drill used to bore a 1″ hole in cast iron.

2. Find the rotational speed of a 30-mm drill used to bore a 30-mm hole in a piece of brass.

3. Find the rotational speed of a 2″ drill used to bore a 2″ hole in a piece of aluminum.

4. Find the rotational speed of a 45-mm milling cutter used to face off machine steel.

5. Find the rotational speed of a 40-mm milling cutter used to side-cut a piece of brass.

6. Find the rotational speed of a 3″ drill used to bore a 3″ hole in a piece of cast iron.

7. Find the rotational speed of a 1.5″ drill used to bore a 1.5″ hole in a piece of machine steel.

8. Find the rotational speed of a 30-mm milling cutter used to side-cut a piece of aluminum.

20–5   DRILL PRESS FEED

The feed of a drill press is the distance the drill penetrates the work in 1 revolution. If the drill revolves 100 times while it penetrates a distance of 1 in., the feed is 0.01 in. per revolution.

■ **Example**

Find the time required to drill through a piece of work 3 in. thick with a drill making 240 rpm if the feed per revolution is 0.007 in.

Solution
$$\text{Penetration in 1 minute} = 240 \times 0.007$$
$$= 1.680 \text{ in.}$$

$$\text{Time to drill through 3 in.} = 3 \div 1.68$$
$$= 1.786 \text{ minutes}$$
$$= 1 \text{ minute 47 seconds}$$

Explanation   Since the drill penetrates 0.007 in. for each turn, in 1 minute it will advance 240 times 0.007 in., or 1.68 in. To drill through 3 in., it will take as many times 1 minute as 1.68 in. is contained in 3 in., or $3 \div 1.68 = 1.786$ minutes.

Practice check: Do exercise 6 on right.

6. Find the time required to drill through a piece of work 4 in. thick with a drill making 180 rpm if the feed per revolution is 0.010.

The answer is in the margin on page 610.

■ Problems 20–5

Using the space provided, find the time required to do the following drill press operations.

1. With a speed of 250 rpm and a feed of 0.018″ per revolution, how long will it take to drill through a piece of cast iron $4\frac{1}{4}$″ thick?

2. A $1\frac{1}{8}$″ drill making 280 rpm is fed 0.016″ per revolution. Find the time required to drill through a piece $3\frac{7}{16}$″ thick.

3. How long will it take to drill 24 holes in a plate $1\frac{1}{8}$″ thick with a drill making 225 rpm and a feed of 0.015″ per revolution? Allow $\frac{1}{2}$ minute for adjusting the plate after each hole is drilled.

4. With a feed of 0.025″ and a speed of 280 rpm, how long will it take to drill through a piece $3\frac{7}{8}$″ thick?

5. At a speed of 300 rpm and a feed of 0.02″, how long will it take to drill through a piece $4\frac{1}{2}$″ thick?

6. With a speed of 250 rpm and a feed of 0.40 mm per revolution, how long will it take to drill through a piece of cast iron 100 mm thick?

7. At a speed of 300 rpm and a feed of 0.50 mm, how long will it take to drill through 90 mm of machine steel?

8. A 30-mm drill at 250 rpm is fed at 0.35 mm per revolution. How long will it take to drill through a piece 85 mm thick?

9. How long will it take to drill 20 holes in a plate 90 mm thick with a drill making 250 rpm and a feed of 0.30 mm per revolution? Allow 40 seconds for adjusting the plate after each hole is drilled.

10. With a feed of 0.35 mm and a speed of 300 rpm, how long will it take to drill through a piece 50 mm thick?

20-6   MILLING MACHINE FEED

The feed of a milling machine is the distance traveled by the table while the cutter makes 1 revolution. If a milling cutter revolves 80 times while the table moves 1 in., the feed is 0.0125 in.

■ **Example 1**

Find the feed per minute of a milling cutter if the feed is 0.02 in. per revolution and the cutter makes 120 rpm.

Solution   Feed per minute = $120 \times 0.02 = 2.4$ in.

Explanation   Since the cutter makes 120 rpm, it will cut 120×0.02 in., or 2.4 in., in 1 minute.

Practice check: Do exercise 7 on the right.

7. Find the feed per minute of a milling cutter if the feed is 0.04 in. per revolution and the cutter makes 150 rpm.

The answer is in the margin on page 611.

■ **Example 2**

How long will it take a cutter to traverse a piece of work 10 in. long if it makes 32 rpm and the feed is 0.025 in.?

Solution   Total revolutions required = $10 \div 0.025 = 400$

$$\text{Time required} = 400 \div 32$$
$$= 12.5 \text{ minutes}$$

Explanation   With a feed of 0.025 in., 40 revolutions of the cutter are required for each inch traversed. Since the cutter makes 32 revolutions in 1 minute, to make 400 revolutions it will take $400 \div 32$ or 12.5 minutes.

Practice check: Do exercise 8 on the right.

8. How long will it take a cutter to traverse a piece of work 35 in. long if it makes 65 rpm and the feed is 0.50?

The answer is in the margin on page 611.

■ **Example 3**

A 2-in. milling cutter makes 210 rpm. If the table feed is 7 in. per minute, find the cutting speed. Find the feed per tooth if the cutter has 14 teeth.

Solution   Cutting speed = circumference $\times$ rpm
$$= 2 \times \tfrac{22}{7} \times \tfrac{1}{12} \times 210 = 110 \text{ ft per minute}$$

Feed per revolution = $7 \div 210 = \tfrac{7}{210} = 0.0333$ in.

Feed per tooth = $0.0333 \div 14 = 0.0024$ in.

Explanation   To find the cutting speed, multiply the circumference by the revolutions per minute. Since the table moves 7 in. in the time that the cutter makes 210 revolutions, in 1 revolution of the cutter the table will move $\tfrac{7}{210}$ of 7 in., or $\tfrac{1}{30}$ in. Since there are 14 teeth in the cutter, the feed per tooth will be $\tfrac{1}{14}$ of the feed per revolution, or $\tfrac{1}{14}$ of 0.0333 in. = 0.0024 in.

Practice check: Do exercise 9 on the right.

9. A 4-in. milling cutter makes 200 rpm. If the table feed is 5 in. per minute, find the cutting speed. Find the feed per tooth if the cutter has 20 teeth.

The answer is in the margin on page 611.

Answer to exercise 6

Penetration in 1 min = 1.8 in.
Time = 2.2 min
or 2 min 12 sec

■ **Example 4**

Compare the result in Example 3 by solving the problem using the simplified formula in Section 20–2. That is, if rpm = CS × 4/diameter of work in inches, then

$$CS = \text{diameter of work in inches} \times \text{rpm}/4$$

or in metric

$$CS = \text{diameter of work in mm} \times \text{rpm}/320$$

Solution

$$\text{Cutting speed} = \text{diameter} \times \text{rpm}/4$$
$$= 2 \times 210/4 = 420/4 = 105 \text{ ft/min}$$

$$\text{Feed per revolution} = 7/120 = 0.0333 \text{ in.}$$

$$\text{Feed per tooth} = 0.0333/14 = 0.0024 \text{ in.}$$

Explanation   We found the cutting speed by using the simplified rpm formula and applying algebra to make the simplified formula into one for cutting speed. The rest of the computation is the same as in Example 3.

■ **Problems 20–6**

Use the space provided to solve the following problems involving the milling machine. Use the simplified formulas for cutting speed.

$$CS = \text{diameter} \times \text{rpm}/4 \quad \text{for inch measures}$$

$$CS = \text{diameter} \times \text{rpm}/320 \quad \text{for metric measures}$$

1. A milling cutter revolves 75 times while the table moves 1″. Find the feed per revolution.

2. A milling cutter making 150 rpm has a feed of 0.02″. What is the feed per minute?

3. What is the feed per revolution of a milling cutter that makes 80 rpm with a table feed of $6\frac{1}{2}''$? What is the feed per tooth if the cutter has 10 teeth?

4. A $\frac{3}{4}''$ end mill makes 210 rpm, and the table feed is $8\frac{1}{2}''$ per minute. Find the cutting speed. Find the feed per tooth if the end mill has 8 teeth.

5. A milling cutter makes 100 rpm with a table feed of $5\frac{3}{8}''$. What is the feed per revolution? What is the feed per tooth if the cutter has 12 teeth?

6. A $\frac{3}{4}''$ cutter with 10 teeth makes 200 rpm with a table feed of $8\frac{3}{4}''$. Find the cutting speed and the feed per tooth.

7. A milling cutter $2\frac{1}{4}''$ in diameter has 12 teeth. If the cutting speed is 75 ft per minute, find the feed per minute when every tooth is allowed to cut a chip 0.005″ thick.

8. A milling cutter 3″ in diameter has 15 teeth. If it makes 48 rpm and the table feed is $2\frac{3}{4}''$ per minute, find the thickness of the chip cut by each tooth.

Answer to exercise 7

Feed per minute = 150 × 0.04
                = 6 in.

Answer to exercise 8

Total revolutions = 70
          Time = 1.077 min
             or 1 min 5 sec

Answer to exercise 9

Cutting speed = 209.5 ft per min
Feed per rev. = 0.025 in.
Feed per tooth = 0.00125 in.

9. A $\frac{7}{8}''$ end mill with 8 teeth has a surface speed of 35 ft per minute. What is the feed per minute if each tooth cuts a chip 0.004″ thick?

10. A milling cutter revolves 80 times while the table moves 25 mm. Find the feed per revolution.

11. A milling cutter making 250 rpm has a feed of 0.60 mm. What is the feed per minute?

12. What is the feed per revolution of a milling cutter that makes 100 rpm with a table feed of 200 mm? What is the feed per tooth if the cutter has 12 teeth?

13. A 20-mm end mill makes 250 rpm, and the table feed is 220 mm per minute. Find the cutting speed. Find the feed per tooth if the cutter has 8 teeth.

14. A milling cutter makes 150 rpm with a table feed of 150 mm per minute. What is the feed per revolution? What is the feed per tooth if the cutter has 10 teeth?

15. A 20-mm cutter with 12 teeth makes 200 rpm with a table feed of 250 mm. Find the cutting speed and the feed per tooth.

20–7   SURFACE SPEED OR RIM SPEED

The *surface speed* or *rim speed* of an emery wheel or flywheel is the number of feet that a point on the rim or circumference of the wheel travels in 1 minute. Every time that a wheel makes 1 complete turn, a point on its rim travels a distance equal to its circumference. The number of feet that a point on the rim will travel in 1 minute is equal to the number of turns that the wheel makes in 1 minute multiplied by its circumference expressed in feet.

■ Example

Find the surface speed of an emery wheel 8 in. in diameter making 240 rpm.

Solution $8 \times \frac{22}{7} \times \frac{1}{12} \times 240 = 503$ ft per minute

Explanation   In each turn of the wheel, a point on its rim travels a distance equal to its circumference, that is, $8 \times \frac{22}{7}$. Divide this product by 12 to express the circumference in feet. Since the wheel turns 240 times in 1 minute, a point on the rim will travel, in 1 minute, 240 times the length of 1 circumference; therefore, multiply the circumference by 240.

Practice check: Do exercise 10 on the left.

10. Find the surface speed of a flywheel 15 in. in diameter making 200 rpm.

The answer is in the margin on page 614.

■ Problems 20–7

Using the space provided below, find the cutting speed in the following problems. Use $\frac{22}{7}$ as the approximation for π when finding the circumference.

1. Find the rim speed of an 8″ emery wheel that is mounted on a 1,760-rpm electric motor.

2. Find the rim speed of a 6″ emery wheel that is mounted on a 3,400-rpm, 220-volt electric motor.

3. Find the surface speed of an emery wheel 12″ in diameter making 210 rpm.

4. An 8′ flywheel makes 200 rpm. Find the rim speed.

5. Find the surface speed of a flywheel 15″ in diameter making 250 rpm.

6. Find the rim speed of a 3″ emery wheel that is mounted on a 4,500-rpm, 220-volt electric motor.

7. A 6′ flywheel makes 180 rpm. Find the rim speed.

8. A 12″ emery wheel is driven by a motor that makes 2,000 rpm. A 3″ pulley on the motor drives an 8″ pulley on the emery wheel spindle. Find the surface speed of the emery wheel in feet per minute.

20–8   FINDING THE REVOLUTIONS PER MINUTE, GIVEN THE DIAMETER AND THE CUTTING SPEED

■ **Example**

At how many rpm should a lathe be run to give a cutting speed of 36 ft per minute when turning a rod $2\frac{1}{2}$ in. in diameter?

Solution   Circumference $= \frac{5}{2} \times \frac{22}{7} \times \frac{1}{12} = \frac{55}{84}$ ft

$$\text{rpm} = \text{cutting speed} \div \text{circumference}$$
$$= 36 \div \frac{55}{84} = 36 \times \frac{84}{55} = 55$$

Explanation   Since in every turn of the piece of work a distance equal to its circumference passes the cutting tool, for 36 ft of the surface of the work to pass the tool, the piece must make as many times one turn as the length of the circumference is contained in 36 ft. The rule is

$$\text{rpm} = \text{cutting speed} \div \text{circumference}$$

Practice check: Do exercise 11 on the right.

11. How many rpm must a 12″ circular saw make to give a cutting speed of 4,000 ft per minute?

The answer is in the margin on page 616.

■ **Problems 20–8**

Using the space provided, find the revolutions per minute required in the following problems. Use $\frac{22}{7}$ as an approximation for π when finding the circumference.

1. What should be the maximum number of rpm of a 6′ flywheel if any point on its rim should travel not more than 1 mile per minute?

2. At how many revolutions per minute should a 6″ gear blank be turned if the cutting speed is to be 42 ft per minute?

3. A certain grindstone may run with a rim speed of 1,200 ft per minute. What should be the rpm if the diameter of the stone is 30″?

4. At how many rpm may a 4′ grindstone be run if it can stand a surface speed of 700 ft per minute?

Answer to exercise 10

Speed = $15 \times \frac{22}{7} \times \frac{1}{12} \times 200$
          = 785.7 ft per min

5. The cutting speed of a piece of work in a lathe is 250 ft per minute. If the diameter is $2\frac{1}{2}''$, what must be its rpm?

6. How many rpm must a 24″ circular saw make to give a cutting speed of 3,600 ft per minute?

7. How many rpm should a steel shaft $1\frac{3}{4}''$ in diameter make if it is to be turned in the lathe with a cutting speed of 30 ft per minute?

8. A crank pin of annealed tool steel $3\frac{3}{4}''$ in diameter is turned in the lathe with a cutting speed of 20 ft per minute. How many rpm must it make?

9. The surface of a cast-iron pulley 12″ in diameter is turned in the lathe with a cutting speed of 40 ft per minute. How many rpm must it make?

10. The cutting speed on a cast-iron cylinder $4\frac{1}{2}''$ in diameter that is bored in a lathe is 42 ft per minute. How many rpm does it make?

20–9   FINDING THE DIAMETER, GIVEN THE RIM SPEED AND THE REVOLUTIONS PER MINUTE

■ **Example**

An emery wheel that is required to have a rim speed of 1 mile per minute is to be placed on a spindle that makes 1,600 rpm. What must be the diameter of the emery wheel?

Solution

$$\text{Circumference} = \frac{\text{rim speed}}{\text{rpm}} = \frac{5,280}{1,600} \times 12 = 39.6 \text{ in.}$$

$$\begin{aligned}\text{Diameter} &= \text{circumference} \div \tfrac{22}{7} \\ &= 39.6 \div \tfrac{22}{7} = 39.6 \times \tfrac{7}{22} \\ &= 12.6 \text{ in.}\end{aligned}$$

Explanation   In a mile there are 5,280 ft. Since in 1,600 turns a point on the rim travels 5,280 ft, in 1 turn it will travel $\frac{1}{1,600}$ of 5,280 ft. But in 1 turn a point on the rim travels a distance equal to the circumference; therefore, dividing the rim speed by the turns made in 1 minute gives the length of the circumference in feet. Multiply by 12 to change it to inches. To find the diameter, divide the circumference by $\frac{22}{7}$, giving 12.6 in. A 12-in. wheel would be used.

Practice check: Do exercise 12 on the right.

■ **Problems 20–9**

Using the space provided, solve the following problems involving rim speed and revolutions per minute. Use $\frac{22}{7}$ as an approximation for π.

1. An emery wheel having an allowable surface speed of 4,500 ft per minute makes 1,500 rpm. What is its diameter?

2. A flywheel making 330 rpm has a rim speed of 1 mile per minute. What is its diameter?

3. What must be the diameter of a circular saw to give a cutting speed of 3,000 ft per minute if it revolves at the rate of 400 rpm?

4. An emery wheel which is required to have a surface speed of 5,000 ft per minute is driven by a 2,000-rpm motor. What must be the diameter of the emery wheel if the motor pulley is 3″ in diameter and the pulley on the emery wheel spindle is 8″ in diameter?

5. If the motor in Problem 4 is used to drive a 12″ emery wheel, what must be the size of the pulley on the emery wheel spindle if a speed of 4,500 ft per minute is required?

6. What is the diameter of a rod that makes 48 rpm and has a cutting speed of 36 ft per minute?

7. What is the diameter of a crank that makes 24 rpm and has a cutting speed of 20 ft per minute?

8. A brass rod is turned in the lathe at 400 rpm with a cutting speed of 75 ft per minute. What is the diameter?

12. A flywheel having an allowable surface speed of 5,200 ft per minute makes 1,400 rpm. What is its diameter?

The answer is in the margin on page 617.

9. The bearings at the ends of a crank shaft are turned in a lathe at 18 rpm with a cutting speed of 30 ft per minute. What is the diameter?

10. The rim speed of a pulley making 50 rpm is 150 ft per minute. What is the diameter of the pulley?

20–10   CUTTING SPEED ON THE PLANER

In figuring the ***actual*** or ***net cutting speed*** on the planer, take into account the time occupied by the idle return stroke. If a planer has a forward stroke of 16 ft per minute and a return stroke of 32 ft per minute, the return stroke will take half as much time as the forward stroke. Assuming the cut to be 16 ft long, the forward stroke will take 1 minute and the return $\frac{1}{2}$ minute. The complete stroke, cut and return, will take $1\frac{1}{2}$ minutes. But since it takes $1\frac{1}{2}$ minutes to cut 16 ft, it will cut in 1 minute a distance equal to $16 \div 1\frac{1}{2} = 16 \times \frac{2}{3} = 10\frac{2}{3}$ ft, which is called the net or actual cutting speed of the planer. That is, the tool in the forward stroke moves at the rate of 16 ft per minute while in contact with the work, but because of the time lost by the idle return stroke, the actual or net result is only an average of $10\frac{2}{3}$ ft cut for every minute that the planer is operating.

■ Example

What is the net or actual cutting speed of a planer making 10 strokes per minute, each 2 ft 6 in. long, if the reverse is 3 to 1? Find the theoretical cutting speed.

13. Find the actual and the theoretical cutting speeds of a planer making 8 strokes per minute. The length of stroke is 30″, and the return stroke is 4 to 1.

The answer is in the margin on page 618.

Solution
$$2 \text{ ft 6 in.} = 2\frac{1}{2} \text{ ft}$$
$$\text{Distance traveled in 1 direction in 1 minute} = 10 \times 2\frac{1}{2} = 25 \text{ ft}$$
$$= \text{net cutting speed}$$

Since the reverse is 3 to 1, the forward stroke will take up $\frac{3}{4}$ of each minute, whereas the return takes $\frac{1}{4}$. That is, it takes the planer $\frac{3}{4}$ of a minute to travel the 25 ft. In 1 minute the theoretical cutting speed would be

$$25 \div \tfrac{3}{4} = 25 \times \tfrac{4}{3} = 33\tfrac{1}{3} \text{ ft}$$

Practice check: Do exercise 13 on the left.

■ Problems 20–10

Use the space provided to find the actual and the theoretical cutting speeds of the following planers.

1. A planer has a forward stroke of 18 ft per minute and a return twice as fast. Find the actual and the theoretical cutting speeds.

2. A planer with a 4-to-1 return makes 12 strokes a minute, each stroke 27″ long. Find the net and the theoretical cutting speeds.

3. Find the actual and the theoretical cutting speeds of a planer making 14 strokes per minute; the length of stroke is 3′6″ and the return stroke 3 to 1.

4. What are the net and the theoretical cutting speeds of a planer having a forward stroke of 24 ft per minute and a return of 2 to 1?

Answer to exercise 12

Circumference = 44.6 in.
Diameter = 14.2 in.

5. With a return of 2 to 1 and a stroke of 4′3″, what are the net and the theoretical cutting speeds of a planer making 8 strokes a minute?

6. A planer with a 3-to-1 return makes 15 strokes per minute. Find the actual and the theoretical cutting speeds if the stroke is 28″ long.

20–11 TIME REQUIRED FOR A JOB ON THE PLANER

■ **Example**

Find the time required for one complete cut on a casting 1 ft 3 in. × 3 ft 6 in. if the cutting speed is 20 ft per minute, the feed $\frac{1}{4}$ in. per stroke, and the return 2 to 1. Allow a clearance of 3 in. at each end.

NOTE: *A feed of $\frac{1}{4}$ in. per stroke means that for each forward stroke, there is a sideways motion of the tool of $\frac{1}{4}$ in.*

Solution Total length of forward stroke = 3 ft 6 in. + clearance
= 3 ft 6 in. + 6 in. = 4 ft 0 in.

Since the cutting speed is 20 ft per minute, whereas the length of the forward stroke is only 4 ft, the forward stroke will take only $\frac{4}{20}$ or $\frac{1}{5}$ of a minute. The return stroke, being twice as fast, will take $\frac{1}{2}$ of $\frac{1}{5}$ or $\frac{1}{10}$ of a minute.

The time required for one complete stroke is $\frac{1}{5} + \frac{1}{10} = \frac{3}{10}$ of a minute. Since the feed is $\frac{1}{4}$ in., four strokes will be required for every inch of width of the casting. Thus,

Total number of strokes = 15 × 4 = 60

Since every complete stroke takes $\frac{3}{10}$ of a minute, 60 strokes will take 60 × $\frac{3}{10}$ = 18 minutes.

Practice check: Do exercise 14 on the right.

14. A planer has a forward movement of 14 ft a minute and a return of 3 to 1. How long will it take for a complete cut on a casting 1′6″ × 4′2″ with a feed of $\frac{1}{8}$″? Allow 3″ at each end for tool clearance.

The answer is in the margin on page 619.

■ **Problems 20–11**

Using the space provided, find the time required for one complete cut in each of the following planer problems.

1. Find the time required for taking a complete cut on a plate $2' \times 3'$ if the cutting speed is 30 ft per minute, the return 4 to 1, and the feed $\tfrac{1}{8}''$. The clearance at each end is $3''$.

2. A planer has a forward movement of 24 ft a minute and a return of 3 to 1. How long will it take for a complete cut on a casting $1'8'' \times 5'6''$ with a feed of $\tfrac{3}{16}''$? Allow $3''$ at each end for tool clearance.

3. How long will it take to finish one face of a casting $2'6'' \times 4'0''$ if the planer has a forward stroke of 20 ft per minute, a return of 3 to 1, a roughing feed of $\tfrac{1}{16}''$, and a finishing feed of $\tfrac{1}{8}''$? Allow $3''$ at each end for the tool to clear.

4. With a forward stroke of 36 ft per minute, a return of 2 to 1, and a feed of $\tfrac{3}{32}''$, how long will it take for one complete cut on a face $15'' \times 3'4''$? Allow $3''$ at each end for tool clearance.

5. Find the time required for finishing a casting $2'6'' \times 6'0''$ on a planer having a forward movement of 40 ft per minute and a return of 2 to 1, if a feed of $\tfrac{1}{8}''$ is used for roughing and $\tfrac{1}{4}''$ for finishing. Clearance for the tool is $3''$ at each end.

20–12   NUMBER OF STROKES PER MINUTE

To find the number of strokes per minute for a given length of stroke and given cutting and return speeds, proceed as in the following example.

■ **Example**

Find the number of strokes per minute a planer must make if the forward or cutting speed is 28 ft per minute, the return 2 to 1, and the length of stroke 8 ft including clearance.

Solution   Time required for forward stroke $= \tfrac{8}{28} = \tfrac{2}{7}$ minute

Time required for return stroke $= \tfrac{1}{2}$ of $\tfrac{2}{7} = \tfrac{1}{7}$ minute

Time required for complete stroke $= \tfrac{1}{7} + \tfrac{2}{7} = \tfrac{3}{7}$ minute

Since each complete stroke takes $\tfrac{3}{7}$ of a minute, in 1 minute the planer will make $1 \div \tfrac{3}{7}$, or $1 \times \tfrac{7}{3} = 2\tfrac{1}{3}$, strokes.

Practice check: Do exercise 15 on the left.

15. A planer has a cutting stroke of 26 ft per minute, a return of 4 to 1, and a length of stroke of $4'4''$. How many strokes will it make in 1 minute?

The answer is in the margin on page 621.

■ Problems 20–12

Use the space provided below to find the number of strokes per minute in the following planer problems.

NOTE: *In each of the following problems, the given length of stroke includes the clearance for the tool at both ends.*

1. Find the number of strokes per minute of a planer having a forward stroke of 24 ft per minute, a return of 2 to 1, and a length of stroke of 6'.

2. How many strokes per minute must a planer make if its forward movement is 30 ft per minute, its return 2 to 1, and its length of stroke 4'?

3. A planer has a cutting stroke of 20 ft per minute, a return of 3 to 1, and a length of stroke of 5'6''. How many strokes will it make in 1 minute?

4. Compute the number of strokes per minute for a cut 6'6'' long if the cutting speed of the planer is 24 ft per minute and the return 2 to 1.

5. The forward movement of a planer is 32 ft per minute, and the return is 2 to 1. Compute the number of strokes per minute required for a stroke 5' long.

6. Find the number of strokes per minute of a planer having a forward stroke of 30 ft per minute, a return of 3 to 1, and a length of stroke of 4'.

7. A planer has a cutting stroke of 25 ft per minute, a return of 2 to 1, and a length of stroke of 5'. How many strokes will it make in 1 minute?

8. The forward movement of a planer is 35 ft per minute and the return is 3 to 1. Compute the number of strokes per minute required for a stroke 6' long.

20–13   CUTTING SPEED OF SHAPERS

The cutting speed of a *geared shaper* is obtained in exactly the same way as the cutting speed of a planer.

The method of computing the cutting speed of a *crank shaper* is illustrated in the following example.

■ **Example**

Find the cutting speed of a shaper making 30 rpm with a reverse of 2 to 1 if the length of stroke is 14 in.

Solution   Length of stroke $= 14$ in. $= 1\frac{1}{6}$ ft

$$\text{Total length of metal cut in 1 minute} = 30 \times 1\tfrac{1}{6} = 35 \text{ ft}$$

$$\text{Cutting speed} = 35 \times \tfrac{3}{2} = 52.5 \text{ ft per minute}$$

Explanation   Since there is 1 stroke for each revolution, there will be 30 strokes per minute. The actual total length of metal cut per minute is 30 times the length of stroke, or $30 \times 1\frac{1}{6} = 35$ ft. That is, 35 feet is the actual net cutting speed. But since the return is twice as fast as the forward stroke, the return takes $\frac{1}{3}$ of the minute, whereas the forward stroke takes $\frac{2}{3}$ of the minute. That is, 35 feet of metal are cut in $\frac{2}{3}$ of a minute. For a whole minute the cutting speed of the shaper is $35 \div \frac{2}{3} = 35 \times \frac{3}{2} = 52.5$ ft per minute.

Practice check: Do exercise 16 on the left.

16. What is the cutting speed of a crank shaper making 24 rpm with a 3-to-1 return if the length of stroke is 16″?

The answer is in the margin on page 622.

■ **Problems 20–13**

Using the space provided, find the cutting speed in the following shaper problems.

1. A crank shaper with a 2-to-1 return makes 28 strokes of 1′ length in 1 minute. Find the cutting speed.

2. Find the cutting speed of a crank shaper making 25 rpm with a reverse of 3 to 1 if the length of stroke is 10″.

3. With a return of 2 to 1 and a speed of 27 rpm, what is the cutting speed of a crank shaper if the length of stroke is 15″?

4. What is the cutting speed of a crank shaper making 32 rpm with a 2-to-1 return if the length of stroke is 14″?

5. A crank shaper having a 3-to-1 return makes 28 rpm, and the length of cut is 9″. What is the cutting speed?

6. A crank shaper with a 3-to-1 return makes 30 strokes of 10″ length in 1 minute. Find the cutting speed.

7. With a return of 2 to 1 and a speed of 24 rpm, what is the cutting speed of a crank shaper if the length of stroke is 9″?

8. A crank shaper having a 3-to-1 return makes 32 rpm, and the length of cut is 15″. What is the cutting speed?

■ Self-Test

Use the space provided to solve the following problems involving cutting speed and feed. Check your answers with the answers in the back of the book. Use $\frac{22}{7}$ for π.

Answer to exercise 15

Forward stroke time $= \frac{1}{6}$ min
Return stroke time $= \frac{1}{4} \times \frac{1}{6}$
$\qquad\qquad = \frac{1}{24}$ min
Total strokes $= 1 \times \frac{24}{5} = 4\frac{4}{5}$

1. Find the cutting speed of a piece of steel $1\frac{1}{4}''$ in diameter turned in a lathe at 200 rpm.

2. An 18-in. flywheel makes 400 rpm. What is its rim speed in feet per minute?

3. A 12″ emery wheel is to have a rim speed of 6,000 ft per minute. What is its rpm?

4. A cut-off silicon carbide blade, 8″ in diameter, is not to exceed 7,000 rpm. What is its maximum rim speed in feet per minute?

5. What is the cutting speed of a 4″ milling cutter making 40 rpm?

6. What must be the diameter of a milling cutter to have a cutting speed of 30 ft per minute when it is run at 12 rpm?

7. What is the actual cutting speed of a planer making 12 strokes per minute, each 2 ft long, if the reverse is 3 to 1?

8. What is the theoretical cutting speed of the planer in Problem 7?

9. With a speed of 400 rpm and a feed of 0.012″ per revolution, how long will it take to drill through 6″ of steel?

10. A milling cutter revolves 100 times while the table moves 1″. What is the feed per revolution?

■ **Chapter Test**

Solve the following problems involving cutting speed and feed.

1. Find the cutting speed of a piece of brass $1\frac{1}{2}''$ in diameter turned in a lathe at 220 rpm.

2. Find the surface speed of an emery wheel $6''$ in diameter making 1,750 rpm.

3. At how many rpm should a lathe be run to give a cutting speed of 40 ft per minute when turning a shaft $3\frac{1}{8}''$ in diameter?

4. A $10''$ emery wheel is required to have a surface speed of 6,000 ft per minute. How fast must the emery wheel turn in rpm?

5. If the rim speed of an emery wheel should be 5,600 ft per minute and the spindle speed is 1,750 rpm, what should be the diameter of the emery wheel?

6. A brass rod is turned in a lathe at 450 rpm with a cutting speed of 80 ft per minute. What is the diameter of the brass rod?

7. A $1\frac{1}{2}''$ hole is drilled in a piece of work with the drill revolving at 280 rpm. What is the cutting speed?

8. Find the cutting speed of a $10''$ milling cutter making 40 rpm.

9. How many rpm must a $2''$ drill make in order to have a cutting speed of 25 ft per minute?

10. What must be the diameter of a milling cutter to have a cutting speed of 30 ft per minute when it is run at 18 rpm?

11. Find the actual cutting speed of a planer making 20 strokes per minute, each 3′6″ long, if the reverse is 3 to 1.

12. Find the theoretical cutting speed of the planer in Problem 11.

13. Find the time required to complete one cut on a casting 3′6″ long and 1′8″ wide if the cutting speed is 20 ft per minute, the feed $\frac{1}{4}$″ per stroke, and the return 2 to 1. Allow 3″ clearance at each end.

14. Find the cutting speed of a shaper making 30 rpm with a reverse of 2 to 1 if the length of the stroke is 16″.

15. A piece of work revolves 80 times while the tool travels $\frac{1}{2}$″. Find the feed.

16. A shaft 16″ long is being turned in a lathe at a speed of 60 rpm and a feed of 20. Find the time required for one cut.

17. With a feed of 24 and a cutting speed of 30 ft per minute, how many inches of a piece of 2″ steel will be turned in 10 minutes?

18. With a feed of 0.025″ and a speed of 300 rpm, how long will it take to drill through a piece $4\frac{1}{2}$″ thick?

19. If the feed were 0.0125″ in Problem 18, how much faster would the drill have to turn in order to drill the piece in the same amount of time?

20. A milling cutter revolves 100 times while the table moves 1″. Find the feed per revolution.

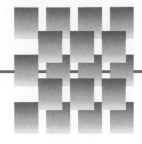

CHAPTER 21

Gears

21–1   SPUR GEARS

Two round steel disks like those shown in Fig. 21–1 will drive each other if pressed close together. This is called a *friction drive* and may be used where little power is to be transmitted. It is quite evident, however, that if any considerable power is required, the two cylindrical disks will slip on each other. To prevent this slippage and make a positive drive, teeth are provided on each disk and corresponding grooves for the teeth to enter. Such cylindrical disks or wheels are called *gear wheels* or *gears.* A pair of gears is shown in Fig. 21–2.

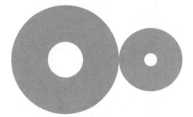

Figure 21–1   **Figure 21–2**

Pitch Circle

The circle representing the disk on which the teeth are built is called the *pitch circle.* It is not visible at all on the actual gear, yet it remains the most important element in the design of the gear (see Figs. 21–3 and 21–4).

Pitch Diameter

The diameter of the pitch circle is called the *pitch diameter* of the gear.

Gear Tooth Parts

The parts of a gear tooth are illustrated in Fig. 21–4.

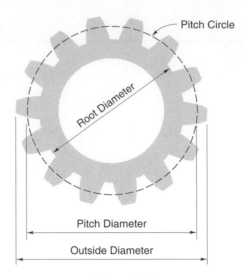

Figure 21–3

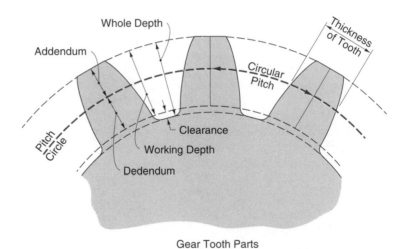

Gear Tooth Parts

Figure 21–4

Addendum, Dendendum, and Clearance

The projection of the gear tooth above the pitch circle is called the ***addendum.*** The part of the tooth below the pitch circle is called the ***dedendum.*** Its depth is equal to that of the addendum plus an additional depth called the ***clearance.***

Whole Depth and Working Depth

The sum of the addendum and the dedendum constitutes the ***whole depth.*** This is also called the ***cutting depth*** because it is the depth of cut required in cutting the gear from the blank. The whole depth minus the clearance is called the ***working depth,*** that is, the depth to which the teeth of the mating gear enter. The part of the working depth below the pitch circle is equal to the depth of the addendum. The working depth is equal to twice the depth of the addendum. The whole depth of a gear tooth is therefore equal to two addenda plus the clearance.

Outside Diameter

The **outside diameter,** or the diameter of the blank, is equal to the pitch diameter plus twice the addendum (see Fig. 21–3).

Root Diameter

The **root diameter,** or the diameter between the bottoms of the teeth, is equal to the outside diameter minus twice the whole depth.

Circular Pitch

The distance P_c between the centers of two adjoining teeth is called the **circular pitch.** It is measured on the circumference of the pitch circle and is equal to that circumference divided by the number of teeth. If we call the diameter of the pitch circle D and the number of teeth N, then

$$\text{Circumference of the pitch circle} = \pi \times D$$

$$\text{Circular pitch} = \frac{\pi \times D}{N} \quad \text{or} \quad P_c = \frac{\pi \times D}{N}$$

Also, since $N \times P_c =$ the pitch circumference, we have

$$\text{Pitch diameter} = \frac{\text{Pitch circumference}}{\pi} \quad \text{or} \quad D = \frac{N \times P_c}{\pi}$$

Also, since $\dfrac{\text{Pitch circumference}}{\text{Circular pitch}} = \text{Number of teeth } (N)$, we have

$$N = \frac{\pi \times D}{P_c}$$

The **thickness of tooth** is half the circular pitch.

■ Example 1

Find the circular pitch of a gear that has 48 teeth and a pitch diameter of 6 in.

Solution
$$P_c = \frac{\text{Pitch circumference}}{\text{Number of teeth}} = \frac{\pi \times D}{N}$$

$$= \frac{3.1416 \times 6}{48} = 0.3927 \text{ in.}$$

Practice check: Do exercise 1 on the right.

■ Example 2

Find the pitch diameter of a gear that has 45 teeth and a circular pitch of 0.3142 in.

Solution
$$\text{Pitch diameter} = \frac{\text{Circumference of pitch circle}}{\pi}$$

$$D = \frac{N \times P_c}{\pi} = \frac{45 \times 0.3142}{3.1416} = 4.5 \text{ in.}$$

Practice check: Do exercise 2 on the right.

1. Find the circular pitch of a gear that has 36 teeth and a pitch diameter of 4 in.

The answer is in the margin on page 629.

2. Find the pitch diameter of a gear that has 24 teeth and a circular pitch of 0.3214 in.

The answer is in the margin on page 629.

■ Example 3

Find the number of teeth in a gear that has a pitch diameter of $12\frac{1}{2}$ in. and a circular pitch of 0.7854 in.

3. Find the number of teeth in a gear that has a pitch diameter of 8.2 in. and a circular pitch of 0.62832 in.

The answer is in the margin on page 630.

Solution

$$\text{Number of teeth} = \frac{\text{Circumference of pitch circle}}{P_c}$$

$$N = \frac{\pi \times D}{P_c} = \frac{3.1416 \times 12.5}{0.7854}$$

$$= 50 \text{ teeth}$$

Practice check: Do exercise 3 on the left.

■ Example 4

Find the thickness of tooth of the gear in Example 3.

4. Find the thickness of tooth if the circular pitch is 0.6328 in.

The answer is in the margin on page 630.

Solution $\text{Thickness of tooth} = \dfrac{P_c}{2} = \dfrac{0.7854}{2} = 0.3927 \text{ in.}$

Practice check: Do exercise 4 on the left.

■ Problems 21–1

Use the space provided below to solve the following problems involving spur gears.

1. Find the circular pitch of a gear that has 60 teeth and a pitch diameter of $7\frac{1}{2}''$.

2. What is the circular pitch of a 70-tooth gear if the pitch diameter is $3\frac{1}{2}''$?

3. A gear that has a pitch diameter of $22''$ has 55 teeth. Find the circular pitch.

4. Find the circular pitch of a 70-tooth gear with a $40''$ pitch diameter.

5. Find the circular pitch of a 36-tooth gear with a $4''$ pitch diameter.

6. A gear has 65 teeth and $0.6283''$ circular pitch. Find the pitch diameter.

7. A gear has 31 teeth and a circular pitch of 1.5708″. Find the pitch diameter.

8. Find the pitch diameter of an 82-tooth gear whose circular pitch is 0.1963″.

Answer to exercise 1

$$P_c = \frac{3.1416 \times 4}{36}$$

$$= 0.349 \text{ in.}$$

9. A gear has 30 teeth and 0.3927″ circular pitch. Find the pitch diameter.

10. Find the pitch diameter of a 45-tooth gear whose circular pitch is 2.0944″.

Answer to exercise 2

$$D = \frac{24 \times 0.3214}{3.1416}$$

$$= 2.5 \text{ in.}$$

11. How many teeth has a gear whose circular pitch is 3.1416″ and whose pitch diameter is 82″?

12. Find the number of teeth on a gear with a pitch diameter of 48″ and a circular pitch of 2.5133″.

13. Find the number of teeth on a gear with a pitch diameter of 4″ and a circular pitch of 0.6283″.

14. Find the number of teeth on a gear with a pitch diameter of $5\frac{1}{2}$″ and a circular pitch of 0.2244″.

15. Find the number of teeth on a gear with a pitch diameter of $2\frac{1}{4}$″ and a circular pitch of 0.2618″.

16. Find the thickness of tooth in the gears in Problems 11–15.

21–2   DIAMETRAL PITCH

In modern gear design, the circular pitch is used very little. Another pitch, called the diametral pitch, is commonly used because of its convenience in computation. When we speak of the pitch of a gear, we mean diametral pitch unless circular pitch is specified. The ***diametral pitch*** is the number of teeth per inch of pitch diameter.

Diametral pitch is not a distance, as pitch is commonly understood. For instance, if a gear has a pitch diameter of 3 in. and 24 teeth, it has $\frac{24}{3}$ or 8 teeth for every inch of its pitch diameter; that is, its diametral pitch is 8. It is called an 8-pitch gear. It will mesh only with other 8-pitch gears. A 60-tooth, 8-pitch gear has a pitch diameter of $\frac{60}{8}$ or $7\frac{1}{2}$ in.; a 40-tooth, 8-pitch gear has a pitch diameter of $\frac{40}{8}$ or 5 in.

Answer to exercise 3

$$N = \frac{3.1416 \times 8.2}{0.62832}$$

$$= 41 \text{ teeth}$$

Answer to exercise 4

$$\text{Thickness} = \frac{0.6328}{2}$$

$$= 0.3164 \text{ in.}$$

Since the diametral pitch means the number of teeth per inch of pitch diameter, the number of teeth on a gear is equal to the pitch diameter times the pitch. A 6-pitch gear with a pitch diameter of $5\frac{1}{2}$ in. will have $5\frac{1}{2} \times 6$ or 33 teeth. This relation between pitch, number of teeth, and pitch diameter simplifies gear computation to a great extent.

Using the letters D for pitch diameter, P for diametral pitch, and N for number of teeth, we have, by definition,

$$\text{Diametral pitch} = \frac{\text{Number of teeth}}{\text{Pitch diameter}} \quad \text{or} \quad P = \frac{N}{D}$$

Also,

$$\text{Number of teeth} = \text{Pitch diameter} \times \text{Pitch} \quad \text{or} \quad N = D \times P$$

Also,

$$\text{Pitch diameter} = \frac{\text{Number of teeth}}{\text{Diametral pitch}} \quad \text{or} \quad D = \frac{N}{P}$$

5. Find the diametral pitch of a gear that has 63 teeth and a pitch diameter of $5\frac{1}{4}$ in.

The answer is in the margin on page 632.

■ Example 1

Find the diametral pitch of a gear that has 72 teeth and a pitch diameter of $4\frac{1}{2}$ in.

Solution $P = \dfrac{N}{D} = \dfrac{72}{4\frac{1}{2}} = 16$. The gear is a 16-pitch gear.

Practice check: Do exercise 5 on the left.

6. What is the pitch diameter of a 15-pitch gear with 42 teeth?

The answer is in the margin on page 632.

■ Example 2

What is the pitch diameter of a 12-pitch gear with 45 teeth?

Solution $D = \dfrac{N}{P} = \dfrac{45}{12} = 3\frac{3}{4}$. The pitch diameter is $3\frac{3}{4}$ in.

Practice check: Do exercise 6 on the left.

7. How many teeth must a 6-pitch gear have if its pitch diameter is 8.5 in.?

The answer is in the margin on page 632.

■ Example 3

How many teeth must a 5-pitch gear have if its pitch diameter is 6.2 in.?

Solution $N = D \times P = 6.2 \times 5 = 31$ teeth. The gear has 31 teeth.

Practice check: Do exercise 7 on the left.

■ Problems 21–2

Using the space provided below, solve the following problems involving diametral pitch and circular pitch.

1. Find the diametral pitch of a 68-tooth gear whose pitch diameter is 17″.

2. Find the diametral pitch of a 40-tooth gear whose pitch diameter is 16″.

3. Find the diametral pitch of a 30-tooth gear whose pitch diameter is $2\frac{1}{2}''$.

4. Find the diametral pitch of a 65-tooth gear whose pitch diameter is $3\frac{1}{4}''$.

5. Find the diametral pitch of a 124-tooth gear whose pitch diameter is $7\frac{3}{4}''$.

6. How many teeth are there on a 12-pitch gear whose pitch diameter is $5\frac{1}{2}''$?

7. How many teeth are there on a 9-pitch gear whose pitch diameter is $11''$?

8. How many teeth are there on a 20-pitch gear whose pitch diameter is $9.1''$?

9. How many teeth are there on a 4-pitch gear whose pitch diameter is $16\frac{3}{4}''$?

10. How many teeth are there on a 16-pitch gear whose pitch diameter is $3\frac{7}{8}''$?

11. Find the pitch diameter of a 70-tooth, 5-pitch gear.

12. Find the pitch diameter of a 48-tooth, 9-pitch gear.

13. Find the pitch diameter of an 18-tooth, 8-pitch gear.

14. Find the pitch diameter of a 45-tooth, $2\frac{1}{2}$-pitch gear.

15. Find the pitch diameter of a 27-tooth, $1\frac{1}{2}$-pitch gear.

Answer to exercise 5

$P = \dfrac{63}{5.25} = 12$. The gear is a 12-pitch gear.

Answer to exercise 6

$D = \frac{42}{16} = 2\frac{4}{8}$. The pitch diameter is $2\frac{4}{8}$ in.

Answer to exercise 7

$N = D \times P = 8.5 \times 6$
   $= 51$ teeth

21–3   PROPORTIONS OF GEAR TEETH

In standard gears, *the addendum is the reciprocal of the pitch;* that is, addendum = 1/pitch. In a 10-pitch gear the addendum is $\frac{1}{10}$ in.; in a 4-pitch gear the addendum is $\frac{1}{4}$ in. The dedendum is equal in depth to the addendum plus the clearance. The *clearance* is usually stated as $\frac{1}{20}$ of the circular pitch.

21–4   RELATION BETWEEN CIRCULAR PITCH AND DIAMETRAL PITCH

It is often necessary to convert circular pitch into diametral pitch, and vice versa. The circular pitch, being part of the circumference of the pitch circle, is found by dividing that circumference by the number of teeth in the gear.

   Let

$$D = \text{Pitch diameter}$$

$$N = \text{Number of teeth}$$

$$P = \text{Pitch}$$

$$P_c = \text{Circular pitch}$$

Then

$$\text{Pitch circumference} = \pi \times \text{Pitch diameter} = \pi \times D$$

$$\text{Circular pitch} = \frac{\text{Pitch circumference}}{\text{Number of teeth}}$$

That is,

$$P_c = \frac{\pi \times D}{N}$$

But

$$N = D \times P$$

Writing this in place of N, we get

$$P_c = \frac{\pi \times D}{P \times D}$$

from which, by cancellation, we get the formula for converting diametral pitch into circular pitch.

$$P_c = \frac{\pi}{P}$$

To convert circular pitch into diametral pitch, we proceed as follows.

$$\text{Diametral pitch} = \frac{\text{Number of teeth}}{\text{Pitch diameter}}$$

That is,

$$P = \frac{N}{D}$$

But

$$N = \frac{\pi \times D}{P_c}$$

Putting this value in place of N in the formula, we get

$$P = \frac{\pi \times D}{P_c \times D}$$

which, by cancellation of D, becomes the formula for converting circular pitch into diametral pitch.

$$P = \frac{\pi}{P_c}$$

■ **Example 1**

Find the circular pitch of a 12-pitch gear.

Solution $P_c = \dfrac{\pi}{P} = \dfrac{3.1416}{12} = 0.2618$ in.

Practice check: Do exercise 8 on the right.

■ **Example 2**

Find the diametral pitch of a gear whose circular pitch is 0.6283 in.

Solution $P = \dfrac{\pi}{P_c} = \dfrac{3.1416}{0.6283} = 5$

Practice check: Do exercise 9 on the right.

8. Find the circular pitch of a 14-pitch gear.

The answer is in the margin on page 635.

9. Find the diametral pitch of a gear whose circular pitch is 0.4488 in.

The answer is in the margin on page 635.

■ **Problems 21–4**

Use the space provided to find the circular pitch of the following gears.

1. 6-pitch gear

2. $2\frac{1}{2}$-pitch gear

3. 10-pitch gear

4. $1\frac{1}{4}$-pitch gear

5. 18-pitch gear

6. 20-pitch gear

7. 1½-pitch gear

8. 3-pitch gear

9. 8-pitch gear

10. 14-pitch gear

Find the diametral pitch of the following gears.

11. A gear whose circular pitch is 1.2566″

12. A gear whose circular pitch is 0.1571″

13. A gear whose circular pitch is 0.3491″

14. A gear whose circular pitch is 1.7952″

15. A gear whose circular pitch is 1.5708″

16. A gear whose circular pitch is $\frac{3}{4}$″

17. A gear whose circular pitch is $1\frac{1}{2}$″

18. A gear whose circular pitch is 0.3927″

19. A gear whose circular pitch is 0.1963″

20. A gear whose circular pitch is 2.5133″

21–5   CLEARANCE

The *clearance,* which is usually stated as $\frac{1}{20}$ of the circular pitch, can also be expressed in terms of the diametral pitch.

$$\text{Clearance} = \tfrac{1}{20} \text{ of } P_c = \tfrac{1}{20} \text{ of } \frac{\pi}{P}$$

$$= \frac{1}{20} \times \frac{3.1416}{P} = \frac{0.157}{P}$$

■ **Example 1**

Find the clearance of a gear whose circular pitch is 0.3491 in.

Solution $\text{Clearance} = \dfrac{P_c}{20} = \dfrac{0.3491}{20} = 0.0175 \text{ in.}$

Practice check: Do exercise 10 on the right.

■ **Example 2**

Find the clearance of a 12-pitch gear.

Solution $\text{Clearance} = \dfrac{0.157}{P} = \dfrac{0.157}{12} = 0.0131 \text{ in.}$

Practice check: Do exercise 11 on the right.

■ **Problems 21–5**

Using the space provided, find the clearance in each of the following gears.

1. 16-pitch gear

2. 12-pitch gear

3. $2\frac{1}{2}$-pitch gear

4. 8-pitch gear

5. $1\frac{3}{4}$-pitch gear

6. $3\frac{1}{4}$-pitch gear

Answer to exercise 8

$P_c = \dfrac{3.1416}{14}$

$= 0.2439$

Answer to exercise 9

$P = \dfrac{3.1416}{0.4488}$

$= 7$

10. Find the clearance of a gear whose circular pitch is 0.4815 in.

The answer is in the margin on page 638.

11. Find the clearance of a 16-pitch gear.

The answer is in the margin on page 638.

7. 5.2-pitch gear

8. A gear whose circular pitch is 1.5708″

9. A gear whose circular pitch is 0.2244″

10. A gear whose circular pitch is 0.6283″

11. A gear whose circular pitch is 0.2618″

12. A gear whose circular pitch is 0.7500″

13. A gear whose circular pitch is 0.3927″

14. A gear whose circular pitch is 0.5236″

15. A gear whose circular pitch is 1.63″

21–6   DEPTH OF TOOTH

$$\text{Whole depth of tooth} = \text{Addendum} + \text{Dedendum}$$

$$= \frac{1}{P} + \frac{1}{P} + \frac{0.157}{P} = \frac{2.157}{P}$$

Since

$$\text{Dedendum} = \text{Addendum} + \text{Clearance}$$

then

$$\text{Whole depth} = \text{Addendum} + \text{Addendum} + \text{Clearance}$$

■ **Example 1**

Find the whole depth of tooth of an 8-pitch gear.

Solution　　Whole depth $= \dfrac{2.157}{P} = \dfrac{2.157}{8} = 0.2696$ in.

Practice check: Do exercise 12 on the right.

■ **Example 2**

Find the whole depth of tooth of a gear whose circular pitch is 0.6283 in.

Solution　　　　　　$P = \dfrac{\pi}{P_c} = \dfrac{3.1416}{0.6283} = 5$

Whole depth $= \dfrac{2.157}{P} = \dfrac{2.157}{5} = 0.4314$ in.

Practice check: Do exercise 13 on the right.

12. Find the whole depth of tooth of a 12-pitch gear.

The answer is in the margin on page 639.

13. Find the whole depth of tooth of a gear whose circular pitch is 0.3927 in.

The answer is in the margin on page 639.

■ **Problems 21–6**

Use the space provided to find the whole depth of tooth in each of the following gears.

1. 16-pitch gear

2. 12-pitch gear

3. $2\frac{1}{2}$-pitch gear

4. 8-pitch gear

5. $1\frac{3}{4}$-pitch gear

6. $3\frac{1}{4}$-pitch gear

7. 5.2-pitch gear

8. A gear whose circular pitch is 1.5708″

9. A gear whose circular pitch is 0.2244″

10. A gear whose circular pitch is 0.6283″

Answer to exercise 10

Clearance $= \dfrac{0.4815}{20}$

$= 0.0241$ in.

Answer to exercise 11

Clearance $= \dfrac{0.157}{16}$

$= 0.0098$ in.

11. A gear whose circular pitch is 0.2618″

12. A gear whose circular pitch is 0.7500″

13. A gear whose circular pitch is 0.3927″

14. A gear whose circular pitch is 0.5236″

15. A gear whose circular pitch is 1.63″

21–7　　OUTSIDE DIAMETER

Outside diameter of a gear = Pitch diameter + 2 × Addendum

$$= D + \frac{1}{P} + \frac{1}{P}$$

$$= D + \frac{2}{P}$$

■ **Example**

Find the outside diameter of an 8-pitch gear whose pitch diameter is $5\frac{1}{2}$ in.

14. Find the outside diameter of a 16-pitch gear whose pitch diameter is $4\frac{3}{4}$ in.

The answer is in the margin on page 641.

Solution　　Outside diameter $= D + \dfrac{2}{P} = 5\frac{1}{2} + \frac{2}{8} = 5\frac{1}{2} + \frac{1}{4} = 5\frac{3}{4}$ in.

Practice check: Do exercise 14 on the left.

■ **Problems 21–7**

Using the space provided, find the outside diameters of the following gears.

1. A 16-pitch gear with a pitch diameter of $8\frac{1}{4}″$

2. A $2\frac{1}{2}$-pitch gear whose pitch diameter is 24″

3. A 16-pitch gear whose pitch diameter is 1.875″

4. A 10-pitch gear whose pitch diameter is 3.1″

Answer to exercise 12

Whole depth $= \dfrac{2.157}{12}$

$\qquad = 0.1798$ in.

5. A 20-pitch gear whose pitch diameter is $9\frac{1}{4}$″

6. A 4-pitch gear whose pitch diameter is 12″

Answer to exercise 13

$P = \dfrac{3.1416}{0.3927}$

$\quad = 8$

Whole depth $= \dfrac{2.157}{8}$

7. A 12-pitch gear with a pitch diameter of $3\frac{1}{4}$″

8. A 9-pitch gear whose pitch diameter is 5.333″

$\qquad = 0.2696$ in.

9. A 14-pitch gear whose pitch diameter is $11\frac{1}{2}$″

10. A 6-pitch gear whose pitch diameter is 3.667″

21–8   PITCH DIAMETER

To find the pitch diameter when the outside diameter and the pitch are given, use the following method.

$$\text{Pitch diameter} = \text{Outside diameter} - 2 \times \text{Addendum}$$

Using the symbol *OD* for outside diameter, we have the formula

$$D = OD - \frac{2}{P}$$

The formula for finding the pitch diameter when the outside diameter and the number of teeth are given is

$$D = \frac{N \times OD}{N + 2}$$

This formula states that we must multiply the outside diameter by the number of teeth and divide that product by the number of teeth plus 2.

■ **Example 1**

Find the pitch diameter of an 8-pitch gear whose outside diameter is $6\frac{1}{4}$ in.

15. Find the pitch diameter of a 16-pitch gear whose outside diameter is $5\frac{3}{4}$ in.

The answer is in the margin on page 642.

16. Find the pitch diameter of a 52-tooth gear whose outside diameter is $6\frac{3}{4}$ in.

The answer is in the margin on page 642.

Solution $D = OD - \dfrac{2}{P} = 6\frac{1}{4} - \frac{2}{8} = 6\frac{1}{4} - \frac{1}{4} = 6$ in.

Practice check: Do exercise 15 on the left.

■ **Example 2**

Find the pitch diameter of a 66-tooth gear whose outside diameter is $8\frac{1}{2}$ in.

Solution $D = \dfrac{N \times OD}{N + 2} = \dfrac{66 \times 8\frac{1}{2}}{66 + 2} = \dfrac{561}{68} = 8\frac{1}{4}$ in.

Practice check: Do exercise 16 on the left.

■ **Problems 21–8**

Use the space provided to find the pitch diameter of the following gears.

1. A gear that has 58 teeth and an outside diameter of $7\frac{1}{2}''$

2. A 40-tooth gear whose outside diameter is 16.8″

3. A gear that has 114 teeth and an outside diameter of 9.667″

4. A 31-tooth gear whose outside diameter is 3.3″

5. A 57-tooth gear whose outside diameter is $14\frac{3}{4}''$

6. An 18-pitch gear whose outside diameter is $5\frac{1}{2}''$

7. A 14-pitch gear whose outside diameter is 3.286″

8. A $2\frac{1}{2}$-pitch gear whose outside diameter is 40.8″

9. A 6-pitch gear whose outside diameter is 15″

10. A 12-pitch gear whose outside diameter is 7.417″

21-9   USE OF FORMULAS

You will notice that frequently there are several formulas for finding a required item. You must select the formula that contains the information given. For instance, to find the pitch diameter of a gear whose outside diameter is $7\frac{3}{4}$ in. and that has

Answer to exercise 14

Outside diameter $= 4\frac{3}{4} + \frac{2}{16}$
$= 4\frac{7}{8}$ in.

Formulas for Gear Computation			
To Find	**Symbol**	**Having**	**Formula**
Pitch	P	Circular pitch	$P = \pi/P_c$
Circular pitch	P_c	Pitch	$P_c = \pi/P$
	P_c	Pitch diameter and number of teeth	$P_c = \dfrac{\pi \times D}{N}$
Pitch	P	Pitch diameter and number of teeth	$P = N/D$
	P	Outside diameter and number of teeth	$P = \dfrac{N + 2}{OD}$
Pitch diameter	D	Number of teeth and pitch	$D = N/P$
	D	Number of teeth and circular pitch	$D = \dfrac{N \times P_c}{\pi}$
	D	Outside diameter and number of teeth	$D = \dfrac{OD \times N}{N + 2}$
	D	Outside diameter and pitch	$D = OD - 2/P$
Outside diameter	OD	Number of teeth and pitch	$OD = \dfrac{N + 2}{P}$
	OD	Pitch diameter and pitch	$OD = D + 2/P$
	OD	Number of teeth and circular pitch	$OD = \dfrac{(N + 2) \times P_c}{\pi}$
	OD	Pitch diameter and circular pitch	$OD = D + 2P_c/\pi$
Number of teeth	N	Pitch diameter and pitch	$N = D \times P$
	N	Pitch diameter and circular pitch	$N = \dfrac{D \times \pi}{P_c}$
Thickness of tooth	T	Circular pitch	$T = P_c/2$
Addendum	Add	Pitch	$Add = 1/P$
	Add	Circular pitch	$Add = P_c/\pi$
Dedendum	Ded	Pitch	$Ded = 1.157/P$
	Ded	Circular pitch	$Ded = 0.3683 P_c$
Clearance	Cl	Circular pitch	$Cl = P_c/20$
	Cl	Pitch	$Cl = 0.157/P$
Whole depth	WD	Pitch	$WD = 2.157/P$
	WD	Circular pitch	$WD = 0.6866 \times P_c$
Working depth	wd	Pitch	$wd = 2/P$
	wd	Circular pitch	$wd = \dfrac{2 \times P_c}{\pi}$

Answer to exercise 15

$D = 5\frac{3}{4} - \frac{1}{8} = 5\frac{5}{8}$ in.

Answer to exercise 16

$D = \dfrac{52 \times 6\frac{3}{4}}{52 + 2} = \dfrac{351}{54}$

$= 6.5$ in.

60 teeth, we select the formula that gives D in terms of OD and N, namely,

$$D = \frac{OD \times N}{N + 2}$$

When the circular pitch is given, it will often be convenient to convert it into diametral pitch and continue the calculation in terms of diametral pitch. The table on page 641 summarizes the formulas for gear computation.

■ **Example**

Find the depth of cut and size of blank for a 6-pitch gear of 80 teeth.

Solution　　The depth of cut is the same as the whole depth.

$$\text{Depth of cut} = \frac{2.157}{P} = \frac{2.157}{6} = 0.3595 \text{ in.}$$

17. Find the depth of cut and size of blank for a 12-pitch gear of 40 teeth.

The answer is in the margin on page 644.

The size of blank is the same as the outside diameter.

$$\text{Size of blank} = \frac{N + 2}{P} = \frac{80 + 2}{6} = \frac{82}{6} = 13.667 \text{ in.}$$

Practice check: Do exercise 17 on the left.

■ **Problems 21–9**

Compute the missing dimensions of the gears in the following chart.

	Pitch	Pitch Diameter	Number of Teeth	Circular Pitch	Addendum	Dedendum	Clearance	Working Depth	Whole Depth	Outside Diameter
1.	10	$2\frac{1}{2}''$	___	___	___	___	___	___	___	___
2.	16	___	25	___	___	___	___	___	___	___
3.	12	___	___	___	___	___	___	___	___	$4\frac{2}{3}''$
4.	___	___	32	___	___	___	___	___	___	$8\frac{1}{2}''$
5.	___	$3.417''$	41	___	___	___	___	___	___	___
6.	___	$3.8''$	___	$0.1571''$	___	___	___	___	___	___
7.	___	___	55	$1.2566''$	___	___	___	___	___	___
8.	___	___	___	$0.3927''$	___	___	___	___	___	$9\frac{1}{2}''$
9.	$1\frac{1}{2}$	___	90	___	___	___	___	___	___	___
10.	14	___	75	___	___	___	___	___	___	___

21–10　RACKS

The proportions of the teeth in a rack are the same as in a gear. The linear pitch of the rack is the same as the circular pitch of the gears that mesh with the rack. (See Fig. 21–5, which shows the tooth proportions for a rack to mesh with a 6-pitch gear.)

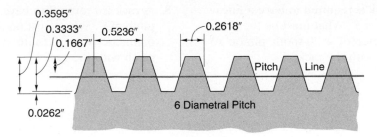

Figure 21–5

■ **Example 1**

Find the linear pitch of a 5-pitch rack.

Solution   Linear pitch = Circular pitch

$$P_c = \frac{\pi}{P} = \frac{3.1416}{5} = 0.6283 \text{ in.}$$

Practice check: Do exercise 18 on the right.

■ **Example 2**

Find the depth of cut for a 12-pitch rack.

Solution   Depth of cut = Whole depth = $\frac{2.157}{P} = \frac{2.157}{12} = 0.1798$ in.

Practice check: Do exercise 19 on the right.

■ **Problems 21–10**

1. Find the whole depth of tooth of a 4-pitch rack.

2. Find the whole depth of tooth of a 6-pitch rack.

3. Find the linear pitch of a 3-pitch rack.

4. Find the linear pitch of a 7-pitch rack.

5. A 6-pitch rack is driven by a 27-tooth pinion. Find the distance traveled for each turn of the pinion.

6. A 9-pitch rack is driven by an 18-tooth pinion. Find the distance traveled for each turn of the pinion.

18. Find the linear pitch of a 6-pitch rack.

The answer is in the margin on page 646.

19. Find the depth of cut for an 8-pitch rack.

The answer is in the margin on page 646.

7. A rack is required to have a linear pitch of $\frac{3}{4}''$. What must be the pitch diameter of a 20-tooth pinion to mesh with the rack?

8. A rack is required to have a linear pitch of $\frac{5}{8}''$. What must be the pitch diameter of a 30-tooth pinion to mesh with the rack?

9. Find the pitch diameter of a 30-tooth pinion to mesh with a rack whose linear pitch is $1\frac{1}{8}''$.

10. Find the pitch diameter of a 23-tooth pinion to mesh with a rack whose linear pitch is $2\frac{1}{4}''$.

21–11   CENTER-TO-CENTER DISTANCE OF GEARS

Since the pitch circles of gears in mesh are supposed to be in contact, the distance from the center of one gear to the center of the other gear is equal to the sum of the pitch radii. In other words, the center-to-center distance of a pair of gears in mesh is equal to half the sum of the pitch diameters (Fig. 21–6).

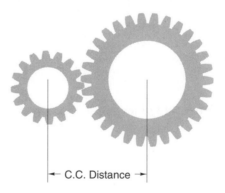

← C.C. Distance →

Figure 21–6

In a rack and pinion, the distance from the center of the pinion to the pitch line of the rack is equal to half the pitch diameter of the pinion (Fig. 21–7).

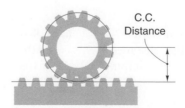

C.C. Distance

Figure 21–7

■ **Example 1**

Two gears in mesh have pitch diameters of 8 in. and $3\frac{1}{4}$ in. Find their center-to-center distance.

Solution   Center-to-center distance $= \dfrac{8 + 3\frac{1}{4}}{2} = \dfrac{11\frac{1}{4}}{2} = 5\frac{5}{8}$ in.

Practice check: Do exercise 20 on the right.

■ **Example 2**

Find the center-to-center distance of a pair of 8-pitch gears having 42 teeth and 75 teeth.

Solution

$$\text{75-tooth gear, } D = \frac{N}{P} = \frac{75}{8} = 9\frac{3}{8} \text{ in.}$$

$$\text{42-tooth gear, } D = \frac{N}{P} = \frac{42}{8} = 5\frac{1}{4} \text{ in.}$$

$$\text{Center-to-center distance} = \frac{9\frac{3}{8} + 5\frac{1}{4}}{2} = \frac{14\frac{5}{8}}{2} = 7\frac{5}{16} \text{ in.}$$

Practice check: Do exercise 21 on the right.

■ **Problems 21–11**

Use the space provided to find the center-to-center distance of the following gears.

1. Find the center-to-center distance of a pair of gears whose pitch diameters are $7\frac{1}{2}''$ and $5\frac{1}{4}''$.

2. A pair of 5-pitch gears have 45 teeth and 62 teeth. Find the distance center to center.

3. Find the distance center to center of a 32-tooth gear and a 100-tooth gear of 12-pitch.

4. A pair of gears of $2\frac{1}{2}$-pitch have 30 teeth and 45 teeth. Find the distance center to center.

5. A 35-tooth, 14-pitch pinion meshes with a rack. Find the distance from the center of pinion to the pitch line of rack.

6. Find the distance from the center of an 8-pitch, 21-tooth pinion to the pitch line of a rack that is driven by this pinion.

20. Two gears in mesh have pitch diameters of 6 in. and $2\frac{1}{2}$ in. Find their center-to-center distance.

The answer is in the margin on page 647.

21. Find the center-to-center distance of a pair of 16-pitch gears having 36 teeth and 68 teeth.

The answer is in the margin on page 647.

Answer to exercise 18

$$P_c = \frac{3.1416}{6}$$

$$= 0.5236 \text{ in.}$$

Answer to exercise 19

Depth of cut $= \dfrac{2.157}{8}$

$$= 0.2696 \text{ in.}$$

7. A pair of gears of 45 teeth and 75 teeth have a center-to-center distance of 12″. Find their pitch diameters.

8. Find the pitch diameters of a pair of gears of 49 teeth and 72 teeth if the distance center to center is 4.321″.

9. Gear *A* has 40 teeth and an outside diameter of 7″; gear *B* has 18 teeth and an outside diameter of 3.333″. What should be the distance center to center when properly adjusted?

10. Find the center-to-center distance of the following pair of gears: gear *A* has 90 teeth and an outside diameter of $7\frac{2}{3}″$; gear *B* has 30 teeth and an outside diameter of $2\frac{2}{3}″$.

21–12   SELECTION OF CUTTERS

Because the shape of the tooth changes with the number of teeth on a gear, cutters are made in sets, usually eight to the set for each pitch. The following table gives the cutter number to select for different numbers of teeth. Theoretically, a different-shaped cutter is required for each number of teeth, but for ordinary work the approximation is close enough if we use the cutters within the limits listed in the table.

Table of Cutters
Use No. 8 gear cutter for cutting gears with 12 or 13 teeth.
Use No. 7 gear cutter for cutting gears with 14 to 16 teeth.
Use No. 6 gear cutter for cutting gears with 17 to 20 teeth.
Use No. 5 gear cutter for cutting gears with 21 to 25 teeth.
Use No. 4 gear cutter for cutting gears with 26 to 34 teeth.
Use No. 3 gear cutter for cutting gears with 35 to 54 teeth.
Use No. 2 gear cutter for cutting gears with 55 to 134 teeth.
Use No. 1 gear cutter for cutting gears with 135 teeth to a rack.

■ **Problems 21–12**

Select cutters for all the gears in the chart of Problem Set 21–9.

21–13   BEVEL GEARS

Bevel gears differ from spur gears in that the teeth are cut on the surface of a cone instead of a cylinder. Figure 21–8 shows two frustums of cones in contact. Such frustums of cones, if pressed firmly together, will drive each other. To secure a

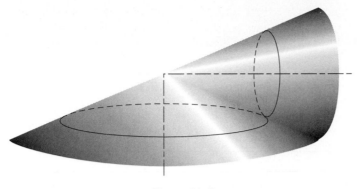

Figure 21–8

Answer to exercise 20

$$\text{Center-to-center distance} = \frac{6 + 2\frac{1}{2}}{2}$$

$$= 4\frac{1}{4} \text{ in.}$$

Answer to exercise 21

68-tooth gear, $D = \frac{68}{16}$
$$= 4\frac{1}{4} \text{ in.}$$
36-tooth gear, $D = \frac{36}{16}$
$$= 2\frac{1}{4} \text{ in.}$$

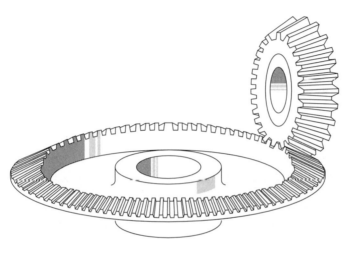

Figure 21–9

positive drive, teeth are cut on the surface of these cones, producing a pair of bevel gears, as shown in Fig. 21–9.

21–14 DEFINITIONS APPLYING TO BEVEL GEARS—PITCH CONES

The cones that form the basis of the bevel gears are called the ***pitch cones,*** and their diameters are the ***pitch diameters*** of the bevel gears. The pitch cones are not visible at all on the finished gears, but they are the most important elements in the design.

The *addendum* is the projection of the tooth above the pitch cone surface. The depth of the tooth below the pitch cone surface is called the *dedendum*. The depth of the dedendum is equal to the depth of the addendum plus the *clearance*.

The terms *addendum, dedendum, clearance, pitch diameter, diametral pitch, number of teeth, circular pitch, thickness of tooth,* and *width of space at pitch line* have the same meaning as in spur gears and are computed in exactly the same way. For further definitions in reference to bevel gears, refer to Fig. 21–10.

Figure 21–10(*a*) shows the pitch cones of a pair of bevel gears in mesh; the axes are at right angles to each other. Figure 21–10(*b*) shows a cross section through

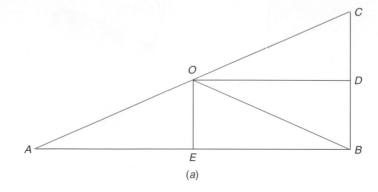

(a)

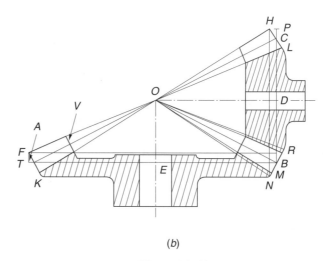

(b)

Figure 21–10

the bevel gears built on these cones. We will call the large one the *gear* and the small one the *pinion*.

AB = Pitch diameter of the gear

AC = Pitch diameter of the pinion

FR = Outside diameter of the gear

HM = Outside diameter of the pinion

$OA = OB = OC$ = Pitch cone radius

Angle EOB = Pitch cone angle of the gear

Angle DOB = Pitch cone angle of the pinion

Angle AOF = Angle COH = Addendum angle
(same for both gear and pinion)

Angle AOK = Angle LOC = Dedendum angle
(same for both gear and pinion)

AT = Angular addendum of gear

CP = Angular addendum of pinion

Angle EOF = Turning angle of gear

Angle DOH = Turning angle of pinion

Angle EOK = Cutting angle of gear

Angle DOL = Cutting angle of pinion

The following relations can be seen from the construction of Fig. 21–10.

Turning angle = Pitch cone angle + Addendum angle

Cutting angle = Pitch cone angle − Dedendum angle

Outside diameter = Pitch diameter + Twice the angular addendum

The following example will illustrate the principles of bevel gear design.

■ **Example**

Compute the dimensions of a pair of 4-pitch bevel gears with shafts at right angles, the gear to have 60 teeth and the pinion 18 teeth.

Solution   To find pitch diameters, use the formulas

$$\text{Pitch diameter of gear} = \frac{N}{P} = \frac{60}{4} = 15 \text{ in.}$$

$$\text{Pitch diameter of pinion} = \frac{N}{P} = \frac{18}{4} = 4\tfrac{1}{2} \text{ in.}$$

To find pitch cone angles, construct the pitch cones by drawing AB = 15 in. at the right angles to $BC = 4\tfrac{1}{2}$ in. Connect A and C, and from the center O of AC, draw OB and the perpendiculars OE and OD (Fig. 21–11).

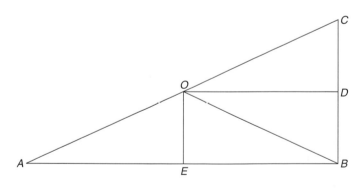

Figure 21–11

Angle BOE is the pitch cone angle of the gear; angle BOD is the pitch cone angle of the pinion.

$$EB - \tfrac{1}{2} \text{ of } AB - \tfrac{1}{2} \text{ of } 15 \text{ in.} = 7\tfrac{1}{2} \text{ in.}$$

$$BD = \tfrac{1}{2} \text{ of } BC = \tfrac{1}{2} \text{ of } 4\tfrac{1}{2} \text{ in.} = 2\tfrac{1}{4} \text{ in.}$$

$$\text{Tangent } BOE = \frac{BE}{OE} = \frac{7\tfrac{1}{2}}{2\tfrac{1}{4}} = 3.33333$$

Therefore, angle $BOE = 73°18' =$ pitch cone angle of gear.

22. Compute the dimensions of a pair of 5-pitch bevel gears with shafts at right angles, the gear to have 45 teeth and the pinion 22 teeth. See Figure 21–11.

The answer is in the margin on page 652.

$$\text{Tangent } BOD = \frac{BD}{OD} = \frac{2\frac{1}{4}}{7\frac{1}{2}} = 0.30000$$

Therefore, angle $BOD = 16°42' =$ pitch cone angle of pinion.

Proof

$$\text{Sum of pitch cone angle} = 73°18' + 16°42' = 90°$$

Practice check: Do exercise 22 on the left.

■ Problems 21–14

Compute the pitch diameters and the pitch cone angles of the following pairs of bevel gears at right angles. Use the space provided to draw the pitch cone diagram in each instance.

1. Gear 60 teeth, pinion 25 teeth, pitch 4

2. Gear 44 teeth, pinion 24 teeth, pitch 8

3. Gear 72 teeth, pinion 36 teeth, pitch 6

4. Gear 35 teeth, pinion 30 teeth, pitch 5

5. Gear 50 teeth, pinion 25 teeth, pitch 10

6. Gear 54 teeth, pinion 30 teeth, pitch 3

7. Gear 100 teeth, pinion 27 teeth, pitch 5

8. Gear 62 teeth, pinion 40 teeth, pitch 8

9. Gear 45 teeth, pinion 36 teeth, pitch 4

10. Gear 75 teeth, pinion 24 teeth, pitch 6

21–15   TOOTH PARTS

Continuing the example of the preceding section and referring to Fig. 21–10,

$$\text{Addendum} = \frac{1}{P} = \frac{1}{4} = 0.25 \text{ in.} = AF \quad \text{in Fig. 21–10}(b)$$

$$\text{Clearance} = \frac{0.157}{P} = \frac{0.157}{4} = 0.0393 \text{ in.}$$

$$\text{Dedendum} = \text{Depth under pitch cone}$$

$$= \text{Addendum} + \text{Clearance}$$

$$= \frac{1}{P} + \frac{0.157}{P} = \frac{1.157}{P} = \frac{1.157}{4} = 0.2893 \text{ in.}$$

Whole depth of tooth at pitch circle
$$= 0.25 + 0.25 + 0.0393 = 0.5393 \text{ in.}$$
$$= FK \quad \text{in Fig. 21–10}(b)$$

$$\text{Circular pitch at pitch line} = \frac{\pi}{P} = \frac{3.1416}{4} = 0.7854 \text{ in.}$$

$$\text{Thickness of tooth} = \text{Space between teeth}$$

$$= \frac{0.7854}{2} = 0.3927 \text{ in.}$$

Practice check: Do exercise 23 on the right.

23. Compute the addendum, clearance, dedendum, whole depth, depth under pitch cone, circular pitch, thickness of tooth, and space between teeth in exercise 22 on page 650.

The answer is in the margin on page 653.

■   **Problems 21–15**

Compute the addendum, clearance, dedendum, whole depth, depth under pitch cone, circular pitch, thickness of tooth, and space between teeth in each of the 10 pairs of bevel gears in Problem Set 21–14.

21–16   PITCH CONE RADIUS

In Fig. 21–11, the pitch cone radius

$$\sin \text{angle } EOA = \frac{AE}{OA}$$

$$OA = \frac{AE}{\sin \text{angle } EOA}$$

$$OA = \frac{7.5}{\sin 73°18'} = \frac{7.5}{0.95782} = 7.83 \text{ in.}$$

Answer to exercise 22

Pitch diameter of gear $= \frac{45}{5}$
$= 9$ in.

Pitch diameter of pinion $= \frac{22}{5}$
$= 4\frac{2}{5}$ in.

$EB = \frac{1}{2}$ of $9 = 4\frac{1}{2}$ in.
$BD = \frac{1}{2}$ of $4\frac{2}{5} = 2\frac{1}{5}$ in.

Tan $BOE = \dfrac{4\frac{1}{2}}{2\frac{1}{5}} = 2.045$

Angle BOE
    $= 64°$
    $=$ Pitch cone angle of gear

Tan $BOD = \dfrac{2\frac{1}{5}}{4\frac{1}{4}} = 0.4889$

Angle BOD
    $= 26°$
    $=$ Pitch cone angle of pinion

---

24. Compute the pitch cone radius, addendum angle, turning angle, dedendum angle, and cutting angle for exercise 22 on page 650.

The answer is in the margin on page 655.

21–17   ADDENDUM ANGLE AND TURNING ANGLE

Since the line *FK* in Fig. 21–10(*b*) is perpendicular to the pitch cone radius, then $FA/FO =$ the tangent of the addendum angle AOF; that is,

$$\text{Tangent of addendum angle} = \frac{\text{Addendum}}{\text{Pitch cone radius}}$$

$$= \frac{0.25}{7.83} = 0.03193$$

$$\text{Addendum angle } AOF = 1°50'$$

The turning angle is equal to the pitch cone angle plus the addendum angle.

$$\text{Turning angle of the gear} = 73°18' + 1°50' = 75°08'$$
$$\text{Turning angle of the pinion} = 16°42' + 1°50' = 18°32'$$

21–18   DEDENDUM ANGLE AND CUTTING ANGLE

$$\text{Tangent of dedendum angle } AOK = \frac{KA}{AO} = \frac{\text{Depth under pitch cone}}{\text{Pitch cone radius}}$$

$$= \frac{0.2893}{7.83} = 0.03695$$

$$\text{Dedendum angle } AOK = 2°07'$$

$$\text{Cutting angle} = \text{Pitch cone angle} - \text{Dedendum angle}$$
$$\text{Cutting angle of gear} = 73°18' - 2°07' = 71°11'$$
$$\text{Cutting angle of pinion} = 16°42' - 2°07' = 14°35'$$

Practice check: Do exercise 24 on the left.

■ **Problems 21–18**

Compute the pitch cone radius, addendum angle, turning angle, dedendum angle, and cutting angle for the 10 pairs of bevel gears in Problem Set 21–14.

21–19   OUTSIDE DIAMETERS

$$\text{Outside diameter} = \text{Pitch diameter} + \text{Twice the angular addendum}$$

Outside Diameter of Gear

In triangle *ATF* in Fig. 21–10(*b*), *AF* is the addendum, 0.25 in.; angle *TAF* is equal to the pitch cone angle of the gear, $EOA = 73°18'$.

$$\text{Angular addendum } AT = AF \times \cos TAF = 0.25 \times \cos 73°18'$$
$$= 0.25 \times 0.28736 = 0.072 \text{ in.}$$

$$\text{Outside diameter of gear} = 15 + 2 \times 0.072 = 15 + 0.144$$
$$= 15.144 \text{ in.}$$

Outside Diameter of Pinion

In triangle *PCH*, *CH* is the addendum, 0.25 in.; angle *PCH* = pitch cone angle of the pinion *DOC*, which is 16°42'.

$$\text{Angular addendum } PC = CH \times \cos \widehat{PCH} = 0.25 \times \cos 16°42'$$
$$= 0.25 \times 0.95782 = 0.239 \text{ in.}$$

$$\text{Outside diameter of pinion} = 4\tfrac{1}{2} + 2 \times 0.239 = 4.5 + 0.478$$
$$= 4.978 \text{ in.}$$

Practice check: Do exercise 25 below on the right.

■ **Problems 21–19**

Compute the outside diameters of the 10 pairs of bevel gears in Problem Set 21–14.

21–20   SELECTING CUTTERS FOR BEVEL GEARS

Bevel gears may be cut on the milling machine with spur gear cutters of the same pitch. The cutters, however, are selected not for the number of teeth on the bevel gear but for an imaginary number of teeth computed as follows.

In Fig. 21–12, showing the pitch cones of the bevel gears under consideration, the line *XY* is drawn perpendicular to the pitch cone radius *OB*; that is, the line *XY* is a prolongation of the edge of tooth *RM* in 21–10(*b*). The center lines of the gear and the pinion are prolonged to meet this line.

$$BY = BO \times \text{Tangent angle } BOY$$
$$= \text{Pitch cone radius} \times \text{Tangent of pitch cone angle of pinion}$$
$$= 7.83 \times \tan 16°42'$$
$$= 7.83 \times 0.30001 = 2.349 \text{ in.}$$

$$BX = BO \times \text{Tangent angle } BOX$$
$$= \text{Pitch cone radius} \times \text{Tangent of pitch cone angle of gear}$$
$$= 7.83 \times \tan 73°18'$$
$$= 7.83 \times 3.33317 = 26.099 \text{ in.}$$

Answer to exercise 23

Addendum $= \tfrac{1}{5}$
$\qquad = 0.2$ in.

Clearance $= \dfrac{0.157}{5}$

$\qquad = 0.0314$ in.

Dedundum
$\qquad = \dfrac{1.157}{5} = 0.2314$ in.
$\qquad = $ Depth under pitch cone

Whole depth
$\qquad = 0.2 + 0.2 + 0.0314$
$\qquad = 0.4314$ in.

Circular pitch $= \dfrac{3.1416}{5}$

$\qquad = 0.62832$ in.

Thickness of tooth
$\qquad = \dfrac{0.62832}{2}$

$\qquad = 0.31416$ in.

$\qquad = $ Space between teeth

25. Compute the outside diameters in exercise 22 on page 650.

The answer is in the margin on page 656.

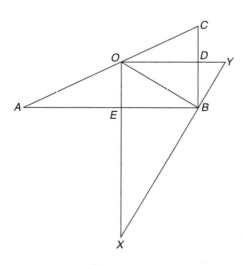

Figure 21–12

BY is the pitch radius of the imaginary pinion for which to select the cutter.

$$\text{Diameter} = 2 \times 2.349 = 4.698 \text{ in.}$$

$$\text{Number of teeth} = \text{Pitch diameter} \times \text{Pitch} = 4.698 \times 4$$
$$= 18.792 \text{ teeth}$$

The pinion is to be cut with a 19-tooth cutter, that is, a No. 6 cutter.
   BX is the pitch radius of the imaginary gear for which to select the cutter.

$$\text{Diameter} = 2 \times 26.1 = 52.2 \text{ in.}$$

26. Select the cutters for exercise 22 on page 650.

The answer is in the margin on page 657.

$$\text{Number of teeth} = \text{Pitch diameter} \times \text{Pitch} = 52.2 \times 4$$
$$= 208.8 \text{ teeth}$$

The gear is to be cut with a 209-tooth cutter, that is, a No. 1 cutter.

Practice check: Do exercise 26 on the left.

■ **Problems 21–20**

Select the cutters for the 10 pairs of bevel gears in Problem Set 21–14.

21–21   DIMENSIONS OF TEETH AT SMALL END

The *width* or *face* of the teeth, *AV* in Fig. 21–10(*b*), is usually made about one-quarter of the pitch cone radius. The dimensions of the teeth at the small end are proportional to their dimensions at the large end. Thus,

Ratio between the small end dimensions and the large end dimensions

$$= \frac{\text{Pitch cone radius at small end}}{\text{Pitch cone radius at large end}}$$

   The pitch cone radius is 7.83 in.; therefore, let us assume a width of face of 2 in., making the pitch cone radius at the small end of the teeth *OV* equal to $7.83 - 2.00 = 5.83$ in. Then the ratio between the small pitch cone radius and the large pitch cone radius is $5.83/7.83 = 0.745$. We can now find the tooth proportions at the small end by multiplying the large end dimensions by the factor 0.745.

27. Compute the addendum, clearance, dedendum, whole depth, circular pitch, thickness of tooth, and space at the small ends of the teeth of the gears in exercise 22 on page 650. Assume a tooth face of approximately $\frac{1}{4}$ of the pitch cone radius but not more than 0.3 of the pitch cone radius.

The answer is in the margin on page 657.

$$\text{Addendum at small end} = 0.745 \times 0.25 = 0.1862 \text{ in.}$$

$$\text{Clearance at small end} = 0.745 \times 0.0393 = 0.0293 \text{ in.}$$

$$\text{Dedendum at small end} = 0.745 \times 0.2893 = 0.2155 \text{ in.}$$

$$\text{Whole depth at small end} = 0.745 \times 0.5393 = 0.4018 \text{ in.}$$

$$\text{Circular pitch at small end} = 0.745 \times 0.7854 = 0.5851 \text{ in.}$$

$$\text{Thickness of tooth at small end} = \text{Space at small end}$$
$$= 0.745 \times 0.3927 = 0.2926 \text{ in.}$$

Practice check: Do exercise 27 on the left.

■ Problems 21–21

Compute the addendum, clearance, dedendum, whole depth, circular pitch, thickness of tooth, and space at the small ends of the teeth of the 10 pairs of bevel gears in Problem Set 21–14. In each instance assume a tooth face of approximately $\frac{1}{4}$ of the pitch cone radius but not more than 0.3 of the pitch cone radius.

21–22 MITER GEARS

When the gear and the pinion are both the same size and their axes are at right angles, they are called ***miter gears.*** The computation in this case is greatly simplified; all elements are the same for both gears.

The pitch cone angle is 45°.

$$\text{Pitch cone radius} = \frac{\text{Pitch radius}}{\sin 45°} = \frac{\text{Pitch diameter}}{2 \times \sin 45°}$$

$$= \frac{\text{Pitch diameter}}{2 \times 0.70711} = \frac{\text{Pitch diameter}}{1.41422}$$

$$= 0.707 \times \text{Pitch diameter}$$

$$\text{Face angle} = 45° + \text{Addendum angle}$$

$$\text{Cutting angle} = 45° - \text{Dedendum angle}$$

$$\text{Angular addendum} = \text{Addendum} \times \cos 45°$$
$$= 0.70711 \times \text{Addendum}$$

$$\text{Number of teeth for which to select cutter} = \frac{\text{Number of teeth}}{\cos 45°}$$

$$= \frac{\text{Number of teeth}}{0.70711}$$

$$= 1.414 \times \text{Number of teeth}$$

The table on page 658 is a summary of all the rules for solving problems involving bevel gears.

■ Self-Test

Use the space provided to solve the following problems involving gears. Check your answers with the answers in the back of the book.

1. Find the circular pitch of a gear that has 64 teeth and a pitch diameter of 8″.

2. Find the pitch diameter of a gear that has 60 teeth and a circular pitch of 0.2356″.

3. Find the number of teeth on a gear with a pitch diameter of 8″ and a circular pitch of 0.5236″.

4. What is the diametral pitch of a gear with 64 teeth and a pitch diameter of 4″?

Answer to exercise 24

Pitch cone radius

$$= OA = \frac{4.5}{\sin 64°}$$

$$= \frac{4.5}{0.8988} = 5.01 \text{ in.}$$

Tan of addendum angle AOF

$$= \frac{0.2}{5.01}$$

$$= 0.0399$$

Addendum angle $AOF = 2°20'$
Gear turning angle $= 64° + 2°20'$
$$= 66°20'$$
Pinion turning angle $= 26° + 2°20'$
$$= 28°20'$$

Tan of dedendum angle AOK

$$= \frac{0.2314}{5.01}$$

$$= 0.04619$$

Dedendum angle $AOK = 2°40'$
Gear cutting angle $= 64° - 2°40'$
$$= 61°20'$$
Pinion cutting angle $= 26° - 2°40'$
$$= 23°20'$$

Answer to exercise 25

Angular addendum *AT*

$$= 0.2 \times \cos 64°$$
$$= 0.2 \times 0.4384$$
$$= 0.088 \text{ in.}$$

Gear outside diameter

$$= 9 + 2 \times 0.08768$$
$$= 9.175 \text{ in.}$$

Angular addendum *PC*

$$= 0.2 \times \cos 26°$$
$$= 0.2 \times 0.8988$$
$$= 0.180 \text{ in.}$$

Pinion outside diameter

$$= 4\tfrac{2}{8} + 2 \times 0.17976$$
$$= 4.4 + 0.35952$$
$$= 4.760 \text{ in.}$$

---

5. What is the pitch diameter of a 16-pitch gear with 48 teeth?

6. What is the circular pitch of an 8-pitch gear?

7. Find the clearance of an 8-pitch gear.

8. Find the whole depth of tooth of a 12-pitch gear.

9. Find the outside diameter of a 12-pitch gear with a pitch diameter of 4″.

10. Two gears in mesh have pitch diameters of 8″ and 6″. Find their center-to-center distance.

■ Chapter Test

Use the space provided to solve the following problems involving gears.

1. Find the circular pitch of a gear that has 64 teeth and a pitch diameter of 6″.

2. Find the pitch diameter of a gear that has 60 teeth and a circular pitch of 0.3927″.

3. Find the number of teeth in a gear that has a pitch diameter of 10″ and a circular pitch of 0.3927″.

4. Find the thickness of tooth of the gear in Problem 3.

5. Find the diametral pitch of a gear that has 64 teeth and a pitch diameter of 4″.

6. What is the pitch diameter of a 16-pitch gear with 128 teeth?

7. How many teeth must an 8-pitch gear have if its pitch diameter is $3\frac{1}{2}''$?

8. What is the circular pitch of a 10-pitch gear?

9. Find the diametral pitch of a gear whose circular pitch is 0.31416".

10. Find the clearance of a gear whose circular pitch is 0.2836".

11. Find the clearance of an 8-pitch gear.

12. Find the whole depth of a 16-pitch gear.

13. Find the outside diameter of an 8-pitch gear whose pitch diameter is $6\frac{3}{4}''$.

14. Find the whole depth of tooth of a 6-pitch rack.

15. Find the center-to-center distance of two gears whose pitch diameters are $6\frac{1}{2}''$ and $4\frac{3}{4}''$.

16. Compute the pitch diameters of a pair of bevel gears at right angles if the gear has 64 teeth, the pinion has 20 teeth, and the pitch is 4.

Answer to exercise 26

Gear:
$BX = BO \times$ tan angle BOX
$\quad = 5.01 \times$ tan 64°
$\quad = 5.01 \times 2.050$
$\quad = 10.271$ in.
Diameter $= 2 \times 10.27$
$\qquad\quad = 20.5$ in.
Number of teeth $= 20.5 \times 5$
$\qquad\qquad\qquad = 102.5$ teeth
Answer: 103 teeth calls for a
$\qquad\quad$ No. 2 cutter.

Pinion:
$BY = BO \times$ tan angle BOY
$\quad = 5.01 \times$ tan 26°
$\quad = 5.01 \times 0.4877$
$\quad = 2.44$ in.
Diameter $= 2 \times 2.44$
$\qquad\quad = 4.88$ in.
Number of teeth $= 4.88 \times 5$
$\qquad\qquad\qquad = 24.4$ teeth
Answer: 24 teeth calls for a
$\qquad\quad$ No. 5 cutter.

Answer to exercise 27

$OV = 5.01 - 1.25$
$\quad\, = 3.76$

Ratio $= \dfrac{3.76}{5.01}$

$\qquad\;\; = 0.75$
Addendum at small end
$\qquad\qquad\qquad = 0.75 \times 0.2$
$\qquad\qquad\qquad = 0.15$ in.
Clearance $= 0.75 \times 0.0314$
$\qquad\qquad = 0.0236$ in.
Dedendum $= 0.75 \times 0.2314$
$\qquad\qquad = 0.174$ in.
Whole depth $= 0.75 \times 0.4314$
$\qquad\qquad\;\; = 0.324$ in.
Circular pitch $= 0.75 \times 0.6283$
$\qquad\qquad\qquad = 0.4712$ in.
Thickness of tooth
$\qquad\qquad\quad = 0.75 \times 0.31416$
$\qquad\qquad\quad = 0.2356$ in.

Table of Rules for Bevel Gears	
To Find	**Rule**
Pitch diameter	Pitch diameter $= N/P$
Pitch cone angles	Tan pitch cone angle, gear $= \dfrac{\text{Pitch diameter, gear}}{\text{Pitch diameter, pinion}}$ Tan pitch cone angle, pinion $= \dfrac{\text{Pitch diameter, pinion}}{\text{Pitch diameter, gear}}$
Pitch cone radius	Pitch cone radius $= \dfrac{\frac{1}{2}\text{-pitch diameter of gear}}{\text{Sine pitch cone angle, gear}}$
Tooth parts, large end. Addendum	Addendum $= 1/P$
Clearance	Clearance $= 0.157/P$
Dedendum or depth below pitch cone	Depth below pitch cone $= 1.157/P$
Whole depth	Whole depth $= 2.157/P$
Circular pitch at pitch line	Circular pitch $= \pi/P$
Thickness of tooth or space at pitch line	Thickness $=$ Space $= P_c/2$
Tooth dimensions at small end	Multiply large end dimensions by the following ratio. $\qquad \dfrac{\text{Pitch cone radius at small end}}{\text{Pitch cone radius at large end}}$
Addendum angle	Tan addendum angle $= \dfrac{\text{Addendum}}{\text{Pitch cone radius}}$
Dedendum angle	Tan dedendum angle $= \dfrac{\text{Dedendum}}{\text{Pitch cone radius}}$
Turning angle	Face angle $=$ Pitch cone angle $+$ Addendum angle
Cutting angle	Cutting angle $=$ Pitch cone angle $\qquad\qquad - $ Dedendum angle
Angular addendum	Gear, angular addendum $=$ Addendum $\qquad\qquad\qquad \times$ Cosine of pitch cone angle of $\qquad\qquad\qquad\quad$ gear Pinion, angular addendum $=$ Addendum $\qquad\qquad\qquad \times$ Cosine of pitch cone angle of $\qquad\qquad\qquad\quad$ pinion
Outside diameter	Outside diameter $=$ Pitch diameter $\qquad\qquad\qquad + $ Two times the angular addendum
Number of teeth for which to select cutter	Number of teeth for which to select the cutter $\qquad = \dfrac{\text{Actual number of teeth}}{\text{Cosine of pitch cone angle}}$

Table 1   Powers, Roots, Circumferences, and Areas   660

Table 2   Natural Trigonometric Functions   685

Table 3   Converting Millimeters to Inches and Decimals   690

Table 4   Decimal and Metric Equivalents of
Fractions of an Inch   691

Table 5   Converting Degrees Fahrenheit to Degrees Celsius   692

Table 6   Converting Degrees Celsius to Degrees Fahrenheit   693

Table 7   Weights of Materials   694

Table 8   National Coarse-Thread Series   695

Table 9   National Fine-Thread Series   696

					No. = Diam.		
No.	Square	Cube	Square Root	Cube Root	Circum.	Area	No.
1	1	1	1.0000	1.0000	3.142	0.7854	1
2	4	8	1.4142	1.2599	6.283	3.1416	2
3	9	27	1.7321	1.4422	9.425	7.0686	3
4	16	64	2.0000	1.5874	12.566	12.5664	4
5	25	125	2.2361	1.7100	15.708	19.6350	5
6	36	216	2.4495	1.8171	18.850	28.2743	6
7	49	343	2.6458	1.9129	21.991	38.4845	7
8	64	512	2.8284	2.0000	25.133	50.2655	8
9	81	729	3.0000	2.0801	28.274	63.6173	9
10	100	1000	3.1623	2.1544	31.416	78.5398	10
11	121	1331	3.3166	2.2240	34.558	95.0332	11
12	144	1728	3.4641	2.2894	37.699	113.097	12
13	169	2197	3.6056	2.3513	40.841	132.732	13
14	196	2744	3.7417	2.4101	43.982	153.938	14
15	225	3375	3.8730	2.4662	47.124	176.715	15
16	256	4096	4.0000	2.5198	50.265	201.062	16
17	289	4913	4.1231	2.5713	53.407	226.980	17
18	324	5832	4.2426	2.6207	56.549	254.469	18
19	361	6859	4.3589	2.6684	59.690	283.529	19
20	400	8000	4.4721	2.7144	62.832	314.159	20
21	441	9261	4.5826	2.7589	65.973	346.361	21
22	484	10648	4.6904	2.8020	69.115	380.133	22
23	529	12167	4.7958	2.8439	72.257	415.476	23
24	576	13824	4.8990	2.8845	75.398	452.389	24
25	625	15625	5.0000	2.9240	78.540	490.874	25
26	676	17576	5.0990	2.9625	81.681	530.929	26
27	729	19683	5.1962	3.0000	84.823	572.555	27
28	784	21952	5.2915	3.0366	87.965	615.752	28
29	841	24389	5.3852	3.0723	91.106	660.520	29
30	900	27000	5.4772	3.1072	94.248	706.858	30
31	961	29791	5.5678	3.1414	97.389	754.768	31
32	1024	32768	5.6569	3.1748	100.531	804.248	32
33	1089	35937	5.7446	3.2075	103.673	855.299	33
34	1156	39304	5.8310	3.2396	106.814	907.920	34
35	1225	42875	5.9161	3.2711	109.956	962.113	35
36	1296	46656	6.0000	3.3019	113.097	1017.88	36
37	1369	50653	6.0828	3.3322	116.239	1075.21	37
38	1444	54872	6.1644	3.3620	119.381	1134.11	38
39	1521	59319	6.2450	3.3912	122.522	1194.59	39

Table 1
Powers, Roots, Circumferences, and Areas

Table 1

No.	Square	Cube	Square Root	Cube Root	No. = Diam.		No.
					Circum.	Area	
40	1600	64000	6.3246	3.4200	125.66	1256.64	40
41	1681	68921	6.4031	3.4482	128.81	1320.25	41
42	1764	74088	6.4807	3.4760	131.95	1385.44	42
43	1849	79507	6.5574	3.5034	135.09	1452.20	43
44	1936	85184	6.6332	3.5303	138.23	1520.53	44
45	2025	91125	6.7082	3.5569	141.37	1590.43	45
46	2116	97336	6.7823	3.5830	144.51	1661.90	46
47	2209	103823	6.8557	3.6088	147.65	1734.94	47
48	2304	110592	6.9282	3.6342	150.80	1809.56	48
49	2401	117649	7.0000	3.6593	153.94	1885.74	49
50	2500	125000	7.0711	3.6840	157.08	1963.50	50
51	2601	132651	7.1414	3.7084	160.22	2042.82	51
52	2704	140608	7.2111	3.7325	163.36	2123.72	52
53	2809	148877	7.2801	3.7563	166.50	2206.18	53
54	2916	157464	7.3485	3.7798	169.65	2290.22	54
55	3025	166375	7.4162	3.8030	172.79	2375.83	55
56	3136	175616	7.4833	3.8259	175.93	2463.01	56
57	3249	185193	7.5498	3.8485	179.07	2551.76	57
58	3364	195112	7.6158	3.8709	182.21	2642.08	58
59	3481	205379	7.6811	3.8930	185.35	2733.97	59
60	3600	216000	7.7460	3.9149	188.50	2827.43	60
61	3721	226981	7.8102	3.9365	191.64	2922.47	61
62	3844	238328	7.8740	3.9579	194.78	3019.07	62
63	3969	250047	7.9373	3.9791	197.92	3117.25	63
64	4096	262144	8.0000	4.0000	201.06	3216.99	64
65	4225	274625	8.0623	4.0207	204.20	3318.31	65
66	4356	287496	8.1240	4.0412	207.35	3421.19	66
67	4489	300763	8.1854	4.0615	210.49	3525.65	67
68	4624	314432	8.2462	4.0817	213.63	3631.68	68
69	4761	328509	8.3066	4.1016	216.77	3739.28	69
70	4900	343000	8.3666	4.1213	219.91	3848.45	70
71	5041	357911	8.4261	4.1408	223.05	3959.19	71
72	5184	373248	8.4853	4.1602	226.19	4071.50	72
73	5329	389017	8.5440	4.1793	229.34	4185.39	73
74	5476	405224	8.6023	4.1983	232.48	4300.84	74
75	5625	421875	8.6603	4.2172	235.62	4417.86	75
76	5776	438976	8.7178	4.2358	238.76	4536.46	76
77	5929	456533	8.7750	4.2543	241.90	4656.63	77
78	6084	474552	8.8318	4.2727	245.04	4778.36	78
79	6241	493039	8.8882	4.2908	248.19	4901.67	79

No.	Square	Cube	Square Root	Cube Root	No. = Diam. Circum.	Area	No.
80	6400	512000	8.9443	4.3089	251.33	5026.55	80
81	6561	531441	9.0000	4.3267	254.47	5153.00	81
82	6724	551368	9.0554	4.3445	257.61	5281.02	82
83	6889	571787	9.1104	4.3621	260.75	5410.61	83
84	7056	592704	9.1652	4.3795	263.89	5541.77	84
85	7225	614125	9.2195	4.3968	267.04	5674.50	85
86	7396	636056	9.2736	4.4140	270.18	5808.80	86
87	7569	658503	9.3274	4.4310	273.32	5944.68	87
88	7744	681472	9.3808	4.4480	276.46	6082.12	88
89	7921	704969	9.4340	4.4647	279.60	6221.14	89
90	8100	729000	9.4868	4.4814	282.74	6361.73	90
91	8281	753571	9.5394	4.4979	285.88	6503.88	91
92	8464	778688	9.5917	4.5144	289.03	6647.61	92
93	8649	804357	9.6437	4.5307	292.17	6792.91	93
94	8836	830584	9.6954	4.5468	295.31	6939.78	94
95	9025	857375	9.7468	4.5629	298.45	7088.22	95
96	9216	884736	9.7980	4.5789	301.59	7238.23	96
97	9409	912673	9.8489	4.5947	304.73	7389.81	97
98	9604	941192	9.8995	4.6104	307.88	7542.96	98
99	9801	970299	9.9499	4.6261	311.02	7697.69	99
100	10000	1000000	10.0000	4.6416	314.16	7853.98	100
101	10201	1030301	10.0499	4.6570	317.30	8011.85	101
102	10404	1061208	10.0995	4.6723	320.44	8171.28	102
103	10609	1092727	10.1489	4.6875	323.58	8332.29	103
104	10816	1124864	10.1980	4.7027	326.73	8494.87	104
105	11025	1157625	10.2470	4.7177	329.87	8659.01	105
106	11236	1191016	10.2956	4.7326	333.01	8824.73	106
107	11449	1225043	10.3441	4.7475	336.15	8992.02	107
108	11664	1259712	10.3923	4.7622	339.29	9160.88	108
109	11881	1295029	10.4403	4.7769	342.43	9331.32	109
110	12100	1331000	10.4881	4.7914	345.58	9503.32	110
111	12321	1367631	10.5357	4.8059	348.72	9676.89	111
112	12544	1404928	10.5830	4.8203	351.86	9852.03	112
113	12769	1442897	10.6301	4.8346	355.00	10028.7	113
114	12996	1481544	10.6771	4.8488	358.14	10207.0	114
115	13225	1520875	10.7238	4.8629	361.28	10386.9	115
116	13456	1560896	10.7703	4.8770	364.42	10568.3	116
117	13689	1601613	10.8167	4.8910	367.57	10751.3	117
118	13924	1643032	10.8682	4.9049	370.71	10935.9	118
119	14161	1685159	10.9087	4.9187	373.85	11122.0	119

Table 1 663

No.	Square	Cube	Square Root	Cube Root	No. = Diam.		No.
					Circum.	Area	
120	14400	1728000	10.9545	4.9324	376.99	11309.7	120
121	14641	1771561	11.0000	4.9461	380.13	11499.0	121
122	14884	1815848	11.0454	4.9597	383.27	11689.9	122
123	15129	1860867	11.0905	4.9732	386.42	11882.3	123
124	15376	1906624	11.1355	4.9866	389.56	12076.3	124
125	15625	1953125	11.1803	5.0000	392.70	12271.8	125
126	15876	2000376	11.2250	5.0133	395.84	12469.0	126
127	16129	2048383	11.2694	5.0265	398.98	12667.7	127
128	16384	2097152	11.3137	5.0397	402.12	12868.0	128
129	16641	2146689	11.3578	5.0528	405.27	13069.8	129
130	16900	2197000	11.4018	5.0658	408.41	13273.2	130
131	17161	2248091	11.4455	5.0788	411.55	13478.2	131
132	17424	2299968	11.4891	5.0916	414.69	13684.8	132
133	17689	2352637	11.5326	5.1045	417.83	13892.9	133
134	17956	2406104	11.5758	5.1172	420.97	14102.6	134
135	18225	2460375	11.6190	5.1299	424.12	14313.9	135
136	18496	2515456	11.6619	5.1426	427.26	14526.7	136
137	18769	2571353	11.7047	5.1551	430.40	14741.1	137
138	19044	2628072	11.7473	5.1676	433.54	14957.1	138
139	19321	2685619	11.7898	5.1801	436.68	15174.7	139
140	19600	2744000	11.8322	5.1925	439.82	15393.8	140
141	19881	2803221	11.8743	5.2048	442.96	15614.5	141
142	20164	2863288	11.9164	5.2171	446.11	15836.8	142
143	20449	2924207	11.9583	5.2293	449.25	16060.6	143
144	20736	2985984	12.0000	2.2415	452.39	16286.0	144
145	21025	3048625	12.0416	5.2536	455.53	16513.0	145
146	21316	3112136	12.0830	5.2656	458.67	16741.5	146
147	21609	3176523	12.1244	5.2776	461.81	16971.7	147
148	21904	3241792	12.1655	5.2896	464.96	17203.4	148
149	22201	3307949	12.2066	5.3015	468.10	17436.6	149
150	22500	3375000	12.2474	5.3133	471.24	17671.5	150
151	22801	3442951	12.2882	5.3251	474.38	17907.9	151
152	23104	3511808	12.3288	5.3368	477.52	18145.8	152
153	23409	3581577	12.3693	5.3485	480.66	18385.4	153
154	23716	3652264	12.4097	5.3601	483.81	18626.5	154
155	24025	3723875	12.4499	5.3717	486.95	18869.2	155
156	24336	3796416	12.4900	5.3832	490.09	19113.4	156
157	24649	3869893	12.5300	5.3947	493.23	19359.3	157
158	24964	3944312	12.5698	5.4061	496.37	19606.7	158
159	25281	4019679	12.6095	5.4175	499.51	19855.7	159

No.	Square	Cube	Square Root	Cube Root	No. = Diam.		No.
					Circum.	Area	
160	25600	4096000	12.6491	5.4288	502.65	20106.2	160
161	25921	4173281	12.6886	5.4401	505.80	20358.3	161
162	26244	4251528	12.7279	5.4514	508.94	20612.0	162
163	26569	4330747	12.7671	5.4626	512.08	20867.2	163
164	26896	4410944	12.8062	5.4737	515.22	21124.1	164
165	27225	4492125	12.8452	5.4848	518.36	21382.5	165
166	27556	4574296	12.8841	5.4959	521.50	21642.4	166
167	27889	4657463	12.9228	5.5069	524.65	21904.0	167
168	28224	4741632	12.9615	5.5178	527.79	22167.1	168
169	28561	4826809	13.0000	5.5288	530.93	22431.8	169
170	28900	4913000	13.0384	5.5397	534.07	22698.0	170
171	29241	5000211	13.0767	5.5505	537.21	22965.8	171
172	29584	5088448	13.1149	5.5613	540.35	23235.2	172
173	29929	5177717	13.1529	5.5721	543.50	23506.2	173
174	30276	5268024	13.1909	5.5828	546.64	23778.7	174
175	30625	5359375	13.2288	5.5934	549.78	24052.8	175
176	30976	5451776	13.2665	5.6041	552.92	24328.5	176
177	31329	5545233	13.3041	5.6147	556.06	24605.7	177
178	31684	5639752	13.3417	5.6252	559.20	24884.6	178
179	32041	5735339	13.3791	5.6357	562.35	25164.9	179
180	32400	5832000	13.4164	5.6462	565.49	25446.9	180
181	32761	5929741	13.4536	5.6567	568.63	25730.4	181
182	33124	6028568	13.4907	5.6671	571.77	26015.5	182
183	33489	6128487	13.5277	5.6774	574.91	26302.2	183
184	33856	6229504	13.5647	5.6877	578.05	26590.4	184
185	34225	6331625	13.6015	5.6980	581.19	26880.3	185
186	34596	6434856	13.6382	5.7083	584.34	27171.6	186
187	34969	6539203	13.6748	5.7185	587.48	27464.6	187
188	35344	6644672	13.7113	5.7287	590.62	27759.1	188
189	35721	6751269	13.7477	5.7388	593.76	28055.2	189
190	36100	6859000	13.7840	5.7489	596.90	28352.9	190
191	36481	6967871	13.8203	5.7590	600.04	28652.1	191
192	36864	7077888	13.8564	5.7690	603.19	28952.9	192
193	37249	7189057	13.8924	5.7790	606.33	29255.3	193
194	37636	7301384	13.9284	5.7890	609.47	29559.2	194
195	38025	7414875	13.9642	5.7989	612.61	29864.8	195
196	38416	7529536	14.0000	5.8088	615.75	30171.9	196
197	38809	7645373	14.0357	5.8186	618.89	30480.5	197
198	39204	7762392	14.0712	5.8285	622.04	30790.7	198
199	39601	7880599	14.1067	5.8383	625.18	31102.6	199

Table 1 665

| No. | Square | Cube | Square Root | Cube Root | No. = Diam. | | No. |
					Circum.	Area	
200	40000	8000000	14.1421	5.8480	628.32	31415.9	200
201	40401	8120601	14.1774	5.8578	631.46	31730.9	201
202	40804	8242408	14.2127	5.8675	634.60	32047.4	202
203	41209	8365427	14.2478	5.8771	637.74	32365.5	203
204	41616	8489664	14.2829	5.8868	640.89	32685.1	204
205	42025	8615125	14.3178	5.8964	644.03	33006.4	205
206	42436	8741816	14.3527	5.9059	647.17	33329.2	206
207	42849	8869743	14.3875	5.9155	650.31	33653.5	207
208	43264	8998912	14.4222	5.9250	653.45	33979.5	208
209	43681	9129329	14.4568	5.9345	656.59	34307.0	209
210	44100	9261000	14.4914	5.9439	659.73	34636.1	210
211	44521	9393931	14.5258	5.9533	662.88	34966.7	211
212	44944	9528128	14.5602	5.9627	666.02	35298.9	212
213	45369	9663597	14.5945	5.9721	669.16	35632.7	213
214	45796	9800344	14.6287	5.9814	672.30	35968.1	214
215	46225	9938375	14.6629	5.9907	675.44	36305.0	215
216	46656	10077696	14.6969	6.0000	678.58	36643.5	216
217	47089	10218313	14.7309	6.0092	681.73	36983.6	217
218	47524	10360232	14.7648	6.0185	684.87	37325.3	218
219	47961	10503459	14.7986	6.0277	688.01	37668.5	219
220	48400	10648000	14.8324	6.0368	691.15	38013.3	220
221	48841	10793861	14.8661	6.0459	694.29	38359.6	221
222	49284	10941048	14.8997	6.0550	697.43	38707.6	222
223	49729	11089567	14.9332	6.0641	700.58	39057.1	223
224	50176	11239424	14.9666	6.0732	703.72	39408.1	224
225	50625	11390625	15.0000	6.0822	706.86	39760.8	225
226	51076	11543176	15.0333	6.0912	710.00	40115.0	226
227	51529	11697083	15.0665	6.1002	713.14	40470.8	227
228	51984	11852352	15.0997	6.1091	716.28	40828.1	228
229	52441	12008989	15.1327	6.1180	719.42	41187.1	229
230	52900	12167000	15.1658	6.1269	722.57	41547.6	230
231	53361	12326391	15.1987	6.1358	725.71	41909.6	231
232	53824	12487168	15.2315	6.1446	728.85	42273.3	232
233	54289	12649337	15.2643	6.1534	731.99	42638.5	233
234	54756	12812904	15.2971	6.1622	735.13	43005.3	234
235	55225	12977875	15.3297	6.1710	738.27	43373.6	235
236	55696	13144256	15.3623	6.1797	741.42	43743.5	236
237	56169	13312053	15.3948	6.1885	744.56	44115.0	237
238	56644	13481272	15.4272	6.1972	747.70	44488.1	238
239	57121	13651919	15.4596	6.2058	750.84	44862.7	239

No.	Square	Cube	Square Root	Cube Root	No. = Diam.		No.
					Circum.	Area	
240	57600	13824000	15.4919	6.2145	753.98	45238.9	240
241	58081	13997521	15.5242	6.2231	757.12	45616.7	241
242	58564	14172488	15.5563	6.2317	760.27	45996.1	242
243	59049	14348907	15.5885	6.2403	763.41	46377.0	243
244	59536	14526784	15.6205	6.2488	766.55	46759.5	244
245	60025	14706125	15.6525	6.2573	769.69	47143.5	245
246	60516	14886936	15.6844	6.2658	772.83	47529.2	246
247	61009	15069223	15.7162	6.2743	775.97	47916.4	247
248	61504	15252992	15.7480	6.2828	779.12	48305.1	248
249	62001	15438249	15.7797	6.2912	782.26	48695.5	249
250	62500	15625000	15.8114	6.2996	785.40	49087.4	250
251	63001	15813251	15.8430	6.3080	788.54	49480.9	251
252	63504	16003008	15.8745	6.3164	791.68	49875.9	252
253	64009	16194277	15.9060	6.3247	794.82	50272.6	253
254	64516	16387064	15.9374	6.3330	797.96	50670.7	254
255	65025	16581375	15.9687	6.3413	801.11	51070.5	255
256	65536	16777216	16.0000	6.3496	804.25	51471.9	256
257	66049	16974593	16.0312	6.3579	807.39	51874.8	257
258	66564	17173512	16.0624	6.3661	810.53	52279.2	258
259	67081	17373979	16.0935	6.3743	813.67	52685.3	259
260	67600	17576000	16.1245	6.3825	816.81	53092.9	260
261	68121	17779581	16.1555	6.3907	819.96	53502.1	261
262	68644	17984728	16.1864	6.3988	823.10	53912.9	262
263	69169	18191447	16.2173	6.4070	826.24	54325.2	263
264	69696	18399744	16.2481	6.4151	829.38	54739.1	264
265	70225	18609625	16.2788	6.4232	832.52	55154.6	265
266	70756	18821096	16.3095	6.4312	835.66	55571.6	266
267	71289	19034163	16.3401	6.4393	838.81	55990.3	267
268	71824	19248832	16.3707	6.4473	841.95	56410.4	268
269	72361	19465109	16.4012	6.4553	845.09	56832.2	269
270	72900	19683000	16.4317	6.4633	848.23	57255.5	270
271	73441	19902511	16.4621	6.4713	851.37	57680.4	271
272	73984	20123648	16.4924	6.4792	854.51	58106.9	272
273	74529	20346417	16.5227	6.4872	857.66	58534.9	273
274	75076	20570824	16.5529	6.4951	860.80	58964.6	274
275	75625	20796875	16.5831	6.5030	863.94	59395.7	275
276	76176	21024576	16.6132	6.5108	867.08	59828.5	276
277	76729	21253933	16.6433	6.5187	870.22	60262.8	277
278	77284	21484952	16.6733	6.5265	873.36	60698.7	278
279	77841	21717639	16.7033	6.5343	876.50	61136.2	279

Table 1 667

No.	Square	Cube	Square Root	Cube Root	No. = Diam.		No.
					Circum.	Area	
280	78400	21952000	16.7332	6.5421	879.65	61575.2	280
281	78961	22188041	16.7631	6.5499	882.79	62015.8	281
282	79524	22425768	16.7929	6.5577	885.93	62458.0	282
283	80089	22665187	16.8226	6.5654	889.07	62901.8	283
284	80656	22906304	16.8523	6.5731	892.21	63347.1	284
285	81225	23149125	16.8819	6.5808	895.35	63794.0	285
286	81796	23393656	16.9115	6.5885	898.50	64242.4	286
287	82369	23639903	16.9411	6.5962	901.64	64692.5	287
288	82944	23887872	16.9706	6.6039	904.78	65144.1	288
289	83521	24137569	17.0000	6.6115	907.92	65597.2	289
290	84100	24389000	17.0294	6.6191	911.06	66052.0	290
291	84681	24642171	17.0587	6.6267	914.20	66508.3	291
292	85264	24897088	17.0880	6.6343	917.35	66966.2	292
293	85849	25153757	17.1172	6.6419	920.49	67425.6	293
294	86436	25412184	17.1464	6.6494	923.63	67886.7	294
295	87025	25672375	17.1756	6.6569	926.77	68349.3	295
296	87616	25934336	17.2047	6.6644	929.91	68813.5	296
297	88209	26198073	17.2337	6.6719	933.05	69279.2	297
298	88804	26463592	17.2627	6.6794	936.19	69746.5	298
299	89401	26730899	17.2916	6.6869	939.34	70215.4	299
300	90000	27000000	17.3205	6.6943	942.48	70685.8	300
301	90601	27270901	17.3494	6.7018	945.62	71157.9	301
302	91204	27543608	17.3781	6.7092	948.76	71631.5	302
303	91809	27818127	17.4069	6.7166	951.90	72106.6	303
304	92416	28094464	17.4356	6.7240	955.04	72583.4	304
305	93025	28372625	17.4642	6.7313	958.19	73061.7	305
306	93636	28652616	17.4929	6.7387	961.33	73541.5	306
307	94249	28934443	17.5214	6.7460	964.47	74023.0	307
308	94864	29218112	17.5499	6.7533	967.61	74506.0	308
309	95481	29503629	17.5784	6.7606	970.75	74990.6	309
310	96100	29791000	17.6068	6.7679	973.89	75476.8	310
311	96721	30080231	17.6352	6.7752	977.04	75964.5	311
312	97344	30371328	17.6635	6.7824	980.18	76453.8	312
313	97969	30664297	17.6918	6.7897	983.32	76944.7	313
314	98596	30959144	17.7200	6.7969	986.46	77437.1	314
315	99225	31255875	17.7482	6.8041	989.60	77931.1	315
316	99856	31554496	17.7764	6.8113	992.74	78426.7	316
317	100489	31855013	17.8045	6.8185	995.88	78923.9	317
318	101124	32157432	17.8326	6.8256	999.03	79422.6	318
319	101761	32461759	17.8606	6.8328	1002.2	79922.9	319

No.	Square	Cube	Square Root	Cube Root	No. = Diam.		No.
					Circum.	Area	
320	102400	32768000	17.8885	6.8399	1005.3	80424.8	320
321	103041	33076161	17.9165	6.8470	1008.5	80928.2	321
322	103684	33386248	17.9444	6.8541	1011.6	81433.2	322
323	104329	33698267	17.9722	6.8612	1014.7	81939.8	323
324	104976	34012224	18.0000	6.8683	1017.9	82448.0	324
325	105625	34328125	18.0278	6.8753	1021.0	82957.7	325
326	106276	34645976	18.0555	6.8824	1024.2	83469.0	326
327	106929	34965783	18.0831	6.8894	1027.3	83981.8	327
328	107584	35287552	18.1108	6.8964	1030.4	84496.3	328
329	108241	35611289	18.1384	6.9034	1033.6	85012.3	329
330	108900	35937000	18.1659	6.9104	1036.7	85529.9	330
331	109561	36264691	18.1934	6.9174	1039.9	86049.0	331
332	110224	36594368	18.2209	6.9244	1043.0	86569.7	332
333	110889	36926037	18.2483	6.9313	1046.2	87092.0	333
334	111556	37259704	18.2757	6.9382	1049.3	87615.9	334
335	112225	37595375	18.3030	6.9451	1052.4	88141.3	335
336	112896	37933056	18.3303	6.9521	1055.6	88668.3	336
337	113569	38272753	18.3576	6.9589	1058.7	89196.9	337
338	114244	38614472	18.3848	6.9658	1061.9	89727.0	338
339	114921	38958219	18.4120	6.9727	1065.0	90258.7	339
340	115600	39304000	18.4391	6.9795	1068.1	90792.0	340
341	116281	39651821	18.4662	6.9864	1071.3	91326.9	341
342	116964	40001688	18.4932	6.9932	1074.4	91863.3	342
343	117649	40353607	18.5203	7.0000	1077.6	92401.3	343
344	118336	40707584	18.5472	7.0068	1080.7	92940.9	344
345	119025	41063625	18.5742	7.0136	1083.8	93482.0	345
346	119716	41421736	18.6011	7.0203	1087.0	94024.7	346
347	120409	41781923	18.6279	7.0271	1090.1	94569.0	347
348	121104	42144192	18.6548	7.0338	1093.3	95114.9	348
349	121801	42508549	18.6815	7.0406	1096.4	95662.3	349
350	122500	42875000	18.7083	7.0473	1099.6	96211.3	350
351	123201	43243551	18.7350	7.0540	1102.7	96761.8	351
352	123904	43614208	18.7617	7.0607	1105.8	97314.0	352
353	124609	43986977	18.7883	7.0674	1109.0	97867.7	353
354	125316	44361864	18.8149	7.0740	1112.1	98423.0	354
355	126025	44738875	18.8414	7.0807	1115.3	98979.8	355
356	126736	45118016	18.8680	7.0873	1118.4	99538.2	356
357	127449	45499293	18.8944	7.0940	1121.5	100098	357
358	128164	45882712	18.9209	7.1006	1124.7	100660	358
359	128881	46268279	18.9473	7.1072	1127.8	101223	359

Table 1

669

| No. | Square | Cube | Square Root | Cube Root | No. = Diam. | | No. |
					Circum.	Area	
360	129600	46656000	18.9737	7.1138	1131.0	101788	360
361	130321	47045881	19.0000	7.1204	1134.1	102354	361
362	131044	47437928	19.0263	7.1269	1137.3	102922	362
363	131769	47832147	19.0526	7.1335	1140.4	103491	363
364	132496	48228544	19.0788	7.1400	1143.5	104062	364
365	133225	48627125	19.1050	7.1466	1146.7	104635	365
366	133956	49027896	19.1311	7.1531	1149.8	105209	366
367	134689	49430863	19.1572	7.1596	1153.0	105785	367
368	135424	49836032	19.1833	7.1661	1156.1	106362	368
369	136161	50243409	19.2094	7.1726	1159.2	106941	369
370	136900	50653000	19.2354	7.1791	1162.4	107521	370
371	137641	51064811	19.2614	7.1855	1165.5	108103	371
372	138384	51478848	19.2873	7.1920	1168.7	108687	372
373	139129	51895117	19.3132	7.1984	1171.8	109272	373
374	139876	52313624	19.3391	7.2048	1175.0	109858	374
375	140625	52734375	19.3649	7.2112	1178.1	110447	375
376	141376	53157376	19.3907	7.2177	1181.2	111036	376
377	142129	53582633	19.4165	7.2240	1184.4	111628	377
378	142884	54010152	19.4422	7.2304	1187.5	112221	378
379	143641	54439939	19.4679	7.2368	1190.7	112815	379
380	144400	54872000	19.4936	7.2432	1193.8	113411	380
381	145161	55306341	19.5192	7.2495	1196.9	114009	381
382	145924	55742968	19.5448	7.2558	1200.1	114608	382
383	146689	56181887	19.5704	7.2622	1203.2	115209	383
384	147456	56623104	19.5959	7.2685	1206.4	115812	384
385	148225	57066625	19.6214	7.2748	1209.5	116416	385
386	148996	57512456	19.6469	7.2811	1212.7	117021	386
387	149769	57960603	19.6723	7.2874	1215.8	117628	387
388	150544	58411072	19.6977	7.2936	1218.9	118237	388
389	151321	58863869	19.7231	7.2999	1221.1	118847	389
390	152100	59319000	19.7484	7.3061	1225.2	119459	390
391	152881	59776471	19.7737	7.3124	1228.4	120072	391
392	153664	60236288	19.7990	7.3186	1231.5	120687	392
393	154449	60698457	19.8242	7.3248	1234.6	121304	393
394	155236	61162984	19.8494	7.3310	1237.8	121922	394
395	156025	61629875	19.8746	7.3372	1240.9	122542	395
396	156816	62099136	19.8997	7.3434	1244.1	123163	396
397	157609	62570773	19.9249	7.3496	1247.2	123786	397
398	158404	63044792	19.9499	7.3558	1250.4	124410	398
399	159201	63521199	19.9750	7.3619	1253.5	125036	399

No.	Square	Cube	Square Root	Cube Root	No. = Diam.		No.
					Circum.	Area	
400	160000	64000000	20.0000	7.3681	1256.6	125664	400
401	160801	64481201	20.0250	7.3742	1259.8	126293	401
402	161604	64964808	20.0499	7.3803	1262.9	126923	402
403	162409	65450827	20.0749	7.3864	1266.1	127556	403
404	163216	65939264	20.0998	7.3925	1269.2	128190	404
405	164025	66430125	20.1246	7.3986	1272.3	128825	405
406	164836	66923416	20.1494	7.4047	1275.5	129462	406
407	165649	67419143	20.1742	7.4108	1278.6	130100	407
408	166464	67917312	20.1990	7.4169	1281.8	130741	408
409	167281	68417929	20.2237	7.4229	1284.9	131382	409
410	168100	68921000	20.2485	7.4290	1288.1	132025	410
411	168921	69426531	20.2731	7.4350	1291.2	132670	411
412	169744	69934528	20.2978	7.4410	1294.3	133317	412
413	170569	70444997	20.3224	7.4470	1297.5	133965	413
414	171396	70957944	20.3470	7.4530	1300.6	134614	414
415	172225	71473375	20.3715	7.4590	1303.8	135265	415
416	173056	71991296	20.3961	7.4650	1306.9	135918	416
417	173889	72511713	20.4206	7.4710	1310.0	136572	417
418	174724	73034632	20.4450	7.4770	1313.2	137228	418
419	175561	73560059	20.4695	7.4829	1316.3	137885	419
420	176400	74088000	20.4939	7.4889	1319.5	138544	420
421	177241	74618461	20.5183	7.4948	1322.6	139205	421
422	178084	75151448	20.5426	7.5007	1325.8	139867	422
423	178929	75686967	20.5670	7.5067	1328.9	140531	423
424	179776	76225024	20.5913	7.5126	1332.0	141196	424
425	180625	76765625	20.6155	7.5185	1335.2	141863	425
426	181476	77308776	20.6398	7.5244	1338.3	142531	426
427	182329	77854483	20.6640	7.5302	1341.5	143201	427
428	183184	78402752	20.6882	7.5361	1344.6	143872	428
429	184041	78953589	20.7123	7.5420	1347.7	144545	429
430	184900	79507000	20.7364	7.5478	1350.9	145220	430
431	185761	80062991	20.7605	7.5537	1354.0	145896	431
432	186624	80621568	20.7846	7.5595	1357.2	146574	432
433	187489	81182737	20.8087	7.5654	1360.3	147254	433
434	188356	81746504	20.8327	7.5712	1363.5	147934	434
435	189225	82312875	20.8567	7.5770	1366.6	148617	435
436	190096	82881856	20.8806	7.5828	1369.7	149301	436
437	190969	83453453	20.9045	7.5886	1372.9	149987	437
438	191844	84027672	20.9284	7.5944	1376.0	150674	438
439	192721	84604519	20.9523	7.6001	1379.2	151363	439

Table 1

No.	Square	Cube	Square Root	Cube Root	No. = Diam.		No.
					Circum.	Area	
440	193600	85184000	20.9762	7.6059	1382.3	152053	440
441	194481	85766121	21.0000	7.6117	1385.4	152745	441
442	195364	86350888	21.0238	7.6174	1388.6	153439	442
443	196249	86938307	21.0476	7.6232	1391.7	154134	443
444	197136	87528384	21.0713	7.6289	1394.9	154830	444
445	198025	88121125	21.0950	7.6346	1398.0	155528	445
446	198916	88716536	21.1187	7.6403	1401.2	156228	446
447	199809	89314623	21.1424	7.6460	1404.3	156930	447
448	200704	89915392	21.1660	7.6517	1407.4	157633	448
449	201601	90518849	21.1896	7.6574	1410.6	158337	449
450	202500	91125000	21.2132	7.6631	1413.7	159043	450
451	203401	91733851	21.2368	7.6688	1416.9	159751	451
452	204304	92345408	21.2603	7.6744	1420.0	160460	452
453	205209	92959677	21.2838	7.6801	1423.1	161171	453
454	206116	93576664	21.3073	7.6857	1426.3	161883	454
455	207025	94196375	21.3307	7.6914	1429.4	162597	455
456	207936	94818816	21.3542	7.6970	1432.6	163313	456
457	208849	95443993	21.3776	7.7026	1435.7	164030	457
458	209764	96071912	21.4009	7.7082	1438.9	164748	458
459	210681	96702579	21.4243	7.7138	1442.0	165468	459
460	211600	97336000	21.4476	7.7194	1445.1	166190	460
461	212521	97972181	21.4709	7.7250	1448.3	166914	461
462	213444	98611128	21.4942	7.7306	1451.4	167639	462
463	214369	99252847	21.5174	7.7362	1454.6	168365	463
464	215296	99897344	21.5407	7.7418	1457.7	169093	464
465	216225	100544625	21.5639	7.7473	1460.8	169823	465
466	217156	101194696	21.5870	7.7529	1464.0	170554	466
467	218089	101847563	21.6102	7.7584	1467.1	171287	467
468	219024	102503232	21.6333	7.7639	1470.3	172021	468
469	219961	103161709	21.6564	7.7695	1473.4	172757	469
470	220900	103823000	21.6795	7.7750	1476.5	173494	470
471	221841	104487111	21.7025	7.7805	1479.7	174234	471
472	222784	105154048	21.7256	7.7860	1482.8	174974	472
473	223729	105823817	21.7486	7.7915	1486.0	175716	473
474	224676	106496424	21.7715	7.7970	1489.1	176460	474
475	225625	107171875	21.7945	7.8025	1492.3	177205	475
476	226576	107850176	21.8174	7.8079	1495.4	177952	476
477	227529	108531333	21.8403	7.8134	1498.5	178701	477
478	228484	109215352	21.8632	7.8188	1501.7	179451	478
479	229441	109902239	21.8861	7.8243	1504.8	180203	479

No.	Square	Cube	Square Root	Cube Root	No. = Diam.		No.
					Circum.	Area	
480	230400	110592000	21.9089	7.8297	1508.0	180956	480
481	231361	111284641	21.9317	7.8352	1511.1	181711	481
482	232324	111980168	21.9545	7.8406	1514.3	182467	482
483	233289	112678587	21.9773	7.8460	1517.4	183225	483
484	234256	113379904	22.0000	7.8514	1520.5	183984	484
485	235225	114084125	22.0227	7.8568	1523.7	184745	485
486	236196	114791256	22.0454	7.8622	1526.8	185508	486
487	237169	115501303	22.0681	7.8676	1530.0	186272	487
488	238144	116214272	22.0907	7.8730	1533.1	187038	488
489	239121	116930169	22.1133	7.8784	1536.2	187805	489
490	240100	117649000	22.1359	7.8837	1539.4	188574	490
491	241081	118370771	22.1585	7.8891	1542.5	189345	491
492	242064	119095488	22.1811	7.8944	1545.7	190117	492
493	243049	119823157	22.2036	7.8998	1548.8	190890	493
494	244036	120553784	22.2261	7.9051	1551.9	191665	494
495	245025	121287375	22.2486	7.9105	1555.1	192442	495
496	246016	122023936	22.2711	7.9158	1558.2	193221	496
497	247009	122763473	22.2935	7.9211	1561.4	194000	497
498	248004	123505992	22.3159	7.9264	1564.5	194782	498
499	249001	124251499	22.3383	7.9317	1567.7	195565	499
500	250000	125000000	22.3607	7.9370	1570.8	196350	500
501	251001	125751501	22.3830	7.9423	1573.9	197136	501
502	252004	126506008	22.4054	7.9476	1577.1	197923	502
503	253009	127263527	22.4277	7.9528	1580.2	198713	503
504	254016	128024064	22.4499	7.9581	1583.4	199504	504
505	255025	128787625	22.4722	7.9634	1586.5	200296	505
506	256036	129554216	22.4944	7.9686	1589.7	201090	506
507	257049	130323843	22.5167	7.9739	1592.8	201886	507
508	258064	131096512	22.5389	7.9791	1595.9	202683	508
509	259081	131872229	22.5610	7.9843	1599.1	202482	509
510	260100	132651000	22.5832	7.9896	1602.2	204282	510
511	261121	133432831	22.6053	7.9948	1605.4	205084	511
512	262144	134217728	22.6274	8.0000	1608.5	205887	512
513	263169	135005697	22.6495	8.0052	1611.6	206692	513
514	264196	135796744	22.6716	8.0104	1614.8	207499	514
515	265225	136590875	22.6936	8.0156	1617.9	208307	515
516	266256	137388096	22.7156	8.0208	1621.1	209117	516
517	267289	138188413	22.7376	8.0260	1624.2	209928	517
518	268324	138991832	22.7596	8.0311	1627.3	210741	518
519	269361	139798359	22.7816	8.0363	1630.5	211556	519

Table 1 673

No.	Square	Cube	Square Root	Cube Root	No. = Diam.		No.
					Circum.	Area	
520	270400	140608000	22.8035	8.0415	1633.6	212372	520
521	271441	141420761	22.8254	8.0466	1636.8	213189	521
522	272484	142236648	22.8473	8.0517	1639.9	214008	522
523	273529	143055667	22.8692	8.0569	1643.1	214829	523
524	274576	143877824	22.8910	8.0620	1646.2	215651	524
525	275625	144703125	22.9129	8.0671	1649.3	216475	525
526	276676	145531576	22.9347	8.0723	1652.5	217301	526
527	277729	146363183	22.9565	8.0774	1655.6	218128	527
528	278784	147197952	22.9783	8.0825	1658.8	218956	528
529	279841	148035889	23.0000	8.0876	1661.9	219787	529
530	280900	148877000	23.0217	8.0927	1665.0	220618	530
531	281961	149721291	23.0434	8.0978	1668.2	221452	531
532	283024	150568768	23.0651	8.1028	1671.3	222287	532
533	284089	151419437	23.0868	8.1079	1674.5	223123	533
534	285156	152273304	23.1084	8.1130	1677.6	223961	534
535	286225	153130375	23.1301	8.1180	1680.8	224801	535
536	287296	153990656	23.1517	8.1231	1683.9	225642	536
537	288369	154854153	23.1733	8.1281	1687.0	226484	537
538	289444	155720872	23.1948	8.1332	1690.2	227329	538
539	290521	156590819	23.2164	8.1382	1693.3	228175	539
540	291600	157464000	23.2379	8.1433	1696.5	229022	540
541	292681	158340421	23.2594	8.1483	1699.6	229871	541
542	293764	159220088	23.2809	8.1533	1702.7	230722	542
543	294849	160103007	23.3024	8.1583	1705.9	231574	543
544	295936	160989184	23.3238	8.1633	1709.0	232428	544
545	297025	161878625	23.3452	8.1683	1712.2	233283	545
546	298116	162771336	23.3666	8.1733	1715.3	234140	546
547	299209	163667323	23.3880	8.1783	1718.5	234998	547
548	300304	164566592	23.4094	8.1833	1721.6	235858	548
549	301401	165469149	23.4307	8.1882	1724.7	236720	549
550	302500	166375000	23.4521	8.1932	1727.9	237583	550
551	303601	167284151	23.4734	8.1982	1731.0	238448	551
552	304704	168196608	23.4947	8.2031	1734.2	239314	552
553	305809	169112377	23.5160	8.2081	1737.3	240182	553
554	306916	170031464	23.5372	8.2130	1740.4	241051	554
555	308025	170953875	23.5584	8.2180	1743.6	241922	555
556	309136	171879616	23.5797	8.2229	1746.7	242795	556
557	310249	172808693	23.6008	8.2278	1749.9	243669	557
558	311364	173741112	23.6220	8.2327	1753.0	244545	558
559	312481	174676879	23.6432	8.2377	1756.2	245422	559

| No. | Square | Cube | Square Root | Cube Root | No. = Diam. | | No. |
					Circum.	Area	
560	313600	175616000	23.6643	8.2426	1759.3	246301	560
561	314721	176558481	23.6854	8.2475	1762.4	247181	561
562	315844	177504328	23.7065	8.2524	1765.6	248063	562
563	316969	178453547	23.7276	8.2573	1768.7	248947	563
564	318096	179406144	23.7487	8.2621	1771.9	249832	564
565	319225	180362125	23.7697	8.2670	1775.0	250719	565
566	320356	181321496	23.7908	8.2719	1778.1	251607	566
567	321489	182284263	23.8118	8.2768	1781.3	252497	567
568	322624	183250432	23.8328	8.2816	1784.4	253388	568
569	323761	184220009	23.8537	8.2865	1787.6	254281	569
570	324900	185193000	23.8747	8.2913	1790.7	255176	570
571	326041	186169411	23.8956	8.2962	1793.9	256072	571
572	327184	187149248	23.9165	8.3010	1797.0	256970	572
573	328329	188132517	23.9374	8.3059	1800.1	257869	573
574	329476	189119224	23.9583	8.3107	1803.3	258770	574
575	330625	190109375	23.9792	8.3155	1806.4	259672	575
576	331776	191102976	24.0000	8.3203	1809.6	260576	576
577	332929	192100033	24.0208	8.3251	1812.7	261482	577
578	334084	193100552	24.0416	8.3300	1815.8	262389	578
579	335241	194104539	24.0624	8.3348	1819.0	263298	579
580	336400	195112000	24.0832	8.3396	1822.1	264208	580
581	337561	196122941	24.1039	8.3443	1825.3	265120	581
582	338724	197137368	24.1247	8.3491	1828.4	266033	582
583	339889	198155287	24.1454	8.3539	1831.6	266948	583
584	341056	199176704	24.1661	8.3587	1834.7	267865	584
585	342225	200201625	24.1868	8.3634	1837.8	268783	585
586	343396	201230056	24.2074	8.3682	1841.0	269701	586
587	344569	202262003	24.2281	8.3730	1844.1	270624	587
588	345744	203297472	24.2487	8.3777	1847.3	271547	588
589	346921	204336469	24.2693	8.3825	1850.4	272471	589
590	348100	205379000	24.2899	8.3872	1853.5	273397	590
591	349281	206425071	24.3105	8.3919	1856.7	274325	591
592	350464	207474688	24.3311	8.3967	1859.8	275254	592
593	351649	208527857	24.3516	8.4014	1863.0	276184	593
594	352836	209584584	24.3721	8.4061	1866.1	277117	594
595	354025	210644875	24.3926	8.4108	1869.3	278051	595
596	355216	211708736	24.4131	8.4155	1872.4	278986	596
597	356409	212776173	24.4336	8.4202	1875.5	279923	597
598	357604	213847192	24.4540	8.4249	1878.7	280862	598
599	358801	214921799	24.4745	8.4296	1881.8	281802	599

Table 1
675

No.	Square	Cube	Square Root	Cube Root	No. = Diam.		No.
					Circum.	Area	
600	360000	216000000	24.4949	8.4343	1885.0	282743	600
601	361201	217081801	24.5153	8.4390	1888.1	283687	601
602	362404	218167208	24.5357	8.4437	1891.2	284631	602
603	363609	219256227	24.5561	8.4484	1894.4	285578	603
604	364816	220348864	24.5764	8.4530	1897.5	286526	604
605	366025	221445125	24.5967	8.4577	1900.7	287475	605
606	367236	222545016	24.6171	8.4623	1903.8	288426	606
607	368449	223648543	24.6374	8.4670	1907.0	289379	607
608	369664	224755712	24.6577	8.4716	1910.1	290333	608
609	370881	225866529	24.6779	8.4763	1913.2	291289	609
610	372100	226981000	24.6982	8.4809	1916.4	292247	610
611	373321	228099131	24.7184	8.4856	1919.5	293206	611
612	374544	229220928	24.7386	8.4902	1922.7	294166	612
613	375769	230346397	24.7588	8.4948	1925.8	295128	613
614	376996	231475544	24.7790	8.4994	1928.9	296092	614
615	378225	232608375	24.7992	8.5040	1932.1	297057	615
616	379456	233744896	24.8193	8.5086	1935.2	298024	616
617	380689	234885113	24.8395	8.5132	1938.4	298992	617
618	381924	236029032	24.8596	8.5178	1941.5	299962	618
619	383161	237176659	24.8797	8.5224	1944.7	300934	619
620	384400	238328000	24.8998	8.5270	1947.8	301907	620
621	385641	239483061	24.9199	8.5316	1950.9	302882	621
622	386884	240641848	24.9399	8.5462	1954.1	303858	622
623	388129	241804367	24.9600	8.5408	1957.2	304836	623
624	389376	242970624	24.9800	8.5453	1960.4	305815	624
625	390625	244140625	25.0000	8.5499	1963.5	306796	625
626	391876	245314376	25.0200	8.5544	1966.6	307779	626
627	393129	246491883	25.0400	8.5590	1969.8	308763	627
628	394384	247673152	25.0599	8.5635	1972.9	309748	628
629	395641	248858189	25.0799	8.5681	1976.1	310736	629
630	396900	250047000	25.0998	8.5726	1979.2	311725	630
631	398161	251239591	25.1197	8.5772	1982.4	312715	631
632	399424	252435968	25.1396	8.5817	1985.5	313707	632
633	400689	253636137	25.1595	8.5862	1988.6	314700	633
634	401956	254840104	25.1794	8.5907	1991.8	315696	634
635	403225	256047875	25.1992	8.5952	1994.9	316692	635
636	404496	257259456	25.2190	8.5997	1998.1	317690	636
637	405769	258474853	25.2389	8.6043	2001.2	318690	637
638	407044	259694072	25.2587	8.6088	2004.3	319692	638
639	408321	260917119	25.2784	8.6132	2007.5	320695	639

No.	Square	Cube	Square Root	Cube Root	No. = Diam.		No.
					Circum.	Area	
640	409600	262144000	25.2982	8.6177	2010.6	321699	640
641	410881	263374721	25.3180	8.6222	2013.8	322705	641
642	412164	264609288	25.3377	8.6267	2016.9	323713	642
643	413449	265847707	25.3574	8.6312	2020.0	324722	643
644	414736	267089984	25.3772	8.6357	2023.2	325733	644
645	416025	268336125	25.3969	8.6401	2026.3	326745	645
646	417316	269586136	25.4165	8.6446	2029.5	327759	646
647	418609	270840023	25.4362	8.6490	2032.6	328775	647
648	419904	272097792	25.4558	8.6535	2035.8	329792	648
649	421201	273359449	25.4755	8.6579	2038.9	330810	649
650	422500	274625000	25.4951	8.6624	2042.0	331831	650
651	423801	275894451	25.5147	8.6668	2045.2	332853	651
652	425104	277167808	25.5343	8.6713	2048.3	333876	652
653	426409	278445077	25.5539	8.6757	2051.5	334901	653
654	427716	279726264	25.5734	8.6801	2054.6	335927	654
655	429025	281011375	25.5930	8.6845	2057.7	336955	655
656	430336	282300416	25.6125	8.6890	2060.9	337985	656
657	431649	283593393	25.6320	8.6934	2064.0	339016	657
658	432964	284890312	25.6515	8.6978	2067.2	340049	658
659	434281	286191179	25.6710	8.7022	2070.3	341084	659
660	435600	287496000	25.6905	8.7066	2073.5	342119	660
661	436921	288804781	25.7099	8.7110	2076.6	343157	661
662	438244	290117528	25.7294	8.7154	2079.7	344196	662
663	439569	291434247	25.7488	8.7198	2082.9	345237	663
664	440896	292754944	25.7682	8.7241	2086.0	346279	664
665	442225	294079625	25.7876	8.7285	2089.2	347323	665
666	443556	295408296	25.8070	8.7329	2092.3	348368	666
667	444889	296740963	25.8263	8.7373	2095.4	349415	667
668	446224	298077632	25.8457	8.7416	2098.6	350464	668
669	447561	299418309	25.8650	8.7460	2101.7	351514	669
670	448900	300763000	25.8844	8.7503	2104.9	352565	670
671	450241	302111711	25.9037	8.7547	2108.0	353618	671
672	451584	303464448	25.9230	8.7590	2111.2	354673	672
673	452929	304821217	25.9422	8.7634	2114.3	355730	673
674	454276	306182024	25.9615	8.7677	2117.4	356788	674
675	455625	307546875	25.9808	8.7721	2120.6	357847	675
676	456976	308915776	26.0000	8.7764	2123.7	358908	676
677	458329	310288733	26.0192	8.7807	2126.9	359971	677
678	459684	311665752	26.0384	8.7850	2130.0	361035	678
679	461041	313046839	26.0576	8.7893	2133.1	362101	679

Table 1 677

No.	Square	Cube	Square Root	Cube Root	No. = Diam.		No.
					Circum.	Area	
680	462400	314432000	26.0768	8.7937	2136.3	363168	680
681	463761	315821241	26.0960	8.7980	2139.4	364237	681
682	465124	317214568	26.1151	8.8023	2142.6	365308	682
683	466489	318611987	26.1343	8.8066	2145.7	366380	683
684	467856	320013504	26.1534	8.8109	2148.9	367453	684
685	469225	321419125	26.1725	8.8152	2152.0	368528	685
686	470596	322828856	26.1916	8.8194	2155.1	369605	686
687	471969	324242703	26.2107	8.8237	2158.3	370684	687
688	473344	325660672	26.2298	8.8280	2161.4	371764	688
689	474721	327082769	26.2488	8.8323	2164.6	372845	689
690	476100	328509000	26.2679	8.8366	2167.7	373928	690
691	477481	329939371	26.2869	8.8408	2170.8	375013	691
692	478864	331373888	26.3059	8.8451	2174.0	376099	692
693	480249	332812557	26.3249	8.8493	2177.1	377187	693
694	481636	334255384	26.3439	8.8536	2180.3	378276	694
695	483025	335702375	26.3629	8.8578	2183.4	379367	695
696	484416	337153536	26.3818	8.8621	2186.6	380459	696
697	485809	338608873	26.4008	8.8663	2189.7	381554	697
698	487204	340068392	26.4197	8.8706	2192.8	382649	698
699	488601	341532099	26.4386	8.8748	2196.0	383746	699
700	490000	343000000	26.4575	8.8790	2199.1	384845	700
701	491401	344472101	26.4764	8.8833	2202.3	385945	701
702	492804	345948408	26.4953	8.8875	2205.4	387047	702
703	494209	347428927	26.5141	8.8917	2208.5	388151	703
704	495616	348913664	26.5330	8.8959	2211.7	389256	704
705	497025	350402625	26.5518	8.9001	2214.8	390363	705
706	498436	351895816	26.5707	8.9043	2218.0	391471	706
707	499849	353393243	26.5895	8.9085	2221.1	392580	707
708	501264	354894912	26.6083	8.9127	2224.3	393692	708
709	502681	356400829	26.6271	8.9169	2227.4	394805	709
710	504100	357911000	26.6458	8.9211	2230.5	395919	710
711	505521	359425431	26.6646	8.9253	2233.7	397035	711
712	506944	360944128	26.6833	8.9295	2236.8	398153	712
713	508369	362467097	26.7021	8.9337	2240.0	399272	713
714	509796	363994344	26.7208	8.9378	2243.1	400393	714
715	511225	365525875	26.7395	8.9420	2246.2	401515	715
716	512656	367061696	26.7582	8.9462	2249.4	402639	716
717	514089	368601813	26.7769	8.9503	2252.5	403765	717
718	515524	370146232	26.7955	8.9545	2255.7	404892	718
719	516961	371694959	26.8142	8.9587	2258.8	406020	719

| No. | Square | Cube | Square Root | Cube Root | No. = Diam. | | No |
					Circum.	Area	
720	518400	373248000	26.8328	8.9628	2261.9	407150	720
721	519841	374805361	26.8514	8.9670	2265.1	408282	721
722	521284	376367048	26.8701	8.9711	2268.2	409416	722
723	522729	377933067	26.8887	8.9752	2271.4	410550	723
724	524176	379503424	26.9072	8.9794	2274.5	411687	724
725	525625	381078125	26.9258	8.9835	2277.7	412825	725
726	527076	382657176	26.9444	8.9876	2280.8	413965	726
727	528529	384240583	26.9629	8.9918	2283.9	415106	727
728	529984	385828352	26.9815	8.9959	2287.1	416248	728
729	531441	387420489	27.0000	9.0000	2290.2	417393	729
730	532900	389017000	27.0185	9.0041	2293.4	418539	730
731	534361	390617891	27.0370	9.0082	2296.5	419686	731
732	535824	392223168	27.0555	9.0123	2299.7	420835	732
733	537289	393832837	27.0740	9.0164	2302.8	421986	733
734	538756	395446904	27.0924	9.0205	2305.9	423138	734
735	540225	397065375	27.1109	9.0246	2309.1	424293	735
736	541696	398688256	27.1293	9.0287	2312.2	425448	736
737	543169	400315553	27.1477	9.0328	2315.4	426604	737
738	544644	401947272	27.1662	9.0369	2318.5	427762	738
739	546121	403583419	27.1846	9.0410	2321.6	428922	739
740	547600	405224000	27.2029	9.0450	2324.8	430084	740
741	549081	406869021	27.2213	9.0491	2327.9	431247	741
742	550564	408518488	27.2397	9.0532	2331.1	432412	742
743	552049	410172407	27.2580	9.0572	2334.2	433578	743
744	553536	411830784	27.2764	9.0613	2337.3	434746	744
745	555025	413493625	27.2947	9.0654	2340.5	435916	745
746	556516	415160936	27.3130	9.0694	2343.6	437087	746
747	558009	416832723	27.3313	9.0735	2346.8	438259	747
748	559504	418508992	27.3496	9.0775	2349.9	439433	748
749	561001	420189749	27.3679	9.0816	2353.1	440609	749
750	562500	421875000	27.3861	9.0856	2356.2	441786	750
751	564001	423564751	27.4044	9.0896	2359.3	442965	751
752	565504	425259008	27.4226	9.0937	2362.5	444146	752
753	567009	426957777	27.4408	9.0977	2365.6	445328	753
754	568516	428661064	27.4591	9.1017	2368.8	446511	754
755	570025	430368875	27.4773	9.1057	2371.9	447697	755
756	571536	432081216	27.4955	9.1098	2375.0	448883	756
757	573049	433798093	27.5136	9.1138	2378.2	450072	757
758	574564	435519512	27.5318	9.1178	2381.3	451262	758
759	576081	437245479	27.5500	9.1218	2384.5	452453	759

Table 1

679

No.	Square	Cube	Square Root	Cube Root	No. = Diam.		No.
					Circum.	Area	
760	577600	438976000	27.5681	9.1258	2387.6	453646	760
761	579121	440711081	27.5862	9.1298	2390.8	454841	761
762	580644	442450728	27.6043	9.1338	2393.9	456037	762
763	582169	444194947	27.6225	9.1378	2397.0	457234	763
764	583696	445943744	27.6405	9.1418	2400.2	458434	764
765	585225	447697125	27.6586	9.1458	2403.3	459635	765
766	586756	449455096	27.6767	9.1498	2406.5	460837	766
767	588289	451217663	27.6948	9.1537	2409.6	462042	767
768	589824	452984832	27.7128	9.1577	2412.7	463247	768
769	591361	454756609	27.7308	9.1617	2415.9	464454	769
770	592900	456533000	27.7489	9.1657	2419.0	465663	770
771	594441	458314011	27.7669	9.1696	2422.2	466873	771
772	595984	460099648	27.7849	9.1736	2425.3	468085	772
773	597529	461889917	27.8029	9.1775	2428.5	469298	773
774	599076	463684824	27.8209	9.1815	2431.6	470513	774
775	600625	465484375	27.8388	9.1855	2434.7	471730	775
776	602176	467288576	27.8568	9.1894	2437.9	472948	776
777	603729	469097433	27.8747	9.1933	2441.0	474168	777
778	605284	470910952	27.8927	9.1973	2444.2	475389	778
779	606841	472729139	27.9106	9.2012	2447.3	476612	779
780	608400	474552000	27.9285	9.2052	2450.4	477836	780
781	609961	476379541	27.9464	9.2091	2453.6	479062	781
782	611524	478211768	27.9643	9.2130	2456.7	480290	782
783	613089	480048687	27.9821	9.2170	2459.9	481510	783
784	614656	481890304	28.0000	9.2209	2463.0	482750	784
785	616225	483736625	28.0179	9.2248	2466.2	483982	785
786	617796	485587656	28.0357	9.2287	2469.3	485216	786
787	619369	487443403	28.0535	9.2326	2472.4	486451	787
788	620944	489303872	28.0713	9.2365	2475.6	487688	788
789	622521	491169069	28.0891	9.2404	2478.7	488927	789
790	624100	493039000	28.1069	9.2443	2481.9	490167	790
791	625681	494913671	28.1247	9.2482	2485.0	491409	791
792	627264	496793088	28.1425	9.2521	2488.1	492652	792
793	628849	498677257	28.1603	9.2560	2491.3	493897	793
794	630436	500566184	28.1780	9.2599	2494.4	495143	794
795	632025	502459875	28.1957	9.2638	2497.6	496391	795
796	633616	504358336	28.2135	9.2677	2500.7	497641	796
797	635209	506261573	28.2312	9.2716	2503.8	498892	797
798	636804	508169592	28.2489	9.2754	2507.0	500145	798
799	638401	510082399	28.2666	9.2793	2510.1	501399	799

No.	Square	Cube	Square Root	Cube Root	No. = Diam. Circum.	No. = Diam. Area	No.
800	640000	512000000	28.2843	9.2832	2513.3	502655	800
801	641601	513922401	28.3019	9.2870	2516.4	503912	801
802	643204	515849608	28.3196	9.2909	2519.6	505171	802
803	644809	517781627	28.3373	9.2948	2522.7	506432	803
804	646416	519718464	28.3549	9.2986	2525.8	507694	804
805	648025	521660125	28.3725	9.3025	2529.0	508958	805
806	649636	523606616	28.3901	9.3063	2532.1	510223	806
807	651249	525557943	28.4077	9.3102	2535.3	511490	807
808	652864	527514112	28.4253	9.3140	2538.4	512758	808
809	654481	529475129	28.4429	9.3179	2541.5	514028	809
810	656100	531441000	28.4605	9.3217	2544.7	515300	810
811	657721	533411731	28.4781	9.3255	2547.8	516573	811
812	659344	535387328	28.4956	9.3294	2551.0	517848	812
813	660969	537367797	28.5132	9.3332	2554.1	519124	813
814	662596	539353144	28.5307	9.3370	2557.3	520402	814
815	664225	541343375	28.5482	9.3408	2560.4	521681	815
816	665856	543338496	28.5657	9.3447	2563.5	522962	816
817	667489	545338513	28.5832	9.3485	2566.7	524245	817
818	669124	547343432	28.6007	9.3523	2569.8	525529	818
819	670761	549353259	28.6182	9.3561	2573.0	526814	819
820	672400	551368000	28.6356	9.3599	2576.1	528102	820
821	674041	553387661	28.6531	9.3637	2579.2	529391	821
822	675684	555412248	28.6705	9.3675	2582.4	530681	822
823	677329	557441767	28.6880	9.3713	2585.5	531973	823
824	678976	559476224	28.7054	9.3751	2588.7	533267	824
825	680625	561515625	28.7228	9.3789	2591.8	534562	825
826	682276	563559976	28.7402	9.3827	2595.0	535858	826
827	683929	565609283	28.7576	9.3865	2598.1	537157	827
828	685584	567663552	28.7750	9.3902	2601.2	538456	828
829	687241	569722789	28.7924	9.3940	2604.4	539758	829
830	688900	571787000	28.8097	9.3978	2607.5	541061	830
831	690561	573856191	28.8271	9.4016	2610.7	542365	831
832	692224	575930368	28.8444	9.4053	2613.8	543671	832
833	693889	578009537	28.8617	9.4091	2616.9	544979	833
834	695556	580093704	28.8791	9.4129	2620.1	546288	834
835	697225	582182875	28.8964	9.4166	2623.2	547599	835
836	698896	584277056	28.9137	9.4204	2626.4	548912	836
837	700569	586376253	28.9310	9.4241	2629.5	550226	837
838	702244	588480472	28.9482	9.4279	2632.7	551541	838
839	703921	590589719	28.9655	9.4316	2635.8	552858	839

Table 1 681

No.	Square	Cube	Square Root	Cube Root	No. = Diam.		No.
					Circum.	Area	
840	705600	592704000	28.9828	9.4354	2638.9	554177	840
841	707281	594823321	29.0000	9.4391	2642.1	555497	841
842	708964	596947688	29.0172	9.4429	2645.2	556819	842
843	710649	599077107	29.0345	9.4466	2648.4	558142	843
844	712336	601211584	29.0517	9.4503	2651.5	559467	844
845	714025	603351125	29.0689	9.4541	2654.6	560794	845
846	715716	605495736	29.0861	9.4578	2657.8	562122	846
847	717409	607645423	29.1033	9.4615	2660.9	563452	847
848	719104	609800192	29.1204	9.4652	2664.1	564783	848
849	720801	611960049	29.1376	9.4690	2667.2	566116	849
850	722500	614125000	29.1548	9.4727	2670.4	567450	850
851	724201	616295051	29.1719	9.4764	2673.5	568786	851
852	725904	618470208	29.1890	9.4801	2676.6	570124	852
853	727609	620650477	29.2062	9.4838	2679.8	571463	853
854	729316	622835864	29.2233	9.4875	2682.9	572803	854
855	731025	625026375	29.2404	9.4912	2686.1	574146	855
856	732736	627222016	29.2575	9.4949	2689.2	575490	856
857	734449	629422793	29.2746	9.4986	2692.3	576835	857
858	736164	631628712	29.2916	9.5023	2695.5	578182	858
859	737881	633839779	29.3087	9.5060	2698.6	579530	859
860	739600	636056000	29.3258	9.5097	2701.8	580880	860
861	741321	638277381	29.3428	9.5134	2704.9	582232	861
862	743044	640503928	29.3598	9.5171	2708.1	583585	862
863	744769	642735647	29.3769	9.5207	2711.2	584940	863
864	746496	644972544	29.3939	9.5244	2714.3	586297	864
865	748225	647214625	29.4109	9.5281	2717.5	587655	865
866	749956	649461896	29.4279	9.5317	2720.6	589014	866
867	751689	651714363	29.4449	9.5354	2723.8	590375	867
868	753424	653972032	29.4618	9.5391	2726.9	591738	868
869	755161	656234909	29.4788	9.5427	2730.0	593102	869
870	756900	658503000	29.4958	9.5464	2733.2	594468	870
871	758641	660776311	29.5127	9.5501	2736.3	595835	871
872	760384	663054848	29.5296	9.5537	2739.5	597204	872
873	762129	665338617	29.5466	9.5574	2742.6	598575	873
874	763876	667627624	29.5635	9.5610	2745.8	599947	874
875	765625	669921875	29.5804	9.5647	2748.9	601320	875
876	767376	672221376	29.5973	9.5683	2752.0	602696	876
877	769129	674526133	29.6142	9.5719	2755.2	604073	877
878	770884	676836152	29.6311	9.5756	2758.3	605451	878
879	772641	679151439	29.6479	9.5792	2761.5	606831	879

No.	Square	Cube	Square Root	Cube Root	No. = Diam.		No.
					Circum.	Area	
880	774400	681472000	29.6648	9.5828	2764.6	608212	880
881	776161	683797841	29.6816	9.5865	2767.7	609595	881
882	777924	686128968	29.6985	9.5901	2770.9	610980	882
883	779689	688465387	29.7153	9.5937	2774.0	612366	883
884	781456	690807104	29.7321	9.5973	2777.2	613754	884
885	783225	693154125	29.7489	9.6010	2780.3	615143	885
886	784996	695506456	29.7658	9.6046	2783.5	616534	886
887	786769	697864103	29.7825	9.6082	2786.6	617927	887
888	788544	700227072	29.7993	9.6118	2789.7	619321	888
889	790321	702595369	29.8161	9.6154	2792.9	620717	889
890	792100	704969000	29.8329	9.6190	2796.0	622114	890
891	793881	707347971	29.8496	9.6226	2799.2	623513	891
892	795664	709732288	29.8664	9.6262	2802.3	624913	892
893	797449	712121957	29.8831	9.6298	2805.4	626315	893
894	799236	714516984	29.8998	9.6334	2808.6	627718	894
895	801025	716917375	29.9166	9.6370	2811.7	629124	895
896	802816	719323136	29.9333	9.6406	2814.9	630530	896
897	804609	721734273	29.9500	9.6442	2818.0	631938	897
898	806404	724150792	29.9666	9.6477	2821.2	633348	898
899	808201	726572699	29.9833	9.6513	2824.3	634760	899
900	810000	729000000	30.0000	9.6549	2827.4	636173	900
901	811801	731432701	30.0167	9.6585	2830.6	637587	901
902	813604	733870808	30.0333	9.6620	2833.7	639003	902
903	815409	736314327	30.0500	9.6656	2836.9	640421	903
904	817216	738763264	30.0666	9.6692	2840.0	641840	904
905	819025	741217625	30.0832	9.6727	2843.1	643261	905
906	820836	743677416	30.0998	9.6763	2846.3	644683	906
907	822649	746142643	30.1164	9.6799	2849.4	646107	907
908	824464	748613312	30.1330	9.6834	2852.6	647533	908
909	826281	751089429	30.1496	9.6870	2855.7	648960	909
910	828100	753571000	30.1662	9.6905	2858.8	650388	910
911	829921	756058031	30.1828	9.6941	2862.0	651818	911
912	831744	758550528	30.1993	9.6976	2865.1	653250	912
913	833569	761048497	30.2159	9.7012	2868.3	654684	913
914	835396	763551944	30.2324	9.7047	2871.4	656118	914
915	837225	766060875	30.2490	9.7082	2874.6	657555	915
916	839056	768575296	30.2655	9.7118	2877.7	658993	916
917	840889	771095213	30.2820	9.7153	2880.8	660433	917
918	842724	773620632	30.2985	9.7188	2884.0	661874	918
919	844561	776151559	30.3150	9.7224	2887.1	663317	919

Table 1 683

No.	Square	Cube	Square Root	Cube Root	No. = Diam.		No.
					Circum.	Area	
920	846400	778688000	30.3315	9.7259	2890.3	664761	920
921	848241	781229961	30.3480	9.7294	2893.4	666207	921
922	850084	783777448	30.3645	9.7329	2896.5	667654	922
923	851929	786330467	30.3809	9.7364	2899.7	669103	923
924	853776	788889024	30.3974	9.7400	2902.8	670554	924
925	855625	791453125	30.4138	9.7435	2906.0	672006	925
926	857476	794022776	30.4302	9.7470	2909.1	673460	926
927	859329	796597983	30.4467	9.7505	2912.3	674915	927
928	861184	799178752	30.4631	9.7540	2915.4	676372	928
929	863041	801765089	30.4795	9.7575	2918.5	677831	929
930	864900	804357000	30.4959	9.7610	2921.7	679291	930
931	866761	806954491	30.5123	9.7645	2924.8	680752	931
932	868624	809557568	30.5287	9.7680	2928.0	682216	932
933	870489	812166237	30.5450	9.7715	2931.1	683680	933
934	872356	814780504	30.5614	9.7750	2934.2	685147	934
935	874225	817400375	30.5778	9.7785	2937.4	686615	935
936	876096	820025856	30.5941	9.7819	2940.5	688084	936
937	877969	822656953	30.6105	9.7854	2943.7	689555	937
938	879844	825293672	30.6268	9.7889	2946.8	691028	938
939	881721	827936019	30.6431	9.7924	2950.0	692502	939
940	883600	830584000	30.6594	9.7959	2953.1	693978	940
941	885481	833237621	30.6757	9.7993	2956.2	695455	941
942	887364	835896888	30.6920	9.8028	2959.4	696934	942
943	889249	838561807	30.7083	9.8063	2962.5	698415	943
944	891136	841232384	30.7246	9.8097	2965.7	699897	944
945	893025	843908625	30.7409	9.8132	2968.8	701380	945
946	894916	846590536	30.7571	9.8167	2971.9	702865	946
947	896809	849278123	30.7734	9.8201	2975.1	704352	947
948	898704	851971392	30.7896	9.8236	2978.2	705840	948
949	900601	854670349	30.8058	9.8270	2981.4	707330	949
950	902500	857375000	30.8221	9.8305	2984.5	708822	950
951	904401	860085351	30.8383	9.8339	2987.7	710315	951
952	906304	862801408	30.8545	9.8374	2990.8	711809	952
953	908209	865523177	30.8707	9.8408	2993.9	713306	953
954	910116	868250664	30.8869	9.8443	2997.1	714803	954
955	912025	870983875	30.9031	9.8477	3000.2	716303	955
956	913936	873722816	30.9192	9.8511	3003.4	717804	956
957	915849	876467493	30.9354	9.8546	3006.5	719306	957
958	917764	879217912	30.9516	9.8580	3009.6	720810	958
959	919681	881974079	30.9677	9.8614	3012.8	722316	959

No.	Square	Cube	Square Root	Cube Root	No. = Diam.		No.
					Circum.	Area	
960	921600	884736000	30.9839	9.8648	3015.9	723823	960
961	923521	887503681	31.0000	9.8683	3019.1	725332	961
962	925444	890277128	31.0161	9.8717	3022.2	726842	962
963	927369	893056347	31.0322	9.8751	3025.4	728354	963
964	929296	895841344	31.0483	9.8785	3028.5	729867	964
965	931225	898632125	31.0644	9.8819	3031.6	731382	965
966	933156	901428696	31.0805	9.8854	3034.8	732899	966
967	935089	904231063	31.0966	9.8888	3037.9	734417	967
968	937024	907039232	31.1127	9.8922	3041.1	735937	968
969	938961	909853209	31.1288	9.8956	3044.2	737458	969
970	940900	912673000	31.1448	9.8990	3047.3	738981	970
971	942841	915498611	31.1609	9.9024	3050.5	740506	971
972	944784	918330048	31.1769	9.9058	3053.6	742032	972
973	946729	921167317	31.1929	9.9092	3056.8	743559	973
974	948676	924010424	31.2090	9.9126	3059.9	745088	974
975	950625	926859375	31.2250	9.9160	3063.1	746619	975
976	952576	929714176	31.2410	9.9194	3066.2	748151	976
977	954529	932574833	31.2570	9.9227	3069.3	749685	977
978	956484	935441352	31.2730	9.9261	3072.5	751221	978
979	958441	938313739	31.2890	9.9295	3075.6	752758	979
980	960400	941192000	31.3050	9.9329	3078.8	754296	980
981	962361	944076141	31.3209	9.9363	3081.9	755837	981
982	964324	946966168	31.3369	9.9396	3085.0	757378	982
983	966289	949862087	31.3528	9.9430	3088.2	758922	983
984	968256	952763904	31.3688	9.9464	3091.3	760466	984
985	970225	955671625	31.3847	9.9497	3094.5	762013	985
986	972196	958585256	31.4006	9.9531	3097.6	763561	986
987	974169	961504803	31.4166	9.9565	3100.8	765111	987
988	976144	964430272	31.4325	9.9598	3103.9	766662	988
989	978121	967361669	31.4484	9.9632	3107.0	768214	989
990	980100	970299000	31.4643	9.9666	3110.2	769769	990
991	982081	973242271	31.4802	9.9699	3113.3	771325	991
992	984064	976191488	31.4960	9.9733	3116.5	772882	992
993	986049	979146657	31.5119	9.9766	3119.6	774441	993
994	988036	982107784	31.5278	9.9800	3122.7	776002	994
995	990025	985074875	31.5436	9.9833	3125.9	777564	995
996	992016	988047936	31.5595	9.9866	3129.0	779128	996
997	994009	991026973	31.5753	9.9900	3132.2	780693	997
998	996004	994011992	31.5911	9.9933	3135.3	782260	998
999	998001	997002999	31.6070	9.9967	3138.5	783828	999

Table 2

685

Table 2
Natural Trigonometric Functions

Degrees	Sin	Cos	Tan	Cot	Sec	Csc	
0°00′	.0000	1.0000	.0000	—	1.000	—	90°00′
10	029	000	029	343.8	000	343.8	50
20	058	000	058	171.9	000	171.9	40
30	.0087	1.0000	.0087	114.6	1.000	114.6	30
40	116	.9999	116	85.94	000	85.95	20
50	145	999	145	68.75	000	68.76	10
1°00′	.0175	.9998	.0175	57.29	1.000	57.30	89°00′
10	204	998	204	49.10	000	49.11	50
20	233	997	233	42.96	000	42.98	40
30	.0262	.9997	.0262	38.19	1.000	38.20	30
40	291	996	291	34.37	000	34.38	20
50	320	995	320	31.24	001	31.26	10
2°00′	.0349	.9994	.0349	28.64	1.001	28.65	88°00′
10	378	993	378	26.43	001	26.45	50
20	407	992	407	24.54	001	24.56	40
30	.0436	.9990	.0437	22.90	1.001	22.93	30
40	465	989	466	21.47	001	21.49	20
50	494	988	495	20.21	001	20.23	10
3°00′	.0523	.9986	.0524	19.08	1.001	19.11	87°00′
10	552	985	553	18.07	002	18.10	50
20	581	983	582	17.17	002	17.20	40
30	.0610	.9981	.0612	16.35	1.002	16.38	30
40	640	980	641	15.60	002	15.64	20
50	669	978	670	14.92	002	14.96	10
4°00′	.0698	.9976	.0699	14.30	1.002	14.34	86°00′
10	727	974	729	13.73	003	13.76	50
20	756	971	758	13.20	003	13.23	40
30	.0785	.9969	.0787	12.71	1.003	12.75	30
40	814	967	816	12.25	003	12.29	20
50	843	964	846	11.83	004	11.87	10
5°00′	.0872	.9962	.0875	11.43	1.004	11.47	85°00′
10	901	959	904	11.06	004	11.10	50
20	929	957	934	10.71	004	10.76	40
30	.0958	.9954	.0963	10.39	1.005	10.43	30
40	987	951	992	10.08	005	10.13	20
50	.1016	948	.1022	9.788	005	9.839	10
6°00′	.1045	.9945	.1051	9.514	1.006	9.567	84°00′
10	074	942	080	9.255	006	9.309	50
20	103	939	110	9.010	006	9.065	40
30	.1132	.9936	.1139	8.777	1.006	8.834	30
40	161	932	169	8.556	007	8.614	20
50	190	929	198	8.345	007	8.405	10
7°00′	.1219	.9925	.1228	8.144	1.008	8.206	83°00′
10	248	922	257	7.953	008	8.016	50
20	276	918	287	7.770	008	7.834	40
30	.1305	.9914	.1317	7.596	1.009	7.661	30
40	334	911	346	7.429	009	7.496	20
50	363	907	376	7.269	009	7.337	10
8°00′	.1392	.9903	.1405	7.115	1.010	7.185	82°00′
10	421	899	435	6.968	010	7.040	50
10	449	894	465	6.827	011	6.900	40
30	.1478	.9890	.1495	6.691	1.011	6.765	30
40	507	886	524	6.561	012	6.636	20
50	536	881	554	6.435	012	6.512	10
9°00′	.1564	.9877	.1584	6.314	1.012	6.392	81°00′
	Cos	Sin	Cot	Tan	Csc	Sec	Degrees

Degrees	Sin	Cos	Tan	Cot	Sec	Csc	
9°00′	.1564	.9877	.1584	6.314	1.012	6.392	**81°00′**
10	593	872	614	197	013	277	50
20	622	868	644	084	013	166	40
30	.1650	.9863	.1673	5.976	1.014	6.059	30
40	679	858	703	871	014	5.955	20
50	708	853	733	769	015	855	10
10°00′	.1736	.9848	.1763	5.671	1.015	5.759	**80°00′**
10	765	843	793	576	016	665	50
20	794	838	823	485	016	575	40
30	.1822	.9833	.1853	5.396	1.017	5.487	30
40	851	827	883	309	018	403	20
50	880	822	914	226	018	320	10
11°00′	.1908	.9816	.1944	5.145	1.019	5.241	**79°00′**
10	937	811	974	066	019	164	50
20	965	805	.2004	4.989	020	089	40
30	.1994	.9799	.2035	4.915	1.020	5.016	30
40	.2022	793	065	843	021	4.945	20
50	051	787	095	773	022	876	10
12°00′	.2079	.9781	.2126	4.705	1.022	4.810	**78°00′**
10	108	775	156	638	023	745	50
20	136	769	186	574	024	682	40
30	.2164	.9763	.2217	4.511	1.024	4.620	30
40	193	757	247	449	025	560	20
50	221	750	278	390	026	502	10
13°00′	.2250	.9744	.2309	4.331	1.026	4.445	**77°00′**
10	278	737	339	275	027	390	50
20	306	730	370	219	028	336	40
30	.2334	.9724	.2401	4.165	1.028	4.284	30
40	363	717	432	113	029	232	20
50	391	710	462	061	030	182	10
14°00′	.2419	.9703	.2493	4.011	1.031	4.134	**76°00′**
10	447	696	524	3.962	031	086	50
20	476	689	555	914	032	039	40
30	.2504	.9681	.2586	3.867	1.033	3.994	30
40	532	674	617	821	034	950	20
50	560	667	648	776	034	906	10
15°00′	.2588	.9659	.2679	3.732	1.035	3.864	**75°00′**
10	616	652	711	689	036	822	50
20	644	644	742	647	037	782	40
30	.2672	.9636	.2773	3.606	1.038	3.742	30
40	700	628	805	566	039	703	20
50	728	621	836	526	039	665	10
16°00′	.2756	.9613	.2867	3.487	1.040	3.628	**74°00′**
10	784	605	899	450	041	592	50
20	812	596	931	412	042	556	40
30	.2840	.9588	.2962	3.376	1.043	3.521	30
40	868	580	994	340	044	487	20
50	896	572	.3026	305	045	453	10
17°00′	.2924	.9563	.3057	3.271	1.046	3.420	**73°00′**
10	952	555	089	237	047	388	50
20	9'16	546	121	204	048	357	40
30	.3007	.9537	.3153	3.172	1.048	3.326	30
40	035	528	185	140	049	295	20
50	062	520	217	108	050	265	10
18°00′	.3090	.9511	.3249	3.078	1.051	3.236	**72°00′**
	Cos	**Sin**	**Cot**	**Tan**	**Csc**	**Sec**	**Degrees**

Table 2
687

Degrees	Sin	Cos	Tan	Cot	Sec	Csc	
18°00'	.3090	.9511	.3249	3.078	1.051	3.236	72°00'
10	118	502	281	047	052	207	50
20	145	492	314	018	053	179	40
30	.3173	.9483	.3346	2.989	1.054	3.152	30
40	201	474	378	960	056	124	20
50	228	465	411	932	057	098	10
19°00'	.3256	.9455	.3443	2.904	1.058	3.072	71°00'
10	283	446	476	877	059	046	50
20	311	436	508	850	060	021	40
30	.3338	.9426	.3541	2.824	1.061	2.996	30
40	365	417	574	798	062	971	20
50	393	407	607	773	063	947	10
20°00'	.3420	.9397	.3640	2.747	1.064	2.924	70°00'
10	448	387	673	723	065	901	50
20	475	377	706	699	066	878	40
30	.3502	.9367	.3739	2.675	1.068	2.855	30
40	529	356	772	651	069	833	20
50	557	346	805	628	070	812	10
21°00'	.3584	.9336	.3839	2.605	1.071	2.790	69°00'
10	611	325	872	583	072	769	50
20	638	315	906	560	074	749	40
30	.3665	.9304	.3939	2.539	1.075	2.729	30
40	692	293	973	517	076	709	20
50	719	283	.4006	496	077	689	10
22°00'	.3746	.9272	.4040	2.475	1.079	2.669	68°00'
10	773	261	074	455	080	650	50
20	800	250	108	434	081	632	40
30	.3827	.9239	.4142	2.414	1.082	2.613	30
40	854	228	176	394	084	595	20
50	881	216	210	375	085	577	10
23°00'	.3907	.9205	.4245	2.356	1.086	2.559	67°00'
10	934	194	270	337	088	542	50
20	961	182	314	318	089	525	40
30	.3987	.9171	.4348	2.300	1.090	2.508	30
40	.4014	159	383	282	092	491	20
50	041	147	417	264	093	475	10
24°00'	.4067	.9135	.4452	2.246	1.095	2.459	66°00'
10	094	124	487	229	096	443	50
20	120	112	522	211	097	427	40
30	.4147	.9100	.4557	2.194	1.099	2.411	30
40	173	088	592	177	100	396	20
50	200	075	628	161	102	381	10
25°00'	.4226	.9063	.4663	2.145	1.103	2.366	65°00'
10	253	051	699	128	105	352	50
20	279	038	734	112	106	337	40
30	.4305	.9026	.4770	2.097	1.108	2.323	30
40	331	013	806	081	109	309	20
50	358	001	841	066	111	295	10
26°00'	.4384	.8988	.4877	2.050	1.113	2.281	64°00'
10	410	975	913	035	114	268	50
20	436	962	950	020	116	254	40
30	.4462	.8949	.4986	2.006	1.117	2.241	30
40	488	936	.5022	1.991	119	228	20
50	514	923	059	977	121	215	10
27°00'	.4540	.8910	.5095	1.963	1.122	2.203	63°00'
	Cos	Sin	Cot	Tan	Csc	Sec	Degrees

Degrees	Sin	Cos	Tan	Cot	Sec	Csc	
27°00′	.4540	.8910	.5095	1.963	1.122	2.203	63°00
10	566	897	132	949	124	190	50
20	592	884	169	935	126	178	40
30	.4617	.8870	.5206	1.921	1.127	2.166	30
40	643	857	243	907	129	154	20
50	669	843	280	894	131	142	10
28°00′	.4695	.8829	.5317	1.881	1.133	2.130	62°00′
10	720	816	354	868	134	118	50
20	746	802	392	855	136	107	40
30	.4772	.8788	.5430	1.842	1.138	2.096	30
40	797	774	467	829	140	085	20
50	823	760	505	816	142	074	10
29°00′	.4848	.8746	.5543	1.804	1.143	2.063	61°00′
10	874	732	581	792	145	052	50
20	899	718	619	780	147	041	40
30	.4924	.8704	.5658	1.767	1.149	2.031	30
40	950	689	696	756	151	020	20
50	975	675	735	744	153	010	10
30°00′	.5000	.8660	.5774	1.732	1.155	2.000	60°00′
10	025	646	812	720	157	1.990	50
20	050	631	851	709	159	980	40
30	.5075	.8616	.5890	1.698	1.161	1.970	30
40	100	601	930	686	163	961	20
50	125	587	969	675	165	951	10
31°00′	.5150	.8572	.6009	1.664	1.167	1.942	59°00′
10	175	557	048	653	169	932	50
20	200	542	088	643	171	923	40
30	.5225	.8526	.6128	1.632	1.173	1.914	30
40	250	511	168	621	175	905	20
50	275	496	208	611	177	896	10
32°00′	.5299	.8480	.6249	1.600	1.179	1.887	58°00′
10	324	465	289	590	181	878	50
20	348	450	330	580	184	870	40
30	.5373	.8434	.6371	1.570	1.186	1.861	30
40	398	418	412	560	188	853	20
50	422	403	453	550	190	844	10
33°00′	.5446	.8387	.6494	1.540	1.192	1.836	57°00′
10	471	371	536	530	195	828	50
20	495	355	577	520	197	820	40
30	.5519	.8339	.6619	1.511	1.199	1.812	30
40	544	323	661	501	202	804	20
50	568	307	703	1.492	204	796	10
34°00′	.5592	.8290	.6745	1.483	1.206	1.788	56°00′
10	616	274	787	473	209	781	50
20	640	258	830	464	211	773	40
30	.5664	.8241	.6873	1.455	1.213	1.766	30
40	688	225	916	446	216	758	20
50	712	208	959	437	218	751	10
35°00′	.5736	.8192	.7002	1.428	1.221	1.743	55°00′
10	760	175	046	419	223	736	50
20	783	158	089	411	226	729	40
30	.5807	.8141	.7133	1.402	1.228	1.722	30
40	831	124	177	393	231	715	20
50	854	107	221	385	233	708	10
36°00′	.5878	.8090	.7265	1.376	1.236	1.701	54°00′
	Cos	Sin	Cot	Tan	Csc	Sec	Degrees

Table 2
689

Degrees	Sin	Cos	Tan	Cot	Sec	Csc	
36°00′	.5878	.8090	.7265	1.376	1.236	1.701	54°00′
10	901	073	310	368	239	695	50
20	925	056	355	360	241	688	40
30	.5948	.8039	.7400	1.351	1.244	1.681	30
40	972	021	445	343	247	675	20
50	995	004	490	335	249	668	10
37°00′	.6018	.7986	.7536	1.327	1.252	1.662	53°00′
10	041	969	581	319	255	655	50
20	065	951	627	311	258	649	40
30	.6088	.7934	.7673	1.303	1.260	1.643	30
40	111	916	720	295	263	636	20
50	134	898	766	288	266	630	10
38°00′	.6157	.7880	.7813	1.280	1.269	1.624	52°00′
10	180	862	860	272	272	618	50
20	202	844	907	265	275	612	40
30	.6225	.7826	.7954	1.257	1.278	1.606	30
40	248	808	.8002	250	281	601	20
50	271	790	050	242	284	595	10
39°00′	.6293	.7771	.8098	1.235	1.287	1.589	51°00′
10	316	753	146	228	290	583	50
20	338	735	195	220	293	578	40
30	.6361	.7716	.8243	1.213	1.296	1.572	30
40	383	698	292	206	299	567	20
50	406	679	342	199	302	561	10
40°00′	.6428	.7660	.8391	1.192	1.305	1.556	50°00′
10	450	642	441	185	309	550	50
20	472	623	491	178	312	545	40
30	.6494	.7604	.8541	1.171	1.315	1.540	30
40	517	585	591	164	318	535	20
50	539	566	642	157	322	529	10
41°00′	.6561	.7547	.8693	1.150	1.325	1.524	49°00′
10	583	528	744	144	328	519	50
20	604	509	796	137	332	514	40
30	.6626	.7490	.8847	1.130	1.335	1.509	30
40	648	470	899	124	339	504	20
50	670	451	952	117	342	499	10
42°00′	.6691	.7431	.9004	1.111	1.346	1.494	48°00′
10	713	412	057	104	349	490	50
20	734	392	110	098	353	485	40
30	.6756	.7373	.9163	1.091	1.356	1.480	30
40	777	353	217	085	360	476	20
50	799	333	271	079	364	471	10
43°00′	.6820	.7314	.9325	1.072	1.367	1.466	47°00′
10	841	294	380	066	371	462	50
20	862	274	435	060	375	457	40
30	.6884	.7254	.9490	1.054	1.379	1.453	30
40	905	234	545	048	382	448	20
50	926	214	601	042	386	444	10
44°00′	.6947	.7193	.9657	1.036	1.390	1.440	46°00′
10	967	173	713	030	394	435	50
20	988	153	770	024	398	431	40
30	.7009	.7133	.9827	1.018	1.402	1.427	30
40	030	112	884	012	406	423	20
50	050	092	942	006	410	418	10
45°00′	.7071	.7071	1.000	1.000	1.414	1.414	45°00′
	Cos	Sin	Cot	Tan	Csc	Sec	Degrees

Table 3
Converting Millimeters to Inches and Decimals

Milli-meters	Inches	Milli-meters	Inches	Milli-meters	Inches	Milli-meters	Inches
1	= .03937	26	= 1.02362	51	= 2.00787	76	= 2.99213
2	= .07874	27	= 1.06299	52	= 2.04724	77	= 3.03150
3	= .11811	28	= 1.10236	53	= 2.08661	78	= 3.07087
4	= .15748	29	= 1.14173	54	= 2.12598	79	= 3.11024
5	= .19685	30	= 1.18110	55	= 2.16535	80	= 3.14961
6	= .23622	31	= 1.22047	56	= 2.20472	81	= 3.18898
7	= .27559	32	= 1.25984	57	= 2.24409	82	= 3.22835
8	= .31496	33	= 1.29921	58	= 2.28346	83	= 3.26772
9	= .35433	34	= 1.33858	59	= 2.32283	84	= 3.30709
10	= .39370	35	= 1.37795	60	= 2.36220	85	= 3.34646
11	= .43307	36	= 1.41732	61	= 2.40157	86	= 3.38583
12	= .47244	37	= 1.45669	62	= 2.44094	87	= 3.42520
13	= .51181	38	= 1.49606	63	= 2.48031	88	= 3.46457
14	= .55118	39	= 1.53543	64	= 2.51969	89	= 3.50394
15	= .59055	40	= 1.57480	65	= 2.55906	90	= 3.54331
16	= .62992	41	= 1.61417	66	= 2.59843	91	= 3.58268
17	= .66929	42	= 1.65354	67	= 2.63780	92	= 3.62204
18	= .70866	43	= 1.69291	68	= 2.67717	93	= 3.66142
19	= .74803	44	= 1.73228	69	= 2.71654	94	= 3.70079
20	= .78740	45	= 1.77165	70	= 2.75591	95	= 3.74016
21	= .82677	46	= 1.81102	71	= 2.79528	96	= 3.77953
22	= .86614	47	= 1.85039	72	= 2.83465	97	= 3.81890
23	= .90551	48	= 1.88976	73	= 2.87402	98	= 3.85827
24	= .94488	49	= 1.92913	74	= 2.91339	99	= 3.89764
25	= .98425	50	= 1.96850	75	= 2.95276	100	= 3.93701

Table 4

691

Table 4
Decimal and Metric Equivalents
of Fractions of an Inch

Fraction	$\frac{1}{32}$ds	$\frac{1}{64}$ths	Decimal	Millimeters	Fraction	$\frac{1}{32}$ds	$\frac{1}{64}$ths	Decimal	Millimeters
		1	.015625	0.3969			33	.515625	13.0969
	1	2	.03125	0.7938		17	34	.53125	13.4938
		3	.046875	1.1906			35	.546875	13.8906
$\frac{1}{16}$	2	4	.0625	1.5875	$\frac{9}{16}$	18	36	.5625	14.2875
		5	.078125	1.9844			37	.578125	14.6844
	3	6	.09375	2.3812		19	38	.59375	15.0812
		7	.109375	2.7781			39	.609375	15.4781
$\frac{1}{8}$	4	8	.125	3.1750	$\frac{5}{8}$	20	40	.625	15.8750
		9	.140625	3.5719			41	.640625	16.2719
	5	10	.15625	3.9688		21	42	.65625	16.6688
		11	.171875	4.3656			43	.671875	17.0656
$\frac{3}{16}$	6	12	.1875	4.7625	$\frac{11}{16}$	22	44	.6875	17.4625
		13	.203125	5.1594			45	.703125	17.8594
	7	14	.21875	5.5562		23	46	.71875	18.2562
		15	.234375	5.9531			47	.734375	18.6531
$\frac{1}{4}$	8	16	.25	6.3500	$\frac{3}{4}$	24	48	.75	19.0500
		17	.265625	6.7469			49	.765625	19.4469
	9	18	.28125	7.1438		25	50	.78125	19.8438
		19	.296875	7.5406			51	.796875	20.2406
$\frac{5}{16}$	10	20	.3125	7.9375	$\frac{13}{16}$	26	52	.8125	20.6375
		21	.328125	8.3344			53	.828125	21.0344
	11	22	.34375	8.7312		27	54	.84375	21.4312
		23	.359375	9.1281			55	.859375	21.8281
$\frac{3}{8}$	12	24	.375	9.5250	$\frac{7}{8}$	28	56	.875	22.2250
		25	.390625	9.9219			57	.890625	22.6219
	13	26	.40625	10.3188		29	58	.90625	23.0188
		27	.421875	10.7156			59	.921875	23.4156
$\frac{7}{16}$	14	28	.4375	11.1125	$\frac{15}{16}$	30	60	.9375	23.8125
		29	.453125	11.5094			61	.953125	24.2094
	15	30	.46875	11.9062		31	62	.96875	24.6062
		31	.484375	12.3031			63	.984375	25.0031
$\frac{1}{2}$	16	32	.5	12.7000	1	32	64	1.	25.4000

Table 5
Converting Degrees Fahrenheit
to Degrees Celsius

32°F = 0.00°C	78°F = 25.56°C	123°F = 50.56°C	168°F = 75.56°C
33°F = 0.56°C	79°F = 26.11°C	124°F = 51.11°C	169°F = 76.11°C
34°F = 1.11°C	80°F = 26.67°C	125°F = 51.67°C	170°F = 76.67°C
35°F = 1.67°C	81°F = 27.22°C	126°F = 52.22°C	171°F = 77.22°C
36°F = 2.22°C	82°F = 27.78°C	127°F = 52.78°C	172°F = 77.78°C
37°F = 2.78°C	83°F = 28.33°C	128°F = 53.33°C	173°F = 78.33°C
38°F = 3.33°C	84°F = 28.89°C	129°F = 53.89°C	174°F = 78.89°C
39°F = 3.89°C	85°F = 29.45°C	130°F = 54.45°C	175°F = 79.45°C
40°F = 4.45°C	86°F = 30.00°C	131°F = 55.00°C	176°F = 80.00°C
41°F = 5.00°C	87°F = 30.56°C	132°F = 55.56°C	177°F = 80.56°C
42°F = 5.56°C	88°F = 31.11°C	133°F = 56.11°C	178°F = 81.11°C
43°F = 6.11°C	89°F = 31.67°C	134°F = 56.67°C	179°F = 81.67°C
44°F = 6.67°C	90°F = 32.22°C	135°F = 57.22°C	180°F = 82.22°C
45°F = 7.22°C	91°F = 32.78°C	136°F = 57.78°C	181°F = 82.78°C
46°F = 7.78°C	92°F = 33.33°C	137°F = 58.33°C	182°F = 83.33°C
47°F = 8.33°C	93°F = 33.89°C	138°F = 58.89°C	183°F = 83.89°C
48°F = 8.89°C	94°F = 34.45°C	139°F = 59.45°C	184°F = 84.45°C
49°F = 9.45°C	95°F = 35.00°C	140°F = 60.00°C	185°F = 85.00°C
50°F = 10.00°C	96°F = 35.56°C	141°F = 60.56°C	186°F = 85.56°C
51°F = 10.56°C	97°F = 36.11°C	142°F = 61.11°C	187°F = 86.11°C
52°F = 11.11°C	98°F = 36.67°C	143°F = 61.67°C	188°F = 86.67°C
53°F = 11.67°C	99°F = 37.22°C	144°F = 62.22°C	189°F = 87.22°C
54°F = 12.22°C	100°F = 37.78°C	145°F = 62.78°C	190°F = 87.78°C
55°F = 12.78°C	101°F = 38.33°C	146°F = 63.33°C	191°F = 88.33°C
56°F = 13.33°C	102°F = 38.89°C	147°F = 63.89°C	192°F = 88.89°C
57°F = 13.89°C	103°F = 39.45°C	148°F = 64.46°C	193°F = 89.45°C
58°F = 14.45°C	104°F = 40.00°C	149°F = 65.00°C	194°F = 90.00°C
59°F = 15.00°C	105°F = 40.56°C	150°F = 65.56°C	195°F = 90.56°C
60°F = 15.56°C	106°F = 41.11°C	151°F = 66.11°C	196°F = 91.11°C
61°F = 16.11°C	107°F = 41.67°C	152°F = 66.67°C	197°F = 91.67°C
62°F = 16.67°C	108°F = 42.22°C	153°F = 67.22°C	198°F = 92.22°C
63°F = 17.22°C	109°F = 42.78°C	154°F = 67.78°C	199°F = 92.78°C
64°F = 17.78°C	110°F = 43.33°C	155°F = 68.33°C	200°F = 93.33°C
65°F = 18.33°C	111°F = 43.89°C	156°F = 68.89°C	201°F = 93.89°C
66°F = 18.89°C	112°F = 44.45°C	157°F = 69.45°C	202°F = 94.45°C
67°F = 19.45°C	113°F = 45.00°C	158°F = 70.00°C	203°F = 95.00°C
68°F = 20.00°C	114°F = 45.56°C	159°F = 70.56°C	204°F = 95.56°C
69°F = 20.56°C	115°F = 46.11°C	160°F = 71.11°C	205°F = 96.11°C
70°F = 21.11°C	116°F = 46.67°C	161°F = 71.67°C	206°F = 96.67°C
71°F = 21.67°C	117°F = 47.22°C	162°F = 72.22°C	207°F = 97.22°C
72°F = 22.22°C	118°F = 47.78°C	163°F = 72.78°C	208°F = 97.78°C
73°F = 22.78°C	119°F = 48.33°C	164°F = 73.33°C	209°F = 98.33°C
74°F = 23.33°C	120°F = 48.89°C	165°F = 73.89°C	210°F = 98.89°C
75°F = 23.89°C	121°F = 49.45°C	166°F = 74.45°C	211°F = 99.45°C
76°F = 24.45°C	122°F = 50.00°C	167°F = 75.00°C	212°F = 100.00°C
77°F = 25.00°C			

Table 6

Table 6
Converting Degrees Celsius to Degrees Fahrenheit

0°C = 32.00°F	26°C = 78.80°F	51°C = 123.80°F	76°C = 168.80°F
1°C = 33.80°F	27°C = 80.60°F	52°C = 125.60°F	77°C = 170.60°F
2°C = 35.60°F	28°C = 82.40°F	53°C = 127.40°F	78°C = 172.40°F
3°C = 37.40°F	29°C = 84.20°F	54°C = 129.20°F	79°C = 174.20°F
4°C = 39.20°F	30°C = 86.00°F	55°C = 131.00°F	80°C = 176.00°F
5°C = 41.00°F	31°C = 87.80°F	56°C = 132.80°F	81°C = 177.80°F
6°C = 42.80°F	32°C = 89.60°F	57°C = 134.60°F	82°C = 179.60°F
7°C = 44.60°F	33°C = 91.40°F	58°C = 136.40°F	83°C = 181.40°F
8°C = 46.40°F	34°C = 93.20°F	59°C = 138.20°F	84°C = 183.20°F
9°C = 48.20°F	35°C = 95.00°F	60°C = 140.00°F	85°C = 185.00°F
10°C = 50.00°F	36°C = 96.80°F	61°C = 141.80°F	86°C = 186.80°F
11°C = 51.80°F	37°C = 98.60°F	62°C = 143.60°F	87°C = 188.60°F
12°C = 53.60°F	38°C = 100.40°F	63°C = 145.40°F	88°C = 190.40°F
13°C = 55.40°F	39°C = 102.20°F	64°C = 147.20°F	89°C = 192.20°F
14°C = 57.20°F	40°C = 104.00°F	65°C = 149.00°F	90°C = 194.00°F
15°C = 59.00°F	41°C = 105.80°F	66°C = 150.80°F	91°C = 195.80°F
16°C = 60.80°F	42°C = 107.60°F	67°C = 152.60°F	92°C = 197.60°F
17°C = 62.60°F	43°C = 109.40°F	68°C = 154.40°F	93°C = 199.40°F
18°C = 64.40°F	44°C = 111.20°F	69°C = 156.20°F	94°C = 201.20°F
19°C = 66.20°F	45°C = 113.00°F	70°C = 158.00°F	95°C = 203.00°F
20°C = 68.00°F	46°C = 114.80°F	71°C = 159.80°F	96°C = 204.80°F
21°C = 69.80°F	47°C = 116.60°F	72°C = 161.60°F	97°C = 206.60°F
22°C = 71.60°F	48°C = 118.40°F	73°C = 163.40°F	98°C = 208.40°F
23°C = 73.40°F	49°C = 120.20°F	74°C = 165.20°F	99°C = 210.20°F
24°C = 75.20°F	50°C = 122.00°F	75°C = 167.00°F	100°C = 212.00°F
25°C = 77.00°F			

Table 7
Weights of Materials

Material	Average Weight in Pounds		Average Weight in Grams per Cm³
	Per Cubic Foot	Per Cubic Inch	
Aluminum	160	0.0924	2.56
Brass	512	0.2960	8.20
Brick, pressed	150		2.40
Brick, common, hard	125		2.00
Cement, American, Rosendale	56		0.90
Cement, Portland	90		1.44
Clay, loose	63		1.01
Coal, broken, loose, Anthracite	54		0.86
Coal, broken, loose, Bituminous	49		0.78
Concrete	154		2.46
Copper	550	0.3184	8.80
Earth, common loam	76		1.22
Earth, packed	95		1.52
Gravel, dry, loose	90 to 106		1.44 to 1.70
Gravel, well shaken	99 to 117		1.58 to 1.87
Gold	1206	0.6975	19.30
Ice	58.7		0.94
Iron, cast	450	0.2600	7.20
Iron, wrought	480	0.2778	7.68
Lead	710	0.4109	11.36
Lime	53		0.85
Masonry, well dressed	165		2.64
Masonry, dry rubble	138		2.21
Mortar, hardened	103		1.65
Nickel	549	0.3177	8.78
Quartz	165		2.64
Sand, dry, loose	90 to 106		1.44 to 1.70
Sand, well shaken	99 to 117		1.58 to 1.87
Silver	657	0.3802	10.51
Snow, freshly fallen	5 to 12		0.08 to 0.19
Snow, wet and compacted	15 to 50		0.24 to 0.80
Steel	490	0.2835	7.84
Stone, gneiss	168		2.67
Stone, granite	170		2.72
Stone, limestone	168		2.69
Stone, marble	168		2.69
Stone, sandstone	151		2.42
Stone, shale	162		2.59
Stone, slate	175		2.80
Tar	62		0.99
Tin	455	0.2632	7.28
Water	62.5		1.00
Wood, dry, ash	38		0.61
Wood, dry, cherry	42		0.67
Wood, dry, chestnut	41		0.66
Wood, dry, elm	35		0.56
Wood, dry, hemlock	25		0.40
Wood, dry, hickory	53		0.85
Wood, dry, lignum vitae	83		1.33
Wood, dry, mahogany	53		0.85
Wood, dry, maple	49		0.78
Wood, dry, oak, white	50		0.80
Wood, dry, oak, other kinds	32 to 45		0.51 to 0.72
Wood, dry, pine, white	25		0.40
Wood, dry, pine, yellow	34 to 45		0.54 to 0.72
Wood, dry, spruce	25		0.40
Wood, dry, sycamore	37		0.59
Wood, dry, walnut, black	38		0.61
Zinc	438	0.2528	7.01

Table 8 695

Table 8
National Coarse-Thread Series

Size	Threads per Inch	Basic Diameters Major Diam. (Inches)	Minor Diam. (Inches)	Cross-section at Root of Thread (Square Inches)	Tap Drill Size Commercial Size $\frac{3}{4}$ Depth of Thread	$\frac{5}{6}$ Depth of Thread
1	64	0.073	0.0538	0.0022	53	0.0561
2	56	.086	.0641	.0031	50	.0667
3	48	.099	.0734	.0041	47	.0764
4	40	.112	.0813	.0050	43	.0849
5	40	.125	.0943	.0067	38	.0979
6	32	.138	.0997	.0075	36	.1042
8	32	.164	.1257	.0120	29	.1302
10	24	.190	.1389	.0145	25	.1449
12	24	.216	.1649	.0206	16	.1709
$\frac{1}{4}$	20	.2500	.1887	.0269	7	.1959
$\frac{5}{16}$	18	.3125	.2443	.0454	F	.2524
$\frac{3}{8}$	16	.3750	.2983	.0678	$\frac{5}{16}$	.3073
$\frac{7}{16}$	14	.4375	.3499	.0933	U	.3602
$\frac{1}{2}$	13	.5000	.4056	.1257	$\frac{27}{64}$	.4167
$\frac{9}{16}$	12	.5625	.4603	.1620	$\frac{31}{64}$	.4723
$\frac{5}{8}$	11	.6250	.5135	.2018	$\frac{17}{32}$	.5266
$\frac{3}{4}$	10	.7500	.6273	.3020	$\frac{21}{32}$	.6417
$\frac{7}{8}$	9	.8750	.7387	.4193	$\frac{49}{64}$	.7547
1	8	1.0000	.8466	.5510	$\frac{7}{8}$	.8647
$1\frac{1}{8}$	7	1.1250	.9497	.6931	$\frac{63}{64}$	.9704
$1\frac{1}{4}$	7	1.2500	1.0747	.8898	$1\frac{7}{64}$	1.0954
$1\frac{3}{8}$	6	1.3750	1.1705	1.0541	$1\frac{7}{32}$	1.1946
$1\frac{1}{2}$	6	1.5000	1.2955	1.2938	$1\frac{11}{32}$	1.3196
$1\frac{3}{4}$	5	1.7500	1.5046	1.7441	$1\frac{9}{16}$	1.5335
2	$4\frac{1}{2}$	2.0000	1.7274	2.3001	$1\frac{25}{32}$	1.7594
$2\frac{1}{4}$	$4\frac{1}{2}$	2.2500	1.9774	3.0212	$2\frac{1}{32}$	2.0094
$2\frac{1}{2}$	4	2.5000	2.1933	3.7161	$2\frac{1}{4}$	2.2294
$2\frac{3}{4}$	4	2.7500	2.4433	4.6194	$2\frac{1}{2}$	2.4794
3	4	3.0000	2.6933	5.6209	$2\frac{23}{32}$	2.7294
$3\frac{1}{4}$	4	3.2500	2.9433	6.7205	$2\frac{31}{32}$	2.9794
$3\frac{1}{2}$	4	3.5000	3.1933	7.9183	$3\frac{7}{32}$	3.2294
$3\frac{3}{4}$	4	3.7500	3.4433	9.2143	$3\frac{15}{32}$	3.4794
4	4	4.0000	3.6933	10.6084	$3\frac{23}{32}$	3.7294

Table 9 National Fine-Thread Series					
		Basic Diameters		Cross-section at Root of Thread (Square Inches)	Tap Drill Size	
Size	Threads per Inch	Major (Inches)	Minor (Inches)		Commercial Size $\frac{3}{4}$ Depth of Thread	$\frac{5}{6}$ Depth of Thread
0	80	0.060	0.0447	0.0015	$\frac{3}{64}$	0.0465
1	72	.073	.0560	.0024	53	.0580
2	64	.086	.0668	.0034	50	.0691
3	56	.099	.0771	.0045	45	.0797
4	48	.112	.0864	.0057	42	.0894
5	44	.125	.0971	.0072	37	.1004
6	40	.138	.1073	.0087	33	.1109
8	36	.164	.1299	.0128	29	.1339
10	32	.190	.1517	.0175	21	.1562
12	28	.216	.1722	.0226	14	.1773
$\frac{1}{4}$	28	.2500	.2062	.0326	3	.2113
$\frac{5}{16}$	24	.3125	.2614	.0524	I	.2674
$\frac{3}{8}$	24	.3750	.3239	.0809	Q	.3299
$\frac{7}{16}$	20	.4375	.3762	.1090	$\frac{25}{64}$	.3834
$\frac{1}{2}$	20	.5000	.4387	.1486	$\frac{29}{64}$	.4459
$\frac{9}{16}$	18	.5625	.4943	.1888	$\frac{33}{64}$	.5024
$\frac{5}{8}$	18	.6250	.5568	.2400	$\frac{37}{64}$	.5649
$\frac{3}{4}$	16	.7500	.6733	.3513	$\frac{11}{16}$	.6823
$\frac{7}{8}$	14	.8750	.7874	.4805	$\frac{13}{16}$	.7977
1	14	1.0000	.8978	.6464	$\frac{15}{16}$	.9227
$1\frac{1}{8}$	12	1.1250	1.0228	.8118	$1\frac{3}{64}$	1.0348
$1\frac{1}{4}$	12	1.2500	1.1478	1.0238	$1\frac{11}{64}$	1.1598
$1\frac{3}{8}$	12	1.3750	1.2728	1.2602	$1\frac{19}{64}$	1.2848
$1\frac{1}{2}$	12	1.5000	1.3978	1.5212	$1\frac{27}{64}$	1.4098

Chapter 1
Problems 1–2, page 2

1. 2,987	**3.** 3,279	**5.** 3,948
7. 2,899	**9.** 3,703	**11.** 3,508
13. 2,449	**15.** 2,542	**17.** 4,105
19. 3,652	**21.** 3,310	**23.** 2,713
25. 3,421	**27.** 17,823	**29.** 39,609
31. 30,745	**33.** 35,464	**35.** 20,923
37. 90,207,937,899	**39.** 464,985,792,605	**41.** 432
43. 118,486	**45.** 258,952	**47.** 834 lb
49. 85,178 tons	**51.** 52 cubic yards	**53.** 34 receptacles
55. 19,123,606 people		

Problems 1–4, page 8

1. 1,514	**3.** 830	**5.** 419
7. 251	**9.** 1,130	**11.** 2,850
13. 2,222	**15.** 652	**17.** 5,457
19. 11,985	**21.** 9,991	**23.** 4,232
25. 12,883	**27.** 1,532	**29.** 10,047
31. 5,805,911 votes; 19,361,278 votes	**33.** $104,012,000	**35.** 1,058,414 units
37. 127,470,455 females; 6,231,037 more females	**39.** 37,635,977 more passengers	

Problems 1–6, page 13

1. 8,520,000	**3.** 66,500	**5.** 6,940
7. 154,000	**9.** 39,300	**11.** 446,000
13. 323,000	**15.** 10,630,000	**17.** 52,700
19. 9,470	**21.** 244,528	**23.** 538,016
25. 145,152	**27.** 257,466	**29.** 417,722
31. 215,020	**33.** 222,306	**35.** 364,186
37. 713,940	**39.** 830,270	**41.** $31,200
43. $715	**45.** 2,856 bolts	**47.** 5,865,696,000,000 miles
49. 132,000 ft		

Problems 1–8, page 19

1. 433	**3.** 175	**5.** 879
7. 537	**9.** 70	**11.** 632
13. 652	**15.** 631	**17.** 692
19. 875	**21.** 889	**23.** 428
25. 257	**27.** 479	**29.** 945
31. 12	**33.** 55	**35.** 24
37. 54	**39.** 62	**41.** $8
43. 80 pieces	**45.** 47 miles/hour	**47.** 25 gallons
49. 13 cents		

Chapter 2
Problems 2–3, page 29

1. proper	**3.** improper	**5.** denominator
7. 6; 8; 10; 24	**9.** 32; 41; 55	

Problems 2–4, page 31

1. 2	**3.** 2	**5.** $1\frac{1}{3}$
7. $5\frac{1}{3}$	**9.** 6	**11.** $1\frac{1}{8}$
13. $6\frac{7}{8}$	**15.** $1\frac{36}{64}$ or $1\frac{9}{16}$	**17.** $7\frac{3}{4}$
19. 5	**21.** $2\frac{2}{4}$ or $2\frac{1}{2}$	**23.** $2\frac{1}{8}$
25. $3\frac{1}{2}$	**27.** $2\frac{5}{10}$ or $2\frac{1}{2}$	**29.** $1\frac{2}{16}$ or $1\frac{1}{8}$
31. $2\frac{7}{32}$	**33.** $1\frac{30}{32}$ or $1\frac{15}{16}$	**35.** 2
37. $2\frac{10}{60}$ or $2\frac{1}{6}$	**39.** 2	**41.** $2\frac{7}{10}$
43. 4	**45.** $1\frac{63}{100}$	**47.** $6\frac{19}{50}$
49. $1\frac{21}{50}$		

Problems 2–5, page 33

1. $\frac{1}{2}$	**3.** $\frac{2}{3}$	**5.** $\frac{3}{4}$
7. $\frac{3}{8}$	**9.** $\frac{1}{4}$	**11.** $\frac{1}{2}$
13. 10	**15.** $\frac{2}{3}$	**17.** $\frac{1}{2}$
19. $\frac{1}{8}$	**21.** $\frac{1}{2}$	**23.** $\frac{5}{6}$
25. $\frac{1}{2}$	**27.** $\frac{5}{8}$	**29.** $\frac{1}{2}$
31. $\frac{1}{4}$	**33.** $\frac{3}{8}$	**35.** $\frac{1}{2}$

Problems 2–6, page 37

1. $1\frac{2}{3}$	**3.** 6	**5.** $\frac{1}{3}$
7. $\frac{15}{64}$	**9.** $4\frac{1}{2}$	**11.** $3\frac{1}{8}$
13. $\frac{4}{7}$	**15.** $\frac{1}{9}$	**17.** $11\frac{7}{15}$
19. $2\frac{49}{256}$	**21.** $\frac{1}{8}$	**23.** $\frac{1}{3}$
25. 3 ft 8 in.	**27.** $197\frac{5}{8}$ cu in.	**29.** $31\frac{39}{64}$ cu in.

Problems 2–7, page 39

1. $\frac{6}{25}$	**3.** $\frac{7}{40}$	**5.** $1\frac{1}{2}$
7. 14	**9.** $18\frac{3}{4}$	**11.** 40

13. $1\frac{19}{32}$ **15.** $10\frac{5}{6}$ **17.** $2\frac{6}{25}$

19. $\frac{2}{3}$ **21.** 3   **23.** $1\frac{1}{3}$

25. $1\frac{19}{105}$ **27.** $\frac{146}{255}$ **29.** $2\frac{8}{9}$

31. $2\frac{12}{29}$ **33.** $3\frac{5}{39}$ **35.** 2

37. $8\frac{45}{88}$ in.   **39.** $13\frac{1}{3}$ cu ft

Problems 2–8, page 41

1. $\frac{8}{45}$ **3.** $1\frac{1}{6}$ **5.** $1\frac{3}{32}$

7. $14\frac{2}{7}$ **9.** $2\frac{1}{4}$ **11.** $\frac{2}{27}$

13. 1   **15.** $\frac{1}{32}$ **17.** $1\frac{9}{56}$

19. 25   **21.** 10   **23.** $\frac{3}{224}$

Problems 2–9, page 43

1. $\frac{2}{4}$ **3.** $\frac{12}{32}$ **5.** $\frac{8}{12}$

7. $\frac{9}{15}$ **9.** $\frac{165}{180}$ **11.** $\frac{138}{144}$

13. $\frac{220}{360}$ **15.** $\frac{36}{64}$ **17.** $\frac{81}{144}$

19. $\frac{45}{60}$ **21.** $\frac{20}{16}$ **23.** $\frac{52}{32}$

Problems 2–10, page 45

1. $\frac{2}{4}, \frac{3}{4}$ **3.** $\frac{3}{6}, \frac{4}{6}$ **5.** $\frac{4}{12}, \frac{3}{12}$

7. $\frac{2}{6}, \frac{5}{6}$ **9.** $\frac{6}{12}, \frac{8}{12}, \frac{9}{12}$ **11.** $\frac{8}{12}, \frac{9}{12}, \frac{10}{12}$

13. $\frac{6}{16}, \frac{5}{16}$ **15.** $\frac{12}{16}, \frac{14}{16}, \frac{5}{16}$ **17.** $\frac{10}{24}, \frac{5}{24}$

19. $\frac{105}{120}, \frac{12}{120}, \frac{70}{120}$ **21.** $\frac{11}{143}, \frac{13}{143}$ **23.** $\frac{10}{70}, \frac{6}{70}, \frac{35}{70}$

25. $\frac{5}{8}, \frac{2}{3}, \frac{5}{7}$ **27.** $\frac{2}{3}$ **29.** yes

Problems 2–11, page 47

1. $1\frac{1}{4}$ **3.** $1\frac{5}{12}$ **5.** $1\frac{7}{24}$

7. $1\frac{5}{16}$ **9.** $1\frac{5}{8}$ **11.** $1\frac{29}{32}$

13. 2   **15.** $2\frac{3}{16}$ **17.** $1\frac{5}{12}$

19. $2\frac{5}{24}$ **21.** $4\frac{13}{48}$ **23.** $2\frac{1}{9}$

25. $1\frac{31}{72}$ **27.** $2\frac{1853}{2940}$ **29.** $\frac{147}{1000}$

Problems 2–12, page 48

1. $7\frac{1}{4}$ **3.** $5\frac{1}{8}$ **5.** $8\frac{5}{6}$

7. $4\frac{15}{16}$ **9.** $12\frac{9}{10}$ **11.** $26\frac{11}{24}$

13. 19   **15.** $14\frac{5}{32}$ **17.** $11\frac{35}{48}$

19. $31\frac{47}{60}$ **21.** $17\frac{3}{4}$ **23.** $18\frac{13}{24}$

25. $39\frac{11}{36}$ **27.** $32\frac{9}{16}$ **29.** $8\frac{7}{64}$ in.

31. $21\frac{13}{64}$ in.   **33.** $33\frac{1}{2}$ hours   **35.** 34 hours

Problems 2–13, page 51

1. $\frac{1}{4}$ **3.** $\frac{1}{4}$ **5.** $\frac{1}{6}$

7. $\frac{9}{16}$ **9.** $\frac{7}{20}$ **11.** $\frac{23}{40}$

13. $\frac{1}{8}$ **15.** $\frac{1}{12}$ **17.** $\frac{9}{32}$ in.

19. $\frac{5}{16}$ in.

Problems 2–14, page 53

1. $4\frac{7}{8}$	**3.** $3\frac{3}{16}$	**5.** $4\frac{3}{8}$
7. $3\frac{5}{8}$	**9.** $4\frac{3}{8}$	**11.** $2\frac{7}{12}$
13. $5\frac{43}{72}$	**15.** $4\frac{1}{2}$	**17.** $1\frac{41}{63}$
19. $1\frac{13}{20}$	**21.** $2\frac{7}{8}$	**23.** $4\frac{13}{16}$
25. $8\frac{15}{16}$	**27.** $7\frac{1}{4}$	**29.** 4
31. $\frac{1}{8}$	**33.** $6\frac{1}{16}$	**35.** $2\frac{23}{32}$
37. $29\frac{7}{8}$	**39.** $9\frac{45}{64}$	**41.** $12\frac{1}{16}$
43. $8\frac{47}{48}$	**45.** $8\frac{1}{32}$ in.	**47.** $\frac{5}{64}$ in.
49. $\frac{11}{64}$ in.		

Problems 2–15, page 56

1. $8\frac{3}{4}$	**3.** $\frac{3}{4}$	**5.** $\frac{28}{53}$
7. $\frac{1}{24}$	**9.** $1\frac{1}{3}$	**11.** $4\frac{1}{6}$
13. $1\frac{1}{2}$	**15.** $\frac{11}{14}$	**17.** $1\frac{2}{5}$
19. $\frac{17}{48}$		

Problems 2–16, page 58

1. $1\frac{1}{2}$	**3.** $1\frac{7}{11}$	**5.** $\frac{21}{32}$
7. $\frac{4}{5}$	**9.** $1\frac{1}{7}$	**11.** $\frac{1}{25}$
13. $\frac{7}{24}$	**15.** 1	**17.** $\frac{5}{6}$
19. $3\frac{1}{5}$	**21.** $\frac{89}{152}$	**23.** $4\frac{9}{10}$

Miscellaneous Problems in Common Fractions, page 59

1. 9 ft $2\frac{13}{16}$ in.	**3.** $1\frac{1}{8}$ in.	**5.** $6\frac{1}{2}$ in.
7. $10\frac{13}{16}$ in.	**9.** $3\frac{3}{64}$ in.	**11.** $6\frac{3}{16}$ in.
13. $\frac{9}{64}$ in.	**15.** $11\frac{23}{32}$ in.	**17.** 3 ft $11\frac{11}{16}$ in.
19. 14 holes	**21.** $23\frac{7}{32}$ in.	**23.** $293\frac{3}{4}$ min = 4 hr $53\frac{3}{4}$ min
25. $1\frac{1}{4}$ hr; \$44.44; \$88.88; \$24.69	**27.** 8 ft $7\frac{1}{8}$ in.	**29.** $1\frac{3}{16}$ in.

Chapter 3
Problems 3–1, page 72

1. $\frac{1}{10}$	**3.** $\frac{1}{1000}$	**5.** $\frac{15}{100}$
7. $\frac{83}{1000}$	**9.** 0.6	**11.** 0.03
13. 1.33	**15.** 10.01	**17.** 0.302
19. 0.1001	**21.** 10.1	**23.** 2.0625
25. 512.0075	**27.** 5.000042	

Problems 3–2, page 75

1. $\frac{1}{2}$	**3.** $\frac{1}{8}$	**5.** $\frac{3}{4}$
7. $\frac{7}{8}$	**9.** $\frac{4}{5}$	**11.** $\frac{1}{16}$
13. $\frac{21}{500}$	**15.** $\frac{1}{32}$	**17.** $\frac{1}{80}$
19. $\frac{77}{80}$	**21.** $\frac{3}{400}$	**23.** $\frac{7}{80}$
25. $\frac{31}{32}$	**27.** $\frac{61}{64}$	**29.** $\frac{15}{16}$

31. $\frac{13}{16}$ **33.** $\frac{7}{8}$ **35.** $\frac{27}{32}$

37. $\frac{9}{64}$ **39.** $\frac{29}{64}$

Problems 3–3, page 78

1. 0.250   **3.** 0.438   **5.** 0.667

7. 0.013   **9.** 0.172   **11.** 0.875

13. 0.833   **15.** 0.083   **17.** 0.004

19. 0.111   **21.** 0.132   **23.** 0.284

25. 0.120   **27.** 0.362   **29.** 0.083

31. 0.050   **33.** 0.007   **35.** 0.748

37. 0.813 lb   **39.** 0.694 sq ft

Problems 3–4, page 80

1. 0.5   **3.** 0.625   **5.** 0.5

7. 0.6875   **9.** 1.4375   **11.** 4.625

13. 2.3125   **15.** 8.3125   **17.** 0.65625

19. 0.3125   **21.** 2.46875   **23.** 3.3125

25. 6.84375   **27.** 8.125   **29.** 4.1875

31. 10.0   **33.** 0.671875   **35.** 0.59375

37. 2.53125   **39.** 3.84375   **41.** 6.828125

43. 9.0   **45.** 4.578125   **47.** 10.078125

Problems 3–5, page 82

1. 0.625 ft   **3.** 0.703 ft   **5.** 0.073 ft

7. 5.979 ft   **9.** 0.188 ft   **11.** 6.302 ft

13. 2.302 ft   **15.** 2.120 ft   **17.** 1.771 ft

19. 3.130 ft   **21.** 18 ft $3\frac{5}{16}$ in.   **23.** 3 ft $3\frac{7}{64}$ in.

25. 6 ft $1\frac{17}{64}$ in.   **27.** 14 ft $4\frac{51}{64}$ in.   **29.** 4 ft $5\frac{11}{64}$ in.

31. 1 ft $2\frac{7}{64}$ in.   **33.** 18 ft $1\frac{19}{32}$ in.   **35.** 4 ft $8\frac{17}{64}$ in.

37. 19 ft $6\frac{35}{64}$ in.   **39.** 16 ft $8\frac{1}{64}$ in.   **41.** $10\frac{59}{64}$ in.

43. $10\frac{3}{8}$ in.   **45.** $2\frac{39}{64}$ in.   **47.** $2\frac{15}{64}$ in.

49. $1\frac{3}{4}$ in.   **51.** $9\frac{27}{32}$ in.   **53.** $22\frac{59}{64}$ in.

55. $14\frac{49}{64}$ in.   **57.** $36\frac{1}{4}$ in.   **59.** $28\frac{5}{16}$ in.

Problems 3–6, page 85

1. 8.6   **3.** 3.6   **5.** 6.075

7. 28.99   **9.** 0.875   **11.** 284.2

13. 1,636.316   **15.** 1,035.49644   **17.** 3.46

19. 3.674   **21.** 3.35725   **23.** 8.9498

25. 6.0211   **27.** 8.2395   **29.** 9.9113

31. 598.704   **33.** 7.784   **35.** 7.7964

37. $30.51   **39.** $2,865.42

Problems 3–7, page 88

1. 391.75   **3.** 42.125   **5.** 0.0904

7. 2.44725   **9.** 0.2618   **11.** 1.9193

13. 3.802	**15.** 10.558	**17.** 2.299
19. 1.695	**21.** 2.089	**23.** 4.765
25. 2.6555	**27.** 0.72425	

Problems 3–8, page 92

1. 2.5	**3.** 50	**5.** 3.125
7. 18.75	**9.** 1,525	**11.** 6,375
13. 0.125	**15.** 0.25	**17.** 0.75
19. 0.72	**21.** 0.3477	**23.** 0.09375
25. 0.2493	**27.** 3.61875	**29.** 0.001922
31. 4.2768	**33.** 5.368848	**35.** 3.796875
37. 4,625.280	**39.** 0.045	**41.** 0.034456
43. 1,073.088	**45.** 2.9580	**47.** 5.64852
49. 19.635	**51.** 0.348; 0.102; 0.094; 3.250; 0.249	**53.** 4.2768; 30.6250; 5.3688; 3.9571; 3.7969
55. 0.16275	**57.** 2.408333	**59.** 7.984833
61. 1,800	**63.** 68.6	**65.** 1,370
67. 6.3		

Problems 3–9, page 96

1. 2.500	**3.** 0.044	**5.** 6.250
7. 0.938	**9.** 0.256	**11.** 0.625
13. 0.063	**15.** 0.625	**17.** 6.250
19. 31.250	**21.** 3.573	**23.** 587.302
25. 4.968	**27.** 4.647	**29.** 2.596
31. 907.000	**33.** 0.075	**35.** 19.753
37. 806.875	**39.** 0.000	**41.** 0.001
43. 0.049	**45.** 1.273	**47.** 0.044
49. 0.166	**51.** 0.084	**53.** 11.877
55. 0.130	**57.** 0.128	**59.** 0.012

Problems 3–10, page 106

Decimal	mm		Decimal	mm
1. 0.03125	0.7938		**3.** 0.09375	2.3812
5. 0.15625	3.9688		**7.** 0.21875	5.5563
9. 0.28125	7.1438		**11.** 0.34375	8.7313
13. 0.40625	10.3188		**15.** 0.46875	11.9063
17. 0.53125	13.4938		**19.** 0.59375	15.0812
21. 0.65625	16.6688		**23.** 0.71875	18.2562
25. 0.78125	19.8438		**27.** 0.84375	21.4312
29. 0.90625	23.0188		**31.** 0.96875	24.6063

	Decimal Fraction	Nearest 16th	Nearest 32nd	Nearest 64th
33.	0.9864	15	31	62
35.	0.5925	9	19	38
37.	0.1183	2	4	8
39.	0.0477	1	2	3
41.	0.4239	7	14	27

	mm	in.	Nearest 32nd
43.	1	0.03937	1
45.	3	0.11811	4
47.	5	0.19685	6
49.	7	0.27559	9
51.	9	0.35433	11
53.	11	0.43307	14
55.	13	0.51181	16
57.	15	0.59055	19
59.	17	0.66929	21
61.	19	0.74803	24

63. 2,997.064   **65.** 9,530.488   **67.** 0.104

69. 13,676.030   **71.** 2,200,914.484   **73.** 0.698 lb/cu in.

75. 9,072 kilograms   **77.** 86,400   **79.** 500 sec

81. 268,083,200,000 cu mi

Miscellaneous Problems in Decimal Fractions, page 109

1. 0.260 lb cast iron; 0.278 lb wrought iron; 0.284 lb steel   **3.** 0.440 cu ft   **5.** 0.138 in.

7. 88.2 lb   **9.** 7.38 in.   **11.** 0.3084 in.

13. 6,768 gal   **15.** $18.63   **17.** $490.40

19. $870   **21.** 42.73 ft/min   **23.** 0.2215 in.

25. 0.3295 in.   **27.** 0.179 in.   **29.** 0.1295 in.

31. 0.0139 in.   **33.** $2\frac{3}{32}$ in.   **35.** 14.79

37. 37.48 in.   **39.** 3.85 mi

Chapter 4
Problems 4–2, page 120

1. 5   **3.** 500   **5.** 50

7. 100   **9.** 7.5   **11.** 2,500

13. 11.25   **15.** 4.35   **17.** 51

19. 0.5   **21.** 17.2 lb copper; 412.9 lb tin; 34.9 lb antimony   **23.** $237.99

25. 257.04 lb copper;
172.37 lb zinc;
2.59 lb lead

27. $10,800 for labor
and materials;
$900 for
supervision;
$3,300 for profit

29. 193.75 lb lead;
16.25 lb tin;
40 lb antimony

31. $39.90

33. $1,289.70

35. $953.97

37. $78.75

39. $584.57

Problems 4–3, page 124

1. 50%

3. 25%

5. 25%

7. 20%

9. 33.33%

11. 50%

13. 16.39%

15. 59.09%

17. 45.45%

19. 40%

21. 85%

23. 13.04%

25. 40% lead;
60% tin

27. 5.21%

29. 3.94%

31. 69.14% lathe;
12.34% milling
machine;
18.52% planer

33. yes

35. 0.91%

37. 0.71%

39. 0.80%

Problems 4–4, page 127

1. 16

3. 20

5. 150

7. 71.25

9. 220

11. 20

13. 10.63

15. 13.64

17. 6.81

19. 6.82

21. 4

23. 2.67

25. 58.68

27. $142.41

29. $4.00

31. 8.81 lb

33. 12.5 hp

35. $707.20

Problems 4–5, page 132

1. $14,940.70

3. 5

5. 121 hits

7. $11,300

9. 140,743,680 sq mi

Problems 4–6, page 133

1. 5.2%

3. 8.5%

5. 23%

7. 13%

9. 158.5% increase

Problems 4–7, page 135

1. 2,800 tons

3. 18.5014 in.

5. 53.57

7. 20 hp

9. 1,286,206

Chapter 5
Problems 5–2, page 142

1. 4 to 1

3. 6 to 5

5. 5 to 7

7. 10 to 13

9. 7 to 3

11. 2 to 1

13. 48 to 41

15. 1 to 11

17. 5 to 1; 1 to 2;
2 to 3; 14 to 9;
3 to 2; 1 to 4;
1 to 5

19. 3 to 1	**21.** 3 to 32	**23.** 48 to 13
25. 9 to 8	**27.** $15 to $25	**29.** 5 to 98
31. 5 to 16	**33.** 603 to 80; gold and aluminum	

Problems 5–3, page 147

1. 12	**3.** 5	**5.** 36
7. 2	**9.** 20	**11.** 1.5
13. $2.23	**15.** 8.07 in.	**17.** 40.5 lb
19. 16 in.	**21.** 0.765 in.	**23.** 1.232 in.
25. 63.62 in.	**27.** 21.72 hp	

Problems 5–4, page 151

1. 0.318 lb/cu in.	**3.** 8,355 lb	**5.** 0.607 g/cu cm
7. $20.40	**9.** 344.1 lb	**11.** 10, 4.34; 50, 21.7
		20, 8.68; 100, 43.4
		30, 13.02; 200, 86.8
		40, 17.36; 300, 130.2

Problems 5–5, page 153

| **1.** 12 lb | **3.** 195 | **5.** 0.213 in. |
| **7.** 1.64333 in. | **9.** 396 lb | **11.** 148 rejects |

Problems 5–6, page 156

| **1.** 29.3° | **3.** 2.0156″ | **5.** 4.6 fish, 2.3 lb each |

Chapter 6
Problems 6–2, page 162

	Starting Number	**Add *n* (Go up *n*)**	**Arrive at**
1.	1	3	4
3.	5	3	8
5.	⁻5	3	⁻2
7.	3	1	2
9.	⁻1	2	⁻3

| **11.** ⁻8° | **13.** 4 | **15.** ⁻13° |

Problems 6–3, page 165

1. 7	**3.** 85	**5.** 1,010
7. $3 \times 3 \times 3$	**9.** $17 \times x \times y \times z$	**11.** second
13. ninth	**15.** fifth	**17.** $2x^2$
19. $m^3 n^2$	**21.** $a \times a \times a \times a \times a$	**23.** $7 \times x \times x \times y \times y$
25. $3 \times x \times x \times x \times x \times x \times x \times y \times y \times z \times z \times z$	**27.** $5a$	**29.** $7a$
31. 5	**33.** x	**35.** 5

Problems 6–4, page 167

1. 20	**3.** 5	**5.** 125
7. 4,800	**9.** 88	**11.** 94
13. 2	**15.** 2	**17.** 10
19. 12	**21.** $3\frac{3}{4}$	**23.** $\frac{3}{4}$
25. 62	**27.** $9\frac{1}{4}$	**29.** $42\frac{7}{8}$

Problems 6–5, page 170

1. $19a$	**3.** $^{-}23abc$	**5.** $31by$
7. $6b$	**9.** ax^2	**11.** $^{-}11ab$
13. $^{-}9a - 7c$	**15.** $^{-}7b - 12d$	**17.** $7r^2 - 9st$

Problems 6–6, page 172

1. $6ax$	**3.** $^{-}6a^2b^2$	**5.** $12x^2y^2$
7. $^{-}4ay$	**9.** $56xy$	**11.** $^{-}12z - 26y$
13. $^{-}2a - 5b$	**15.** $21x + 24y$	**17.** $13x^2 - 5xy^2$
19. $^{-}17a^2 - a$		

Problems 6–7, page 174

1. 19	**3.** 5	**5.** $2a + b$
7. $3a - b$	**9.** $4a + b$	**11.** 1
13. $6ab + 2a^2$	**15.** $2b$	**17.** $^{-}x^2 - y$
19. $^{-}m + 3n$		

Problems 6–8, page 177

1. a^2	**3.** a^6	**5.** $15b^3$
7. $^{-}12a^3$	**9.** $30a^4$	**11.** $^{-}6x^6$
13. $6a^3c - 9a^2bc - 3a^2c^2$	**15.** $15a^2 - 38ab + 24b^2$	**17.** $8a^2 + 22ab + 15b^2$
19. $4x^2 - 9y^2$	**21.** $49a^2 - 42ab + 9b^2$	**23.** $9x^2 + 24xy + 16y^2$

Problems 6–9, page 180

1. $3x^2$	**3.** $2x^2y^3$	**5.** $^{-}8x^2$
7. $^{-}9x^2y$	**9.** $12ab + 9a^2b^2$	**11.** $^{-}7b^2y^3 - 6b^3y^4$
13. $x + 3$	**15.** $x - y$	**17.** $x - y$
19. 1	**21.** $4b + 7$	

Problems 6–10, page 185

1. $x = 10$	**3.** $x = 5$	**5.** $x = b/a$
7. $x = (d - b)/(a - c)$	**9.** $x = 10$	**11.** $x = (ab + ac)/5$
13. $x = {}^{\pm}6$	**15.** $h = V/lw$	**17.** $I = E/R; R = E/I$
19. $L = (33,000 \times \text{hp})/PAN$	**21.** $L = R \times CM/10.4;$ $CM = 10.4 \times L/R$	**23.** $R = \sqrt{A/\pi}$
25. $b = ad/c; c = ad/b$	**27.** $F = 9C/5 + 32$	**29.** $b = 6M/h^2;$ $h = \sqrt{6M/b}$
31. $r = (1/2)\sqrt{S/\pi}$	**33.** $N = OD/P_c - 2$	**35.** $R = \sqrt[3]{3V/4\pi}$

37. $d = \sqrt{A/0.7854}$

39. Eff pwr =
App pwr × pwr factor;
App pwr =
Eff pwr/pwr factor

41. $L = BHP \times 33{,}000/2\pi NW$;
$N = BHP \times 33{,}000/2\pi LW$;
$W = BHP \times 33{,}000/2\pi LN$

43. Amperes = Watts/
Volts;
Volts = Watts/
Amperes

45. $S = A/0.866$

47. $s = f/1.732$

49. $s = \sqrt{A/0.433}$

Chapter 7
Problems 7–1, page 193

	Length	Width	Estimated Area	Computed Area
1.	5′	4′	20 sq ft	20 sq ft
3.	5′3″	4′9″	25 sq ft	24.9 sq ft
5.	8′9″	6′6″	54 sq ft	56.9 sq ft
7.	7′4″	5′3″	35 sq ft	38.5 sq ft
9.	5′8″	4′4″	24 sq ft	24.6 sq ft
11.	6′9″	3′3″	21 sq ft	21.9 sq ft

13. 37.5″

15. 3,564 sq ft

17. 10.11 sq ft

19. 1,880.67 sq yd

21. 1,120 pupils

23. 1.41 sq ft

25. 8′

27. 3′6″;
3′6″

29. $10'2\frac{1}{2}''$;
1′9″

31. 360 sq ft

33. 130.6 sq ft

35. 453.12 sq ft

37. $425.37

39. $336.00

Problems 7–2, page 198

	Length	Width	Perimeter
1.	15′	10′	50′
3.	10′	6′8″	33′4″
5.	9′9″	5′4″	30′2″
7.	$2'6\frac{1}{2}''$	3′7″	12′3″

9. 12.4 yd

11. 900 ft

Problems 7–3, page 199

	Width	Length	Area
1.	5′	3′	15 sq ft
3.	4′8″	6′6″	$30\frac{1}{3}$ sq ft
5.	4′6″	3′4″	15 sq ft
7.	3′6″	2′9″	$9\frac{5}{8}$ sq ft
9.	8′6″	12′8″	$107\frac{2}{3}$ sq ft

11. 15′

13. 125′

15. 10′6″

17. 124′6″

19. 4′4″

21. 4 times

23. 4 times

25. $5\frac{4}{9}$ times

27. $\frac{9}{16}$ times

29. quadrupled

Problems 7–4, page 202

1. 36

3. 144

5. 3,721

7. 15,625

9. 3,600

11. 7

13. 9

15. 12

17. 100

19. 70

21. 36 sq in.

23. 40 ft

Problems 7–5, page 204

1. 85

3. 94

5. 45

7. 65

9. 63

11. 35

13. 37

15. 73

17. 34

19. 23

Problems 7–6, page 206

1. 4.36

3. 60.41

5. 1.414

7. 29.09

9. 167.76

11. 26.717

13. 0.522

15. 5.448

17. 1.904

19. 2.52

Problems 7–7, page 208

1. 15

3. 35

5. 16

7. 17

9. 18

11. 1.2

13. 4.43

15. 6.19

17. 68

19. 12.7

21. 2.934

23. 31.62 in.

25. 2.236 in.

27. 5.92 ft

Problems 7–8, page 211

1. 6.08

3. 13.748

5. 21.401

7. 63.29

9. 1.661

11. 9.034

13. 10.455

15. 0.653

17. 0.299

19. 0.0249

21. 3.122

23. 1.010

25. 0.645

27. 0.6508

29. 0.3391

Problems 7–9, page 216

	a	*b*	*c*
1.	5″	12″	13″
3.	7″	24″	25″
5.	1′4″	2′6″	2′10″
7.	1′2″	4′	4′2″
9.	10″	2′0″	2′2″
11.	$10\frac{1}{2}″$	3′0″	$3′1\frac{1}{2}″$

13. 1.41 in.; 2.82 in.;   **15.** $6\frac{1}{8}$ in.   **17.** $11'\frac{35''}{64}$
4.23 in.

19. $27'7\frac{33''}{64}$

Problems 7–10, page 221

	Side a, Altitude	Side b, Base	Side c, Hypotenuse	Area
1.	5″	11″	12.1″	27.5 sq in.
3.	$3\frac{1}{2}''$	$5\frac{1}{2}''$	6.5″	9.6 sq in.
5.	8″	1′3″	1′5″	60 sq in.
7.	$1'4\frac{1}{2}''$	$2'1\frac{1}{4}''$	30.2″	208.3 sq in.
9.	3′	$10\frac{1}{2}''$	$3'1\frac{1}{2}''$	189 sq in.
11.	1.732″	1.000″	2.000″	0.866 sq in.
13.	1″	1.4″	1.7″	0.7 sq in.
15.	5.1″	8.66″	10″	22.1 sq in.
17.	9″	3.75″	$9\frac{3}{4}''$	16.88 sq in.
19.	3′	1′3″	3.25′	$1\frac{7}{8}$ sq ft

Problems 7–11, page 222

	Altitude to the Base	Area
1.	3″	12 sq in.
3.	24″	168 sq in.
5.	1.7′	1.7 sq ft
7.	27.8″	347.5 sq in.
9.	22″	211.2 sq in.
11.	8.0″	153.6 sq in.
13.	7.03′	18.45 sq ft
15.	7.9′	88.88 sq ft
17.	17.32′	381.04 sq ft
19.	13 yd	85.8 sq yd

Problems 7–12, page 224

1. 37.5 sq in.   **3.** 16.2 sq in.   **5.** 187.5 sq in.

7. 42.43 sq in.   **9.** 3.73 sq in.

Test on Areas of Triangles, page 230

1. 2.6 sq ft   **3.** 74.7 sq ft   **5.** 1,397.5 sq ft

7. 438.3 sq in.   **9.** 45.25 sq in.   **11.** 3.05 sq in.

Problems 7–14, page 235

	Angle A	Angle B	Angle C
1.	30°	60°	90°
3.	20°	80°	80°
5.	27°10′	52°20′	100°30′
7.	40°45′	98°30′	40°45′
9.	75°50′	25°28′	78°42′
11.	40°27′10″	83°20′30″	56°12′20″
13.	33°50′9″	129°49′18″	16°20′33″
15.	8°29′40″	100°59′35″	70°30′45″

NOTE: 10°37′45″ means 10 degrees 37 min 45 sec.

Chapter 8
Problems 8–2, page 246

	Side	Altitude	Area
1.	10″	8.66″	43.30 sq in.
3.	$4\frac{1}{2}″$	3.90″	8.77 sq in.
5.	6.25 cm	5.41 cm	16.91 cm²
7.	9.75 ft	8.444 ft	41.162 sq ft
9.	5.78″	5″	14.43 sq in.
11.	4.04″	$3\frac{1}{2}″$	7.07 sq in.
13.	8.95″	7.75″	34.66 sq in.
15.	9.53 cm	$8\frac{1}{4}$ cm	39.32 cm²

Problems 8–3, page 249

	Side	Diagonal	Area
1.	7 m	9.90 m	49 m²
3.	$2\frac{1}{2}″$	3.53″	6.25 sq in.
5.	3.82 cm	5.40 cm	14.59 cm²
7.	20.1 m	28.42 m	404.01 m²
9.	6.36 cm	9 cm	40.5 cm²
11.	3.01″	$4\frac{1}{4}″$	9.03 sq in.
13.	2.16″	3.05″	4.65 sq in.
15.	100 ft	70.7 ft	4,998 sq ft

Problems 8–4, page 252

Side	Distance across Flats	Diagonal	Area
1. 1″	1.73″	2″	2.60 sq in.
3. 2½″	4.33″	5″	16.24 sq in.
5. 7.25 m	12.56 m	14.5 m	136.56 m²
7. 3.17″	5½″	6.35″	26.20 sq in.
9. 6.12 cm	10.6 cm	12.243 cm	97.3 cm²
11. 3 cm	5.20 cm	6 cm	23.40 cm²
13. 2.62″	4.55″	5.25″	17.92 sq in.
15. 1.1375 m	1.97 m	2.275 m	3.362 m²

Problems 8–5, page 255

1. 18.92 sq in.   **3.** 14.48 in.   **5.** 57.27 sq in.

Side	Distance across Flats	Diagonal	Area
7. 5½″	13.28″	14.37″	146.05 sq in.
9. 1.04″	2½″	2.71″	5.18 sq in.
11. 2.01″	4.85″	5¼″	19.49 sq in.
13. 4.75″	11.47″	12.41″	108.93 sq in.

Practice Problems on the Use of the Table of Constants, page 255

15. 13.02 sq in.   **17.** 2.44″   **19.** 7.96 sq in.

21. 9.53″   **23.** 28.13 sq in.   **25.** 6.53″

27. 3.09 sq in.   **29.** 2.47″   **31.** 1.95″

33. 0.73″   **35.** 1.12 m

Problems 8–8, page 264

Average Width	Area
1. 12 cm	96 cm²
3. 7.25″	54.375 sq in.
5. 6.625″	41.075 sq in.
7. 20.25″	303.75 sq in.
9. 6.045″	36.3909 sq in.

11. 195.5 sq in.   **13.** 132.6 sq in.   **15.** 441 sq ft

17. 3.11 sq ft   **19.** 24.66 in.; 4.97 sq ft   **21.** 15.20 in.; 2.11 sq ft

Problems 8–9, page 267

1. 5.3 sq ft   **3.** 331.4 sq ft   **5.** 342.8 sq ft

7. 526.6 ft   **9.** 140.4 sq ft

Problems 8–11, page 269

1. 5″   **3.** $l = 8''$;   **5.** $1\frac{3}{8}''$
                          $w = 7''$;
                          $h = 4\frac{1}{2}''$

7. **(a)** $13\frac{1}{2}$ in.   **9.** $\frac{1''}{4} = 1'$
    (b) $4\frac{7}{8}$ in.
    (c) $31\frac{1}{8}$ in.
    (d) $8\frac{13}{16}$ in.
    (e) $10\frac{1}{2}$ in.

Problems 8–12, page 271

1. $l = 3\frac{7}{24}$ in.   **3.** Beginning at left   **5.** Long diameter =
    $w = 1\frac{3}{4}$ in.   edge: 2′0″, 1′8″, 1′6″,   4 in.; short
                                         4′6″, 1′6″, 1′8″, 2′0″,   diameter = $2\frac{9}{16}$ in.
                                         8′0″

Problems 8–13, page 274

1. 2.86 sq ft   **3.** 50.62 sq in.   **5.** 24.5 sq ft

Problems 8–15, page 276

1. 43.98″   **3.** 6.28″   **5.** 6.68″

7. 10.21″   **9.** 424.12′   **11.** 62.832 cm

13. 10.2102′   **15.** $13'2\frac{1}{16}''$ **17.** $7'3\frac{15}{16}''$

19. $33\frac{1}{2}$ ft

Problems 8–17, page 279

1. 3.82 m   **3.** 1.19″   **5.** 3.50″

7. 6.50″   **9.** 2.72″   **11.** 210.08′

13. 28.01″   **15.** 16.55′

Problems 8–18, page 281

	Radius	Estimated Area	Computed Area
1.	1″	3 sq in.	3.14 sq in.
3.	32.5″	3,000 sq in.	3,318.32 sq in.
5.	$1\frac{9}{16}''$	7 sq in.	7.67 sq in.
7.	0.925″	3 sq in.	2.69 sq in.
9.	$2'10\frac{1}{2}''$	27 sq ft	25.97 sq ft

11. 3.31 lb   **13.** 11.04 sq in.   **15.** 102 sq in.

Problems 8–19, page 283

1. 19.64 sq in.   **3.** 4,071.51 sq ft   **5.** 63.62 sq in.

7. 5.47 sq in.   **9.** 30.68 sq ft   **11.** 65.36 sq ft

Problems 8–21, page 286

1. 11.28 m	**3.** 2.72″	**5.** 6.24″
7. 6.257′	**9.** 10.3″	**11.** 3.16″
13. 0.80″	**15.** 12.62″	

Problems 8–22, page 288

1. 10.09″	**3.** 4.31″	**5.** 7.30″
7. 6.49 cm	**9.** 0.399″	**11.** 1.78″
13. 3.5″	**15.** 12″	

Problems 8–23, page 290

1. 7.21″	**3.** 3.67″	**5.** 7.91″
7. 2.24″	**9.** 13	

Problems 8–25, page 293

1. 56.55 sq in.	**3.** 786.19 sq ft	**5.** $9,882.78
7. 16,709.40 sq ft	**9.** 117.81 sq in.	

Problems 8–27, page 296

1. **(a)** 4.71 in.	**3.** **(a)** 36.9 ft	**5.** **(a)** 10.35 in.
(b) 10.60 sq in.	**(b)** 415.3 sq ft	**(b)** 84.11 sq in.
7. **(a)** 9.42 in.	**9.** **(a)** 2.36 in.	**11.** **(a)** 44.55 cm
(b) 56.55 sq in.	**(b)** 10.60 sq in.	**(b)** 334.12 cm^2
13. **(a)** 45.76 in.	**15.** **(a)** 31.77 ft	
(b) 260.27 sq in. or 1.8 sq ft	**(b)** 82.07 sq ft	

Problems 8–28, page 300

1. 21.13 sq in.	**3.** 1.02 sq in.	**5.** 9.62 sq in.
7. 10.29 sq in.	**9.** 0.65″	**11.** 4.33′
13. 2.82″	**15.** 0.22″	

Problems 8–29, page 302

1. 9.83 sq in.	**3.** 14.10 sq in.	**5.** 7.53 cm^2
7. 0.64 sq in.		

Problems 8–30, page 305

	a	*b*	Area	Perimeter
1.	4″	2.5″	31.4 sq in.	20.96″
3.	3.25″	2.125″	21.70 sq in.	17.25″
5.	2.1″	1.3″	8.58 sq in.	10.97″
7.	32.5′	21.5′	2,196.19 sq ft	173.13′
9.	8.25″	7.4″	191.80 sq in.	49.24″

11. 27.49 sq ft

Chapter 9
Problems 9–3, page 313

1. 3,456 cu in.	**3.** 140 cu in.	**5.** 301.22 cu in.
7. 4,603.5 cu in.	**9.** 84.2 gal	**11.** 216.5 cu in.
13. 328.5 cu in.	**15.** 162.4 cu in.	**17.** 221.9 cu in.
19. 4,888.35 cu in.	**21.** 59.6 cu ft	**23.** 13,242.7 gal
25. 4,544.9 gal	**27.** 430 lb	**29.** 10,227 lb

Problems 9–4, page 317

1. 5′0″	**3.** 6.00″	**5.** 5′8.67″
7. 7.27″	**9.** 4′7″	

Problems 9–5, page 319

1. 262.50 sq in.	**3.** 16.20″	**5.** 5.42″
7. 7′3.17″	**9.** 12.13″	

Problems 9–6, page 320

1. 4.71 sq ft	**3.** 44.44 sq ft	**5.** 9.33 sq ft
7. 63 sq in.	**9.** 11.04 sq ft	**11.** $2,851.35
13. 128.47 sq ft	**15.** 397.83 sq ft	

Problems 9–8, page 323

1. 1.23 cu ft	**3.** 1.81 cu ft	**5.** 216.5 cu in.
7. 4.21 cu ft	**9.** 84.46 cu in.	**11.** 6.12 cu ft
13. 246.5 cu yd	**15.** 11.9 lb	

Problems 9–9, page 326

1. 81.6 sq in.	**3.** 123.2 sq in.	**5.** 165.8 in.
7. 15.87 ft	**9.** 139.8 sq ft	

Problems 9–10, page 329

1. 10.20 cu ft	**3.** 164.9 cu in.	**5.** 3.04 gal
7. 2,752.8 cu ft	**9.** 24.15 cu yd	

Problems 9–11, page 331

1. 4.92″	**3.** 5.92″	**5.** 8.70″
7. 20.2″	**9.** 14.9″	

Problems 9–12, page 333

1. 1,541.9 sq ft	**3.** 22.21 sq ft	**5.** 3.99 sq ft

Problems 9–13, page 335

1. 904.8 cu in.	**3.** 4.18 cu in.	**5.** 1,767.2 cu in.
7. 0.221 cu in.	**9.** 44,602.34 cu ft	**11.** 348.46 cu in.
13. 1,072.33 cu ft	**15.** 8.18 cu in.	**17.** 1,596.3 cu in.
19. 8,181.3 cu ft	**21.** 0.350 cu in.	**23.** 303.1 cu in.
25. 13.81 cu in.		

Problems 9–15, page 338

1. 3.14 sq in.   **3.** 7.06 sq in.   **5.** 31,416 sq in.
7. 6.15 sq in.   **9.** 1,520.5 sq ft   **11.** 19.63 sq ft
13. 4,536.5 sq in.   **15.** 5.93 sq in.   **17.** 490.9 sq in.
19. 9,331.3 sq in.

Problems 9–16, page 340

1. 5.66 cu in.   **3.** 11.07 cu in.   **5.** 7.94 cu in.
7. 4.63 cu in.   **9.** 10.01 cu in.

Problems 9–17, page 343

1. 5.40 cu in.   **3.** 5.30 cu ft   **5.** 4.19 gal

Problems 9–18, page 346

1. 207 lb   **3.** 412 lb   **5.** 435 lb
7. 4.01 lb   **9.** 5.04 lb

Problems 9–19, page 347

1. 94.5 lb   **3.** 78 lb   **5. (a)** 1 to 20.48
                                                  (b) 1 to 9.18
                                                  (c) 1 to 10.38
                                                  (d) 1 to 30.1
                                                  (e) 1 to 3.2

7. 89.6 lb   **9.** 134 to 73

Problems 9–20, page 350

1. 8 fbm   **3.** $9\frac{1}{3}$ fbm   **5.** 10 fbm
7. 6 fbm   **9.** $26\frac{2}{3}$ fbm   **11.** 5 fbm
13. $10\frac{2}{3}$ fbm   **15.** 8 fbm   **17.** $13\frac{1}{3}$ fbm
19. 10 fbm   **21.** 4,779 fbm   **23.** 3,960 fbm
25. 138,096 fbm   **27.** 816 fbm   **29.** 741 fbm

Problems 9–21, page 352

1. 495 fbm   **3.** 781 fbm   **5.** 1,027 fbm
7. 344 fbm   **9.** 3,704 fbm

Chapter 10
Problems 10–1, page 361

1. (a) 1 mm   **3.** No   **5. (c)** 27 m
7. (a) 25 mm   **9.** $8\frac{1}{3}$ min   **11.** 9 mm
13. 23 mm   **15.** 35 mm   **17.** 32 cm
19. 3,500 mm or 3.5 m

Problems 10–2, page 364

1. 0.0050 m   **3.** 630 m   **5.** 0.276 m
7. 47.635 m   **9.** 50 m   **11.** 172.8 cm
13. 78.5 cm   **15.** 21.7 mm   **17.** 21.2 cm
19. 25.4 mm

Problems 10–3, page 367

1. 1.765 dm^2	**3.** 0.862 cm^2	**5.** 7,520 cm^2
7. 5,000 mm^2	**9.** 56 ha	**11.** 16.25 m^2
13. 181.5 cm^2	**15.** 33.33 m	**17.** 21.6 cm^2
19. 150 mm^2	**21.** 1,161.5 cm^2	**23.** 0.274 km^2
25. 46.98 mm^2	**27.** 32.19 m^2	**29.** 56.59 cm^2

Problems 10–4, page 372

1. 0.0035 m^3	**3.** 0.487 cm^3	**5.** 560 l
7. 0.000547345 m^3	**9.** 6.7 mm^3	**11.** 24,192 cm^3
13. 883.6 cm^3	**15.** 35.8 m^3	**17.** 273.8 m^3
19. 116.2 l	**21.** 3,455.8 l	**23.** 204 mm
25. 21.2 mm	**27.** 730 cm^2	**29.** 86 m^3

Problems 10–5, page 376

1. 42 g	**3.** 4.58 g	**5.** 0.078 g
7. 0.087 g	**9.** 90 g	**11.** 30.05 g
13. 7 g	**15.** 933.25 g	**17.** 675.75 kg
19. 2.63 kg		

Problems 10–6, page 378

	mm	cm	m
1.	101.6	10.16	0.10
3.	203.2	20.32	0.20
5.	114.3	11.43	0.11
7.	0.79	0.08	0.0
9.	692.15	69.22	0.69

11. 134.3 cm	**13.** 3.2 mm	**15.** 7.98 cm
17. 10.8 cm	**19.** 4.2 m	

Problems 10–7, page 382

	Length (cm)	Width (cm)	Area
1.	7.6	5.1	38.7 cm^2
3.	8.9	7.6	67.7 cm^2
5.	8.3	5.1	41.9 cm^2
7.	35.6	22.9	812.9 cm^2
9.	41.3	38.1	1,572.6 cm^2

11. 241.9 cm^2	**13.** 73.6 m^2	**15.** 1,040 children
17. 31 m^2	**19.** 42 m^2	

Problems 10–8, page 384

1. 1,660 m²	**3.** 382.00 m²	**5.** 0.81 m²
7. 258 cm²	**9.** 191.9 cm²	**11.** 0.78 ha
13. 87.81 ares	**15.** 2,023 ha	

Problems 10–9, page 386

	Length	Width	Height	Volume
1.	10.16 cm	7.62 cm	12.70 cm	983.22 cm³
3.	6.35 cm	6.35 cm	7.62 cm	307.26 cm³
5.	30.48 cm	20.32 cm	22.86 cm	14,158.40 cm³

	Diameter	Height	Volume
7.	10.16 cm	7.62 cm	617.78 cm³
9.	8.89 cm	6.35 cm	394.16 cm³
11.	30.48 cm	20.32 cm	14,826.7 cm³

13. 23,662.8 cm³	**15.** 141.5 *l*	**17.** 9.9 m³
19. 4.5 m³		

Problems 10–10, page 389

1. 3.97 kg	**3.** 5.45 kg	**5.** 624.3 g
7. 2 kg	**9.** 249.7 metric tons	**11.** varies

Problems 10–11, page 391

	°F	°C
1.	115°	46.1°
3.	210°	98.9°
5.	0°	−17.8°
7.	113°	45°
9.	194°	90°

11. 40.6°C	**13.** Probably a light jacket	**15.** −40°C

Chapter 12
Problems 12–1, page 422

1. 0.008 in.	**3.** 0.292 in.	**5.** 0.130 in.
7. 0.047 in.	**9.** 0.152 in.	

Thimble Is Between	Coinciding Line on Thimble
11. 0.100 and 0.125	9
13. 0.200 and 0.225	18
15. 0.400 and 0.425	1
17. On 0.250	0
19. 0.650 and 0.675	6

Problems 12–2, page 424

1. 0.1413 in.	**3.** 0.0092 in.	**5.** 0.0397 in.
7. 0.6342 in.	**9.** 0.8576 in.	

Thimble Is Between	Gage Line Is Between Thimble Lines	Coinciding Vernier Line
11. 0.975 and 1.000	6 and 7	5
13. 0.100 and 0.125	4 and 5	3
15. 0.200 and 0.225	10 and 11	8
17. 0.000 and 0.025	7 and 8	1
19. 0.175 and 0.200	12 and 13	5

Problems 12–3, page 426

1. 0.141 in.	**3.** 0.268 in.	**5.** 0.981 in.
7. 0.113 in.	**9.** 0.099 in.	

Index Is Between Lines	Coinciding Vernier Line
11. 0.0 and 0.025	20
13. 0.150 and 0.175	23
15. 0.900 and 0.925	8
17. On 0.125	0
19. On 0.750	0

Problems 12–5, page 428

1. 12°15′	**3.** 11°57′	**5.** 27°06′
7. 125°57′	**9.** 3°51′	

Index Is Between	Coinciding Vernier Line
11. 27°30′ and 28°	24
13. 45°30′ and 46°	15
15. 52°00′ and 52°30′	3
17. 9°00′ and 9°30′	27
19. 21°30′ and 22°00′	3

Problems 12–7, page 430

1. 25.62

3. 50.01

5. 54.58

7. 31.16

9. 73.34

Chapter 14
Problems 14–3, page 457

1. 0.3420

3. 0.4226

5. 0.5000

7. 1.000

9. 1.732

11. 0.7642

13. 0.7133

15. 0.5544

17. 0.3772

19. 0.7951

21. 0.5038

23. 0.8711

25. 0.8133

27. 0.3789

29. 2.186

31. 0.1616

33. 0.9684

35. 0.6706

37. 0.9305

39. 0.7055

Problems 14–5, page 459

1. 31°30′

3. 38°30′

5. 27°30′

7. 45°

9. 48°10′

11. 16°30′

13. 57°

15. 86°

17. 50°

19. 41°20′

21. 28°06′

23. 8°20′

25. 30°24′

27. 42°45′

29. 58°46′

31. 44°58′

33. 47°56′

35. 45°44′

37. 39°35′

39. 78°33′

Problems 14–7, page 466

1. $B = 66°$;
$b = 15.72$ in.;
$c = 17.21$ in.

3. $B = 19°20′$;
$b = 4$ in.;
$c = 12.08$ in.

5. $B = 82°$;
$a = 6.32$ in.;
$c = 45.45$ in.

7. $A = 23°40′$;
$a = 24.30$ cm;
$b = 55.46$ cm

9. $B = 58°04′$;
$a = 11.48$ m;
$b = 18.42$ m

11. $A = 40°36′$;
$B = 49°24′$;
$c = 11.52$ in.

13. $A = 38°40′$;
$B = 51°20′$;
$c = 20.49$ in.

15. $A = 55°$;
$B = 35°$;
$c = 48.83$ ft

17. $A = 35°32′$;
$B = 54°28′$;
$c = 34.41$ mm

19. $A = 34°22'$;
$B = 55°38'$;
$c = 29.68$ cm

21. $A = 46°57'$;
$B = 43°03'$;
$b = 3.55$ in.

23. $A = 42°06'$;
$B = 47°54'$;
$a = 83.14$ ft

25. $A = 63°22'$;
$B = 26°38'$;
$a = 14.16$ in.

27. $A = 37°40'$
$B = 52°20'$;
$b = 18.52$ cm

29. $A = 40°$;
$B = 50°$;
$b = 53.62$ m

Problems 14–8, page 470

	Base	Sides	Altitude	Base Angles	Vertex Angle
1.	12″	10″	8″	53°08′	73°44′
3.	8″	8″	6.93″	60°	60°
5.	18 m	12.73 m	9 m	45°	90°
7.	27.71 cm	16 cm	8 cm	30°	120°
9.	9.4″	10.8″	9.7″	64°15′	51°30′
11.	18.2″	15.5″	12.6″	54°10′	71°40′
13.	24.5 cm	19.8 cm	15.6 cm	51°52′	76°16′
15.	41.5 cm	32.8 cm	25.4 cm	50°45′	78°30′

Miscellaneous Problems Using Trigonometry, page 470

1. Base = 41.43 in.;
base angles =
34°03′;
apex angle =
111°54′

3. Base angles =
77°45′24″; apex
angle = 24°29′12″;
sides = 10.23 cm

5. Base angles =
70°49′; base =
1.64 in.; altitude =
2.36 in.

7. 3°35′

9. 0.3655 in.

11. 2°17′20″

13. 4.243 in.

15. 33°41′30″

17. 70°28′; 7.96 cm

19. 0.414 in.

21. (a) 0.5742 in.,
(b) 0.3263 in.,
(c) 2.1137 in.,
(d) 1.8417 in.

23. 5.613 in.

25. 3.016 in.

27. Tongue = 3.342 in.;
groove = 1.462 in.

29. (a) 0.2792 in.,
(b) 0.5863 in.

Chapter 15
Problems 15–7, page 484

1. 91,875 lb

3. 2.34″

5. 2.48″

7. 10,938 lb

9. 6.32″ × 6.32″

Problems 15–8, page 487

1. 0.375″

3. 0.5625″

5. 219 lb

Problems 15–9, page 489

1. 72,120 lb

3. 12 rivets

5. 45,000 lb

7. (a) 3 rivets
   (b) 8 rivets
   (c) 5 rivets
   (d) 14 rivets
   (e) 2 rivets

9. (a) 72,184 lb
   (b) 94,248 lb
   (c) 122,720 lb
   (d) 82,504 lb

Chapter 16

Problems 16–1, page 496

1. 500 ft-lb

3. 18.18 hp

5. 20,250 ft-lb

Problems 16–2, page 497

1. 172.6 hp

3. 365.4 hp

5. 426.3 hp

Problems 16–3, page 499

1. 26.3 hp

3. 46.9 hp

5. 2.5 hp

7. 15.6 hp

Problems 16–4, page 500

1. 9.64 bhp

3. 1.52 bhp

5. 1.49 bhp

Problems 16–5, page 502

1. 0.34 hp

3. 0.40 hp

5. 1:2

7. 1.1 kW

9. 704 watts

11. 106.7 hp

Problems 16–6, page 505

1. 6 hp

3. 2.49 hp

5. 77.4%

7. 233 volts

9. 90.9 amperes

11. 83.3%

Chapter 17

Problems 17–3, page 513

1. $\frac{3}{8}$ in.; $\frac{3}{64}$ in.; $\frac{9}{16}$ in.

3. 1.362 in.

5. 0.624 in.

7. 0.431 in.

9. 0.500 in.

11. 0.601 in.

13. 1.15 in; 1.105 in.

15. 0.913 in.

17. 0.904 in.

19. 0.0938 in.

21. 0.708 in.

Problems 17–4, page 521

1. 2°23′; 3°00′; 3°34′

3. 25°03′

5. 1.049 in.

7. 1.469 in.

9. 0.998 in.

11. 43°11′

Problems 17–5, page 526

1. $\frac{7}{16}$ in.

3. 0.065 in.

5. 0.806 in.

7. 0.215 in.

9. 0.003 in.

Problems 17–6, page 529

1. (a) $61°30'$; **(b)** $47°30'$;   **3.** $68°12'$ **5.** $81°47'$; $87°22'$
  (c) $55°00'$, $25°00'$;
  (d) $75°00'$;
  (e) $56°15'$, $50°00'$

Problems 17–7, page 532

1. $\frac{4}{8}$ in.; $\frac{6}{10}$ in.; $\frac{6}{8}$ in.   **3.** Left, $3\frac{7}{16}$ in.;   **5.** $\frac{9}{16}$ in.
                         right, $1\frac{1}{8}$ in.

7. Beyond range   **9.** $\frac{5}{8}$ in.   **11.** $\frac{1}{64}$ in.

Chapter 18
Problems 18–1, page 537

1. 5 revolutions   **3.** 147 rpm   **5.** 64 teeth

Problems 18–3, page 540

1. 60 rpm   **3.** 60 teeth   **5.** 24, 44
7. $87\frac{1}{2}$ rpm

Problems 18–4, page 544

1. 315 rpm   **3.** 42 teeth   **5.** $46\frac{7}{8}$ rpm
7. $166\frac{2}{3}$ rpm   **9.** $A = 32$; $B = 40$;
                        $C = 24$; $D = 48$

Problems 18–6, page 549

1. $61\frac{7}{8}$ rpm   **3.** 463 rpm

Problems 18–7, page 551

1. 48 rpm   **3.** 224 rpm   **5.** $A = 10$ in.; $B = 20$ in.;
                                               $C = 16$ in.; $D = 12$ in.

7. 1,050 rpm

Chapter 19
Problems 19–2, page 560

1. $\frac{2}{7}$ in.   **3.** $\frac{1}{5}$ in.   **5.** $\frac{2}{11}$ in.
7. $\frac{4}{11}$ **9.** $2\frac{1}{2}$ threads   **11.** $3\frac{1}{5}$ threads
13. $2\frac{2}{7}$ threads   **15.** 14 threads

Problems 19–3, page 562

1. $\frac{1}{12}$ in.; $\frac{1}{12}$ **3.** $\frac{1}{4}$ in.   **5.** $\frac{3}{8}$ in.
7. $\frac{3}{16}$ in.   **9.** $\frac{3}{4}$ in.   **11.** $\frac{1}{4}$ in.
13. $\frac{1}{16}$ in.   **15.** $\frac{3}{64}$ in.

Problems 19–5, page 565

1. 0.0433 in.   **3.** 0.0619 in.   **5.** 0.0787 in.
7. 0.1082 in.   **9.** 0.1443 in.   **11.** 0.0456 in.
13. 0.0662 in.   **15.** 0.05773 in.   **17.** 0.03936 in.
19. 0.0289 in.

Problems 19–7, page 566

1. 0.5773 in. **3.** 0.4949 in. **5.** 0.3849 in.
7. 0.2165 in. **9.** 0.1082 in. **11.** 0.866 in.
13. 0.1237 in. **15.** 0.33 in. **17.** 0.1732 in.
19. 0.0787 in. **21.** 0.1634 in. **23.** 0.4182 in.
25. 0.7451 in. **27.** 1.2113 in. **29.** 1.8651 in.
31. 0.2838 in. **33.** 0.572 in. **35.** 3.6713 in.
37. 1.9351 in. **39.** 2.6923 in.

Problems 19–9, page 569

1. $\frac{15}{64}$ in. **3.** $\frac{3}{8}$ in. **5.** $\frac{35}{64}$ in.
7. $\frac{23}{32}$ in. **9.** $1\frac{1}{8}$ in. **11.** $\frac{19}{64}$ in.
13. $1\frac{49}{64}$ in. **15.** $\frac{3}{16}$ in. **17.** $\frac{23}{64}$ in.
19. $\frac{17}{32}$ in. **21.** 0.0307 **23.** 0.0383
25. 0.0472 **27.** 0.0558 **29.** 0.0682
31. 0.1227 **33.** 0.0876 **35.** 0.0361
37. 0.0292 **39.** 0.0256

Problems 19–11, page 571

1. 0.1534 **3.** 0.1227 **5.** 0.1022
7. 0.0876 **9.** 0.0682 **11.** 0.2454
13. 0.1753 **15.** 0.0722 **17.** 0.0584
19. 0.0511 **21.** 0.2444 **23.** 0.3499
25. 0.4603 **27.** 0.6273 **29.** 0.57597
31. 1.17052 **33.** 0.67783 **35.** 0.8796

Problems 19–14, page 573

1. 0.0928 in. **3.** 0.0650 in. **5.** 0.1181 in.
7. 0.0650 in. **9.** 0.1181 in. **11.** 0.2165 in.
13. 0.1850 in. **15.** 0.4001 in. **17.** 0.8376 in.
19. 1.4902 in. **21.** 2.4252 in.

Problems 19–16, page 575

1. F, 0.2524 in. **3.** $\frac{40}{64}$ in.; 0.7547 in. **5.** $1\frac{7}{32}$ in.; 1.1946 in.
7. $3\frac{49}{64}$ in.; 3.7294 in. **9.** $2\frac{1}{32}$ in.; 2.0094 in. **11.** 0.0278 in.
13. 0.0078 in. **15.** 0.0104 in.

Problems 19–18, page 577

1. 0.1767 in. **3.** 0.21 in. **5.** 0.26 in.
7. 0.177 in. **9.** 0.6428 in. **11.** 3.6567 in.
13. 0.49 in. **15.** $\frac{19}{32}$ in. **17.** $\frac{13}{16}$ in.
19. $2\frac{1}{4}$ in. **21.** $1\frac{1}{2}$ in.

Problems 19–20, page 580

	(a) Root Diam. of Screw	(b) Top Flat of Screw	(c) Bottom Flat of Screw	(d) Tap Drill Size	(e) Bottom Flat of Tap	(f) Top Flat of Tap	(g) Outside Diam. of Tap
1.	1.7300	0.0927	0.0875	1.7500	0.0875	0.0875	2.02
3.	1.8193	0.1059	0.1007	1.8393	0.1007	0.1007	2.145
5.	0.8372	0.0530	0.0478	0.8572	0.0478	0.0478	1.02
7.	0.6300	0.03707	0.3187	0.6500	0.03187	0.03187	0.770
9.	0.7967	0.03089	0.02569	0.8167	0.02569	0.02569	0.92

Problems 19–22, page 582

	Double Depth (in.)	Root Diameter (in.)	Drill Size (in.)	Width of Tool Point (in.)
1.	0.0256	0.0925	$\frac{7}{64}$	0.0025
3.	0.0512	0.1850	$\frac{13}{64}$	0.0049
5.	0.0768	0.3956	$\frac{13}{32}$	0.0074
7.	0.1278	0.7383	$\frac{49}{64}$	0.0123
9.	0.1790	1.0021	$1\frac{1}{32}$	0.0172

Problems 19–25, page 584

1. 0.0534 in.	**3.** 0.1423 in.	**5.** 0.3659 in.
7. 0.5086 in.	**9.** 1.5904 in.	**11.** 0.5711 in.
13. $\frac{13}{32}$ in.	**15.** $\frac{55}{64}$ in.	**17.** $1\frac{5}{16}$ in.
19. $2\frac{7}{32}$ in.	**21.** $4\frac{13}{16}$ in.	**23.** 0.0114 in.
25. 0.0172 in.	**27.** 0.0229 in.	**29.** 0.0343 in.
31. 0.0523		

Problems 19–28, page 586

1. 0.0296 in.	**3.** 0.0571 in.	**5.** 0.19 in.; 0.26 in.
7. 0.40 in.; 0.55 in.	**9.** 1.00 in.; 1.25 in.	**11.** 0.051 in.
13. 0.117 in.		

Problems 19–29, page 590

1. 24, 32	**3.** 24, 36	**5.** 24, 60
7. 22 threads	**9.** 80 teeth	**11.** 48, 24
13. 64, 24	**15.** 96	

Problems 19–30, page 592

1. 64, 36	**3.** 48, 44	**5.** 32, 88
7. 96, 44	**9.** 64, 21	

Problems 19–31, page 594

1. $24 \times \frac{36}{48} \times 60$

3. $24 \times \frac{36}{48} \times 72$

5. $28 \times \frac{48}{56} \times 92$

7. 24 and 48

9. $64 \times \frac{24}{48} \times 20$

11. $108 \times \frac{24}{48} \times 24$

13. $72 \times \frac{96}{32} \times 24$

Problems 19–32, page 597

1. $60 \times \frac{36}{24} \times 127$

3. $80 \times \frac{48}{24} \times 127$

5. $120 \times \frac{60}{24} \times 127$

Chapter 20

Problems 20–2, page 603

1. 360 rpm

3. 80 rpm

5. 210 rpm

7. 480 rpm

Problems 20–3, page 605

1. 5.625 min

3. 3.75 min

5. 1.6 min

7. 1.38 min

9. 1.875 min

Problems 20–4, page 606

1. 270 rpm

3. 400 rpm

5. 400 rpm

7. 240 rpm

Problems 20–5, page 608

1. 0.94 min

3. $19\frac{1}{2}$ min

5. 0.75 min

7. 0.6 min

9. 36.7 min

Problems 20–6, page 610

1. 0.013 min

3. 0.0812 in.; 0.0081 in.

5. 0.054 in.; 0.004 in.

7. 8 in.

9. 5.12 in.

11. 150 mm/min

13. 15.625 m/min; 0.11 mm

15. 12.5 mm/min; 0.104 mm

Problems 20–7, page 612

1. 3,687.6 ft/min

3. 660 ft/min

5. 982.14 ft/min

7. 3,394.29 ft/min

Problems 20–8, page 613

1. 280 rpm

3. 152.7 rpm

5. 382 rpm

7. 65 rpm

9. 13 rpm

Problems 20–9, page 615

1. 11.45 in.

3. 28.6 in.

5. 4.19 in.

7. 3.18 in.

9. 6.36 in.

Problems 20–10, page 616

1. 12 ft/min; 18 ft/min

3. 49 ft/min; $65\frac{1}{3}$ ft/min

5. 34 ft/min; 51 ft/min

Problems 20–11, page 617

1. 28 min   **3.** 3 hr 36 min   **5.** 1 hr 28 min

Problems 20–12, page 619

1. $2\frac{2}{9}$ strokes   **3.** $2\frac{8}{11}$ strokes   **5.** $4\frac{4}{15}$ strokes
7. $3\frac{1}{3}$ strokes

Problems 20–13, page 620

1. 42 ft/min   **3.** 51 ft/min   **5.** 28 ft/min
7. 27 ft/min

Chapter 21
Problems 21–1, page 628

1. 0.3927″   **3.** 1.2566″   **5.** 0.3491″
7. 15.5″   **9.** $3\frac{3}{4}$″   **11.** 82 teeth
13. 20 teeth   **15.** 27 teeth

Problems 21–2, page 630

1. 4   **3.** 12   **5.** 16
7. 99 teeth   **9.** 67 teeth   **11.** 14″
13. $2\frac{1}{4}$″   **15.** 18″

Problems 21–4, page 633

1. 0.5236″   **3.** 0.3142″   **5.** 0.1745″
7. 2.0944″   **9.** 0.3927″   **11.** $2\frac{1}{2}$
13. 9   **15.** 2   **17.** 2.0944
19. 16

Problems 21–5, page 635

1. 0.0098″   **3.** 0.0628″   **5.** 0.0897″
7. 0.0302″   **9.** 0.0112″   **11.** 0.0131″
13. 0.0196″   **15.** 0.0815″

Problems 21–6, page 637

1. 0.1348″   **3.** 0.8628″   **5.** 1.2326″
7. 0.4148″   **9.** 0.1541″   **11.** 0.1798″
13. 0.2696″   **15.** 1.0785″

Problems 21–7, page 638

1. $8\frac{3}{8}$″   **3.** 2″   **5.** 9.35″
7. 3.4167″   **9.** 11.6429″

Problems 21–8, page 640

1. $7\frac{1}{4}$″   **3.** $9\frac{1}{2}$″   **5.** $14\frac{1}{4}$″
7. 3.143″   **9.** 14.667″

Problems 21–9, page 642

	P	D	N	P_c	Add.	Ded.	Cl.	Work. Depth	Whole Depth	Outside Diam.
1.	10	$2\frac{1}{2}$	25	0.3142	0.1000	0.1157	0.0157	0.2000	0.2157	2.7000
3.	12	$4\frac{1}{2}$	54	0.2618	0.0833	0.0964	0.0131	0.1667	0.1798	$4\frac{2}{3}$
5.	12	3.417	41	0.2618	0.0833	0.0964	0.0131	0.1667	0.1798	3.584
7.	2.5	22	55	1.2566	0.4000	0.4628	0.0628	0.8000	0.8628	22.8
9.	$1\frac{1}{2}$	60	90	2.0944	0.6667	0.7714	0.1047	1.3333	1.4380	61.3333

Problems 21–10, page 643

1. 0.5393″   **3.** 1.0472″   **5.** 14.1372″

7. 4.7746″   **9.** 10.743″

Problems 21–11, page 645

1. 6.375″   **3.** 5.5″   **5.** $1\frac{1}{4}$″

7. 9″; 15″   **9.** 4.833″

Problems 21–12, page 646

1. No. 5   **3.** No. 3   **5.** No. 3

7. No. 2   **9.** No. 2

Problems 21–14, page 650

	Pitch Diam. of Gear	Pitch Diam. of Pinion	Pitch Cone Angle of Gear	Pitch Cone Angle of Pinion
1.	15 in.	$6\frac{1}{4}$ in.	67°23′	22°37′
3.	12 in.	6 in.	63°26′	26°34′
5.	5 in.	$2\frac{1}{2}$ in.	63°26′	26°34′
7.	20 in.	5.4 in.	74°53′	15°07′
9.	$11\frac{1}{4}$ in.	9 in.	51°20′	38°40′

Problems 21–15, page 651

	Add.	Cl.	Ded.	Whole Depth	Depth under Pitch Cone	Circ. Pitch	Thickness
1.	0.25	0.0393	0.2893	0.5393	0.2893	0.7854	0.3927
3.	0.1667	0.0261	0.1928	0.3595	0.1928	0.5236	0.2618
5.	0.1	0.0157	0.1157	0.2157	0.1157	0.3142	0.1571
7.	0.2	0.0314	0.2314	0.4314	0.2314	0.6283	0.3142
9.	0.25	0.0393	0.2893	0.5393	0.2893	0.7854	0.3927

Problems 21–18, page 652

	Pitch Cone Radius	Add. Angle	Turn. Gear	Angle Pinion	Ded. Angle	Cut. Gear	Angle Pinion
1.	8.125	1°46′	69°09′	24°23′	2°02′	65°21′	20°35′
3.	6.7082	1°26′	64°52′	28°00′	1°39′	61°47′	24°55′
5.	2.7956	2°03′	65°29′	28°37′	2°22′	61°04′	24°12′
7.	10.3581	1°06′	75°59′	16°13′	1°17′	73°36′	13°50′
9.	7.2035	1°59′	53°19′	40°39′	2°18′	49°02′	36°22′

Problems 21–19, page 653

	Outside Diameter	
	Gear	**Pinion**
1.	15.1923	6.7116
3.	12.1491	6.2981
5.	5.0894	2.6789
7.	20.1043	5.7862
9.	11.5624	9.3904

Problems 21–20, page 654

	Gear		Pinion	
	No. Teeth	**Cutter No.**	**No. Teeth**	**Cutter No.**
1.	156	1	28	4
3.	161	1	41	3
5.	112	2	28	4
7.	384	1	28	4
9.	73	2	47	3

Problems 21–21, page 655

	Assumed Face Dimension	Add.	Cl.	Ded.	Whole Depth	Circ. Pitch	Thickness of Tooth Space at End
1.	2	0.1885	0.0296	0.2181	0.4065	0.5922	0.2961
3.	$1\frac{1}{2}$	0.1293	0.0203	0.1496	0.2789	0.4063	0.2032
5.	$\frac{3}{4}$	0.0732	0.0115	0.0847	0.1578	0.2297	0.1149
7.	2	0.1614	0.0253	0.1867	0.3481	0.5070	0.2535
9.	$1\frac{3}{4}$	0.1895	0.0298	0.2193	0.4088	0.5953	0.2977

◼◼◼◼ ANSWERS TO SELF-TESTS

Chapter 1

1. 11,576   **2.** 15,888   **3.** 479   **4.** 3,440,596   **5.** 5,490,000
6. 21,772,089   **7.** 13   **8.** 25   **9.** $170,558,830,458   **10.** 9,828,898,692 bills

Chapter 2

1. $\frac{7}{8}$ **2.** $7\frac{1}{4}$ **3.** $\frac{5}{16}$ **4.** $1\frac{7}{8}$ **5.** $\frac{8}{15}$
6. $2\frac{5}{8}$ **7.** $\frac{2}{3}$ **8.** $1\frac{2}{3}$ **9.** 1   **10.** $\frac{10}{27}$

Chapter 3

1. 0.627   **2.** 2.102   **3.** 4.704   **4.** 5.81   **5.** 2.4
6. 20.21   **7.** 10.6   **8.** 1.838   **9.** 1.065   **10.** 73.5

Chapter 4

1. 50%   **2.** 33.33%   **3.** 7.5   **4.** 20%   **5.** 200
6. $2,640   **7.** 0.8   **8.** 1   **9.** 3%   **10.** $184.40

Chapter 5

1. 4 to 3   **2.** 1 to 4   **3.** 4 to 3   **4.** 5 to 6   **5.** 1 to 2
6. 1.08 ohms   **7.** 1 to 2   **8.** $60; $40   **9.** 7 lb   **10.** $7,800

Chapter 6

1. 1   **2.** 11   **3.** 3   **4.** 15   **5.** 336
6. $9ab^2$ **7.** $8x^2y$ **8.** $6x^2 - 11xy - 10y^2$ **9.** $x + y$ **10.** $a = \dfrac{3V}{\pi r^2}$

Chapter 7

1. 80 sq in.   **2.** 36 in.   **3.** 12 in.   **4.** 13 in.   **5.** 1,600
6. 33 in.   **7.** 10 in.   **8.** 51°   **9.** 21 sq in.   **10.** 43.3 sq in.

Chapter 8

1. 5.20 in.   **2.** 18 sq in.   **3.** 3.46 in.   **4.** 18.85 in.   **5.** 78.54 sq in.
6. 14.14 in.   **7.** 0.59 sq in.   **8.** 0.81 sq in.   **9.** 94.25 sq in.   **10.** 16.02 in.

Chapter 9

1. 160 cu in.   **2.** 63.58 cu in.   **3.** 33.49 cu in.   **4.** 25.62 sq in.   **5.** 113.04 cu in.
6. 7.4 cu in.   **7.** 1,089 lb   **8.** 180 lb   **9.** 672 fbm   **10.** 168 fbm

Chapter 10
1. 4,800 m   **2.** 30 cm^2 **3.** 3,663 cm^3 **4.** 9 mg   **5.** 63.8 cm

6. 348.4 cm^3 **7.** 16.13 cm^2 **8.** 3,540 cm^3 **9.** 22.7 kg   **10.** 37.8°

Chapter 12
1. 0.165   **2.** 0.731

Thimble Is between	Gage Line Is between Lines	Coinciding Vernier Line
3. 0.275 and 0.300	7 and 8	3
4. 0.425 and 0.450	12 and 13	5

5. 0.804   **6.** 0.039

Index Is between	Coinciding Vernier Line
7. 51°30′ and 52°	14
8. 13°00′ and 13°30′	19

9. 75.53   **10.** 41.36

Chapter 14
1. 0.5   **2.** 50°   **3.** 0.6428   **4.** 24.9″   **5.** 19.1″

6. 40.6 mm   **7.** 20.3 mm   **8.** 120°   **9.** 70.3 mm   **10.** 713.7 mm^2

Chapter 15
1. 42,500 lb   **2.** 3.57 in.   **3.** 325 lb   **4.** 6 rivets   **5.** 7,031 lb

Chapter 16
1. 0.5 hp   **2.** 960 ft-lb   **3.** 228.4 hp   **4.** 182.8 bhp   **5.** 1.2 kW

6. 0.5 hp   **7.** 5.5 amperes   **8.** 87%   **9.** 0.86 hp   **10.** 1.1 hp

Chapter 17
1. 0.24 in.   **2.** 0.4 in.   **3.** 0.0625 in.   **4.** 0.375 in.   **5.** 0.25 in.

6. 0.028 in.   **7.** 0.336 in.   **8.** 0.28 in.   **9.** 3°34′   **10.** 0°54′

Chapter 18
1. 90 rpm   **2.** 40 teeth   **3.** 90 rpm   **4.** 7,000 rpm   **5.** 3 rpm

6. 1,312.5 rpm   **7.** 4.4 in.   **8.** 583 rpm; 729 rpm; 862 rpm   **9.** No   **10.** Clockwise

Chapter 19
1. $\frac{1}{13}$-in. pitch   **2.** $\frac{1}{4}$ in.   **3.** $\frac{1}{10}$ in.   **4.** 0.096 in.   **5.** 0.068 in.

6. 0.185 in.   **7.** 0.201 or 7 drill   **8.** 0.573 in.   **9.** 0.057 in.   **10.** $\frac{2}{3}$

Chapter 20

1. 65.5 ft/min

2. 1,886 ft/min

3. 1,909 rpm

4. 14,667 ft/min

5. 42 ft/min

6. 9.5 in.

7. 24 ft/min

8. 32 ft/min

9. 1.25 min

10. 0.01 in.

Chapter 21

1. 0.3927 in.

2. 4.5 in.

3. 48 teeth

4. 16 pitch

5. 3 in.

6. 0.3927 in.

7. 0.0196 in.

8. 0.1798 in.

9. 4.167 in.

10. 7 in.

ANSWERS TO CHAPTER TESTS

Chapter 1 Test

1. 882,509	**2.** 1,259,819	**3.** 1,600,052	**4.** 1,381,769	**5.** 1,296,929
6. 1,748,132	**7.** 1,602,033	**8.** 1,921,446	**9.** 92,310	**10.** 75,242
11. 30,770	**12.** 33,558	**13.** 60,605	**14.** 138,669	**15.** 8,075
16. 1,870	**17.** 3,357,620	**18.** 1,311,981	**19.** 55,223	**20.** 2,795,544
21. 3,407,040	**22.** 747,504	**23.** 2,644,992	**24.** 555,390	**25.** 71
26. 97	**27.** 34	**28.** 67	**29.** 52	**30.** 62
31. 25	**32.** 69	**33.** \$355,703,113; 7,098,746,642 coins	**34.** 99,920 more licensed drivers	**35.** 36,583 more units

Chapter 2 Test

1. $1\frac{1}{4}$	**2.** $2\frac{1}{15}$	**3.** $3\frac{1}{8}$	**4.** $4\frac{3}{8}$	**5.** $2\frac{5}{32}$
6. $6\frac{13}{16}$	**7.** $4\frac{27}{32}$	**8.** $8\frac{47}{64}$	**9.** $\frac{1}{12}$	**10.** $\frac{7}{8}$
11. $3\frac{1}{16}$	**12.** $\frac{5}{8}$	**13.** $9\frac{1}{16}$	**14.** $\frac{5}{16}$	**15.** $\frac{7}{32}$
16. $1\frac{51}{64}$	**17.** $\frac{3}{8}$	**18.** $1\frac{5}{16}$	**19.** $1\frac{7}{8}$	**20.** $11\frac{7}{8}$
21. $7\frac{31}{32}$	**22.** $14\frac{1}{16}$	**23.** $36\frac{2}{3}$	**24.** $\frac{3}{16}$	**25.** $\frac{3}{8}$
26. $3\frac{1}{7}$	**27.** 7	**28.** $16\frac{1}{8}$	**29.** $\frac{10}{13}$	**30.** $\frac{1}{5}$
31. $1\frac{15}{19}$	**32.** $\frac{24}{31}$	**33.** $\frac{35}{66}$	**34.** $\frac{5}{9}$	**35.** $\frac{26}{33}$
36. $\frac{5}{16}$				

Chapter 3 Test

1. 0.638	**2.** 1.757	**3.** 1.451	**4.** 7.262	**5.** 5.035
6. 4.682	**7.** 7.006	**8.** 18.610	**9.** 8.312	**10.** 9.370
11. 1.000	**12.** 0.237	**13.** 1.703	**14.** 7.303	**15.** 6.095
16. 8.518	**17.** 87.292	**18.** 0.339	**19.** 6.293	**20.** 50.380
21. 0.388	**22.** 0.119	**23.** 0.002	**24.** 3.698	**25.** 10.350
26. 8.039	**27.** 0.078	**28.** 0.039	**29.** 25.428	**30.** \$277.20
31. 0.063	**32.** 25.600	**33.** 0.014	**34.** 2.679	**35.** 1.904
36. 6.135	**37.** 196.078	**38.** 100.000	**39.** \$1.25	**40.** \$81.95

Chapter 4 Test

1. 12.5%	**2.** 25%	**3.** 75%	**4.** 66.67%	**5.** 16.67%
6. 62.5%	**7.** 25	**8.** 91.44	**9.** 57.15	**10.** 0.225
11. 33.33%	**12.** 50%	**13.** 15%	**14.** 6.67%	**15.** 1,600%
16. $5\frac{1}{3}$	**17.** 180	**18.** 90	**19.** 342	**20.** 3.1416
21. $6\frac{1}{4}\%$	**22.** 20 workers			

Chapter 5 Test
1. 3 to 1   **2.** 6 to 5   **3.** 1 to 4   **4.** 5 to 11   **5.** 8 to 3
6. 3 to 2   **7.** 2 to 1   **8.** 7 to 6   **9.** 3 to 1   **10.** $36 and $24
11. 6 to 5   **12.** 1 to 52   **13.** 5.420 ohms   **14.** 1.125 in.   **15.** 20 in.
16. 16 lb   **17.** 4 ft 5 in.   **18.** 228 lb   **19.** $287.50   **20.** $15,886

Chapter 6 Test
1. 1   **2.** $4x$ if $x > 0$; ^-4x if $x < 0$ **3.** $^-8,190$ **4.** $14\ mn^2$ **5.** $^-3xy^3$
6. $16a^4 - b^2$ **7.** $4xy$ **8.** $r = \sqrt{A/\pi}$ **9.** $b = 2A/a$ **10.** $R = E/I$
11. $x = {}^\pm 4$ **12.** $(F - 32)\frac{5}{9}$ **13.** $c = \sqrt{E/m}$ **14.** $d = \sqrt[3]{6V/\pi}$ **15.** $R = P/I^2$; $I = \sqrt{P/R}$

Chapter 7 Test
1. 157.5 sq in.   **2.** 600 sq in.   **3.** 785.375 sq in.   **4.** $13\frac{1}{2}''$ **5.** $1'7\frac{1}{2}''$
6. 86 in.   **7.** 7 ft $6\frac{1}{2}$ in.   **8.** 9 in.   **9.** 8 in.   **10.** 729
11. 1,056.25   **12.** 15   **13.** 93   **14.** 14.14 in.   **15.** 51 in.
16. 20 in.; 48 in.   **17.** 24 in.; 60 in.   **18.** 24 in.   **19.** 511.5 sq in.   **20.** 73.5 sq in.
21. 252.02 sq in.   **22.** 300 sq in.   **23.** 270.6 sq in.   **24.** 80°   **25.** 84°41′

Chapter 8 Test
1. 7.79″   **2.** 83.09 sq in.   **3.** 32 sq in.   **4.** 17.32″   **5.** 52.26″
6. 96 sq in.   **7.** $3\frac{3}{4}''$ **8.** 32.94″   **9.** 9.55″   **10.** 23.76 sq in.
11. 35.68″   **12.** 7.07″   **13.** 34.56 sq in.   **14.** 5.24″   **15.** 13.09 sq in.
16. 112.5 sq in.   **17.** 225 sq in.   **18.** 22.12 sq in.   **19.** 117.81 sq in.   **20.** 40.05″

Chapter 9 Test
1. 3,410 cu in.   **2.** 763.8 cu in.   **3.** 11.2 cu ft   **4.** 519.4 lb   **5.** 15″
6. 11.1″   **7.** 942 sq in.   **8.** 1,059.8 cu in.   **9.** 768 cu in.   **10.** 459.2 sq in.
11. 410.1 sq in.   **12.** 321.4 cu in.   **13.** 4,188.8 cu in.   **14.** 452 sq in.   **15.** 12.3 cu in.
16. 104.2 lb   **17.** 12.5 gal   **18.** 3,519 lb   **19.** 21.8 lb   **20.** 7.3 lb
21. 26.6 fbm   **22.** 6,480 fbm   **23.** 16 fbm   **24.** 356 fbm   **25.** $658.60

Chapter 10 Test
1. 160.8 cm   **2.** 1.15 cm   **3.** 2.8 m   **4.** 71.4 cm   **5.** 116.25 m²
6. $491.40   **7.** 38 m   **8.** 182.25 cm²   **9.** 3,175 cm²   **10.** 3,810 cm²
11. 6.1 cm²   **12.** 38,448 cm³   **13.** 21.2 m³   **14.** 144 l **15.** 720 g
16. 5.25 kg   **17.** 137.2 cm   **18.** 12.1 cm   **19.** 23 m²   **20.** 96.8 cm²
21. 428.7 cm²   **22.** 1,029.1 cm³   **23.** 2 kg   **24.** 4.4°C   **25.** 158°F

Chapter 12 Test
1. Thimble between 0.200 and 0.225; coinciding line on thimble, 22   **2.** On 0.625; coinciding line on thimble, 0   **3.** 0.7062   **4.** 0.2585
5. Index between lines 0.300 and 0.325; coinciding vernier line, 15   **6.** On 0.375; coinciding vernier line, 0   **7.** 50°06′   **8.** 18°44′
9. 93.24   **10.** 21.79

Chapter 14 Test

1. 0.7046   **2.** 0.4292   **3.** 14°30′   **4.** 75°08′   **5.** 35.49 in.
6. 19.76 in.   **7.** 50°12′   **8.** 31.24 in.   **9.** 20.57 in.   **10.** 20.57 in.
11. 25 mm   **12.** 10.44 in.   **13.** 35°39′   **14.** 40°   **15.** 192 cm^2

Chapter 15 Test

1. 2.92 in.   **2.** 60,937.5 lb   **3.** 0.54 in.   **4.** 300 lb   **5.** 15 rivets
6. 35,328 lb

Chapter 16 Test

1. 24,000 ft-lb   **2.** 0.58 hp   **3.** 1.82 hp   **4.** 153 hp   **5.** 25.4 hp
6. 0.9-hp increase   **7.** 60.9 bhp   **8.** 0.4 hp   **9.** 6.8 amperes   **10.** 0.26 hp
11. 1.35 amperes   **12.** 5 hp   **13.** 0.7 hp   **14.** 94%   **15.** 76%

Chapter 17 Test

1. 0.6 in.   **2.** 0.0729 in.   **3.** 0.375 in.   **4.** 2.095 in.   **5.** 1.1719 in.
6. $\frac{1}{24}$ in. or 0.0416 in.   **7.** 3°33′   **8.** 2°24′   **9.** complement   **10.** one-half

Chapter 18 Test

1. 75 rpm   **2.** 85$\frac{5}{7}$ rpm   **3.** 90 teeth   **4.** 44 rpm   **5.** 50 rpm
6. 75 rpm   **7.** 225 rpm   **8.** 90 rpm   **9.** 20 rpm   **10.** 3 to 1
11. 2$\frac{2}{3}$ rpm   **12.** 2,625 rpm   **13.** For example, 4 in. and 1$\frac{3}{4}$ in.   **14.** 3,400 rpm; 2,975 rpm; 2,550 rpm

15.

	Driver @ 3,400 rpm		
	4 in.	**3$\frac{1}{2}$ in.**	**3 in.**
4 in.	3,400 rpm	2,975 rpm	2,550 rpm
3$\frac{1}{2}$ in.	3,885$\frac{5}{7}$ rpm	3,400 rpm	2,914$\frac{2}{7}$ rpm
3 in.	4,533$\frac{1}{3}$ rpm	3,966$\frac{2}{3}$ rpm	3,400 rpm

Chapter 19 Test

1. $\frac{1}{20}$ in.   **2.** $\frac{1}{13}$ in.; $\frac{1}{13}$ in.   **3.** 0.0433 in.   **4.** 0.1443 in.   **5.** 0.1634 in.
6. $\frac{15}{16}$ in.   **7.** 0.0500 in.   **8.** 0.1850 in.   **9.** 0.1767 in.   **10.** 0.044 in.
11. 24 : 80   **12.** 64 : 36

Chapter 20 Test

1. 86$\frac{3}{7}$ ft/min   **2.** 2,750 ft/min   **3.** 49 rpm   **4.** 2,300 rpm   **5.** 12.4 in.
6. 0.68 in.   **7.** 110 ft/min   **8.** 105 ft/min   **9.** 47.7 rpm   **10.** 6.36 in.
11. 70 ft   **12.** 93$\frac{1}{3}$ ft   **13.** 24 min   **14.** 60 ft/min   **15.** 160 rev/in.
16. 5$\frac{1}{3}$ min   **17.** 23.8 in.   **18.** 36 sec   **19.** 600 rpm   **20.** 0.01 in./rev

Chapter 21 Test

1. 0.2945 in.   **2.** 7.5 in.   **3.** 80 teeth   **4.** 0.19635 in.   **5.** 16-pitch gear
6. 8 in.   **7.** 28 teeth   **8.** 0.31416 in.   **9.** 10 in.   **10.** 0.01418 in.
11. 0.0196 in.   **12.** 0.1348 in.; 103   **13.** 7 in.   **14.** 0.3595 in.   **15.** 5$\frac{5}{8}$ in.
16. Pitch diameter of gear = 16 in.; pitch diameter of pinion = 5 in.

■ ■ ■ ■ INDEX

Absolute value, 163
Acme 29-degree screw thread, 578
Actual dimension of object, 257
Acute angle, 232
Addendum, of bevel gear, 647
  of spur gear, 625
Addition, decimals, 84
  fractions, 36
  mixed numbers, 46
  whole numbers, 1
Algebra, 161
  addition, 168
  division, 178
  grouping symbols, 173
  multiplication, 175
  subtraction, 171
American National thread, 571
American Standard, pipe thread, 584
  taper series, 512
Angles in triangles, 232
Angular addendum, 652
Arc, circle, 295
Area, of circle, 282
  of composite figures, 195
  of ellipse, 303
  of figure drawn to certain scale, 271
  of isosceles triangle, 222
  of rectangle, 192
  of ring sections, 292
  of scalene triangle, 223
  of sphere, 388
  of trapezoid, 262
  of triangle, 219
  table of, 660
  units of, 191, 366
Averages, 153
Axioms, 181

Bar graphs, 412
Base, percent, 119
  rule, 128
Base, of prism, 318
Bending stress, 482
Bevel gears, 646
  rules, 655
Bisect, angle, 434
  arc, 434
  line, 433
Board feet, 348
Board measure, 348
  rule, 348
Brake horsepower, 499
Brake, Prony, 499
Brown and Sharpe taper, 512
  worm thread, 580

Calculator, addition (whole numbers),
    6
  division, 22
  multiplication, 17
  subtraction, 11
  to change fractions to decimals, 99
  to find: angle measure, 235, 461
    area of circle, 284
    average of group, 155
    base, 134
    circumference of circle, 278
    percentage, 131
    polygon problems, 259
    proportion problems, 150
    rate, 133
    sine, cosine, tangent, 459
    square root problems, 210
  to solve: decimal fraction problems,
    99

Calculator, *continued*
    triangle problems, 226
    volume of sphere, 337
Caliper, vernier, 424
Castings, weight, 346
Celsius, 390
Center of circle, 440
Chord, 301
Circle, 275
  arc of, 295
  area of, 280
  central angle of, 295
  circumference of, 275
  circumscribed, 297
  diameter of, 279
  inscribed, 297
  radius of, 275
  sectors, 295
  segments, 301
  short method of comparing areas,
    291
Circle graphs, 409
Circular pitch, 627
Circumference, 276
Coefficient, 165
Common denominator, 44
Complex fractions, 57
Composite plane figures, areas of,
    266
Composite solid figures, volume of,
    341
Compound gearing, 541, 593
Compound rest, 527
Compression, 481
Compressive stress, 481
Cone, 322
    frustum, 327

Cone, *continued*
  lateral surface, 324, 332
  volume of, 322
Constants, use of, 255
  for hexagons, octagons, squares,
     triangles, 244
Conversion of dimensions, 81
Cosecant of angle, 456
Cosine of angle, 456
  table of, 685
Cotangent of angle, 456
  table of, 685
Counting numbers, 28
Cross-section paper, graph, 397
Cubic figures, weight, 344
Cubic meter, 371
Cutters, for bevel gears, 653
  for spur gears, 646
  table of, 646
Cutting feed, on drill press, 606
  on lathe, 604
  on milling machine, 606
Cutting speed, crank shaper, 619
  drill press, 606
  geared shaper, 619
  lathe, 601
  milling machine, 606
  planer, 616
Cylinder, 311. *See also* Prism
  base, 318
  height, 316
  hollow, 313
  lateral surface, 319
  volume of, 311

Decimal fractions, 71
  addition, 84
  changing to common, 74
  division, 95
  multiplication, 90
  subtraction, 88
Decimal point, 71
Decimals to metric, 691
Dedendum, bevel gear teeth, 647
  spur gear teeth, 625
Denominator, 27
Diameter of circle, 275
  finding, 285
Diametral pitch, 629
Difference, 7
Dissimilar terms, 168
Dividend, 17
Division, algebra, 178
  decimals, 95
  fractions, 38
  mixed numbers, 39
Divisor, 17
Double shear, rivets, 488
Dressed lumber, allowance for, 348

Dressed stock lumber, 348
Drill press, cutting speed, 606
  feed on, 606
Driven gears, 594
Driving gears, 535

Efficiency of machines, 503
Elastic limit, 482
Electrical power, 501
Electric horsepower, 501
Ellipse, area of, 303
  construction of, 451
  perimeter of, 304
English measure to metric measure,
  379
English to metric conversions, area,
  381
  temperature, 390
  volume, 385
  weight, 388
Equations, definition, 181
Equilateral triangles, 219, 243
  constants for, 244
Equivalent fractions, 35
Exponent, 164, 176
Extremes of proportion, 144

Factor of safety, 483
Factors, 164
Fahrenheit, 390
Flat ring, formula, 292
Flooring, measure, 351
Followers or driven gears, 535
Foot-pound, work, 495
Formulas for plane figures, summary,
  305
  for solids, 354
Fractions,
  common, 27
    addition, 46
    changing, 32, 74
    complex, 57
    denominator, 27
    division, 38
    equivalent, 44
    improper, 28
    lowest terms, 32
    multiplication, 35
    numerator, 27
    proper, 27
    reducing, 30
    simplification, 35
    square root of, 208
    subtraction, 50
    terms, 27
  decimal, addition, 84
    changing to common, 74
    division, 95
    multiplication, 90

Fractions, *continued*
    reducing to common, 74
    subtraction, 88
Friction drive, spur gears, 625
Frustum, definition, 327
  height, 330
Frustum of cone, lateral surface, 332
  volume, 327
Frustum of pyramid, lateral surface,
  332
  volume of, 327
Frustums of pyramids and cones,
  formula, 354
Functions, trigonometric, 455
  tables of, 685

Gas engine, horsepower of, 498
Gear compound, use of, 593
  for screw cutting, 593
Gear computation, formulas, 641
Gears, 535, 625, 646
  bevel, 548, 647, 655
  clearance, 635
  cutting angle, 652
  idlers, 539
  spur, 625
  trains, 535, 548
  wheels, 625
Geometry applications, 433
Graphs, 397
  causal relation, 399
  combined, 402
  no causal relation, 398
Grouping symbols, 173

Hectare, unit of area, 366
Height, frustum, 330
  prism, 316
Hero's Formula, 223
Hexagon, 249
  constants for, 244
  construction of, 443
  inscribed in circle, 443
  regular, 249
Horsepower, brake, 499
  electrical, 501
  of gas engine, 498
  of steam engine, 497
Hypotenuse, 213

Idlers, 539
  effect, 539
Improper fractions, 28
  changing to whole or mixed
    numbers, 30
Index, root, 165
Indicated horsepower, 497
Input, 503
Inverse ratio, 141, 536

Isosceles trapezoid, 262
Isosceles triangles, 219, 468
   area of, 222

Joints, riveted, 487

Kilogram, 375
Kilowatt, 501

Lateral surface, cones, 324

   frustum of cone, 332
   frustum of pyramid, 332
   prisms, 319
   pyramids, 324
Lathe, cutting feed on, 604
   cutting speed on, 601
Lathe gearing, cutting screw threads, 587
Law of exponents, division, 178
   multiplication, 175
Law of signs, division, 178
   multiplication, 175
Lead of screw, 562
Least common multiple, 44
Lengths, metric, 363
Like terms, 168
Lumber, dressed stock, 348
   rough stock, 348
   standard lengths and widths, 349

Major axis, ellipse, 303
Mean (average), 153
Means (proportion), 145
Mechanical efficiency of machines, 503
Members of equation, 181
Meter, 359
Metric measure, 359
   linear measure, 363
   square measure, 366
   standard screw threads, 580
   temperature measure, 390
   threads, cutting of, 596
   volume measure, 371
   weight measure, 375
Micrometer, 421
   ten-thousandths, 423
Millimeters to inches-decimals, 690
Milling machine cutting feed, 606
   cutting speed, 606
Minor axis, ellipse, 303
Minor diameter, screw threads, 566
Minuend, 7
Miter gears, 655
Mixed number, 28
   addition, 48
   division, 39
   multiplication, 11

Mixed number, *continued*
   square root of, 204
   subtraction, 52
Morse taper, 512
Multiplicand, 12
Multiplication, algebra, 175
   fractions, common, 35
   fractions, decimal, 88
   mixed numbers, 36
   whole numbers, 11
Multiplier, 12

Negative numbers, 161
Numerator, 27

Obtuse angle, 232
Octagon, 252
   constants for, 244
   construction of, 445
   inscribed in circle, 445
   inscribed in square, 445
   regular, 252
Origin, graphs, 397
Output, 503
Outside diameter, gear, 627

Parallelogram, 262
Pentagon, 446
Percentage, 119
Perimeter, rectangle, 197
   ellipse, 303
Perpendiculars, 435
Pinion, 648
Pipes, pressure in, 486
Pipe threads, American Standard, 584
Pitch cone, 647
   angle, 648
   radius, 651
Pitch diameter, of bevel gears, 647
   of spur gears, 625
Pitch of gear, circular, 627
   diametral, 629
Pitch of screw thread, 559
Place value, 1
Plane figures, formulas, 305
Planer, cutting speed, 616
Planimeter, 428
Polygon, 243
Power, 164, 495
   electrical, 501
   horsepower, 497
   unit of, 501
Prism, 311
   base, 318
   formula, 354
   height, 316
   lateral surface, 319
   volume of, 311
Problems, all four operations, 55

Prony brake, 499
Proper fractions, 27
Proportion, 144
Protractor, 426
   vernier, 427
Pulley trains, 549
Pyramid, 322
   frustum, 327
   lateral surface, 324, 332
   slant height, 324
   volume of, 322
Pythagorean theorem, 214

Quadrilateral, 262
Quadruple thread, 562
Quotient, 17

Racks, 642
Radius of circle, 275
Rate, percent, 119
   rule, 123
Ratio, 141
Ratio, gear, 540
Ratio, inverse, 141, 536
   reduction to lowest terms, 141
Reciprocal, 38
Rectangle, area, 191
   formula, 306
Reducing a fraction, 30
Rhomboid, 262
Rhombus, 262
Right angle, 232
   prism, 311
   triangle, 212, 219, 429
      sides, formula, 220
      solution, 462
Rim speed, 612
Ring, volume, 340
Ring sections, area, 292
Riveted joints, 487
Rivets, bearing value of, 487
   shearing value of, 487
Root diameter of threads, 564
Roots, 165
Rough stock, lumber, 348
Round off rule, 76

Safety factor, 483
Scale, 269
   area, 273
Scalene triangles, 219
   area of, 222
Screw, double-threaded, 562
   lead, 562
   quadruple, 562
   single-threaded, 562
   triple-threaded, 562
Screw threads, 559
   definitions, 564

Secant of angle, 456
   table of, 685
Sector of circle, 295
Segment of circle, 301
Shaper, cutting speed on, 619
Sharp V-thread, 564
Shear, 482
   of rivets, 487
Shearing stress, 487
   value, 487
Similar terms, 168
Sine of angle, 456
   table of, 685
Single shear, 488
Slant height (pyramid), 325
Solid ring, formula, 340, 354
Solids, 311
   formulas for, 354
Solution of equations, 181
Speed of gears, rule, 536
Sphere, 334
   formula, 354
   surface area of, 338
   volume of, 334
Spur gears, 548, 625
   circular pitch, 627
   clearance, 625
   diametral pitch, 629
   root diameter, 627
   whole depth, 626
Square, constants for, 244
   inscribed in circle, 440
Square foot, inch, yard, 191
Square meter, 366
Square root applications, 212
   of fraction, 208
   list of, 202, 660
   of mixed numbers, 204
   of whole numbers, 203
Squares, 247
Squares of numbers, 201
   table of, 660
Square thread, 576
Steam engine, horsepower, 497
Steep taper, 512
Strength, of materials, 481
   ultimate, 482
Stress and strain, 481
Stresses, units of, 482
Substitution, 167
Subtraction, algebra, 171
   decimals, 88
   fractions, 50
   mixed numbers, 52
   whole numbers, 7
Subtrahend, 7
Surfaces, area of, 191
Surface speed, 612

Tables,
   cosines, 685
   cotangents, 685
   cutters, gear, 646
   decimals to metric, 691
   degrees Celsius to Fahrenheit, 693
   degrees Fahrenheit to Celsius, 692
   linear measure, metric, 363
   millimeters to inches-decimals, 690
   National Coarse-Thread Series, 695
   National Fine-Thread Series, 696
   natural trigonometric functions, 660
   percentage equivalents, 120
   powers, roots, circumferences, areas,
      660
   secants, 685
   sines, 685
   square measure, metric, 366
   squares of numbers, 660
   tangents, 685
   threads, National Coarse, 695
   threads, National Fine, 696
   volume measure, metric, 371
   weight measure, metric, 375
   weights of materials, 694
   working stresses, 483
Tangent of angle, 454
   to circle, 447
   table of, 660
Tap drill, 567
   Acme thread, 578
   American National thread, 571
   for metric thread, 580
   square threads, 576
   V-threads, 564
   Whitworth thread, 582
Taper, 509
   American Standard, 512
   angle, 519
   Brown and Sharpe, 512
   compound rest, 527
   computing, 509
   definition, 509
   methods of turning by, 523
   Morse, 512
   tailstock offset, 523
   taper attachment, 531
   $\frac{3}{4}$ in. per ft series, 512
Tensile stress, 481
Tension, 482
Ten-thousandth micrometer, 423
Terms, algebraic, 168
   of fraction, 27
   ratio, 141
Thread cutting, 587
Threads, 560
   Acme, 578
   American National, 571

   Brown and Sharpe, 580
   definition of parts of, 564
   fractional, 591
   metric, 580
   pipe, 584
   Sharp V, 564
   square, 576
   unified, 568
   Whitworth, 582
Torsion, 482
Transposition of terms, 181
Trapezium, 262
Trapezoid, 262
Triangle, equilateral, 244
   circumscribed, 298
   inscribed, 298
Triangles, 212, 219
   angles in, 232
   area of, 219, 222
   equilateral, 219, 245
   isosceles, 219
   right, 219
   scalene, 219, 223
Trigonometric functions, 455
   tables of, 685–689
Triple-thread screw, 562

Ultimate strength, 482
Unified thread, 568
Unit stress, 482
   allowable, table of, 484
Unlike terms, 168

Vernier caliper, 424
   protractor, 426
   scale, 423
Volume, composite surface figures, 341
   cylinder, 311
   prism, 311
   pyramid, 322
   ring, 340
   sphere, 334
V-thread, 564

Watt, 501
Weight of materials, 344, 694
Whitworth threads, 582
   radius of tool point, 583
Work, 495
   unit of, 495
Working unit stresses, 483
Worm gear, 547

X-axis, 397

Y-axis, 397